ADVANCES IN
SCATTERING
AND BIOMEDICAL
ENGINEERING

EDITORS

DIMITRIOS I. FOTIADIS ◆ CHRISTOS V. MASSALAS

University of Ioannina, Greece

Proceedings of the
Sixth International Workshop

ADVANCES IN
SCATTERING
AND BIOMEDICAL
ENGINEERING

Tsepelovo, Greece   18 – 21 September 2003

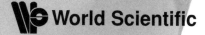

World Scientific

NEW JERSEY • LONDON • SINGAPORE • BEIJING • SHANGHAI • HONG KONG • TAIPEI • CHENNAI

Published by

World Scientific Publishing Co. Pte. Ltd.

5 Toh Tuck Link, Singapore 596224

USA office: Suite 202, 1060 Main Street, River Edge, NJ 07661

UK office: 57 Shelton Street, Covent Garden, London WC2H 9HE

British Library Cataloguing-in-Publication Data
A catalogue record for this book is available from the British Library.

ISBN 981-238-924-5

Printed by FuIsland Offset Printing (S) Pte Ltd, Singapore

Preface

The present volume of proceedings consists of the papers presented during the 6th International Workshop on Scattering Theory and Biomedical Engineering held in the Monastery of Rogovos (Zagori), Epirus, Greece, on 18–21 September 2003. The Workshop is organized every two years by the University of Ioannina, the University of Patras, the National Technical University of Athens, the Institute of Chemical Engineering and High Temperature Chemical Processes and the Biomedical Research Institute.

The aim of the Workshop is to provide an overview of the "hot topics" in Scattering Theory and Biomedical Technology and to bring together young researchers and senior scientists and at the same time to create a forum for exchange of new scientific ideas. All the invited speakers, who are recognized both as eminent in their field but more importantly as "stimulating" speakers, presented their latest achievements.

It was really nice to see both old faces and new ones, and to notice strong collaborations formulating and new friendships emerging, both inside and outside the meeting room. The selection of the venue of the Workshop, the quite monastery of Rogovos, was made in order to give emphasis to our respect to the local culture. This could be an answer to the environment of globalization, because globalization cannot grasp anything from us unless we let it go.

The Workshop Organizing Committee would like to thank all the authors for their contributions, the University of Ioannina, the National Technical University of Athens, the Institute of Chemical Engineering and High Temperature Chemical Processes, the Biomedical Research Institute, EMPHASIS S.A. and ANCO S.A. for their financial support, and for the organization of the workshop Ms. Evi Kalabakioti and Ms. Eirini Zisaki.

Dimitrios I. Fotiadis and C.V. Massalas
University of Ioannina

Ioannina, February 2004

Contents

SCATTERING THEORY

ON THE ELASTIC SCATTERING PROBLEM FROM CUBIC ANISOTROPIC INCLUSIONS

K. A. ANAGNOSTOPOULOS AND A. CHARALAMBOPOULOS

Department of Material Science and Engineering, University of Ioannina, 45110 Ioannina, Greece

The present work addresses the problem of a rigorous spectral analysis of the elasticity equations governing the elastic behavior of an anisotropic material. The analysis is based on a suitable diagonalization process finishing up with the decoupling of linear elasticity coupled differential system of equations and a plane wave expansion of the sought solutions provided that the anisotropic medium exhibits cubic symmetry. The theoretical results thus obtained are then exploited towards the analytic solution of the elastic scattering problem from a cubic anisotropic inclusion embedded in an isotropic medium. Computed results based on the numerical implementation of the analytical scheme demonstrate the effect of the anisotropy on the scattering process.

1. Introduction

Wave propagation and scattering of elastic waves in anisotropic media constitute areas of increased scientific interest since they are closely related to a great number of specific applications. Examples include the nondestructive testing, the characterization of material properties and the seismic theory. The majority of the models developed to analyze scattering processes in elastic media is concerned with isotropic materials due to obvious simplifications that such a consideration offers. However, the most part of real materials dispose more or less an anisotropic behavior and this fact renders the development of a scattering model incorporating such media in one of major concern. The present works aims at contributing towards this primary objective.

From a mathematical point of view, the success of an analytical approach to the anisotropic elastic scattering problem is strongly dependent on the anisotropy level of the investigated medium. Examples of scattering problems in two dimensions involving transversely isotropic media are [1-2] and [3] where scattering by a strip-like crack and a cavity is investigated, respectively. An analytic approach to the three-dimensional scattering problem by transversely isotropic cylinders is presented in [4-5] and by a cylindrical shell in [6]. On the other hand, when it comes to materials with a lower degree of symmetry, the investigation of the wave propagation and scattering phenomena is based upon numerical methods such as the finite

3

element method (FEM) [7], the elastodynamic finite integration technique (EFIT) [8], the boundary element method (BEM) [9] and the strip element method (SEM) [10].

The rest of this work is organized as follows. A spectral analysis of the equations governing the elastic behavior of a cubic anisotropic material is realized in Sec. 2. Employing an appropriate diagonalization procedure and adopting a plane wave expansion of the sought solutions decouples the differential system of elasticity equations. By the end of this section, a spectral representation for the elastic displacement field in the cubic anisotropic medium is in our disposal. In Sec. 3, the two-dimensional transmission scattering problem of time-harmonic elastic plane waves from a cubic anisotropic inclusion embedded in an isotropic background medium is stated and our approach towards its analytical solution is briefly discussed. Numerical examples based on the numerical implementation of the proposed scheme are presented in Sec. 4, exemplifying the effect of the specific anisotropic character on the scattering mechanism.

2. Spectral Analysis of the Equation of Cubic Elasticity

The equations of elasticity describe the vibration of an elastic medium in terms of an elastic displacement vector $\mathbf{U}: G \to R^2$ and a stress tensor (τ_{jk}). Assuming the usual tensor summation convention, stress and strain are connected by means of Hooke's law

$$\tau_{jk} = C_{jkmn} U_{mn}, \ j,k,m,n = 1,2, \tag{1}$$

while the components of the strain tensor in terms of the displacement field are

$$U_{mn} = \frac{1}{2}(\partial_m U_n + \partial_n U_m), \ m,n = 1,2, \tag{2}$$

where ∂_m indicates the partial derivative with respect to the Cartesian coordinate $x_m, m = 1,2$. The elastic moduli $C_{jkmn}, \ j,k,m,n = 1,2$, which characterize the stiffness properties of the material are real valued, bounded and measurable functions on G and satisfy the following symmetry relations

$$C_{jkmn} = C_{mnjk} = C_{kjmn}, \ j,k,m,n = 1,2. \tag{3}$$

In addition, the elastic moduli share coercivity properties expressed by the relation

$$\exists c > 0, \ \xi_{ij} \in C, \ \xi_{ij} = \xi_{ji}, \ \xi_{jk} C_{jkmn} \bar{\xi}_{mn} \geq c \sum_{j,k} |\xi_{jk}|^2, \tag{4}$$

where the bar denotes complex conjugation.

Adopting the terminology presented in [11], we use the Sommerfeld symbolism for the stresses and strains defining

$$\alpha_1 := \tau_{11}, \ \alpha_2 := \tau_{22}, \ \alpha_3 := \tau_{12}$$
$$\varepsilon_1 := U_{11}, \ \varepsilon_2 := U_{22}, \ \varepsilon_3 := 2U_{12},$$

thus obtaining the alternative form of Hooke's law

$$\alpha_j = S_{jk}\varepsilon_k, \ j,k = 1,2,3 \qquad (5)$$

where we have introduced the 3×3 symmetric positive definite stiffness matrix

$$S = \begin{pmatrix} C_{1111} & C_{1122} & C_{1112} \\ . & C_{2222} & C_{2212} \\ . & . & C_{1212} \end{pmatrix}. \qquad (6)$$

We remark that the cubic medium under consideration is completely characterized by three independent stiffness coefficients $C_{1111}, C_{1122}, C_{1212}$, while the remaining elastic moduli are assigned the values $C_{1112} = C_{2212} = 0$ and $C_{2222} = C_{1111}$. Therefore matrix $S = [S_{jk}]$ can also be written as follows

$$S = \begin{pmatrix} S_{11} & S_{12} & 0 \\ S_{12} & S_{11} & 0 \\ 0 & 0 & S_{33} \end{pmatrix}, \qquad (7)$$

where the specific values of its entries in terms of the elastic moduli are obvious. The stiffness entry S_{33} actually corresponds to the relative shear stiffness S_{44} in terms of the Voigt's contracted notation but we keep with this terminology for notational convenience. It is also noted that Eqs. (5) with S given by Eq. (7) correspond to the two-dimensional stress-strain relations of the cubic material in coordinates aligned with principal material directions. The degenerate case of an isotropic medium is obtained in case that $S_{11} = S_{12} + 2S_{33}$. Moreover, the coercivity property (4) implies that $S_{ij} > 0$, $i, j = 1,2$, and $S_{11} > S_{12}$.

Hence the equations of elasticity, whose componentwise form in the case that no external forces apply on region G is

$$\rho \ddot{U}_j - \partial_k \tau_{jk} = 0, \ j = 1,2, \qquad (8)$$

where ρ is the material density and the dot symbol indicates time differentiation, obtain the form

$$\rho \ddot{U} - D'SDU = 0, \qquad (9)$$

where we have used the generalized gradient operator

$$D = \begin{pmatrix} \partial_1 & 0 \\ 0 & \partial_2 \\ \partial_2 & \partial_1 \end{pmatrix}. \tag{10}$$

and the prime denotes the transposed operator. Formally, Eq. (9) is a wave equation of the form

$$\rho\ddot{u} - \mathrm{div}A\,\mathrm{grad}\mathbf{u} = \mathbf{0}, \tag{11}$$

expressing in condensed form the equations of motion of the elastic medium. Assuming that the anisotropic medium undergoes harmonic oscillations of the form $e^{-i\omega t}$ with angular frequency ω, Eq. (9) reduces to the homogeneous coupled system of equations of elasticity

$$S_{11}\partial_1^2 U_1 + (S_{12} + S_{33})\partial_{12} U_2 + S_{33}\partial_2^2 U_1 + \rho\omega^2 U_1 = 0$$
$$S_{33}\partial_1^2 U_2 + (S_{12} + S_{33})\partial_{12} U_1 + S_{11}\partial_2^2 U_2 + \rho\omega^2 U_2 = 0. \tag{12}$$

Decoupling this differential system is equivalent to consider the following eigenvalue-eigenvectors problem

$$\begin{pmatrix} S_{11}\partial_1^2 + S_{33}\partial_2^2 + \rho\omega^2 - \lambda & (S_{12} + S_{33})\partial_{12} \\ (S_{12} + S_{33})\partial_{12} & S_{33}\partial_1^2 + S_{11}\partial_2^2 + \rho\omega^2 - \lambda \end{pmatrix} \begin{pmatrix} U_1^{(\lambda)} \\ U_2^{(\lambda)} \end{pmatrix} = \begin{pmatrix} 0 \\ 0 \end{pmatrix}, \tag{13}$$

where λ stands for a suitable differential operator to be determined shortly and $\left(U_1^{(\lambda)} \quad U_2^{(\lambda)}\right)^T$ is the corresponding eigenvector. Demanding from the determinant of the matrix in Eq. (13) to vanish we obtain

$$\lambda^2 - \{(S_{11} + S_{33})(\partial_1^2 + \partial_2^2) + 2\rho\omega^2\}\lambda +$$
$$(S_{11}\partial_1^2 + S_{33}\partial_2^2 + \rho\omega^2)(S_{33}\partial_1^2 + S_{11}\partial_2^2 + \rho\omega^2) - (S_{12} + S_{33})^2\partial_1^2\partial_2^2 = 0. \tag{14}$$

After straightforward manipulations the discriminant of the above quadratic equation in λ is found to be $d = (S_{11} - S_{33})^2\Delta^2 - 4\tilde{S}\partial_1^2\partial_2^2$, where

$$\tilde{S} = (S_{11} + S_{12})(S_{11} - S_{12} - 2S_{33}), \tag{15}$$

and Δ is the well known Laplace operator. It is to be noticed that dealing with an isotropic medium implies the vanishing of the parameter $\tilde{S}$ and this fact renders $\tilde{S}$ the suitable quantity representing the declination of the cubic medium from the isotropic one. Solving the operational quadratic equation (14), we determine the eigenvalue operators $\lambda_i, i = 1,2$,

$$\lambda_{1,2} = \frac{(S_{11} + S_{33})\Delta + 2\rho\omega^2 \pm \sqrt{d}}{2}, \tag{16}$$

which in the isotropic case become $\lambda_1 = S_{11}\Delta + \rho\omega^2$ and $\lambda_2 = S_{33}\Delta + \rho\omega^2$, referring to the well known Helmholtz differential operators concerning the longitudinal and transverse elastic waves, respectively.

Let $U = (U_1 \quad U_2)^T$ be a solution of system (12). Then the field U is represented as

$$U = PV, \tag{17}$$

where

$$P = \begin{pmatrix} U_1^{(\lambda_1)} & U_1^{(\lambda_2)} \\ U_2^{(\lambda_1)} & U_2^{(\lambda_2)} \end{pmatrix}, \tag{18}$$

is the eigenvector matrix and $V = (V_1 \quad V_2)^T$ is an auxiliary field whose components $V_i, i = 1,2$ are solutions of the decoupled equations

$$\lambda_1 V_1 = 0 \Leftrightarrow \left\{ (S_{11} + S_{33})\Delta + 2\rho\omega^2 + \sqrt{d} \right\} V_1 = 0, \text{(19.a)}$$

$$\lambda_2 V_2 = 0 \Leftrightarrow \left\{ (S_{11} + S_{33})\Delta + 2\rho\omega^2 - \sqrt{d} \right\} V_2 = 0. \text{(19.b)}$$

In order to determine the functions $V_i, i = 1,2$, we first exploit the commuting property of $\lambda_i, i = 1,2$ which assures that $\lambda_1\lambda_2 V_i = 0, i = 1,2$. Defining the operator L by

$$L := \left[(S_{11} + S_{33})\Delta + 2\rho\omega^2 \right]^2 - d, \tag{20}$$

it can be proved that the determination of the sought functions V_i, $i=1,2$ satisfying $LV_i, i=1,2$ reduces to the solution of the fourth order scalar differential equation $L\Psi=0$. Indeed, for every Ψ satisfying $L\Psi = 0$, we easily verify that $W_1 = \lambda_2\Psi$ (res. $W_2 = \lambda_1\Psi$) satisfies the equation $\lambda_1 W_1 = 0$ (res. $\lambda_2 W_2 = 0$) and therefore the pair $(W_1 \quad W_2)^T$ is a suitable candidate pair of functions for the sought pair of solutions $(V_1 \quad V_2)^T$. We postpone the treatment of the equation $L\Psi = 0$ and proceed with the implementation of the transformation induced by Eq. (17). Adopting the reasonable symbol correspondence $\lambda_1 \Leftrightarrow (+)$ and $\lambda_2 \Leftrightarrow (-)$ we find from (13) that

$$\left[\frac{S_{11} - S_{33}}{2}\left(\partial_1^2 - \partial_2^2\right) \mp \frac{\sqrt{d}}{2} \right] U_1^{(\pm)} + (S_{12} + S_{33})\partial_1\partial_2 U_2^{(\pm)} = 0. \tag{21}$$

Consequently, the matrix P given by (18) can be selected as follows

$$P = \begin{pmatrix} (S_{12} + S_{33})\partial_1\partial_2 & -(S_{12} + S_{33})\partial_1\partial_2 \\ \frac{S_{11} - S_{33}}{2}\left(\partial_2^2 - \partial_1^2\right) + \frac{\sqrt{d}}{2} & \frac{S_{11} - S_{33}}{2}\left(\partial_1^2 - \partial_2^2\right) + \frac{\sqrt{d}}{2} \end{pmatrix} A, \tag{22}$$

where A is an arbitrary operator commuting with the differential operators appeared in the entries of matrix P (actually A can be considered as part of a redefined vector V). The transformation (17) now reads

$$U_1 = (S_{12} + S_{33})\partial_1\partial_2 A(V_1 - V_2), \quad (23.\text{a})$$

$$U_2 = \frac{S_{11} - S_{33}}{2}(\partial_2^2 - \partial_1^2)A(V_1 - V_2) + \frac{\sqrt{d}}{2}A(V_1 + V_2). \quad (23.\text{b})$$

Given that $V_1 = \lambda_2\Psi$ and $V_2 = \lambda_1\Psi$, where Ψ solves $L\Psi = 0$ and selecting $A = (\sqrt{d})^{-1}/(S_{12} + S_{33})$ we obtain the differential representation

$$U_1 = -\partial_1\partial_2\Psi, \quad (24.\text{a})$$

$$U_2 = \frac{1}{(S_{12} + S_{33})}(S_{11}\partial_1^2 + S_{33}\partial_2^2 + \rho\omega^2)\Psi. \quad (24.\text{b})$$

It is easily verified that the pair (24) satisfies the coupled differential system (12) with $\Psi \in \ker L$.

What remains is the investigation of the fourth order scalar differential equation $L\Psi = 0$. We seek solutions expressed via plane waves of the form $e^{-ik\hat{\mathbf{k}}\cdot\mathbf{r}}$ where $\mathbf{r} = (x_1, x_2)$ is the position vector and $\hat{\mathbf{k}} = (\cos u, \sin u)$, $u \in [0, 2\pi)$ is the plane wave propagation vector. Forcing the plane waves to belong to $\ker L$ we find that

$$\{S_{11}S_{33}\Delta^2 + \rho\omega^2(S_{11} + S_{33})\Delta + \tilde{S}\partial_1^2\partial_2^2 + \rho^2\omega^4\}(e^{-ik\hat{\mathbf{k}}\cdot\mathbf{r}}) = 0 \Rightarrow$$

$$(S_{11}S_{33} + \tilde{S}\cos^2 u\sin^2 u)k^4 - \rho\omega^2(S_{11} + S_{33})k^2 + \rho^2\omega^4 = 0. \quad (25)$$

Solving the dispersion equation (25) in terms of the wavenumber k we obtain

$$k^2 = k_\pm^2(u) = \rho\omega^2\frac{(S_{11} + S_{33}) \pm \sqrt{(S_{11} - S_{33})^2 - 4\tilde{S}\cos^2 u\sin^2 u}}{2(S_{11}S_{33} + \tilde{S}\cos^2 u\sin^2 u)}, \quad (26)$$

observing that the wave numbers $k_\pm(u)$ are functions of the propagation direction cosines. The established constraints on the values of the elastic moduli assure that all the quantities appearing in Eq. (26) are well defined. Hence the solution Ψ of $L\Psi = 0$ can be expanded in terms of the above constructed plane waves as follows

$$\Psi(\mathbf{r}) = \int_0^{2\pi}\left[A_+(u)e^{-ik_+(u)\hat{\mathbf{k}}\cdot\mathbf{r}} + A_-(u)e^{-ik_-(u)\hat{\mathbf{k}}\cdot\mathbf{r}}\right]du, \quad (27)$$

with initially arbitrary coefficients $A_\pm(u)$ to be determined from the boundary conditions of the problem. Exploiting a Fourier type expansion for the functions $A_\pm(u)$, the solution $\Psi(\mathbf{r})$ is represented as

$$\Psi(\mathbf{r}) = \sum_{m \in Z} A_m^- \frac{1}{2\pi i^m} \int_0^{2\pi} e^{-ik_-(u)\hat{\mathbf{k}}\cdot\mathbf{r}} e^{imu} \, du + \sum_{m \in Z} A_m^+ \frac{1}{2\pi i^m} \int_0^{2\pi} e^{-ik_+(u)\hat{\mathbf{k}}\cdot\mathbf{r}} e^{imu} \, du$$

(28)

where $k_\pm(u)$ are given by the square roots of the r.h.s. of Eq. (26) and $A_m^\pm$ are the Fourier expansion coefficients. It is noticed that this representation in the isotropic limit reduces to the usual spectral analysis of Helmholtz equation solutions in interior domains containing the origin O in polar coordinates [12]. In view of Eq. (28), the Cartesian components of the sought elastic field $\mathbf{U} = (U_1, U_2)$ given by Eqs. (24) obtain the following spectral representation

$$U_1 = \frac{1}{4i} \sum_\pm \sum_{m \in Z} A_m^\pm \left\{ \Theta_{m+2}^\pm(\mathbf{r}) - \Theta_{m-2}^\pm(\mathbf{r}) \right\}, \quad (29.a)$$

$$U_2 = \sum_\pm \sum_{m \in Z} A_m^\pm \left\{ \frac{\rho\omega^2}{(S_{12} + S_{33})} \Phi_m^\pm(\mathbf{r}) - \Theta_m^\pm(\mathbf{r}) \right.$$

$$\left. + \frac{(S_{33} - S_{11})}{4(S_{12} + S_{33})} \left(\Theta_{m+2}^\pm(\mathbf{r}) + \Theta_{m-2}^\pm(\mathbf{r}) \right) \right\}, \quad (29.b)$$

where the functions $\Phi_m^\pm(\mathbf{r})$ and $\Theta_m^\pm(\mathbf{r})$ constituting the basis of this representation are defined by the following integral expressions

$$\Phi_m^\pm(\mathbf{r}) = \frac{1}{2\pi i^m} \int_0^{2\pi} e^{-ik_\pm(u)\hat{\mathbf{k}}\cdot\mathbf{r}} e^{imu} \, du, \quad (30)$$

$$\Theta_m^\pm(\mathbf{r}) = \frac{1}{2\pi i^m} \int_0^{2\pi} k_\pm^2(u) e^{-ik_\pm(u)\hat{\mathbf{k}}\cdot\mathbf{r}} e^{imu} \, du. \quad (31)$$

3. The Transmission Scattering Problem

In the sequel, our previous results concerning the spectral representation of the elastic displacement field in a cubic symmetric anisotropic material are exploited in order to solve a specific boundary value problem (BVP), namely the elastic transmission scattering problem. In particular, we consider the two-dimensional scattering problem of the elastic plane wave $\mathbf{u}^{inc}(\mathbf{r})$ from a cubic symmetric scatterer occupying region G which is characterized by the fourth-order stiffness tensor $\widetilde{C}$ and the material density ρ_{in} and is embedded in an isotropic background medium with Lame constants λ, μ and density ρ_{ex}. The plane wave is either longitudinally (P-wave) or transversely (S-wave) polarized and propagates at an angle k_{inc} with respect to the principal material axis x_1. The interference of the incident plane wave with the anisotropic scatterer leads to the creation of the scattered field $\mathbf{u}^{sc}(\mathbf{r}) = \mathbf{u}_p^{sc}(\mathbf{r}) + \mathbf{u}_s^{sc}(\mathbf{r})$, $\mathbf{r} \in \mathbb{R}^2 \setminus \overline{G}$ consisting of a longitudinal and a transverse part, which propagates with wavenumber $k_p^{ex} = \omega\sqrt{\rho_{ex}/(\lambda + 2\mu)}$ and $k_s^{ex} = \omega\sqrt{\rho_{ex}/\mu}$, respectively, and the transmitted field $\mathbf{u}^{in}(\mathbf{r})$, $\mathbf{r} \in G$. For the statement of the transmission scattering problem in mathematical terms we refer to [13].

The analytical solution of this well-posed BVP is reached by performing a Navier eigenvectors expansion of the scattered displacement field [14], utilizing the spectral representation (29) for the description of the elastic field in the interior of the cubic anisotropic scatterer and requiring the satisfaction of the transmission boundary conditions-assuring displacement and stress continuity on scatterer's surface-for the involved (scattered, transmitted and incident) displacement fields. The details of the described solvability process can be found in [15]. Solving the underlying scattering problem by means of a suitable numerical scheme finally leads to the complete determination of the transmitted and the scattered field. In particular, the radial $g_r(\phi)$ and tangential $g_\phi(\phi)$ scattering amplitudes, which characterize the far-field behavior of the scattered field through the representation

$$\mathbf{u}_\infty^{sc}(\mathbf{r}) = \frac{e^{ik_p^{ex}r}}{\sqrt{r}}\,g_r(\phi)\hat{\mathbf{r}} + \frac{e^{ik_s^{ex}r}}{\sqrt{r}}\,g_\phi(\phi)\hat{\phi} + O(r^{-3/2}), \qquad (32)$$

are then available.

4. Numerical Results-Discussion

In this section we present some computed results based on the numerical implementation of the proposed analytical scheme. We examine the scattering by an infinite cylinder of circular cross-section filled with a cubic anisotropic medium under plane wave excitation of either longitudinal or transverse type. For the isotropic background medium we assume the values $\lambda = \mu = 1.0\,[\times 10^{10}\,\text{Nt/m}^2]$ for the Lame constants and the density $\rho_{ex} = 1.0\,[\times 10^3\,\text{Kgr/m}^3]$, for the cubic material the density is

$\rho_{in} = 2.0 \, [\times 10^3 \, \text{Kgr/m}^3]$ and $k_p^{ex} a = 1.0$ is the dimensionless frequency, where a is the radius of the circular cross-section. Moreover, we use the cubic anisotropy ratio $A = 2S_{33} / (S_{11} - S_{12})$, which is also referred to as the dimensionless Zener anisotropy factor, with the degree of anisotropy measured by the deviation of A from the value $A = 1.0$, valid for an isotropic medium. In addition, we select angles of incidence in the range of 0 to $\pi/2$, including $\pi/4$, which coincides with a symmetry plane of the cubic material. We remind that the cubic material in the 2D-case has four planes of symmetry whose normal are on the coordinate axes and on the coordinate planes making an angle $\pi/4$ with the coordinate axes.

In Figs. 1-3, the angular distribution of the radial scattering amplitude $|g_r(\phi)|$ is shown for several angles of longitudinal P-incidence k_{inc} and varying anisotropy factor A. In the case of $k_{inc} = 0$ the effect of the anisotropy is pronounced in both the forward and backscattering direction for any value of A. We also observe that the distributions are symmetric with respect to the direction of incidence since the latter coincides with a plane of material symmetry. For $\pi/4$ incidence, the symmetry is also preserved but the effect of the anisotropy on the computed quantities is weaker, especially in the forward scattering direction, for any value of A. Finally, for an angle of incidence which does not coincide with a plane of symmetry, e.g. $\pi/9$, the symmetry w.r.t. the direction of incidence is no longer preserved, except, of course, when $A = 1.0$ corresponding to isotropy.

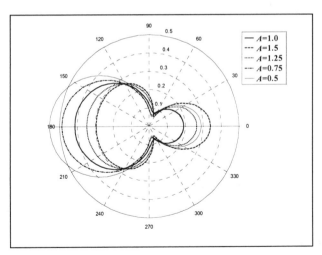

Figure 1. Distribution of the radial scattering amplitude $|g_r|$ for P-incidence with k_{inc}=0.

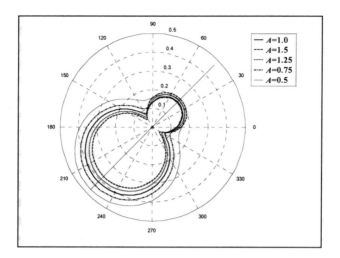

Figure 2. Distribution of the radial scattering amplitude $|g_r|$ for P-incidence with $k_{inc}=\pi/4$.

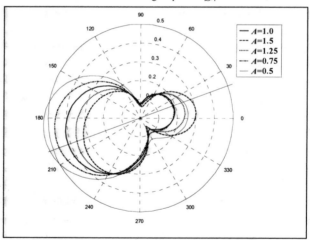

Figure 3. Distribution of the radial scattering amplitude $|g_r|$ for P-incidence with $k_{inc}=\pi/9$

In Figs. 4-6, the magnitude of the tangential scattering amplitude $|g_\phi(\phi)|$ for transverse S-incidence at several angles is examined. In this case, $k_{inc} = 0$ has almost no effect on the computed quantity for any value of the anisotropy factor. In contrast, a $\pi/4$ incidence is shown to be the one that reveals the anisotropy and up to a point its measure. Our previous remarks relatively to the symmetry of the distributions w.r.t. the angle of incidence are also valid here as one may observe from the case of $\pi/9$ incidence.

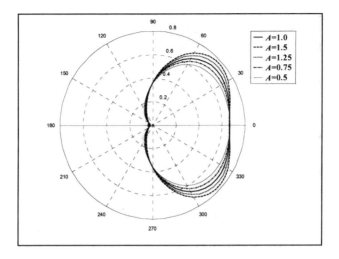

Figure 4. Distribution of the tangential scattering amplitude $|g_\varphi|$ for S-incidence with $k_{inc}=0$.

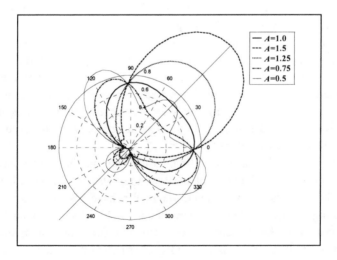

Figure 5. Distribution of the tangential scattering amplitude $|g_\varphi|$ for S-incidence with $k_{inc}=\pi/4$.

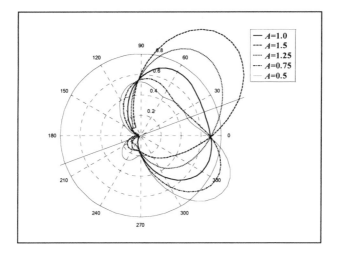

Figure 6. Distribution of the tangential scattering amplitude $|g_\varphi|$ for S-incidence with $k_{inc}=\pi/9$.

We conclude with a few comments on a 'mixed' case, in the sense that the incidence is of the transverse type, i.e. an S-wave, and the radial scattering amplitude $|g_r(\phi)|$ is observed. The corresponding illustrations can be found in [15]. In this case, for $k_{inc}=0$ the distributions almost coincide for any value of A thus providing a clear indication that such a combination of incidence-observation is unsuitable for detecting the presence of the anisotropy. Similarly, in the case of $k_{inc} = \pi/4$ the effect of the anisotropy, as far as the magnitudes of the computed quantities are concerned, is significantly smaller compared with the corresponding case where the tangential scattering amplitude was observed. In agreement with previously examined examples, a loss of symmetry is again observed for an intermediate incidence of $\pi/9$.

The presented results clearly indicate that the information on the specific anisotropic character of the scattering structure is systematically encoded in the far-field scattered wave. We have also obtained results concerning boundary values of the transmitted field, which are in qualitative agreement with the presented ones [15]. To conclude, an appropriate interpretation of the observation results referring to either far-field or near-field quantities for suitably chosen types and angles of incidence seems to offer the possibility of recovering some special features of the scatterer's anisotropic behavior.

References

1. J. Mattsson, Report 1994: **1**, Div. Mech., Chalmers Univ. Tech., Goteborg, Sweden (1994).
2. J. Mattsson, Report 1995: **3**, Div. Mech., Chalmers Univ.Tech., Goteborg, Sweden (1995).
3. R. K. N. D. Rajapakse and D. Gross, *Wave Motion* **21**, 231-252 (1995).
4. F. Honarvar and A.N. Sinclair, *J. Acoust. Soc. Am.* **100**, 57-63 (1996).
5. A. J. Niklasson and S. K. Datta, *Wave Motion* **27**, 169-185 (1998).
6. J.-Y. Kim and J.-G Ih, *Appl. Acoust.* **64**, 1187-1204 (2003).
7. W. Lord, R. Ludwig and Z. You, *J. Nondestr. Eval.* **9**, 129-143 (1990).
8. P. Fellinger, R. Marklein, K. J. Langenberg and S. Klaholz, *Wave Motion* **21**, 47-66 (1995).
9. C.-Y. Wang, J. D. Achenbach and S. Hirose, *Int. J. Solids Structures* **33**, 3843-3864 (1996).
10. G. R. Liu, J. D. Achenbach, *J. Appl. Mech.* **62**, 607-613 (1995).
11. R. Leis, *Initial Boundary Value Problems in Mathematical Sciences*, Wiley, New York (1986).
12. P. M. Morse and H. Feshbach, *Methods of Theoretical Physics*, Vol. II, McGraw-Hill, New York (1953).
13. A. Charalambopoulos, *Inverse Problems* **67**, 149-170 (2002).
14. A. Ben-Menahem and S. J. Singh, *Seismic Waves and Sources*, Springer, New York (1981).
15. K. A. Anagnostopoulos and A. Charalambopoulos, Orthotropic Elastic Materials: Spectral Analysis of the Governing Equations and Solution of the Associated Scattering Problem, submitted for publication (2003).

ACCURACY ANALYSIS AND OPTIMIZATION OF THE METHOD OF AUXILIARY SOURCES (MAS) FOR SCATTERING FROM A DIELECTRIC CIRCULAR CYLINDER

HRISTOS T. ANASTASSIU AND DIMITRA I. KAKLAMANI

Institute of Communication and Computer Systems
School of Electrical and Computer Engineering
National Technical University of Athens
Heroon Polytechniou 9
GR-15780 Zografou, Athens
GREECE
e-mail:hristosa@esd.ece.ntua.gr

This paper presents a rigorous error analysis of the Method of Auxiliary Sources (MAS), when the latter is applied to the solution of the electromagnetic scattering problem involving a circular, dielectric cylinder of infinite length. The MAS matrix is shown to be analytically invertible via advanced eigenvalue analysis, and thus an exact expression for the discretization error can be derived. Large discretization errors are proven to be partly associated with internal resonances of the scattering body. Furthermore, an exact, analytical formula for the condition number of the linear system is extracted, and its properties are utilized to explain many aspects of the irregular behavior of the computational error, resulting from numerical matrix inversion. Finally, the optimal locations of both sets of auxiliary sources (inner and outer) are determined, on the grounds of error minimization.

1. Introduction

This is essentially the second part of an extensive work investigating the behavior of the Method of Auxiliary Sources (MAS) [1]-[5] when the latter is applied to electromagnetic scattering from non-metallic objects. A companion paper [5] was related to a perfectly conducting (PEC) circular cylinder, coated with a thin dielectric layer, modelled as a single entity by the Standard Impedance Boundary Condition (SIBC). In this paper, the case of a dielectric circular cylinder is examined. The same analytical method will be invoked, i.e. eigenvalue evaluation, diagonalization and subsequent analytical inversion of the MAS matrix. The mathematics for the dielectric cylinder is far more complicated than in the PEC [4] or the SIBC [5] case, because two, instead of one, sets of auxiliary sources (AS's) are involved in the solution, one inside and the other outside the scatterer, as described in [2,3]. The appropriate boundary conditions for both the electric and the magnetic field must be satisfied. Analytical inversion of the MAS matrix yields exact expressions for the fields radiated by the AS's, and also an analytical expression for the midpoint (MP) error, like in the PEC and SIBC cases. Minimization of the MP error facilitates the determination of the optimal location of the AS's, which is the principal question in any MAS formulation. A $e^{j\omega t}$ time convention is assumed and suppressed throughout the paper.

2. Analytical Considerations (TM Polarization)

We assume a dielectric, infinite, circular cylinder of radius b characterized by complex relative permittivity ε_r and relative permeability equal to 1. The dielectric is assumed to be linear, homogeneous and isotropic. The structure is illuminated by a plane wave impinging from a direction, determined by polar angle ϕ^i (see Fig. 1). The polarization of the plane wave is assumed to be transverse magnetic (TM) with respect to the cylinder axis z.

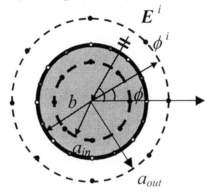

Fig. 1. Geometry of the problem. Black, white and gray bullets represent Auxiliary Sources (AS's), Collocation Points (CP's) and Midpoints (MP's) respectively. The gray disk corresponds to the dielectric cylinder.

To construct the MAS solution [2,3], two separate, fictitious auxiliary surfaces S^{in} and S^{out} are defined, both conformal to the actual boundary S. The first surface S^{in} is located inside the dielectric scatterer, and hence has a circular cross section of radius $a_{in} < b$. The second surface S^{out} is located outside the dielectric scatterer, and hence its radius is $a_{out} > b$ (see Fig. 1). A number of N AS's, in the form of elementary electric currents, are located on each one of S^{in} or S^{out}, radiating elementary electric fields, proportional to the two-dimensional Green's function. The AS's on S^{in} radiate outside the scatterer, whereas AS's on S^{out} radiate inside it. Matching the boundary condition (equality of the electric and magnetic tangential fields) at $M = N$ collocation points (CP's) on the z-plane projection of the scatterer surface, and using the addition theorem for cylindrical functions [6,7], yields the MAS square linear system. This system can finally be written, in a compact, block matrix form as

$$
\begin{bmatrix} [U] & [V] \\ [W] & [Y] \end{bmatrix} \begin{Bmatrix} \{C\} \\ \{D\} \end{Bmatrix} = \begin{Bmatrix} \{A^i\} \\ \{B^i\} \end{Bmatrix} \tag{1}
$$

where $[U], [V], [W], [Y]$ are $N \times N$ square matrices with elements given by

$$
U_{mn} \equiv -\frac{j}{4} \sum_{l=-\infty}^{\infty} J_l(kb) H_l^{(2)}(ka_{out}) \exp\{-jl(\phi_m - \phi_n)\} \tag{2}
$$

$$
V_{mn} \equiv \frac{j}{4} \sum_{l=-\infty}^{\infty} J_l(k_0 a_{in}) H_l^{(2)}(k_0 b) \exp\{-jl(\phi_m - \phi_n)\} \tag{3}
$$

$$
W_{mn} \equiv -\frac{j\sqrt{\varepsilon_r}}{4} \sum_{l=-\infty}^{\infty} \dot{J}_l(kb) H_l^{(2)}(ka_{out}) \exp\{-jl(\phi_m - \phi_n)\} \tag{4}
$$

$$
Y_{mn} \equiv \frac{j}{4} \sum_{l=-\infty}^{\infty} J_l(k_0 a_{in}) \dot{H}_l^{(2)}(k_0 b) \exp\{-jl(\phi_m - \phi_n)\} \tag{5}
$$

In (2)-(5), $\phi_m \equiv m \cdot 2\pi/N$, $m=1,\ldots,N$, k is the wavenumber inside the dielectric, k_0 is the free-space wavenumber, $J_l(\bullet)$ is the Bessel function of l order, $H_l^{(2)}(\bullet)$ is the Hankel function of l order and second kind, and the dot over the cylindrical functions denotes differentiation with respect to the *entire* argument. Furthermore, $\{A^i\}, \{B^i\}$ are $N \times 1$ column vectors, with

$$
A_n^i \equiv E_0 \exp\{jk_0 b\cos(\phi^i - \phi_n)\} \tag{6}
$$

$$
B_n^i \equiv jE_0 \cos(\phi^i - \phi_n) \exp\{jk_0 b\cos(\phi^i - \phi_n)\} \tag{7}
$$

representing the samples of the incident fields, and finally, $\{C\}, \{D\}$ are $N \times 1$ column vectors containing the unknown AS weights.

To derive an expression for the boundary condition error, in a way similar to [4,5], (1) must be inverted analytically. Such an inversion is feasible, given that each one of $[U], [V], [W], [Y]$ is diagonalizable, on the basis of the method described in [4,5], [7]. Indeed, invoking the diagonalization scheme of [4,5], it can be shown that

$$[U] = [G][D_u][G]^{-1}, \ [V] = [G][D_v][G]^{-1} \tag{8}$$

$$[W] = [G][D_w][G]^{-1}, \ [Y] = [G][D_y][G]^{-1} \tag{9}$$

where $[D_u], [D_v], [D_w], [D_y]$ are diagonal matrices containing the eigenvalues of $[U], [V], [W], [Y]$ respectively, given by ($q = 1,...,N$)

$$u_q = -\frac{jN}{4} \sum_{s=-\infty}^{\infty} J_{q+sN}(kb) H_{q+sN}^{(2)}(ka_{out}) \tag{10}$$

$$v_q = \frac{jN}{4} \sum_{s=-\infty}^{\infty} J_{q+sN}(k_0 a_{in}) H_{q+sN}^{(2)}(k_0 b) \tag{11}$$

$$w_q = -\frac{j\sqrt{\varepsilon_r} N}{4} \sum_{s=-\infty}^{\infty} \dot{J}_{q+sN}(kb) H_{q+sN}^{(2)}(ka_{out}) \tag{12}$$

$$y_q = \frac{jN}{4} \sum_{sl=-\infty}^{\infty} J_{q+sN}(k_0 a_{in}) \dot{H}_{q+sN}^{(2)}(k_0 b) \tag{13}$$

and $[G]$ is the eigenvector square matrix (common for all $[U], [V], [W], [Y]$), defined by

$$[G] \equiv [\{g\}_1, \{g\}_2, ..., \{g\}_N] \tag{14}$$

where

$$\{g\}_q \equiv \frac{1}{\sqrt{N}} [\exp\{-jq\phi_1\}, \exp\{-jq\phi_2\}, ..., \exp\{-jq\phi_N\}]^T \tag{15}$$

are the normalized eigenvectors, identical to the PEC [4] or SIBC [5] cases. Using (8)-(15) the square matrix in (1) can be written as

$$[Z] \equiv \begin{bmatrix} [U] & [V] \\ [W] & [Y] \end{bmatrix} = \begin{bmatrix} [G] & [0] \\ [0] & [G] \end{bmatrix} \begin{bmatrix} [D_u] & [D_v] \\ [D_w] & [D_y] \end{bmatrix} \begin{bmatrix} [G]^{-1} & [0] \\ [0] & [G]^{-1} \end{bmatrix} \tag{16}$$

where $[0]$ is the $N \times N$ null matrix, and therefore its inverse is

$$[Z]^{-1} = \begin{bmatrix} [G] & [0] \\ [0] & [G] \end{bmatrix} \begin{bmatrix} [D_y] & -[D_v] \\ -[D_w] & [D_u] \end{bmatrix} \begin{bmatrix} [P]^{-1} & [0] \\ [0] & [P]^{-1} \end{bmatrix} \begin{bmatrix} [G]^{-1} & [0] \\ [0] & [G]^{-1} \end{bmatrix} \tag{17}$$

where

$$[P] \equiv [D_u][D_y] - [D_w][D_v] \tag{18}$$

is the block matrix "determinant" (the proof of (17) is straightforward, by showing directly that $[Z]^{-1}[Z] = [Z][Z]^{-1} = [I]$, i.e. that both products equal the identity matrix). Hence, inversion of (1) in view of (17) yields an analytic expression for the unknown weights.

Suppose, now, that we are interested in calculating the boundary condition error at points of the boundary surface with azimuth angles equal to $\phi_m + \tilde{\phi}$, where $0 \le \tilde{\phi} \le 2\pi/N$. Obviously the choice $\tilde{\phi} = \pi/N$ corresponds to the midpoints (MP's) between the CP's (see Fig. 1). The MP normalized error in the boundary condition can be defined, in a manner analogous to [4,5] by

$$e(a_{in}, a_{out}, b, N) \equiv \frac{\left\| \begin{Bmatrix} \{\tilde{A}^{rad}\} \\ \{\tilde{B}^{rad}\} \end{Bmatrix} - \begin{Bmatrix} \{\tilde{A}^i\} \\ \{\tilde{B}^i\} \end{Bmatrix} \right\|_2}{\left\| \begin{Bmatrix} \{A^i\} \\ \{B^i\} \end{Bmatrix} \right\|_2} \tag{19}$$

where $\{\tilde{A}^i\}\{\tilde{B}^i\}$ are the incident fields evaluated at the MP's, and $\|\bullet\|_2$ is the standard 2-norm. It is easy to show, that when the MP's coincide with the CP's, i.e. when $\tilde{\phi} = 0$, the error in (19) vanishes, as expected. The error also vanishes in the limit as $N \to \infty$, verifying the convergence properties of MAS, just like in the PEC [4] and the SIBC [5] cylinders. In the general case, (19) can be evaluated explicitly after a considerable amount of tedious algebra. The final result for the normalized error can be written as

$$e(a_{in}, a_{out}, b, N) = \sqrt{\frac{\displaystyle\sum_{m=1}^{N} |e_{m1}|^2 + \sum_{m=1}^{N} |e_{m2}|^2}{\frac{3}{2} N E_0^2}} \tag{20}$$

where

$$e_{m1} \equiv \frac{1}{N} \sum_{n=1}^{N} \sum_{p=1}^{N} \exp\{-j(m-n)\phi_p\}\left(l_p^{(11)} A_n^i + l_p^{(12)} B_n^i\right) - \tilde{A}_m^i \qquad (21)$$

$$e_{m2} \equiv \frac{1}{N} \sum_{n=1}^{N} \sum_{p=1}^{N} \exp\{-j(m-n)\phi_p\}\left(l_p^{(21)} A_n^i + l_p^{(22)} B_n^i\right) - \tilde{B}_m^i \qquad (22)$$

whereas $\phi_p \equiv p \cdot 2\pi/N$ and

$$l_q^{(11)} \equiv \frac{\tilde{u}_q y_q - \tilde{v}_q w_q}{u_q y_q - v_q w_q}, \qquad l_q^{(12)} \equiv \frac{\tilde{v}_q u_q - \tilde{u}_q v_q}{u_q y_q - v_q w_q} \qquad (23)$$

$$l_q^{(21)} \equiv \frac{\tilde{w}_q y_q - \tilde{y}_q w_q}{u_q y_q - v_q w_q}, \qquad l_q^{(22)} \equiv \frac{\tilde{y}_q u_q - \tilde{w}_q v_q}{u_q y_q - v_q w_q} \qquad (24)$$

while the tilde sign denotes evaluation at the MP's, similarly to [4,5]. To achieve the highest possible accuracy for the MAS solution, e in (20), must be minimized by choosing the most appropriate a_{in} and a_{out} for given b and N.

Like in the PEC [4] or the SIBC [5] cases, the analytical expression for the boundary condition error reveals the occurrence of resonance effects, i.e. situations where a poor choice of the AS's location may cause very high errors. In the dielectric case, it follows from (20)-(24) that a resonance occurs when

$$u_q y_q - v_q w_q = 0 \qquad (25)$$

After some algebraic manipulation, it turns out that (25) is equivalent to $J_q(k_0 a_{in}) = 0$, which is identical with the PEC [4] and SIBC [5] cases, and is the only condition that may result in resonance effects. Hence, in the dielectric cylinder case, like in all other situations, it is the location of the *interior* AS's that should be carefully chosen to avoid any poor behavior of the method.

To fully assess the accuracy of the numerical MAS solution, it is important to investigate the behavior of the matrix condition number, since the latter largely determines the significance of the computational (round-off) errors. Unlike in the PEC [4] or the SIBC [5] cases, the MAS square matrix for the dielectric cylinder, given in the left hand side of (1), is not normal. Therefore, the condition number κ_2 cannot be determined by the ratio of its eigenvalues, but only through its singular values μ_q, i.e.

$$\kappa_2[\mathbf{Z}] = \frac{\max_q\{\mu_q\}}{\min_q\{\mu_q\}} = \frac{\max_q\{\sqrt{\lambda_q}\}}{\min_q\{\sqrt{\lambda_q}\}} \qquad (26)$$

where λ_q are the eigenvalues of the matrix $[Z]^*[Z]$ (the asterisk denotes the complex transpose). Due to the complexity of the expression for the condition number, any asymptotic estimates that were feasible in [4,5] for other types of boundary conditions, are not obviously derivable in the dielectric case. However, it is possible to determine the situations when (26) can blow up. Indeed, a substantial amount of tedious algebra finally yields that this situation is equivalent to $J_q(k_0 a_{in}) = 0$, which is exactly the case when the boundary condition error approaches infinity (see (25) and the discussion following it). This property is strongly reminiscent of the PEC [4] and SIBC [5] cases. Finally, using the appropriate asymptotics, it can be shown that the condition number approaches infinity for very large N, as expected.

3. Numerical Results and Discussion

To validate the expressions derived, direct comparisons were performed between the analytical error given in (20)-(24) and the computational error calculated by a LU decomposition of the MAS matrix, in a manner similar to [4,5].

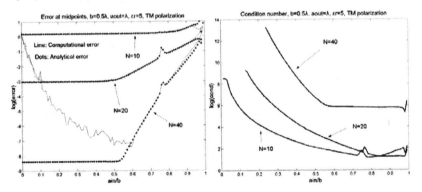

Fig. 2. Midpoint error plots as a function of a_{in}/b for $b=0.5\lambda$, $a_{out}=\lambda$, $\varepsilon_r=5$, TM incidence and various numbers of unknowns N.

Fig. 3. Condition number a_{in}/b for $b=0.5\lambda$, $a_{out}=\lambda$, $\varepsilon_r=5$, TM incidence and various numbers of unknowns N.

Fig. 2 presents the comparison for a geometry with $b=0.5\lambda$, $a_{out}=\lambda$, $\varepsilon_r=5$, TM incidence and a_{in} varying from 0 to b. The vertical axis maps the base 10 logarithm of the error. Three sets of curves are plotted, for $N=10$, 20 and 40. The parameters were chosen to be similar to the PEC [4] and SIBC [5] cases, to compare the MAS behavior for analogous geometries with different materials.

The overall behavior of the error is indeed reminiscent of both PEC and SIBC cylinders. Discrepancies between the analytical and computational errors for small a_{in} are due to large condition numbers, plotted in Fig. 3. Discrepancies for large a_{in} are due to the coincidence of the AS's with the CP's [4,5]. Resonance locations are again observed (a protrusion between $a_{in}/b=0.7$ and 0.8) and are associated with the zeros of the Bessel functions, as discussed in Section 2, as well as in [4,5]. What is particular to the dielectric cylinder, though, is the existence of an additional parameter, namely the radius of the outer auxiliary surface a_{out}. It can be demonstrated that the MAS error is practically independent of the S^{out} location (unless a_{out} is too close to b), meaning that S^{out} can be arbitrarily chosen without risking degradation of the method's accuracy. These results on the error behavior for varying radii of S^{in} and S^{out} completely agree with the empirical observations in [2], where no mathematical proofs were given.

References

1. Popovidi R. S and Z. S. Tsverikmazashvili, "Numerical Study of a Diffraction Problem by a Modified Method of Non-Orthogonal Series", *Journal of Applied Mathematics and Mathematical Physics*, Moscow, 1977. Translated from Russian by D. E. Brown, *Zh. vychisl. Mat. mat. Fiz.*, vol. 17, no. 2, pp. 384-393, 1977.
2. Leviatan Y. and A. Boag, "Analysis of Electromagnetic Scattering from Dielectric Cylinders Using a Multifilament Current Model", *IEEE Transactions on Antennas and Propagation*, vol. AP-35, no. 10, pp. 1119-1127, Oct. 1987.
3. Kaklamani, D. I. and H. T. Anastassiu, "Aspects of the Method of Auxiliary Sources (MAS) in Computational Electromagnetics", *IEEE Antennas and Propagation Magazine,* vol. 44, no. 3, pp. 48-64, June 2002.
4. Anastassiu, H. T., D. G. Lymperopoulos and D. I. Kaklamani, "Accuracy Analysis of the Method of Auxiliary Sources (MAS) for Scattering by a Perfectly Conducting Cylinder", *2003 IEEE International Symposium on Electromagnetic Compatibility (EMC)*, Istanbul, Turkey, May 11-16, 2003.
5. Anastassiu, H. T. and D. I. Kaklamani, "Accuracy Analysis and Optimization of the Method of Auxiliary Sources (MAS) for Scattering from an Impedance Circular Cylinder", *6th International Workshop on Mathematical Methods in Scattering Theory and Biomedical Engineering*, Tsepelovo, Greece, Sept. 18-21, 2003.

6. Hochstadt, H., *The Functions of Mathematical Physics,* Dover, New York, 1986.
7. Warnick, K. F. and W. C. Chew, "Accuracy of the Method of Moments for Scattering by a Cylinder", *IEEE Trans. on Microwave Theory and Techniques,* vol. MTT-48, no. 10, pp. 1652-1660, Oct. 2000.

ACCURACY ANALYSIS AND OPTIMIZATION OF THE METHOD OF AUXILIARY SOURCES (MAS) FOR SCATTERING FROM AN IMPEDANCE CIRCULAR CYLINDER

HRISTOS T. ANASTASSIU AND DIMITRA I. KAKLAMANI

Institute of Communication and Computer Systems
School of Electrical and Computer Engineering
National Technical University of Athens
Heroon Polytechniou 9
GR-15780 Zografou, Athens
GREECE
e-mail:hristosa@esd.ece.ntua.gr

The purpose of this paper is a rigorous accuracy assessment of the Method of Auxiliary Sources (MAS), when the latter is applied to electromagnetic, plane wave scattering from a circular, perfectly conducting cylinder, coated with a thin dielectric layer, and modelled by the Standard Impedance Boundary Condition (SIBC). An analytical inversion of the MAS matrix is possible via eigenvalue analysis, and an exact expression for the discretization error is therefore derived. Large discretization error values are shown to be partly associated with internal resonances of the scattering geometry. Furthermore, exact and asymptotic formulas for the condition number of the linear system are also extracted, showing an exponential growth with the number of unknowns. Finally, the fundamental MAS question, i.e. the optimal location of the auxiliary sources, is resolved on the grounds of error minimization.

1. Introduction

The Method of Auxiliary Sources [1]-[5] is a useful alternative to standard integral equation techniques, such as the Moment Method (MoM). Its salient features include low computational complexity [4], straightforward algorithmic structure and considerable physical insight. However, MAS is still not as popular as MoM, since the latter is often more reliable for the extraction of reference data. This is mainly due to MAS's limited robustness, which stems from the ambiguity related to the location of the Auxiliary Sources (AS's). It has been observed that poor AS's positioning often leads to an inexplicable, irregular behavior of the numerical solution, namely slow convergence rates, or unacceptably high boundary condition errors.

Very recently, a rigorous investigation into the MAS accuracy for scattering from a perfectly conducting (PEC) circular cylinder has been performed [5]. It was demonstrated that for this specific geometry, the MAS matrix can be inverted analytically, through spectral analysis and matrix diagonalization. The eigenvalues and eigenvectors were evaluated using a technique [6] based on the addition theorem of cylindrical functions [7, p.229]. The main outcome of this approach was the derivation of an exact expression for the boundary condition error, as well as for the system condition number, which finally led to the determination of the optimal AS's location.

Although the results in [5] deciphered the MAS behavior to a great extent for scattering from a PEC, cylindrical surface, they cannot be immediately applied to non-metallic or arbitrarily shaped boundaries. Nevertheless, it is well known that MAS capabilities can be exploited in many more cases [3], including dielectric circular, coated circular [2] and impedance square cylinders [4]. Given all the aforementioned complications, deep understanding of the MAS accuracy characteristics is absolutely necessary, for the most generic scatterers possible. The purpose of this work is the MAS error estimation for scattering from impedance circular cylinders.

A $e^{j\omega t}$ time convention is assumed and suppressed throughout the paper.

2. Analytical Considerations (TM Polarization)

We assume a PEC, infinite, circular cylinder of radius b_m, coated with a thin dielectric layer of thickness d. The dielectric coating is characterized by complex relative permittivity ε_r and relative permeability μ_r and is assumed linear, homogeneous and isotropic. The composite structure can be modelled in a compact fashion, by invoking the Standard Impedance Boundary Condition (SIBC) [4,8].

To simulate the composite scatterer, a cylinder S of radius $b=b_m+d$ is initially defined (see Fig. 1). According to SIBC, the total electric E and magnetic field H on S are related to each other through

$$\hat{n} \times E = \zeta \hat{n} \times (\hat{n} \times H) \tag{1}$$

where $\hat{n}$ is the outward pointing, normal unit vector to the surface S, ζ is the surface impedance given by

$$\zeta = j\zeta_1 \tan(k_1 d) \tag{1a}$$

with $\quad \zeta_1 \equiv \sqrt{(\mu_0\mu_r)/(\varepsilon_0 \varepsilon_r)}$ (coating intrinsic impedance) and $k_1 \equiv \omega\sqrt{\mu_0\varepsilon_0\mu_r\varepsilon_r}$ (wavenumber inside the coating).

The structure is illuminated by a plane wave impinging from a direction, determined by polar angle ϕ^i (see Fig. 1). The polarization of the plane wave is assumed to be transverse magnetic (TM) with respect to the cylinder axis z. The incident electric field at an arbitrary point (ρ, ϕ) is therefore expressed by

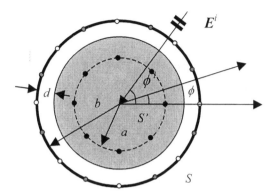

Fig. 1. Geometry of the problem. Black, white and gray bullets represent Auxiliary Sources (AS's), Collocation Points (CP's) and Midpoints (MP's) respectively. The gray disk corresponds to the PEC cylinder, coated by the dielectric layer, colored white.

$$E_z^i = E_0 \exp\{jk_0 \rho \cos(\phi^i - \phi)\} \qquad (2)$$

where k_0 is the free space wavenumber and E_0 is the amplitude of the incident electric field.

To construct the MAS solution [1]-[5], a fictitious auxiliary surface S' of radius $a<b$ is defined, conformal to the actual boundary S. A number of N AS's, in the form of elementary electric currents, are located on S', each one of which radiates an elementary electric field, proportional to the two-dimensional Green's function. Matching the boundary condition (1) at $M = N$ collocation points (CP's) on the z-plane projection of the scatterer surface yields the MAS square linear system. To determine the elements of the MAS matrix and the right hand side of the system, (1) must be properly processed. Given the cylindrical geometry of the problem, (1) is easily transformed into

$$E_z^s - \frac{\zeta}{j\omega\mu_0}\frac{\partial E_z^s}{\partial \rho} = -E_z^i + \frac{\zeta}{j\omega\mu_0}\frac{\partial E_z^i}{\partial \rho} \qquad (3)$$

where E_z^s is the z-directed, scattered electric field, which is due to the assembly of the radiating AS's, expressed as

$$E_z^s(\rho) = -\frac{j}{4}\sum_{n=1}^{N} c_n H_0^{(2)}(k_0|\rho - \rho_n'|) \tag{4}$$

where $\rho(\rho,\phi)$ is an arbitrary observation point, $\rho_n'(\rho_n',\phi_n')$ is the location of the nth AS, c_n is the unknown weight of the nth AS, and $H_0^{(2)}(\bullet)$ is the Hankel function of zero order and second kind. For brevity, we define the left-hand side of (3) as the scattered **combined field** F_z^s, and the negative of the right hand side of (3) as the incident **combined field** F_z^i. Applying the addition theorem for cylindrical functions [6,7 p.229], the Hankel function in (4) is conveniently written as (for $\rho > a$)

$$H_0^{(2)}(k_0|\rho - \rho_n'|) = \sum_{l=-\infty}^{\infty} J_l(k_0 a) H_l^{(2)}(k_0 \rho)\exp\{-jl(\phi - \phi_n')\} \tag{5}$$

where $J_l(\bullet)$ is the Bessel function of order l and $H_l^{(2)}(\bullet)$ is the Hankel function of order l and second kind. Applying (3) at M CP's on S, and invoking (4) and (5) yields the pertinent MAS linear system in the form

$$[Z]\{C\} = -\{F^i\} \tag{6}$$

where $\{C\}$ is a column vector of length N, consisting of the unknown source weights c_n, and $\{F^i\}$ is a column vector consisting of the incident combined field, sampled at the CP's, i.e.

$$F_m^i \equiv E_0\left[1 - \frac{\zeta}{\zeta_0}\cos(\phi^i - \phi_m)\right]\exp\{jk_0 b\cos(\phi^i - \phi_m)\}, \quad m = 1,\dots,M \tag{7}$$

where $\phi_m \equiv m\cdot 2\pi/M$ is the azimuth angle of the mth CP (see Fig. 1) and $\zeta_0 \equiv \sqrt{\mu_0/\varepsilon_0}$ (free space intrinsic impedance). Furthermore, the elements of the interaction matrix $[Z]$ are given by

$$Z_{mn} = -\frac{j}{4}\sum_{l=-\infty}^{\infty} J_l(k_0 a)\left[H_l^{(2)}(k_0 b) - \frac{\zeta}{j\zeta_0}\dot{H}_l^{(2)}(k_0 b)\right]\exp\{-jl(\phi_m - \phi_n)\} \tag{8}$$

where $\dot{H}_l^{(2)}(\bullet)$ is the derivative of the Hankel function of order l and second kind, the dot denoting differentiation with respect to the *entire* argument. For the circular cylinder case, (6) can be solved analytically, via diagonalization and subsequent inversion of $[Z]$, i.e. via a procedure described in detail in [5]. The eigenvalues and eigenvectors of $[Z]$ are evaluated according to a method described in [5,6], adapted to the present problem. After some long manipulation, the eigenvalues of $[Z]$ are finally given by

$$\lambda_q = -\frac{jN}{4} \sum_{s=-\infty}^{\infty} J_{q+sN}(k_0 a)\left[H_{q+sN}^{(2)}(k_0 b) - \frac{\zeta}{j\zeta_0} \dot{H}_{q+sN}^{(2)}(k_0 b)\right] \qquad q = 1,...,N \quad (9)$$

whereas the normalized eigenvectors are identical to the PEC case [5] and are given by

$$\{g\}_q = \frac{1}{\sqrt{N}}\left[\exp\{-jq\phi_1\}, \exp\{-jq\phi_2\},...,\exp\{-jq\phi_N\}\right]^T \qquad (10)$$

Therefore, $[Z]$ can be diagonalized and inverted in a standard manner.

Suppose, now, that we are interested in calculating the boundary condition error at points of the outer surface with azimuth angles equal to $\phi_m + \tilde{\phi}$, where $0 \le \tilde{\phi} \le 2\pi/N$. Obviously the choice $\tilde{\phi} = \pi/N$ corresponds to the midpoints (MP's) between the CP's (see Fig. 1). Following a procedure similar to [5], the error e, defined in the mean square sense over the scatterer surface, is finally given by

$$e(a,b,N,\zeta) = \left(\frac{1}{N}\sum_{m=1}^{N}|e_m|^2\right)^{\frac{1}{2}}\left(1 + \left|\frac{\zeta}{2\zeta_0}\right|^2\right)^{-\frac{1}{2}} \qquad (11)$$

where

$$e_m \equiv \frac{1}{N}\sum_{n=1}^{N}\sum_{q=1}^{N}\frac{\tilde{\lambda}_q}{\lambda_q}\exp\left\{-j(m-n)\phi_q + jk_0 b\cos(\phi^i - \phi_n)\right\}\left[1 - \frac{\zeta}{\zeta_0}\cos(\phi^i - \phi_n)\right] -$$

$$- \exp\left\{jk_0 b\cos(\phi^i - \phi_m - \tilde{\phi})\right\}\left[1 - \frac{\zeta}{\zeta_0}\cos(\phi^i - \phi_m - \tilde{\phi})\right] \qquad (12)$$

$\phi_p \equiv p \cdot 2\pi/N$ and

$$\tilde{\lambda}_q \equiv -\frac{jN}{4}\sum_{s=-\infty}^{\infty} J_{q+sN}(k_0 a)\left[H_{q+sN}^{(2)}(k_0 b) - \frac{\zeta}{j\zeta_0}\dot{H}_{q+sN}^{(2)}(k_0 b)\right]\exp\left\{-j(q+sN)\tilde{\phi}\right\} \quad (13)$$

To achieve the highest possible accuracy for the MAS solution, e in (11), which is a function of a, b, N and ζ, must be minimized by choosing the most appropriate a for given b, N and ζ.

Finally, an expression for the system condition number is derivable from the analysis above. Since the eigenvector set of $\{g\}_q$ is a complete basis for the space of N-dimensional complex vectors, it follows that $[Z]$ is normal, and hence the condition number κ_2 in the $\|\cdot\|_2$ norm can be given by

$$\kappa_2(a,b,N,\zeta) = \frac{\max_q\left(\left|\sum_{s=-\infty}^{\infty} J_{q+sN}(k_0 a)\left[H^{(2)}_{q+sN}(k_0 b) - \frac{\zeta}{j\zeta_0}\dot{H}^{(2)}_{q+sN}(k_0 b)\right]\right|\right)}{\min_q\left(\left|\sum_{s=-\infty}^{\infty} J_{q+sN}(k_0 a)\left[H^{(2)}_{q+sN}(k_0 b) - \frac{\zeta}{j\zeta_0}\dot{H}^{(2)}_{q+sN}(k_0 b)\right]\right|\right)} \quad (14)$$

To draw useful conclusions from (14), it is interesting to extract an asymptotic estimate. On the basis of the methodology described in [6], an asymptotic expression of (14), for very large numbers of unknowns N, or for very large external radius b, is given by

$$\kappa_2 \sim \frac{1}{h_\lambda}\left(\frac{a}{b}\right)^{-\frac{k_0 b}{2h_\lambda}}\left|\frac{1+\frac{\zeta}{\zeta_0}}{1+\frac{\zeta}{2\zeta_0 h_\lambda}}\right|, \qquad h_\lambda \equiv \frac{2\pi b}{\lambda N} \quad (15)$$

where λ is the wavelength. Like in the PEC case [5], (15) reveals an exponential growth of the condition number for large numbers of unknowns, which is an inherent drawback of MAS with respect to its computational efficiency. In the numerical results section the behavior of the condition number will be investigated in depth, and recommendations for circumventing its effects will be given.

3. Numerical Results and Discussion

Fig. 2 presents an error plot for a coated PEC cylinder of total radius $b=0.5\lambda$, for TM incidence, and for a varying between 0 and b. Three sets of curves are plotted, for a number of unknowns equal to $N=10$, 20 and 40. The dielectric coating has a depth equal to $d=0.05\lambda$, and relative permittivity equal to $\varepsilon_r=3-j10$. The horizontal axis maps the ratio a/b and the vertical axis maps the logarithm of the error. The analytical error was computed by (11), (12), and the computational error resulted from numerical inversion of the MAS linear system (LU decomposition). Although not identical in values, the overall behavior of both the analytical and the computational plots are very similar to the PEC case [5]. The geometrical parameters and the numbers of unknowns were deliberately chosen to be equal to one of the PEC cases studied in [5], allowing immediate comparison, and assessment of the coating effects. The irregular behavior of the computational error for small values of a is explained by the high values of the condition number, plotted in Fig. 3, which also strikingly resembles the PEC case. It is emphasized that the MAS solution, as a discrete approximation of a continuous problem, is only indirectly responsible for the irregular behavior of the computational error for small radii a, since the algorithm structure introduces bad system conditioning. However, the discrete nature of the method gives rise only to the analytical error, given by (11), (12).

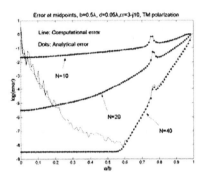
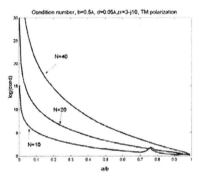

Fig. 2. Midpoint error plots for $b=0.5\lambda$, $d=0.05\lambda$ $\varepsilon_r=3$-$j10$, TM incidence and various numbers of unknowns.

Fig. 3. Matrix condition number for $b=0.5\lambda$, $d=0.05\lambda$, $\varepsilon_r=3$-$j10$, TM incidence and various numbers of unknowns.

The resonance effects, related to the zeros of the Bessel function, are represented by a protrusion between $a/b=0.7$ and 0.8. This "bump" corresponds to the first zero of $J_0(\bullet)$, which is equal to $j_{0,1} \cong 2.405$ [9, p. 409]. The argument of $J_0(k_0 a)$ equals $j_{0,1}$ when $a/b \cong 0.7655$, which is the precise location of the bump in the plot. In general, if $j_{q,n}$ is the nth root of $J_q(\bullet)$, error protrusions are expected for

$$0 < \frac{a}{b} = \frac{j_{q,n}}{k_0 b} \leq 1 \tag{16}$$

Given the properties of the error examined in this analysis, it is very interesting to address the fundamental question of MAS, i.e. what is the optimal location of the AS's, to obtain the most accurate solution possible. Evidently, what was proposed in [5] for the PEC case is valid for the impedance surface, too. Therefore, the most accurate MAS computational solution for the impedance cylinder illuminated by a TM wave, is obtained by setting a/b as small as possible, provided that the condition number of the system does not exceed a fixed threshold. This threshold is determined solely by the arithmetic precision, used in the calculations. Additionally, $k_0 a$ should not lie in the vicinity of any root of any Bessel function of integer order, to avoid resonance effects. A similar analysis for the TE polarization shows that the same rule holds for that case, too.

References

1. Popovidi R. S and Z. S. Tsverikmazashvili, "Numerical Study of a Diffraction Problem by a Modified Method of Non-Orthogonal Series", *Journal of Applied Mathematics and Mathematical Physics*, Moscow, 1977. Translated from Russian by D. E. Brown, *Zh. vychisl. Mat. mat. Fiz.*, vol. 17, no. 2, pp. 384-393, 1977.
2. Leviatan Y., Am. Boag and Al. Boag, "Analysis of Electromagnetic Scattering from Dielectrically Coated Conducting Cylinders Using a Multifilament Current Model", *IEEE Transactions on Antennas and Propagation*, vol. AP-36, no. 11, pp. 1602-1607, Nov. 1988.
3. Kaklamani, D. I. and H. T. Anastassiu, "Aspects of the Method of Auxiliary Sources (MAS) in Computational Electromagnetics", *IEEE Antennas and Propagation Magazine*, vol. 44, no. 3, pp. 48-64, June 2002.
4. Anastassiu, H. T., D. I. Kaklamani, D. P. Economou and O. Breinbjerg, "Electromagnetic Scattering Analysis of Coated Conductors with Edges Using the Method of Auxiliary Sources (MAS) in Conjunction with the Standard Impedance Boundary Condition (SIBC)", *IEEE Trans. on Antennas and Propagation*, vol. AP-50, no. 1, pp. 59-66, Jan. 2002.
5. Anastassiu, H. T., D. G. Lymperopoulos and D. I. Kaklamani, "Accuracy Analysis of the Method of Auxiliary Sources (MAS) for Scattering by a Perfectly Conducting Cylinder", *2003 IEEE International Symposium on Electromagnetic Compatibility (EMC)*, Istanbul, Turkey, May 11-16, 2003.
6. Warnick, K. F. and W. C. Chew, "Accuracy of the Method of Moments for Scattering by a Cylinder", *IEEE Trans. Microwave Theory and Techniques*, vol. MTT-48, no. 10, pp. 1652-1660, Oct. 2000.
7. Hochstadt, H., *The Functions of Mathematical Physics*, Dover, New York, 1986.

8. T. B. A. Senior and J. L. Volakis, *Approximate Boundary Conditions in Electromagnetics,* IEE Press, 1994.

9. Abramowitz, A. and I. E. Stegun, *Handbook of Mathematical Functions,* Dover, New York, 1972.

ON THE SCATTERING OF SPHERICAL ELECTROMAGNETIC WAVES BY A PENETRABLE CHIRAL OBSTACLE

C. ATHANASIADIS

University of Athens,
Department of Mathematics, Panepistimiopolis,
GR 157 84 Athens, Greece

P. A. MARTIN

Colorado School of Mines,
Department of Mathematical and Computer Sciences,
Golden CO 80401-1887, USA

A. SPYROPOULOS, I. G. STRATIS

University of Athens,
Department of Mathematics, Panepistimiopolis,
GR 157 84 Athens, Greece

The problem of scattering of spherical electromagnetic waves by a bounded chiral obstacle is considered. General scattering theorems, relating the far–field patterns due to scattering of waves from a point source put in any two different locations (the reciprocity principle, the optical theorem, etc), and mixed scattering relations (relating the scattered fields due to a point source and a plane wave) are established. Further, in the case of a spherical chiral scatterer, the exact Green's function and the electric far–field pattern of the problem are constructed.

1. Introduction

The interaction of an incident wavefield with a bounded 3-dimensional obstacle is a classic problem in scattering theory. The vast majority of the literature is concerned with incident plane waves. However, in some recent papers Dassios and his co–workers, as well as three and all of the present authors, see [1], [2], [3], [5], have studied incident waves generated by a point source in the vicinity of the scatterer. Point sources have been used by Potthast [8] to solve standard inverse problems. For related work by other authors see the bibliographies of all the previous references. In this work we consider the problem of scattering of spherical electromagnetic waves by a bounded chiral obstacle. General scattering theorems, relating the far–field patterns due to scattering of waves from a point source put in any two different locations (the reciprocity principle, the optical theorem, etc), and mixed scattering relations (relating the scattered fields due to a point source and a plane wave) are established. Further, in the case of a *spherical* chiral scatterer, the exact Green's function and the electric far–field pattern of the problem are constructed, using spherical vector wave functions. These results generalize related properties of the problem where the scatterer is achiral [1], [3].

2. Formulation

Let Ω_c be a bounded three–dimensional obstacle with a smooth closed boundary S, the scatterer. The exterior Ω is an infinite homogeneous isotropic achiral medium with electric permittivity ε, magnetic permeability μ and conductivity $\sigma = 0$. The scatterer Ω_c is filled with a chiral homogeneous isotropic medium with corresponding electromagnetic parameters ε_c, μ_c and chirality measure β_c. The parameters ε and μ are assumed to be real constants and ε_c, μ_c and β_c to be complex constants.

We consider an incident spherical electromagnetic wave due to a point source located at a point with position vector $\boldsymbol{a}$, with respect to the origin O. Suppressing the time dependence $\mathrm{e}^{-\mathrm{i}\omega t}$, ω being the angular frequency, the incident wave $(\boldsymbol{E}_a^i, \boldsymbol{H}_a^i)$ has the form [1], [3]

$$
\begin{aligned}
\boldsymbol{E}_a^i(\boldsymbol{r}; \widehat{\boldsymbol{p}}) &= \tfrac{a\,\mathrm{e}^{-\mathrm{i}ka}}{\mathrm{i}k} \, \nabla \times \left(\tfrac{\mathrm{e}^{\mathrm{i}k|\boldsymbol{r}-\boldsymbol{a}|}}{|\boldsymbol{r}-\boldsymbol{a}|} \, \widehat{\boldsymbol{a}} \times \widehat{\boldsymbol{p}} \right), \\
\boldsymbol{H}_a^i(\boldsymbol{r}; \widehat{\boldsymbol{p}}) &= \tfrac{1}{\mathrm{i}k\eta} \, \nabla \times \boldsymbol{E}_a^i(\boldsymbol{r}; \widehat{\boldsymbol{p}}),
\end{aligned}
\tag{1}
$$

where $\widehat{\boldsymbol{p}}$ is a constant unit vector with $\widehat{\boldsymbol{p}} \cdot \boldsymbol{a} = 0$, $\eta = \sqrt{\mu/\varepsilon}$ is the intrinsic impedance, $k = \omega\sqrt{\varepsilon\mu} > 0$ is the free–space wave number, and $a = |\boldsymbol{a}|$. Physically $(\boldsymbol{E}_a^i, \boldsymbol{H}_a^i)$ represents the field generated by a magnetic dipole with dipole moment $\widehat{\boldsymbol{a}} \times \widehat{\boldsymbol{p}}$; see p.163 of [4], or p.23 of [5]. The coefficient $a\,\mathrm{e}^{-\mathrm{i}ka}/(\mathrm{i}k)$ in (1) assures that when the point source recedes to infinity the spherical wave reduces to a plane electric wave with direction of propagation $-\widehat{\boldsymbol{a}}$ and polarization $\widehat{\boldsymbol{p}}$. The total exterior electric field $\boldsymbol{E}_a^t$ is given by

$$
\boldsymbol{E}_a^t(\boldsymbol{r}; \widehat{\boldsymbol{p}}) = \boldsymbol{E}_a^i(\boldsymbol{r}; \widehat{\boldsymbol{p}}) + \boldsymbol{E}_a^s(\boldsymbol{r}; \widehat{\boldsymbol{p}}), \quad \boldsymbol{r} \in \Omega \setminus \{\boldsymbol{a}\},
\tag{2}
$$

where $\boldsymbol{E}_a^s(\boldsymbol{r}; \widehat{\boldsymbol{p}})$ is the scattered electric field, which is assumed to satisfy the Silver–Müller radiation condition

$$
\lim_{r \to \infty} (\widehat{\boldsymbol{r}} \times \nabla \times \boldsymbol{E}_a^s + \mathrm{i}kr\boldsymbol{E}_a^s) = \boldsymbol{0},
\tag{3}
$$

uniformly in all directions $\widehat{r} \in S^2$, where S^2 is the unit sphere.

The behaviour of $\boldsymbol{E}_a^{\text{s}}$ in the radiation zone is given by

$$\boldsymbol{E}_a^{\text{s}}(\boldsymbol{r}) = h_0(kr)\,\boldsymbol{g}_a(\widehat{\boldsymbol{r}}) + O(r^{-2}), \quad r \to \infty, \tag{4}$$

where $h_0(x) = \mathrm{e}^{\mathrm{i}x}/(\mathrm{i}x)$ is the spherical Hankel function of the first kind and order zero, and $\boldsymbol{g}_a(\widehat{\boldsymbol{r}})$ is the electric far–field pattern.

The total exterior electric field solves the equation

$$\nabla \times \nabla \times \boldsymbol{E}_a^{\text{t}} - k^2 \boldsymbol{E}_a^{\text{t}} = \boldsymbol{0} \quad \text{in } \Omega. \tag{5}$$

We note that the incident electric field satisfies the radiation condition (3), and hence the total electric field also satisfies (3).

The incident electromagnetic waves are transmitted into the chiral scatterer. Let $\boldsymbol{E}_a^{\text{c}}$ be the total electric field in the interior. Then $\boldsymbol{E}_a^{\text{c}}$ satisfies [6]

$$\nabla \times \nabla \times \boldsymbol{E}_a^{\text{c}} - 2\beta_c\gamma^2\nabla \times \boldsymbol{E}_a^{\text{c}} - \gamma^2 \boldsymbol{E}_a^{\text{c}} = \boldsymbol{0} \quad \text{in } \Omega_c , \tag{6}$$

where $\gamma^2 = k_c^2(1-k_c^2\beta_c^2)^{-1}$ and $k_c^2 = \omega^2\varepsilon_c\mu_c$. On the surface of the scatterer we have the following transmission conditions:

$$\left.\begin{array}{r}\widehat{\boldsymbol{n}} \times \boldsymbol{E}_a^{\text{t}} = \widehat{\boldsymbol{n}} \times \boldsymbol{E}_a^{\text{c}} \\ \widehat{\boldsymbol{n}} \times \nabla \times \boldsymbol{E}_a^{\text{t}} = \mathcal{B}_1\widehat{\boldsymbol{n}} \times \nabla \times \boldsymbol{E}_a^{\text{c}} + \mathcal{B}_2\widehat{\boldsymbol{n}} \times \boldsymbol{E}_a^{\text{c}}\end{array}\right\} \quad \text{on } S \tag{7}$$

where $\mathcal{B}_1 = (\mu/\mu_c)k_c^2/\gamma^2$ and $\mathcal{B}_2 = -(\mu/\mu_c)\beta_c k_c^2$.

3. The general scattering theorem

In the sequel, for an incident time–harmonic spherical wave $\boldsymbol{E}_a^{\text{i}}(\boldsymbol{r};\widehat{\boldsymbol{p}})$ due to a point source located at $\boldsymbol{a}$, we will denote the total field in Ω, the scattered field and the far–field pattern by writing $\boldsymbol{E}_a^{\text{t}}(\boldsymbol{r};\widehat{\boldsymbol{p}})$, $\boldsymbol{E}_a^{\text{s}}(\boldsymbol{r};\widehat{\boldsymbol{p}})$ and $\boldsymbol{g}_a(\widehat{\boldsymbol{r}};\widehat{\boldsymbol{p}})$, respectively, indicating the dependence on the position $\boldsymbol{a}$ of the point source and the polarization $\widehat{\boldsymbol{p}}$. Also, the total electric field in Ω_c will be denoted by $\boldsymbol{E}_a^{\text{c}}(\boldsymbol{r};\widehat{\boldsymbol{p}})$.

We are interested in relations between these fields. We consider a point source at $\boldsymbol{a}$ with polarization $\widehat{\boldsymbol{p}}_1$ and another point source at $\boldsymbol{b}$ with polarization $\widehat{\boldsymbol{p}}_2$. For a shorthand notation, we use

$$\left\{\boldsymbol{E}, \boldsymbol{E}'\right\}_S = \int_S \left[(\widehat{\boldsymbol{n}} \times \overline{\boldsymbol{E}})\cdot(\nabla \times \boldsymbol{E}') - (\widehat{\boldsymbol{n}} \times \boldsymbol{E}')\cdot(\nabla \times \overline{\boldsymbol{E}})\right] ds,$$

where the overbar denotes complex conjugation.

Let S_r denote a large sphere of radius r, surrounding the points $\boldsymbol{a}$ and $\boldsymbol{b}$, and let $S_{a,\epsilon} = \{\boldsymbol{r} \in \mathbb{R}^3 : |\boldsymbol{a} - \boldsymbol{r}| = \epsilon\}$ surrounding the point $\boldsymbol{a}$. Then we have the following Lemma from [1].

Lemma 1. *Let $\boldsymbol{E}_a^i(\boldsymbol{r};\widehat{\boldsymbol{p}}_1)$ be a point source at $\boldsymbol{a}$. Let $\boldsymbol{E}_b^i(\boldsymbol{r};\widehat{\boldsymbol{p}}_2)$ be a point source at $\boldsymbol{b}$, with corresponding scattered field $\boldsymbol{E}_b^s(\boldsymbol{r};\widehat{\boldsymbol{p}}_2)$ and far–field pattern $\boldsymbol{g}_b(\widehat{\boldsymbol{r}};\widehat{\boldsymbol{p}}_2)$. Then*

$$\lim_{\epsilon \to 0} \left\{ \boldsymbol{E}_a^i(\,\cdot\,;\widehat{\boldsymbol{p}}_1), \boldsymbol{E}_b^s(\,\cdot\,;\widehat{\boldsymbol{p}}_2) \right\}_{S_{a,\epsilon}} - \lim_{r \to \infty} \left\{ \boldsymbol{E}_a^i(\,\cdot\,;\widehat{\boldsymbol{p}}_1), \boldsymbol{E}_b^s(\,\cdot\,;\widehat{\boldsymbol{p}}_2) \right\}_{S_r} =$$

$$\frac{4\pi a}{ik}\,\widehat{\boldsymbol{p}}_1 \cdot \boldsymbol{G}_b(\boldsymbol{a};\widehat{\boldsymbol{p}}_2),$$

where

$$\boldsymbol{G}_b(\boldsymbol{a};\widehat{\boldsymbol{p}}_2) = e^{ika}\boldsymbol{a} \times \left[\nabla \times \boldsymbol{E}_b^s(\boldsymbol{a};\widehat{\boldsymbol{p}}_2) - \frac{ik}{2\pi}\int_{S^2} \widehat{\boldsymbol{r}} \times \boldsymbol{g}_b(\widehat{\boldsymbol{r}};\widehat{\boldsymbol{p}}_2)\,e^{ik\widehat{\boldsymbol{r}}\cdot\boldsymbol{a}}ds(\widehat{\boldsymbol{r}}) \right]$$

is a spherical far–field pattern generator.

Now, the general scattering theorem, [1], for spherical electric waves scattered by a chiral obstacle is formulated as follows.

Theorem 2. *For any two point–source locations in Ω, $\boldsymbol{a}$ and $\boldsymbol{b}$, and for any polarizations, $\widehat{\boldsymbol{p}}_1$ and $\widehat{\boldsymbol{p}}_2$, we have*

$$\widehat{\boldsymbol{p}}_1 \cdot \boldsymbol{G}_b(\boldsymbol{a};\widehat{\boldsymbol{p}}_2) + \widehat{\boldsymbol{p}}_2 \cdot \overline{\boldsymbol{G}_a(\boldsymbol{b};\widehat{\boldsymbol{p}}_1)} + \frac{1}{2\pi}\int_{S^2} \boldsymbol{g}_b(\widehat{\boldsymbol{r}};\widehat{\boldsymbol{p}}_2) \cdot \overline{\boldsymbol{g}_a(\widehat{\boldsymbol{r}};\widehat{\boldsymbol{p}}_1)}\,ds(\widehat{\boldsymbol{r}})$$
$$= \mathcal{E}_{a,b}(\widehat{\boldsymbol{p}}_1;\widehat{\boldsymbol{p}}_2) \tag{8}$$

where

$$\mathcal{E}_{a,b}(\widehat{\boldsymbol{p}}_1;\widehat{\boldsymbol{p}}_2) = \frac{k}{2\pi}\Bigg\{ \mathrm{Im}\,(\mathcal{B}_1)\int_{\Omega_c} \left(\nabla \times \overline{\boldsymbol{E}_a(\boldsymbol{r};\widehat{\boldsymbol{p}}_1)}\right) \cdot \left(\nabla \times \boldsymbol{E}_b(\boldsymbol{r};\widehat{\boldsymbol{p}}_2)\right)dv$$

$$- \mathrm{Im}\,(\mathcal{B}_1\gamma^2)\int_{\Omega_c} \overline{\boldsymbol{E}_a(\boldsymbol{r};\widehat{\boldsymbol{p}}_1)} \cdot \boldsymbol{E}_b(\boldsymbol{r};\widehat{\boldsymbol{p}}_2)\,dv$$

$$+ i[\beta_c\gamma^2\mathcal{B}_1 + \mathrm{Re}\,(\mathcal{B}_2)]\int_{\Omega_c} \overline{\boldsymbol{E}_a(\boldsymbol{r};\widehat{\boldsymbol{p}}_1)} \cdot \left(\nabla \times \boldsymbol{E}_b(\boldsymbol{r};\widehat{\boldsymbol{p}}_2)\right)dv$$

$$- i[\overline{\beta_c\gamma^2\mathcal{B}_1} + \mathrm{Re}\,(\mathcal{B}_2)]\int_{\Omega_c} \boldsymbol{E}_b(\boldsymbol{r};\widehat{\boldsymbol{p}}_2) \cdot \left(\nabla \times \overline{\boldsymbol{E}_a(\boldsymbol{r};\widehat{\boldsymbol{p}}_1)}\right)dv \Bigg\}.$$

Proof. In view of the relations $\boldsymbol{E}_\alpha^t = \boldsymbol{E}_\alpha^i + \boldsymbol{E}_\alpha^s$, $\alpha = a, b$, we have

$$\{\boldsymbol{E}_a^t, \boldsymbol{E}_b^t\} = \{\boldsymbol{E}_a^i, \boldsymbol{E}_b^i\} + \{\boldsymbol{E}_a^i, \boldsymbol{E}_b^s\} + \{\boldsymbol{E}_a^s, \boldsymbol{E}_b^i\} + \{\boldsymbol{E}_a^s, \boldsymbol{E}_b^s\}. \tag{9}$$

We use the transmission conditions (7) and apply the divergence theorem in Ω_c; this gives

$$\{\boldsymbol{E}_a^t, \boldsymbol{E}_b^t\} = \frac{4\pi i}{k}\mathcal{E}_{a,b}(\widehat{\boldsymbol{p}}_1;\widehat{\boldsymbol{p}}_2). \tag{10}$$

Since $\overline{\boldsymbol{E}_a^{\mathrm{i}}}$ and $\boldsymbol{E}_b^{\mathrm{i}}$ are entire solutions of (5), the vector Green's second theorem gives

$$\{\boldsymbol{E}_a^{\mathrm{i}}, \boldsymbol{E}_b^{\mathrm{i}}\} = 0. \tag{11}$$

For the other terms in (9), we consider two small spheres, S_{a,ϵ_1} and S_{b,ϵ_2}, centred at $\boldsymbol{a}$ and $\boldsymbol{b}$ with radii ϵ_1 and ϵ_2, respectively, with $S_{a,\epsilon_1} \cap S_{b,\epsilon_2} = \emptyset$, as well as a large sphere S_R centred at the origin, surrounding the whole system of the scatterer and the two small spheres. Since $\overline{\boldsymbol{E}_a^{\mathrm{i}}}$ and $\boldsymbol{E}_b^{\mathrm{s}}$ are solutions of (5) for $\boldsymbol{r} \neq \boldsymbol{a}, \boldsymbol{b}$, the vector Green's second theorem gives

$$\{\boldsymbol{E}_a^{\mathrm{i}}, \boldsymbol{E}_b^{\mathrm{s}}\} = \{\boldsymbol{E}_a^{\mathrm{i}}, \boldsymbol{E}_b^{\mathrm{s}}\}_{S_R} - \{\boldsymbol{E}_a^{\mathrm{i}}, \boldsymbol{E}_b^{\mathrm{s}}\}_{S_{a,\epsilon_1}} - \{\boldsymbol{E}_a^{\mathrm{i}}, \boldsymbol{E}_b^{\mathrm{s}}\}_{S_{b,\epsilon_2}}. \tag{12}$$

The last term in (12) is zero because $\overline{\boldsymbol{E}_a^{\mathrm{i}}}$ and $\boldsymbol{E}_b^{\mathrm{s}}$ are regular solutions of (5) in the interior of S_{b,ϵ_2}. Then letting $R \to \infty$ and $\epsilon_1 \to 0$, using Lemma 1, we obtain

$$\{\boldsymbol{E}_a^{\mathrm{i}}, \boldsymbol{E}_b^{\mathrm{s}}\} = \frac{4\pi\mathrm{i}}{k} \widehat{\boldsymbol{p}}_1 \cdot \boldsymbol{G}_b(\boldsymbol{a}; \widehat{\boldsymbol{p}}_2). \tag{13}$$

As $\{\boldsymbol{E}_a^{\mathrm{s}}, \boldsymbol{E}_b^{\mathrm{i}}\} = -\overline{\{\boldsymbol{E}_a^{\mathrm{i}}, \boldsymbol{E}_b^{\mathrm{s}}\}}$, we easily deduce that

$$\{\boldsymbol{E}_a^{\mathrm{s}}, \boldsymbol{E}_b^{\mathrm{i}}\} = \frac{4\pi\mathrm{i}}{k} \widehat{\boldsymbol{p}}_2 \cdot \overline{\boldsymbol{G}_a(\boldsymbol{b}; \widehat{\boldsymbol{p}}_1)}. \tag{14}$$

Finally, in view of the regularity of $\overline{\boldsymbol{E}_a^{\mathrm{s}}}$ and $\boldsymbol{E}_b^{\mathrm{s}}$ in the region exterior to S, we have

$$\{\boldsymbol{E}_a^{\mathrm{s}}, \boldsymbol{E}_b^{\mathrm{s}}\} = \{\boldsymbol{E}_a^{\mathrm{s}}, \boldsymbol{E}_b^{\mathrm{s}}\}_{S_R}. \tag{15}$$

Then, letting $R \to \infty$, we pass to the radiation zone and thus using (4) we get

$$\{\boldsymbol{E}_a^{\mathrm{s}}, \boldsymbol{E}_b^{\mathrm{s}}\} = \frac{2\mathrm{i}}{k} \int_{S^2} \overline{\boldsymbol{g}_a(\widehat{\boldsymbol{r}}; \widehat{\boldsymbol{p}}_1)} \cdot \boldsymbol{g}_b(\widehat{\boldsymbol{r}}; \widehat{\boldsymbol{p}}_2)\, ds(\widehat{\boldsymbol{r}}). \tag{16}$$

Substituting (10), (11), (13), (14), and (16) in (12) gives (8), and the theorem is proved.

4. Reciprocity

In [5] a reciprocity relation for spherical waves scattered by an achiral obstacle has been proved. The same relation also holds for a penetrable chiral scatterer.

Theorem 3. *For any two point–source locations in Ω, $\boldsymbol{a}$ and $\boldsymbol{b}$, for any polarizations, $\widehat{\boldsymbol{p}}_1$ and $\widehat{\boldsymbol{p}}_2$, and for a penetrable chiral scatterer, we have*

$$h_0(ka)\,(\widehat{\boldsymbol{b}} \times \widehat{\boldsymbol{p}}_2) \cdot (\nabla \times \boldsymbol{E}_a^{\mathrm{s}}(\widehat{\boldsymbol{b}}; \widehat{\boldsymbol{p}}_1)) = h_0(kb)\,(\widehat{\boldsymbol{a}} \times \widehat{\boldsymbol{p}}_1) \cdot (\nabla \times \boldsymbol{E}_b^{\mathrm{s}}(\widehat{\boldsymbol{a}}; \widehat{\boldsymbol{p}}_2)) \tag{17}$$

Proof. Using the transmission conditions (7) and applying the divergence theorem in Ω_c, we obtain $\{\overline{\boldsymbol{E}_a^{\mathrm{t}}}, \boldsymbol{E}_b^{\mathrm{t}}\}_S = 0$. Also, using some asymptotics we get $\{\overline{\boldsymbol{E}_a^{\mathrm{s}}}, \boldsymbol{E}_b^{\mathrm{s}}\}_S = 0$. Now $\{\overline{\boldsymbol{E}_a^{\mathrm{i}}}, \boldsymbol{E}_b^{\mathrm{s}}\}_S = 0$, since the corresponding integral on the large sphere S_R vanishes due to the asymptotic form (4) and $\boldsymbol{E}_a^{\mathrm{i}}(\widehat{\boldsymbol{r}}; \widehat{\boldsymbol{p}}_1) = h_0(kr) \boldsymbol{g}_a^{\mathrm{i}}(\widehat{\boldsymbol{r}}; \widehat{\boldsymbol{p}}_1) + O(r^{-2})$ as $r \to \infty$, where $\boldsymbol{g}_a^{\mathrm{i}}(\widehat{\boldsymbol{r}}; \widehat{\boldsymbol{p}}_1) = \mathrm{i}k a e^{-\mathrm{i}ka(1+\widehat{\boldsymbol{r}}\cdot\boldsymbol{a})} (\widehat{\boldsymbol{r}} \times (\boldsymbol{a} \times \widehat{\boldsymbol{p}}_1))$. Combining the above we finally obtain (17).

5. The optical theorem

We define the scattering and absorption cross–sections due to a point source at $\boldsymbol{a}$, [5], as

$$\sigma_a^{\mathrm{s}} = \frac{1}{k^2} \int_{S^2} |\boldsymbol{g}_a(\widehat{\boldsymbol{r}}; \widehat{\boldsymbol{p}})|^2 \, ds(\widehat{\boldsymbol{r}}) \quad \text{and} \quad \sigma_a^{\mathrm{a}} = \frac{1}{k} \operatorname{Im} \int_S \widehat{\boldsymbol{n}} \cdot (\boldsymbol{E}_a^{\mathrm{t}} \times \nabla \times \overline{\boldsymbol{E}_a^{\mathrm{t}}}) \, ds,$$

respectively, and the extinction cross–section by $\sigma_a^{\mathrm{e}} = \sigma_a^{\mathrm{s}} + \sigma_a^{\mathrm{a}}$. If we put $\boldsymbol{a} = \boldsymbol{b}$ and $\widehat{\boldsymbol{p}}_1 = \widehat{\boldsymbol{p}}_2 = \widehat{\boldsymbol{p}}$ in Theorem 2, we obtain

$$\sigma_a^{\mathrm{s}} = -4\pi k^{-2} \operatorname{Re} \left[\widehat{\boldsymbol{p}} \cdot \boldsymbol{G}_a(\boldsymbol{a}; \widehat{\boldsymbol{p}}) \right] + 2\pi k^{-2} \mathcal{E}_{a,a}(\widehat{\boldsymbol{p}}; \widehat{\boldsymbol{p}}). \tag{18}$$

From the above definitions and (9) we have

$$\sigma_a^{\mathrm{a}} = -2\pi k^{-2} \mathcal{E}_{a,a}(\widehat{\boldsymbol{p}}; \widehat{\boldsymbol{p}}). \tag{19}$$

Hence, adding (18) and (19), the definition (8) gives

$$\sigma_a^{\mathrm{e}} = -4\pi k^{-2} \operatorname{Re} \left[\widehat{\boldsymbol{p}} \cdot \boldsymbol{G}_a(\boldsymbol{a}; \widehat{\boldsymbol{p}}) \right]. \tag{20}$$

The value of $\mathcal{E}_{a,a}(\widehat{\boldsymbol{p}}; \widehat{\boldsymbol{p}})$ is given in Theorem 2; it depends on the scatterer's properties.

6. Mixed scattering relations

Let $\boldsymbol{E}^{\mathrm{i}}(\boldsymbol{r}; \widehat{\boldsymbol{d}}, \widehat{\boldsymbol{p}}) = \widehat{\boldsymbol{p}} \exp\{\mathrm{i}k\widehat{\boldsymbol{d}} \cdot \boldsymbol{r}\}$ be an incident time–harmonic plane electric wave, where the unit vector $\widehat{\boldsymbol{d}}$ describes the direction of propagation and the unit vector $\widehat{\boldsymbol{p}}$ gives the polarization. We will indicate the dependence of the total field in Ω, the total field in Ω_c, the scattered field and the electric far–field pattern on the incident direction $\widehat{\boldsymbol{d}}$ and the polarization $\widehat{\boldsymbol{p}}$ by writing $\boldsymbol{E}^{\mathrm{t}}(\boldsymbol{r}; \widehat{\boldsymbol{d}}, \widehat{\boldsymbol{p}})$, $\boldsymbol{E}^{-}(\boldsymbol{r}; \widehat{\boldsymbol{d}}, \widehat{\boldsymbol{p}})$, $\boldsymbol{E}^{\mathrm{s}}(\boldsymbol{r}; \widehat{\boldsymbol{d}}, \widehat{\boldsymbol{p}})$ and $\boldsymbol{g}(\widehat{\boldsymbol{r}}; \widehat{\boldsymbol{d}}, \widehat{\boldsymbol{p}})$, respectively.

Here, we consider mixed situations, and relate fields due to one spherical electric wave $\boldsymbol{E}_a^{\mathrm{i}}(\boldsymbol{r}; \widehat{\boldsymbol{p}}_1)$ and one plane electric wave $\boldsymbol{E}^{\mathrm{i}}(\boldsymbol{r}; -\widehat{\boldsymbol{b}}, \widehat{\boldsymbol{p}}_2)$; we do this by letting $b \to \infty$ in our previous results.

Using the asymptotic forms $|\boldsymbol{r} - \boldsymbol{a}| = r - \widehat{\boldsymbol{r}} \cdot \boldsymbol{a} + O(r^{-1})$ and $|\boldsymbol{r} - \boldsymbol{a}|^{-1} = r^{-1} + O(r^{-2})$, we can easily show that for the spherical electric wave

(1) we have $\lim_{b\to\infty} E_b^i(r;\widehat{p}) = E^i(r;-\widehat{b},\widehat{p})$, that is the spherical electric wave, when the point source goes to infinity, reduces to a plane electric wave with direction of propagation $-\widehat{b}$ and polarization $\widehat{p}$. Similarly, we have $E_b^t(r;\widehat{p}) \to E^t(r;-\widehat{b},\widehat{p})$, $E_b^s(r;\widehat{p}) \to E^s(r;-\widehat{b},\widehat{p})$ and $g_b(\widehat{r};\widehat{p}) \to g(\widehat{r};-\widehat{b},\widehat{p})$ as $b \to \infty$.

Next, let $b \to \infty$ in Lemma 1 to give the following result.

Lemma 4. *Let $E_a^i(r;\widehat{p}_1)$ be an incident spherical electric wave and let $E^i(r;-\widehat{b},\widehat{p}_2)$ be an incident plane electric wave. Then*

$$\lim_{\varepsilon\to 0}\left\{E_a^i(\,\cdot\,;\widehat{p}_1), E^s(\,\cdot\,;-\widehat{b},\widehat{p}_2)\right\}_{S_{a,\varepsilon}} - \lim_{r\to\infty}\left\{E_a^i(\,\cdot\,;\widehat{p}_1), E^s(\,\cdot\,;-\widehat{b},\widehat{p}_2)\right\}_{S_r}$$

$$= \frac{4\pi a}{ik}\widehat{p}_1 \cdot G(a;-\widehat{b},\widehat{p}_2)$$

where

$$G(a;-\widehat{b},\widehat{p}_2) = \lim_{b\to\infty} G_b(a;\widehat{p}_2)$$

$$= e^{ika}a \times \left[\nabla \times E^s(a;-\widehat{b},\widehat{p}_2)\right.$$

$$\left. - \frac{ik}{2\pi}\int_{S^2} \widehat{r} \times g(\widehat{r};-\widehat{b},\widehat{p}_2)\, e^{ik\widehat{r}\cdot a}ds(\widehat{r})\right]$$

is a plane far–field pattern generator.

For the generators $G_b(a;\widehat{p}_2)$ and $G(a;-\widehat{b},\widehat{p}_2)$ we have the following limiting values [8].

Theorem 5. *For two incident point–source electric waves, $E_a^i(r;\widehat{p}_1)$ and $E_b^i(r;\widehat{p}_2)$, we have*

$$\lim_{a\to\infty} G_b(a;\widehat{p}_2) = g_b(-\widehat{a};\widehat{p}_2) \tag{21}$$

and

$$\lim_{a\to\infty} G(a;-\widehat{b},\widehat{p}_2) = g(-\widehat{a};-\widehat{b},\widehat{p}_2). \tag{22}$$

We can now let $b \to \infty$ in the general scattering theorem, Theorem 1. The proof of the following result is similar to that of Theorem 2 and is omitted for the sake of brevity.

Theorem 6. *Let $E_a^i(r;\widehat{p}_1)$ be an incident spherical electric wave and let $E^i(r;-\widehat{b},\widehat{p}_2)$ be an incident plane electric wave. Then*

$$\widehat{p}_1 \cdot G(a;-\widehat{b},\widehat{p}_2) + \widehat{p}_2 \cdot \overline{g_a(\widehat{b};\widehat{p}_1)} + \frac{1}{2\pi}\int_{S^2} g(\widehat{r};-\widehat{b},\widehat{p}_2) \cdot \overline{g_a(\widehat{r};\widehat{p}_1)}\, ds(\widehat{r})$$

$$= \mathcal{M}_a(-\widehat{b}; \widehat{p}_1, \widehat{p}_2),$$

where

$$\mathcal{M}_a(-\widehat{b}; \widehat{p}_1, \widehat{p}_2) = \lim_{b \to \infty} \mathcal{E}_{a,b}(\widehat{p}_1; \widehat{p}_2)$$

$$= \frac{k}{2\pi} \left\{ \operatorname{Im}(\mathcal{B}_1) \int_{\Omega_c} \left(\nabla \times \overline{E_a(r; \widehat{p}_1)} \right) \cdot \left(\nabla \times E(r; -\widehat{b}, \widehat{p}_2) \right) dv \right.$$

$$- \operatorname{Im}(\gamma^2 \mathcal{B}_1) \int_{\Omega_c} \overline{E_a(r; \widehat{p}_1)} \cdot E(r; -\widehat{b}, \widehat{p}_2) \, dv$$

$$+ i[\beta_c \gamma^2 + \operatorname{Re}(\mathcal{B}_2)] \int_{\Omega_c} E_a(r; \widehat{p}_1) \cdot \left(\nabla \times E(r; -\widehat{b}, \widehat{p}_2) \right) dv$$

$$\left. - i[\overline{\beta_c \gamma^2 \mathcal{B}_1} + \operatorname{Re}(\mathcal{B}_2)] \int_{\Omega_c} \left(\nabla \times E_a(r; \widehat{p}_1) \right) \cdot E(r; -\widehat{b}, \widehat{p}_2) \, dv \right\}.$$

To conclude, we note that we also have

$$\lim_{a \to \infty} \lim_{b \to \infty} G_b(a; \widehat{p}_2) = \lim_{b \to \infty} \lim_{a \to \infty} G_b(a; \widehat{p}_2) = g(-\widehat{a}; -\widehat{b}, \widehat{p}_2).$$

This can be used to verify that the known scattering relations for plane–wave incidence [4], [5] are recovered when $a \to \infty$ and $b \to \infty$. Furthermore, (22) and the reciprocity principle for plane waves [5] give the following limiting property:

$$\lim_{a \to \infty} \widehat{p}_1 \cdot G(a; -\widehat{b}, \widehat{p}_2) = \lim_{b \to \infty} \widehat{p}_2 \cdot G(-b; \widehat{a}, \widehat{p}_1).$$

7. Exact Green's function for a chiral dielectric sphere

Consider a spherical scatterer of radius a. Take spherical polar coordinates (r, θ, ϕ) with the origin at the centre of the sphere, so that the point source is at $r = r_0$, $\theta = 0$, and so that the polarization vector $\widehat{p}$ is in the x–direction. Thus, $r_0 = r_0 \widehat{z}$ and $\widehat{b} = \widehat{x}$, where $\widehat{x}$ and $\widehat{z}$ are unit vectors in the x and z directions, respectively.

Using spherical vector wave functions, and in particular (13.3.68), (13.3.69), (13.3.70) of [7] we obtain the following expansion for the incident field, see [3]

$$E_{r_0}^{\text{inc}}(r; \widehat{x}) = \frac{i}{h_0(kr_0)} \sum_{n=1}^{\infty} \frac{2n+1}{n(n+1)} \left\{ h_n(kr_0) N_{e1n}^1(r) - \widetilde{h}_n(kr_0) M_{o1n}^1(r) \right\}$$

for $r < r_0$, where $h_n \equiv h_n^{(1)}$ is a spherical Hankel function, $\widetilde{h}_n(x) = x^{-1} h_n(x) + h_n'(x) = x^{-1}[x h_n(x)]'$, and $M_{\sigma 1n}^1$ and $N_{\sigma 1n}^1$, for $n = 1, 2, \ldots$ and $\sigma = e$ or o are the spherical vector wave functions of the first kind.

The scattered field due to a chiral sphere does not have the same ϕ–dependence as the incident wave, so it has the following general form, [6], p. 394,

$$E^s_{r_0}(r;\hat{x})=\frac{i}{h_0(kr_0)} \sum_{n=1}^{\infty} \frac{2n+1}{n(n+1)}\{a_n N^3_{e1n}(r) - b_n M^3_{o1n}(r) + c_n M^3_{e1n}(r) - d_n N^3_{o1n}(r)\},$$

for $r > a$. In order to evaluate the electric field in the interior of the chiral sphere, we consider the Bohren decomposition, [6], and for $r \in \Omega_c$, write

$$E^c(r) = Q_L(r)-\sqrt{\mu_c/\varepsilon_c}\,Q_R(r) \quad \text{and} \quad H^c(r) = -i\sqrt{\varepsilon_c/\mu_c}\,Q_L(r)+Q_R(r),$$

where $Q_L(r)$ and $Q_R(r)$ are Beltrami fields, satisfying $\nabla \times Q_L = \gamma_L Q_L$ and $\nabla \times Q_R = -\gamma_R Q_R$. We employ the following expansions for the Beltrami fields, [6], p. 395

$$Q_L(r) = \frac{i}{h_0(kr_0)} \sum_{n=1}^{\infty} \frac{2n+1}{n(n+1)}\{A_n \left[M^1_{o1n}(r) + N^1_{o1n}(r)\right] + B_n \left[M^1_{e1n}(r) + N^1_{e1n}(r)\right]\}$$

$$Q_R(r) = \frac{i}{h_0(kr_0)} \sum_{n=1}^{\infty} \frac{2n+1}{n(n+1)}\{C_n \left[M^1_{o1n}(r) - N^1_{o1n}(r)\right] + D_n \left[M^1_{e1n}(r) - N^1_{e1n}(r)\right]\}$$

Using the transmission conditions on $r = a$, we obtain

$$a_n = \frac{S_{nL}W_{nR} + S_{nR}W_{nL}}{V_{nL}W_{nR} + V_{nR}W_{nL}}, \quad b_n = \frac{T_{nL}V_{nR} + T_{nR}V_{nL}}{V_{nL}W_{nR} + V_{nR}W_{nL}}$$

$$c_n = -\frac{\tilde{h}_n(kr_0)}{h_n(kr_0)}d_n = \frac{\tilde{h}_n(kr_0)}{h_n(kr_0)} \frac{S_{nL}V_{nR} - S_{nR}V_{nL}}{V_{nL}W_{nR} + V_{nR}W_{nL}}$$

where, for $A = L, R$, and $\tilde{j}_n(x) = x^{-1}j_n(x) + j'_n(x) = x^{-1}\left[xj_n(x)\right]'$,

$$W_{nA} = \tilde{h}_n(ka)j_n(\gamma_A a) - (\eta/\eta_c)\,h_n(ka)\tilde{j}_n(\gamma_A a)$$

$$V_{nA} = h_n(ka)\tilde{j}_n(\gamma_A a) - (\eta/\eta_c)\,\tilde{h}_n(ka)j_n(\gamma_A a)$$

$$S_{nA} = h_n(kr_0)\left[j_n(ka)\tilde{j}_n(\gamma_A a) - (\eta/\eta_c)\,\tilde{j}_n(ka)j_n(\gamma_A a)\right]$$

$$T_{nA} = \tilde{h}_n(kr_0)\left[(\eta/\eta_c)\,j_n(ka)\tilde{j}_n(\gamma_A a) - \tilde{j}_n(ka)j_n(\gamma_A a)\right],$$

$$A_n = \frac{1}{2j_n(\gamma_L a)}\left\{-h_n(ka)\left[\frac{\eta_c}{\eta}c_n + b_n\right] - \tilde{h}_n(kr_0)j_n(ka)\right\}$$

$$B_n = \frac{1}{2j_n(\gamma_L a)}\left\{h_n(ka)\left[\frac{\eta_c}{\eta}a_n + c_n\right] + \frac{\eta_c}{\eta}h_n(kr_0)j_n(ka)\right\}$$

$$C_n = \frac{1}{2i\eta_c j_n(\gamma_R a)} \left\{ h_n(ka) \left[-\frac{\eta_c}{\eta} d_n + b_n \right] + \widetilde{h}_n(kr_0) j_n(ka) \right\}$$

$$D_n = \frac{1}{2i\eta_c j_n(\gamma_R a)} \left\{ h_n(ka) \left[\frac{\eta_c}{\eta} a_n - c_n \right] + \frac{\eta_c}{\eta} h_n(kr_0) j_n(ka) \right\}$$

and $\eta_c = \sqrt{\mu_c/\varepsilon_c}$ is the interior intrinsic impedance.

Let us calculate the electric far–field pattern. Since $h_n(x) \sim (-i)^n h_0(x)$ and $h'_n(x) \sim (-i)^{n-1} h_0(x)$ as $x \to \infty$, and using (13.3.68) and (13.3.69) of [7] we find that

$$\begin{aligned} M^3_{\sigma 1 n}(r) &= \sqrt{n(n+1)} \, h_n(kr) \, C_{\sigma 1 n}(\widehat{r}) \\ &\sim \sqrt{n(n+1)} \, (-i)^n \, h_0(kr) \, C_{\sigma 1 n}(\widehat{r}) \end{aligned}$$

and

$$\begin{aligned} N^3_{\sigma 1 n}(r) &= n(n+1) \, (kr)^{-1} h_n(kr) \, P_{\sigma e 1 n}(\widehat{r}) + \sqrt{n(n+1)} \, \widetilde{h}_n(kr) \, B_{\sigma 1 n}(\widehat{r}) \\ &\sim \sqrt{n(n+1)} \, (-i)^{n-1} \, h_0(kr) \, B_{\sigma 1 n}(\widehat{r}) \end{aligned}$$

as $kr \to \infty$, where $C_{\sigma 1 n}(\widehat{r})$, $P_{\sigma 1 n}(\widehat{r})$ and $B_{\sigma 1 n}(\widehat{r})$ can be found e.g. in the Appendix of [3]. Therefore for the electric far–field pattern, we have

$$g_{r_0}(r; \widehat{x}) = -\frac{1}{h_0(kr_0)} \sum_{n=1}^{\infty} \frac{(2n+1)(-i)^n}{\sqrt{n(n+1)}} \left\{ a_n \, B_{e 1 n}(\widehat{r}) + i b_n \, C_{o 1 n}(\widehat{r}) + \right.$$
$$\left. i c_n \, C_{e 1 n}(\widehat{r}) + d_n \, B_{o 1 n}(\widehat{r}) \right\}.$$

References

1. C. Athanasiadis, P. A. Martin, A. Spyropoulos, I. G. Stratis: J. Math. Phys. **43** (2002) 5683–5697.
2. C. Athanasiadis, P. A. Martin, I. G. Stratis: IMA J. Appl. Math. **66** (2001) 539–549.
3. C. Athanasiadis, P. A. Martin, I. G. Stratis: Z. Angew. Math. Mech. **83** (2003) 129–136.
4. D. Colton, R. Kress: Inverse Acoustic and Electromagnetic Scattering Theory (2nd ed.). Springer, 1998.
5. G. Dassios, R. Kleinman: Low Frequency Scattering. Clarendon Press, 2000.
6. A. Lakhtakia: Beltrami Fields in Chiral Media. World Scientific, 1994.
7. P. M. Morse, H. Feshbach: Methods of Theoretical Physics, Part II. McGraw–Hill, 1953.
8. R. Potthast: Point Sources and Multipoles in Inverse Scattering Theory. Chapman & Hall CRC, 2001.

COMPUTATIONAL COMPLEXITY ANALYSIS OF MAS/MMAS AND COMPARISON WITH MOM FOR SCATTERING AND RADIATION PROBLEMS

GEORGIOS K. AVDIKOS, HRISTOS T. ANASTASSIU,
DIMITRA I. KAKLAMANI AND NIKOLAOS K. UZUNOGLU

Institute of Communication and Computer Systems
School of Electrical and Computer Engineering
National Technical University of Athens
Heroon Polytechniou 9
GR-15780 Zografou, Athens
GREECE
e-mail: gaudi@ece.ntua.gr

This paper presents a computational complexity analysis of two numerical techniques, namely MAS (Method of Auxiliary Sources) and MMAS (Modified Method of Auxiliary Sources), when applied to scattering and radiation problems. Due to their inherent merits, these methods are more attractive than the conventional Moment Method (MoM) for a wide variety of problems. To estimate the methods' computational cost, the number of multiplications required for the system matrix fill is initially calculated. Furthermore, the algorithmic cost of the matrix inversion is quantified and added to the matrix filling complexity. The analysis is applied to various objects, such as a perfectly conducting (PEC) parallelepiped, a PEC sphere and a microstrip patch antenna. In analogous MoM computations, performed for comparison purposes, an implementation utilizing typical basis functions is considered. In all cases examined, the complexity superiority of MAS and MMAS versus MoM is fully investigated. It is concluded that MAS and MMAS are potentially more efficient techniques than MoM under certain conditions, and are thus suitable for the treatment of problems not easily tractable by the latter, such as scattering from large-size targets.

1. Introduction

Electromagnetic (EM) interactions with arbitrarily shaped objects can be characterized via the solution of an integral equation, whose unknown quantity is the induced current. Since the integral equations involved are rarely solvable by exact techniques, approximate numerical methods are routinely used for their solution. A popular procedure of this type in Electromagnetics is the method of moments (MoM) [1]. In this technique, the PEC scatterer's surface is initially divided into a number of triangular or quadrilateral patches. The surface electric current is most often represented by roof-top functions for quadrilateral patches and the Rao-Wilton-Glisson (RWG) functions for triangular patches [2]. Fairly small patches (typically, over 100 patches per square wavelength) must be used to yield a sufficiently accurate representation of the currents. However, such a discretization may lead to an exceedingly high numbers of unknowns for a large scatterer, making MoM impractical for numerous real life problems. One of the remedies to this issue is to employ higher order basis functions [3], whose calculation, though, increases the computational cost of the algorithm.

Given the low efficiency of MoM for large bodies, alternative techniques have been receiving a lot of attention recently. Among these, the Method of Auxiliary Sources (MAS) [4]-[6] is generally considered highly promising. It possesses significant advantages with respect to numerical stability, computational accuracy, and ease of implementation. The EM boundary-value problems is *not* formulated in terms of continuous equivalent surface currents flowing on surfaces –where the corresponding boundary conditions are enforced- but in terms of discrete fictitious currents, the "Auxiliary Sources" (AS's), located at some distance away from the physical boundaries. Especially for thin or open structures, where the inevitable proximity of source and observation points causes numerical instabilities, a modified version of MAS (MMAS [7,8]) is much more accurate. MMAS is based on the use of current and charge densities on the auxiliary surface, distributed on a canonical grid, where the dyadic Green's function is calculated in a finite difference sense.

Although both MAS and MMAS have been empirically proven to be faster methods than MoM for an extensive class of problems, very few previous studies have been performed [6] focused on the comparison of their computational complexities. In this paper, we initially explore the complexity expressions for plane wave scattering from an arbitrary three-dimensional conducting scatterer for MoM, MAS and MMAS. Additionally, the MAS and MMAS expressions for several specific geometries (PEC sphere, PEC parallelepiped, and microstrip patch antenna) are investigated, and their respective computational cost is estimated, including matrix fill and inversion procedures. Several plots depict the complexity behavior of all three methods as a function of the number of unknowns, revealing the relative computational efficiencies of each.

A $e^{j\omega t}$ time convention is assumed and suppressed throughout the paper.

2. Analytical Considerations for PEC Scatterers

2.1. *MoM Complexity*

We assume a three-dimensional PEC scatterer with a smooth surface S, illuminated by a known external field E^{inc}, lying inside an infinite homogeneous and linear space with dielectric permittivity ε and magnetic permeability μ.

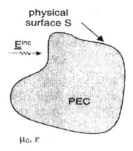

Fig. 1. Scattering from a 3D PEC object.

In standard MoM, a set of basis functions is chosen to approximate the induced current J on S. A standard set of these functions was originally introduced by Rao *et al.* [2] and is defined in Fig. 2, where $A_n^\pm$ denotes the area of the two triangles forming the dihedral patch. Current J is expressed as a weighted superposition of the basis functions, namely as $J \cong \sum_{n=1}^{N} I_n f_n(r)$ where the coefficients I_n are to be determined. Following the Galerkin procedure for the EFIE, where testing and basis functions are chosen to be identical, we obtain the well-known matrix equation $[Z]\{I\}=\{V\}$. The expressions for the elements of $[Z]$ and $\{V\}$ are given in [2], and they reduce to surface integrals of the scalar Green's function, the most typical one of which can be numerically evaluated by

$$\int_0^1 \int_0^{1-\eta} \frac{e^{-jkR^P}}{R^P} d\xi d\eta \cong \sum_{i=1}^{K} \sum_{j=1}^{K} \frac{e^{-jkR^{P(i,j)}}}{R^{P(i,j)}} \Delta\xi_i \Delta\eta_j \qquad (1)$$

where k is the wavenumber, ξ, η are the familiar normalized area co-ordinates on each triangle, R^P is the source-observation distance and K is the numerical integration order (number of subintervals per integration interval).

Based on (1), it is possible to count the number of multiplications required for the evaluation of all elements in $[Z]$. Adding the number of multiplications for the solution of the MoM linear system via Gauss elimination [9], the total complexity of the method, when involving N_{MoM} unknowns, becomes

$$\#\text{MoM} \cong N_{MoM}^2\left(110K^2 + 4K + 2\right) + 1/6\left(2N_{MoM}^3 + 9N_{MoM}^2 + 5N_{MoM}\right) \qquad (2)$$

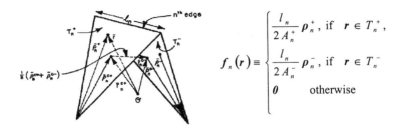

$$f_n(r) \equiv \begin{cases} \dfrac{l_n}{2\,A_n^{+}}\,\boldsymbol{\rho}_n^{+}, & \text{if } \ r \in T_n^{+}, \\[2mm] \dfrac{l_n}{2\,A_n^{-}}\,\boldsymbol{\rho}_n^{-}, & \text{if } \ r \in T_n^{-} \\[2mm] 0 & \text{otherwise} \end{cases}$$

Fig. 2. The Rao-Wilton-Glisson linear surface basis functions [2].

2.2. MAS complexity

In MAS applications pertaining to PEC scatterers, the auxiliary sources (AS's) are distributed on fictitious surfaces, conformal to the physical boundaries (see Fig. 3). The electric field of the nth AS located at r_n' is proportional to the components of the dyadic Green's function, namely

$$E_n(r) = E_n \left(\overline{\overline{I}} + \frac{\nabla\nabla}{k^2} \right) \frac{\exp\{-jk|r - r_n'|\}}{4\pi|r - r_n'|} \cdot \hat{u}_n \qquad (3)$$

where E_n is the corresponding unknown weight, I is the unit dyad and $\hat{u}_n$ is the unit vector along the AS current direction. The total scattered field is expressed as a superposition of all fields given in (3), and the weights are determined via the solution of a linear system, constructed through imposition of the boundary conditions at a set of collocation points [4]-[6]. Obviously, the computational complexity of MAS is proportional to the number of multiplications required for the evaluation of (3), in a convenient coordinate system attached to the geometry. Careful calculation of the complexity finally yields

$$\#\text{MAS} \cong (180+C)N_{MAS}^2 + 1/6\left(2N_{MAS}^3 + 9N_{MAS}^2 + 5N_{MAS}\right) \qquad (4)$$

where N_{MAS} is the number of AS's, and C represents the operations required for the calculation of (3) in the appropriate coordinates. Specific application of (4) to a PEC parallelepiped implies $C=0$, and to a sphere $C=30$.

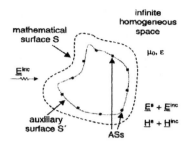

Fig. 4. The AS's are located on a auxiliary surface S' enclosed by the physical surface S.

2.3. MMAS complexity

MMAS is a recently developed variant of MAS [7,8], particularly suitable for the accurate scattering characterization of thin or open structures. MMAS avoids the complications caused by the high order singularities of the dyadic Green's function in (3), by approximating the derivatives through a finite difference scheme, adapted to a local coordinate system, which is attached to the geometry in a conformal fashion. Thus, the MMAS computational complexity differs from MAS basically in the finite difference calculations. The pertinent cost expression is finally estimated as

$$\#\text{MMAS} \cong \left(330 + A + B\right)N_{MMAS}^2 + 1/6\left(2N_{MMAS}^3 + 9N_{MMAS}^2 + 5N_{MMAS}\right) \quad (5)$$

where A and B correspond to the operations number required for the evaluation of the metric coefficients of the surface-tangential coordinates in the aforementioned system. Specific application of (5) to a PEC parallelepiped is valid for $A=B=0$, and to a sphere for $A+B=20$.

3. The planar microstrip patch antenna

A typical geometry of a planar microstrip patch antenna is depicted in Fig. 5. Given the existence of the dielectric substrate, the mathematical analysis of its radiation properties is much more complex than in the PEC cases addressed above. To solve the problem with MoM, the Volume Integral Equation (VIE) is employed, and elementary basis functions are utilized, based on a cubical or tetrahedral discretization scheme [10]. The metallic elements of the antenna are modeled as in Section 2, and their interaction with the dielectric is taken into account. If N_1, N_2, N_3 and N_4 are the number of the unknowns for the dielectric, patch, ground and strip line respectively, and assuming sufficiently smooth behavior of the potentials [10], the total complexity of MoM becomes

$$\#\text{MoM} \cong \left(86K^3 + 86K^2 + 12K + 4\right)N_1^2 + \left(504K^2 + 6K + 2\right)\left(N_2^2 + N_3^2 + N_4^2\right) +$$
$$+ \left(86K^2 + 6K + 2\right)\left(N_1N_2 + N_1N_3 + N_1N_4 + N_2N_3 + N_2N_4 + N_3N_4\right) +$$
$$+ 1/6\left(2N_{tot}^3 + 9N_{tot}^2 + 5N_{tot}\right) \quad , \qquad N_{tot} \equiv N_1 + N_2 + N_3 + N_4 \qquad (6)$$

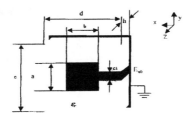

Fig. 5. A microstrip patch antenna.

where K is the order of numerical integration. Analogous estimates for MAS and MMAS yield

$$\#\text{MAS} \cong 210N_{MAS}^2 + 1/6\left(2N_{MAS}^3 + 9N_{MAS}^2 + 5N_{MAS}\right) \qquad (7)$$

$$\#\text{MMAS} \cong 450N_{MMAS}^2 + 1/6\left(2N_{MMAS}^3 + 9N_{MMAS}^2 + 5N_{MMAS}\right) \qquad (8)$$

4. Numerical Results and Discussion

Several graphs are plotted, comparing the complexities of MAS, MMAS and MoM, for the parallelepiped, sphere and the microstrip antenna. The calculations were performed assuming $N_{MAS} = N_{MMAS} = aN_{MoM}$. Evidently, for $a \leq 1$, (2), (4) and (5) yield that MoM is always less efficient than either MAS or MMAS, therefore it is only interesting to investigate the case when $a > 1$. For the parallelepiped we set $a=2$, whereas for the sphere and the microstrip antenna we set $a=3$. The ratios #MAS/#MoM and #MMAS/#MoM are plotted in Fig. 6, 7 and 8 for a PEC parallelepiped, a PEC sphere and a microstrip patch antenna respectively. For the microstrip antenna also, Fig. 9 depicts a direct complexity comparison between MAS and MMAS, showing the expected higher computational cost of MMAS. The behavior for a small number of unknowns N_{MoM} is nearly linear while for large N_{MoM} the curves reach the asymptote #(M)MAS/#MoM = a^3, as it is evident from (2), (4) and (5). The main conclusion drawn from the plots is that there always exists a threshold of

N_{MoM}, under which MAS or MMAS are more efficient than MoM, even if they require a denser discretization.

The results in this section are valid for MoM numerical integration based on the trapezoidal rule and for matrix inversion via Gaussian elimination. For other integration or inversion schemes, the estimates must be modified accordingly.

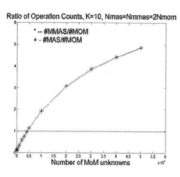

Fig. 6. Operations ratio for a parallelepiped.

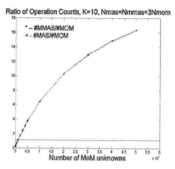

Fig. 7. Operations ratio for a sphere.

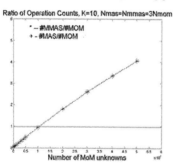

Fig. 8. Count ratio for a microstrip patch antenna.

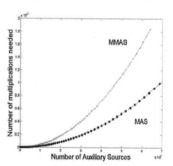

Fig. 9. Operations required vs. number of AS's.

5. Summary and Conclusions

In this paper the computational complexities of three numerical techniques, namely MoM, MAS and MMAS were estimated and compared to each other. General, three-dimensional PEC scatterers were considered, and the results were further particularized to a parallelepiped and a sphere. The radiation analysis of a microstrip patch antenna was also examined. Computational cost comparisons showed that for the same discretization density, both MAS and MMAS are always more efficient than MoM. Moreover, even if MoM requires fewer unknowns than MAS or MMAS, MoM is still less efficient under a given threshold of the number of unknowns.

References

1. R. F. Harrington, *Field Computation by Moment Method*, New York: Macmillan, 1968.
2. S. M. Rao, D. R. Glisson, and A. W. Wilton, "Electromagnetic scattering by surfaces of arbitrary shape," *IEEE Transactions on Antennas and Propagation*, vol. 30, pp. 409–418, May 1982.
3. S. Wandzura, "Electric current basis functions for curved surfaces", *Electromagnetics*, vol. 12, no. 1, pp. 77-91, Jan. 1992.
4. Leviatan Y. and A. Boag, "Analysis of Electromagnetic Scattering from Dielectric Cylinders Using a Multifilament Current Model", *IEEE Transactions on Antennas and Propagation*, vol. AP-35, no. 10, pp. 1119-1127, Oct. 1987.
5. Kaklamani, D. I. and H. T. Anastassiu, "Aspects of the Method of Auxiliary Sources (MAS) in Computational Electromagnetics", *IEEE Antennas and Propagation Magazine,* vol. 44, no. 3, pp. 48-64, June 2002.
6. Anastassiu, H. T., D. I. Kaklamani, D. P. Economou and O. Breinbjerg, "Electromagnetic Scattering Analysis of Coated Conductors with Edges Using the Method of Auxiliary Sources (MAS) in Conjunction with the Standard Impedance Boundary Condition (SIBC)", *IEEE Transactions on Antennas and Propagation,* vol. AP-50, no. 1, pp. 59-66, Jan. 2002.
7. Kaklamani D. I., Anastassiu H. T. and P. Shubitidze, "Analysis of Conformal Patch Arrays via a Modified Method of Auxiliary Sources (MMAS*), Proceedings of the Millenium Conference on Antennas and Propagation (AP2000)*, Davos, Switzerland, 9-14 April 2000.
8. F. Shubitidze, H. T. Anastassiu and D. I. Kaklamani, "An Improved Accuracy Version of the Method of Auxiliary Sources for Computational Electromagnetics", accepted by *IEEE Transactions on Antennas and Propagation* (to appear Feb. 2004).
9. P. G. Ciarlet, *Introduction to Numerical Linear Algebra and Optimization.* Cambridge, U.K.: Cambridge Univ. Press, 1991.
10. D. H. Schaubert, D. R. Wilton, and A. W. Glisson, "A tetrahedral modelling method for electromagnetic scattering by arbitrary shaped inhomogeneous dielectric bodies," *IEEE Transactions on Antennas and Propagation*, vol. 32, pp.77–85, Jan. 1984.

ON THE DIRECT AND INVERSE SCATTERING PROBLEM FROM A SPHERICAL INCLUSION IMBEDDED IN A CYLINDRICAL ELASTIC BOUNDED DOMAIN

K. BAGANAS AND A. CHARALAMBOPOULOS

Department of Materials Science and Engineering,
University of Ioannina,
GR45110, Ioannina, Greece

G. D. MANOLIS

Department of Civil Engineering,
Aristotle University,
GR54006, Thessaloniki, Greece

In this paper we study the theoretical part behind a non-destructive testing (NDT) method for the identification of a spherical inclusion inside a bounded cylindrical elastic medium. The latter is considered to be linear, homogeneous, isotropic and bounded in 3-D space while the inclusion contains an ideal fluid. We apply a normal, harmonic and uniform pressure load on the one tranverse cavity surface and express the dispacement fields inside and outside the inclusion in terms of Navier eigenvector expansions. The rest of the exterior surface is assumed to be traction free. We solve the problem above and investigate the displacement field on the outer surface of the cylinder. Suitable numerical proccessing is performed at the final stage of the theoretical model. In the inverse problem regime, we show that knowledge of the displacement field on the lateral cylinder surface leads to determination of the inclusions position as well as an estimation of its size.

1. Introduction

Problems involving bounded domains in 2D and 3D with defects or inhomogeities that have been treated by analytical methods have been restricted to cases where simple, similar geometries comprise the problem at hand, i.e., cylindrical objects in a cylindrical body [1], spherical bodies within spherical substrate [2], etc. In such cases, the relevant fields are expanded in series that use the same complete basis of eigenfunctions. Then, by using addition theorems referring to the same geometry and manipulating the equations that arise from the boundary conditions, the problem under discussion degenerates into an NxN algebraic system of the type $\mathbf{AX=B}$ where $\mathbf{X}$ includes the unknown expansion coefficients of the underlying fields. Matrix $\mathbf{A}$ is accurately extrapolated and its coefficients are simple expressions or series of well-known functions.

In problems with dissimilar geometries the coefficients of the $\mathbf{A}$ and $\mathbf{B}$ matrices are expressed as multiple infinite series of projections (surface integrals) on the characteristic boundary surfaces of the problem. Numerical treatment for such expressions is a difficult task, since the truncation and round-off errors that accumulate because of multiple iterative loops cause unavoidable errors, and careful numerical implementation is required in order to produce satisfactory results. An efficient approach to the problem is the application of T-matrix methods (or null-field approach) [3, 4]. Also, Lee and Mal [5] use a volume integral technique to handle multiscattering phenomena in elastodynamics. Some representative techniques for studying the interference of harmonic elastic waves with (i) inclusions in half-spaces and (ii) circular holes with or without edge cracks in an infinite plate can be found in [6, 7] and [8], respectively.

In this work, we investigate a boundary value problem with dissimilar geometries. More precisely, we look at the general framework of spectral analysis of BVPs, extending the use of Navier eigenvectors [9, 15] to problems consisting of a spherical inclusion (the spherical geometry) containing an ideal fluid, arbitrarily placed inside a cylindrical elastic body (the cylindrical geometry), where both are bounded in Euclidean 3D space. This model simulates concrete specimens with one fluid (or void, as a special case) spherical inclusion. That investigate forced vibrations of this system, although one could study the free vibrations as well by determining the eigenfrequencies and the corresponding eigenvectors. Actually, the latter analysis would also work for inclusion detection. In fact, these two approaches are connected, since given eigenproperties of the system, forced vibrations can be obtained by superposition, although this implementation is numerically more demanding. In this work, we prefer to face the excitation problem directly, in order to avoid solution of several intermediate homogeneous problems, giving priority to detection mechanisms based on outside loads instead of free vibrations. Finally, the excitation process has an addition degree of freedom; namely, the source position itself.

The displacement field at the cylinder's lateral surfaces provides useful information for identifying both position and size of the inclusion. Specifically, the inclusion's position is clearly identifiable through simple inspection, while only a good estimate of its diameter can be attained.

2. Problem Formulation

We consider a cylindrical body V of radius a and length d characterized completely by its Lamé's constants λ, μ, and its density ρ. Inside this structure a spherical inclusion is centered at an arbitrary point O, having radius b and containing an ideal fluid with density ρ_f and sound propagation velocity c_f. A normal, time-harmonic uniform pressure load is applied on the surface element S_s. The system responds to this excitation and this dynamical process is described by the displacement field $\mathbf{u}(\mathbf{r},t) = \mathbf{u}(\mathbf{r})\,e^{-i\omega t}$ developed in the elastic medium as well as the velocity potetial field $\Phi(\mathbf{r},t) = \Phi(\mathbf{r})\,e^{-i\omega t}$ in the perfect fluid region.

The displacement field $\mathbf{u}(\mathbf{r})$ satisfies the time-reduced Navier equation

$$c_s'\nabla'^2\mathbf{u}(\mathbf{r}') + (1 - c_s'^2)\nabla'\nabla'\cdot\mathbf{u}(\mathbf{r}') + \Omega^2\mathbf{u}(\mathbf{r}') = 0, \tag{1}$$

where $\mathbf{r}' = \frac{\mathbf{r}}{a}$, $\Omega = \frac{\omega a}{c_p}$, $\nabla' = a\nabla$, $c_s' = \frac{c_s}{c_p}$, $c_p = \left(\frac{\lambda+2\mu}{\rho}\right)^{1/2}$, $c_s = \left(\frac{\mu}{\rho}\right)^{1/2}$.

Also, the motion of an inviscid and irrotational fluid is destribed by the velocity potential $\Phi(\mathbf{r},t)$ whose time-independent part satisfies the Helmholtz equation

$$\nabla'^2\Phi(\mathbf{r}') + k_f'^2\Phi(\mathbf{r}') = 0, \tag{2}$$

where $c_f' = \frac{c_f}{c_p}$ and $k_f' = \frac{\Omega}{c_f'}$. The pressure, $P(\mathbf{r},t)$, in the fluid is found from the velocity potential as

$$P'(\mathbf{r}') = \frac{P(\mathbf{r}')}{r} = i\Omega\rho_f'\frac{1}{c_s^2}\Phi(\mathbf{r}'), \tag{3}$$

where $\rho_f' = \rho_f/\rho$. The displacement field $\mathbf{u}^f(\mathbf{r}')$ and the resulting velocity $\mathbf{v}^f(\mathbf{r}') = -i\Omega\mathbf{u}^f(\mathbf{r}')$ are recovered easily from the potential field through the equation $\mathbf{u}^f(\cdot) = \frac{i}{\Omega}\nabla'\Phi(\cdot)$.

The curved surface S_a and flat surfaces S_b, S_c are traction free

$$\mathbf{Tu}(\mathbf{r}') = 0, \ \mathbf{r}' \in S', \tag{4}$$

where $\mathbf{T} = 2\mu'\hat{\mathbf{n}}\cdot\nabla' + \lambda'\hat{\mathbf{n}}\nabla'\cdot + \mu'\hat{\mathbf{n}}\times\nabla'\times$, stands for the dimensionless surface traction operator, S' is the external surface of the cylinder with the exception of the "source" region S_s: $S' = S_a \cup S_b \cup (S_c\backslash S_s)$, $\hat{\mathbf{n}}$ is the unit outward normal vector and $(\lambda',\mu') = (\lambda/\mu, 1)$.

The excitation field is described as $\mathbf{P}_s(\mathbf{r}') = -P_s'\hat{\mathbf{n}}(\mathbf{r}')$, $\mathbf{r}' \in S_s$, where P_s' denotes its magnitude. Thus, $\mathbf{Tu}(\mathbf{r}') = -P_s'\hat{\mathbf{n}}(\mathbf{r}')$, for $\mathbf{r}' \in S_s$.

Considering now the inclusion's surface, the conditions of pressure and normal displacement continuity must hold

$$\mathbf{Tu}\,(\mathbf{r}') = -P'\,(\mathbf{r}')\hat{\mathbf{r}}, \ \mathbf{r}' \in S_f, \tag{5}$$

$$\hat{\mathbf{r}} \cdot \mathbf{u}\,(\mathbf{r}') = \hat{\mathbf{r}} \cdot \mathbf{u}^f\,(\mathbf{r}'), \ \mathbf{r}' \in S_f, \tag{6}$$

where $\mathbf{u}^f$ denotes the displacement field in the inclusion.

3. The Direct Problem Solution

In the elastic region, the displacement field in written in terms of the complete basis of Navier eigenvectors $\{\mathbf{L}_n^{m,l}, \mathbf{M}_n^{m,l}, \mathbf{N}_n^{m,l}\}$ [9, 15]. The expressions $\mathbf{TL}_n^{m,l}\,(\hat{\mathbf{r}})$, $\mathbf{TM}_n^{m,l}\,(\hat{\mathbf{r}})$ and $\mathbf{TN}_n^{m,l}\,(\hat{\mathbf{r}})$ that result when the traction operator $\mathbf{T}$ is applied to the Navier eigenvector expansions are explicitly given in [2].

The displacement field in the fluid is expressed as

$$\mathbf{u}^f\,(\mathbf{r}') = \frac{i}{c_f'} \sum_{n=0}^{\infty} \sum_{m=-n}^{n} \left[\begin{array}{c} c_n^m j_n'(k_f'r')\mathbf{P}_n^m(\hat{\mathbf{r}})+ \\ c_n^m j_n(k_f'r')\frac{\sqrt{n(n+1)}}{r'}\mathbf{B}_n^m(\hat{\mathbf{r}}) \end{array} \right], \tag{7}$$

and the pressure takes the form

$$P'(\mathbf{r}') = i\Omega\rho_f'\frac{1}{c_s^2} \sum_{n=0}^{\infty} \sum_{m=-n}^{n} \left[c_n^m j_n(k_f'r')P_n^m(\cos\theta)e^{im\phi} \right].$$

The application of the boundary conditions at S_f gives at $\mathbf{r}' \in S_f$, where $r' = r_f' = b/a$:

1) $\mathbf{Tu}\,(\mathbf{r}') = -P'\,(\mathbf{r}')\hat{\mathbf{r}} \Leftrightarrow$

$$\sum_{l=1}^{2} [a_n^{m,l} A_n^l\,(k_p';r_f') + \gamma_n^{m,l} D_n^l\,(k_s';r_f')] = -i\Omega\rho_f'\frac{1}{c_s^2}c_n^m j_n\,(k_f'r_f'), \tag{8}$$

$$\sum_{l=1}^{2} [a_n^{m,l} B_n^l\,(k_p';r_f') + \gamma_n^{m,l} E_n^l\,(k_s';r_f')] = 0, \tag{9}$$

$$\sum_{l=1}^{2} [\beta_n^{m,l} C_n^l\,(k_s';r_f')] = 0, \tag{10}$$

2) $\hat{\mathbf{r}} \cdot \mathbf{u}\,(\mathbf{r}') = \hat{\mathbf{r}} \cdot \mathbf{u}^f\,(\mathbf{r}') \Leftrightarrow$

$$\sum_{l=1}^{2} \left[a_n^{m,l} g_n^l\,(k_p'r_f') + \gamma_n^{m,l} n\,(n+1)\frac{g_n^l\,(k_s'r_f')}{k_s'r_f'} \right] = \frac{i}{c_f'}c_n^m j_n'\,(k_f'r_f'), \tag{11}$$

wherethe $A_n^l(k_p'; r')$, $B_n^l(k_p'; r')$, $C_n^l(k_s'; r')$, $D_n^l(k_s'; r')$ and $E_n^l(k_s'; r')$ expressions can also be found in [2].

Solving the system of equations (10-13) we obtain

$$\beta_n^{m,2} = -\frac{C_n^1\left(k_s'; r_f'\right)}{C_n^2\left(k_s'; r_f'\right)}\beta_n^{m,1} \text{ for } n \geqslant 1, \; a_0^{0,2} = c_0^0 F_{12}, \; a_0^{0,1} = \frac{F_2}{F_1}c_0^0, \quad (12)$$

$$a_n^{m,1} = \sum_{l=1}^{2}\gamma_n^{m,l} F_{10}\left(l; n\right), \; a_n^{m,2} = -\sum_{l=1}^{2}\gamma_n^{m,l} F_{11}\left(l; n\right), \; n \geqslant 1, \quad (13)$$

where the coefficients F_i, $i = 1..12$ are appropriate known functions. Inserting now these expressions into eq.(7), $\mathbf{u}\left(\mathbf{r}'\right)$ becomes

$$\mathbf{u}\left(\mathbf{r}'\right) = c_0^0\left(F_{12}\mathbf{L}_0^{0,2} + \frac{F_2}{F_1}\mathbf{L}_0^{0,1}\right) + \sum_{n=1}^{\infty}\sum_{m=-n}^{n}\left\{\beta_n^{m,1}\left(\mathbf{M}_n^{m,1} - \frac{C_n^1}{C_n^2}\mathbf{M}_n^{m,2}\right) + \right.$$

$$\left. \sum_{l=1}^{2}\gamma_n^{m,l}\left(F_{10}\mathbf{L}_n^{m,1} - F_{11}\mathbf{L}_n^{m,2} + \mathbf{N}_n^{m,l}\right)\right\}. \quad (14)$$

This substitution implies more convenient handling of the rest boundary conditions.

The expansion must satisfy the remaining boundary conditions on the cylinder's surfaces. We have to adapt the above representation to a coordinate system matching the cylindrical geometry. This is necessary in case the inclusion is not centered along the symmetry axis z'. Then, use of translational addition theorems for Hansen's vectors give rise to a new expression of $\mathbf{u}(\mathbf{r}')$ in terms of spherical vector wave functions, with reference to origin O' ($\mathbf{r}_0' = (r_0', \theta_0, \phi_0)$) lying on the axis of symmetry z' of the elastic cylinder. We therefore consider the expressions derived by Cruzan in [12] and concern the $A\left(r, s; m, n, l\right)$, $A_{rs}^{mn}\left(m, n, l\right)$, $B_{rs}^{mn}\left(m, n, l\right)$ coefficients, and the $\mathbf{L}_n^{m,l}$, $\mathbf{M}_n^{m,l}$, $\mathbf{N}_n^{m,l}$. The displacement field according to O'

is finally expressed as

$$\mathbf{u}\left(\widetilde{\mathbf{r}}'\right) = c_0^0 \sum_{s=0}^{\infty} \sum_{r=-s}^{s} \left[F_{12} A\left(r, s; 0, 0, 2\right) \mathbf{L}'_{rs2} + \frac{F_2}{F_1} A\left(r, s; 0, 0, 1\right) \mathbf{L}'_{rs1} \right] +$$

$$\sum_{n=1}^{\infty} \sum_{m=-n}^{n} \left\{ \beta_n^{m,1} \sum_{s=0}^{\infty} \sum_{r=-s}^{s} \left(\begin{bmatrix} A_{rs}^{mn}\left(m, n, 1\right) \mathbf{M}'_{rs1} + \\ B_{rs}^{mn}\left(m, n, 1\right) \mathbf{N}'_{rs1} \end{bmatrix} - \atop \frac{C_n^1}{C_n^2} \begin{bmatrix} A_{rs}^{mn}\left(m, n, 2\right) \mathbf{M}'_{rs2} + \\ B_{rs}^{mn}\left(m, n, 2\right) \mathbf{N}'_{rs2} \end{bmatrix} \right) + \tag{15}$$

$$\sum_{l=1}^{2} \gamma_n^{m,l} \sum_{s=0}^{\infty} \sum_{r=-s}^{s} \left(\begin{matrix} F_{10} A\left(r, s; m, n, 1\right) \mathbf{L}'_{rs1} - F_{11} A\left(r, s; m, n, 2\right) \mathbf{L}'_{rs2} + \\ \left[A_{rs}^{mn}\left(m, n, l\right) \mathbf{N}'_{rsl} + B_{rs}^{mn}\left(m, n, l\right) \mathbf{M}'_{rsl} \right] \end{matrix} \right) \right\}.$$

The next step, after inserting $\mathbf{u}(\widetilde{\mathbf{r}}')$ into eqs.(4-5) and expressing $[\mathbf{Tu}\left(\widetilde{\mathbf{r}}'\right)]$ in terms of $\left(\widetilde{r}', \widetilde{\theta}, \widetilde{\phi}\right)$, is to transform the O' spherical coordinates to cylindrical ones. In order then to obtain an algebraic system of linearly independent equations for the unknown coefficients $c_0^0, \beta_n^{m,1}$ and $\gamma_n^{m,l}$, we use a set of basis functions, say $\{f_n(\widetilde{\phi}, \widetilde{z}')\}$, as follows. Consider, for example, the term $[\cdot]_{\widetilde{z}'} = 0$; obtain from the boundary conditions on the S_a surface. Then, the following surface integrals over $\widetilde{\phi}'$ and $\widetilde{z}'$:
$\int f_i \left(\widetilde{\phi}, \widetilde{z}'\right) [\cdot]_{\widetilde{z}'} \, d\widetilde{\phi} d\widetilde{z}' = 0$, $i = 1, 2, ..., N$, form a system of N linearly independent equations in $c_0^0, \beta_n^{m,1}, \gamma_n^{m,l}$, where N is a truncation parameter for the series solution; i.e., N implies that $n = 1 .. N_0$, while $N_0 = N_0(N)$.

Finally we collect these systems of equations, so as to form a linear system $\mathbf{AX=B}$, where $\mathbf{X}$ denotes the column vector of the unknown coefficients $c_0^0, \beta_n^{m,1}, \gamma_n^{m,l}$, $\mathbf{A}$ comprises integrals over all of the functions appeared into the brackets (by that we mean what is enclosed in $[\cdot]_{\widetilde{z}'}$ into the surface integrals) , and $\mathbf{B}$ is the "source" vector, due to the pressure load. Thus, $\mathbf{X=A^{-1}B}$ and the solution is

$$\mathbf{u}\left(\widetilde{\mathbf{r}}'\right) = c_0^0 \sum_{s=0}^{N_1} \sum_{r=-s}^{s} \left[F_{12} A\left(r, s; 0, 0, 2\right) \mathbf{L}'_{rs2} + \frac{F_2}{F_1} A\left(r, s; 0, 0, 1\right) \mathbf{L}'_{rs1} \right] +$$

$$\sum_{n=1}^{N_0} \sum_{m=-n}^{n} \left\{ \beta_n^{m,1} \sum_{s=0}^{N_2} \sum_{r=-s}^{s} \left(\begin{bmatrix} A_{rs}^{mn}\left(m, n, 1\right) \mathbf{M}'_{rs1} + \\ B_{rs}^{mn}\left(m, n, 1\right) \mathbf{N}'_{rs1} \end{bmatrix} - \atop \frac{C_n^1}{C_n^2} \begin{bmatrix} A_{rs}^{mn}\left(m, n, 2\right) \mathbf{M}'_{rs2} + \\ B_{rs}^{mn}\left(m, n, 2\right) \mathbf{N}'_{rs2} \end{bmatrix} \right) + \tag{16}$$

$$\sum_{l=1}^{2} \gamma_n^{m,l} \sum_{s=0}^{N_3} \sum_{r=-s}^{s} \left(\begin{matrix} F_{10} A\left(r, s; m, n, 1\right) \mathbf{L}'_{rs1} - F_{11} A\left(r, s; m, n, 2\right) \mathbf{L}'_{rs2} + \\ \left[A_{rs}^{mn}\left(m, n, l\right) \mathbf{N}'_{rsl} + B_{rs}^{mn}\left(m, n, l\right) \mathbf{M}'_{rsl} \right] \end{matrix} \right) \right\},$$

where N_1, N_2, N_3 are suitable truncation parameters so as the corresponding finite series sufficiently well approximate their own limits.

4. Numerical Results

We present results concerning the axisymmetric case where the inclusion and the pressure load lie on the symmetry axis of the cylinder. The results corresponds to material properties analogous to those of concrete test speciments and the fluid is considered to be water.

Figures 1-2 represent the dispacement field on the S_a surface for two different inclusion positions. The load frequency is 250 KHz.

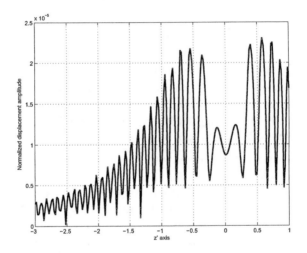

Figure 1.   Normalized displacement field on S_a; the inclusion is placed at $z' = 0$.

References

1. A. Charalambopoulos, D.I. Fotiadis, C.V. Massalas, Int. J. Eng. Sc., 36, 711 (1998).
2. A. Charalambopoulos, G. Dassios, D.I. Fotiadis, C.V. Massalas, Comp. Math. Model., 27(2), 81 (1998).
3. S. Olsson, J. Acoust. Soc. Am., 93, 2479 (1993).
4. S. Olsson, J. Acoust. Soc. Am., 88, 515 (1990).
5. J.K. Lee, A. Mal, Appl. Math. Comp., 67, 135 (1995).

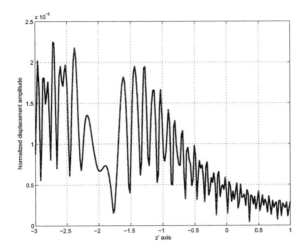

Figure 2.   Normalized displacement field on S_a; the inclusion is placed at $z' = -2$.

6. A.B. Doyum and F. Erdogan, J. Sound Vibr., 147(1), 13 (1991).
7. S. Nintcheu Fata, B. B. Guzina, M. Bonnet, IABEM 2002, Conf. Proc., Univ. Texas Publ., Austin (2002).
8. Z. Chang, A. Mal, Mech. Mater., 31, 197 (1999).
9. W.W. Hansen, Phys. Rev., 47, 139 (1935).
10. B. Friedman, J. Russek, Quart. Appl. Math., 12, 13 (1954).
11. S. Stein, Quart. Appl. Math., 19, 15 (1961).
12. O. R. Cruzan, Quart. Appl. Math., 20, 33 (1962).
13. A.K. Mal, S.K. Bose, Proc. Cambr. Phil. Soc., 76, 587 (1974).
14. J. A. Stratton, Electrom. Th., N.Y. (1941).
15. P.M. Morse and H. Feshbach, Meth. Theor. Phys., 2, N.Y. (1953).

MIXED BOUNDARY VALUE PROBLEMS IN INVERSE ELECTROMAGNETIC SCATTERING *

F. CAKONI AND D. COLTON

Department of Mathematical Sciences
University of Delaware,
Newark, Delaware 19711, USA
E-mail: cakoni@math.udel.edu, colton@math.udel.edu

We consider the scattering of time harmonic incident plane waves by partially coated perfect conductors and screens (including the case of no coating, i.e. the scatterer is a perfect conductor). Of particular interest to us is the inverse problem of determining the shape of surface impedance from a knowledge of the far field pattern of the electric field. Our analysis is based on a study of mixed boundary value problems for Maxwell's equations .

1. Introduction

In this paper we will survey recent results we have obtained on the inverse electromagnetic scattering problem for partially coated perfect conductors and screens. In particular, we are interested in obtaining the shape of the scatterer and material properties of the coating (if it exists) from a knowledge of the time harmonic incident field at a fixed frequency and the electric far field pattern for incident and observation directions on the unit sphere. The difficulty in solving such problems lies in the fact that the material properties of the coating, if it exists, is a priori unknown nor is the connectivity or dimensionality of the scatterer. Such problems are ideally suited to the linear sampling method in inverse scattering theory [7] and in this paper we will outline how the inverse scattering problems of the above type can be solved using this method. For further details we refer the reader to our papers [2], [3], [4] and [5].

---

*This work is supported in part by grants F49620-02-1-0071 and F49620-02-1-0353 from the Air Force Office of Scientific Research

2. The Inverse Scattering Problem for Partially Coated Obstacles

Let $D \subset \mathbb{R}^3$ be a bounded region with boundary Γ such that $D_e := \mathbb{R}^3 \backslash \overline{D}$ is connected. Each simply connected piece of D is assumed to be a Lipschitz curvilinear polyhedron. We assume that the boundary $\Gamma = \Gamma_D \cup \Pi \cup \Gamma_I$ is split into two disjoint parts Γ_D and Γ_I having Π as their possible common boundary and denote by ν the unit outward normal defined almost everywhere on Γ.

The direct scattering problem for the scattering of a time harmonic electromagnetic plane wave by a partially coated obstacle D is to find an electric field E and a magnetic field H such that

$$\operatorname{curl} E - ikH = 0$$
$$\operatorname{curl} H + ikE = 0 \tag{1}$$

in $\mathbb{R}^3 \setminus \overline{D}$ and on the boundary Γ satisfy

$$\nu \times E = 0 \qquad \text{on} \qquad \Gamma_D$$
$$\nu \times \operatorname{curl} E - i\lambda(x)(\nu \times E) \times \nu = 0 \qquad \text{on} \qquad \Gamma_I \tag{2}$$

where $\lambda(x) \geq \lambda_0 > 0$ is the surface impedance, $\lambda \in L_\infty(\Gamma_I)$,

$$E = E^i + E^s, \qquad H = H^i + H^s \tag{3}$$

and

$$\lim_{r \to \infty} (\operatorname{curl} E^s \times x - ikr E^s) = 0 \tag{4}$$

uniformly in $\hat{x} = x/|x|$ where $r = |x|$. The incident field E^i, H^i is, in general, an entire solution of Maxwell's equations in $\mathbb{R}^3$. In particular we consider electromagnetic plane waves given by

$$E^i(x) := \tfrac{i}{k} \operatorname{curl} \operatorname{curl} p \, e^{ikx \cdot d}$$
$$H^i(x) := \operatorname{curl} p \, e^{ikx \cdot d} \tag{5}$$

where $k > 0$ is the wave number, d is a unit vector giving the direction of propagation and p is the polarization vector. Letting

$$L_t^2(\Gamma) : = \left\{ u \in \left(L^2(\Gamma)\right)^3 : \nu \cdot u = 0 \text{ on } \Gamma \right\}$$
$$H(\operatorname{curl}, D) := \{ u \in \left(L^2(D)\right)^3 : \operatorname{curl} u \in \left(L^2(D)\right)^3 \}$$

it is shown in [4] that there exists a unique solution to (1)-(5) such that E is in $H_{loc}(\mathrm{curl}, \mathbb{R}^3 \setminus \overline{D}) \cap \{u : \nu \times u|_{\Gamma_I} \in L_t^2(\Gamma_I)\}$.

In addition to the scattering problem (1)-(5), we will also need to consider the *interior* mixed boundary value problem of finding a solution to

$$\mathrm{curl}\, E - ikH = 0$$

$$\mathrm{curl}\, H + ikE = 0$$

(6)

in D such that

$$\nu \times E = f \qquad \text{on} \qquad \Gamma_D$$

$$\nu \times \mathrm{curl}\, E - i\lambda(x)(\nu \times E) \times \nu = h \qquad \text{on} \qquad \Gamma_I$$

(7)

for a $f = \nu \times u|_{\Gamma_D}$ where $u \in H(\mathrm{curl}, D)$ such that $\nu \times u|_{\Gamma_I} \in L_t^2(\Gamma_I)$ and $h \in L_t^2(\Gamma_I)$. If $\Gamma_I \neq \emptyset$ it was shown in [4] that there exists a unique solution to (6)-(7) such that E is in $H(\mathrm{curl}, D) \cap \{u : \nu \times u|_{\Gamma_I} \in L_t^2(\Gamma_I)\}$.

We now turn our attention to the inverse scattering problem associated with the direct scattering problem (1)-(5). To this end, we note that for $k > 0$ fixed the radiating solution E^s of (1)-(5) has the asymptotic behavior

$$E^s(x) = \frac{e^{ik|x|}}{|x|} \left\{ E_\infty(\hat{x}, d, p) + O\left(\frac{1}{|x|}\right) \right\}$$

(8)

as $|x| \to \infty$, where $E_\infty(\hat{x}, d, p)$ is the electric far field pattern and the dependence on k has been suppressed. It can easily be shown [8] that $\hat{x} \cdot E_\infty(\hat{x}) = 0$. the *inverse scattering problem* we are concerned with is to determine D and λ from $E_\infty(\hat{x}, d, p)$ for $\hat{x}, d \in \Omega = \{x : |x| = 1\}$, $p \in \mathbb{R}^3$ and k fixed. We note that it suffices to only consider $\hat{x}, -d \in \Omega_0 \subset \Omega$, i.e. limited aperture far field data and, since E^s depends linearly on p, we only need to consider three linearly independent polarizations. However, for the sake of simplicity we will restrict our attention to $\hat{x}, d \in \Omega$ and $p \in \mathbb{R}^3$.

Theorem 2.1. *D and λ are uniquely determined by $E_\infty(\hat{x}, d, p)$ for $\hat{x}, d \in \Omega$ and $p \in \mathbb{R}^3$.*

Proof. The proof follows from a slight generation of the results of [9]. $\square$

Having established the uniqueness of our inverse scattering problem, the next step is to reconstruct D (without of course knowing λ a priori). To this end we will use the linear sampling method. We first define the far field operator $F : L^2_t(\Omega) \to L^2_t(\Omega)$ by

$$(Fg)(\hat{x}) := \int_\Omega E_\infty(\hat{x}, d, g(d)) ds(d), \qquad \hat{x} \in \Omega \qquad (9)$$

and note that F is a compact linear operator. Let $E_{e,\infty}$ be the electric far field pattern of the electric dipole

$$E_e(x, z, q) : = \tfrac{i}{k} \mathrm{curl}_x \, \mathrm{curl}_x \, q \, \Phi(x, z)$$

$$(10)$$

$$H_e(x, z, q) : = \mathrm{curl}_x \, q \, \Phi(x, z)$$

where

$$\Phi(x, z) := \frac{1}{4\pi} \frac{e^{ik|x-z|}}{|x - z|} \qquad (11)$$

and $q \in \mathbb{R}^3$, $z \in \mathbb{R}^3$, are the polarization and the source location, respectively, of the electric dipole. The *far field equation* is

$$Fg(\hat{x}) = E_{e,\infty}(\hat{x}, z, q). \qquad (12)$$

In general, no solution exists to the far field equation. However, we can assert the existence of an approximate solution which has, as a function of z, growth properties which allow us to determine D. In the theorem which follows E_g denotes the electric field of an *electromagnetic Herglotz pair* defined by [8]

$$E_g(x) = \int_\Omega e^{ikx \cdot d} g(d) \, ds(d)$$

$$(13)$$

$$H_g(x) = \tfrac{1}{ik} \mathrm{curl} E_g(x)$$

where $g \in L^2_t(\Omega)$.

Theorem 2.2. *Assume that $\Gamma_I \neq \emptyset$ and $\lambda > 0$. Then if F is the electric far field operator (9) corresponding to the direct scattering problem (1)-(5), we have that*

(1) *If $z \in D$ then for every $\epsilon > 0$ there exists a solution $g_\epsilon(\cdot, z) = g_\epsilon(\cdot, z, q) \in L^2_t(\Omega)$ satisfying the inequality*

$$\|Fg_\epsilon(\cdot, z) - E_{e,\infty}(\cdot, z, q)\|_{L^2_t(\Omega)} < \epsilon.$$

Moreover this solution satisfies

$$\lim_{z \to \Gamma} \|E_{g_\epsilon(\cdot, z)}\|_X = \infty, \quad and \quad \lim_{z \to \Gamma} \|g_\epsilon(\cdot, z)\|_{L^2_t(\Omega)} = \infty.$$

(2) *If $z \in D_e$ then for every $\epsilon > 0$ and $\delta > 0$ there exists a solution*
$g_{\delta,\epsilon}(\cdot, z) = g_{\delta,\epsilon}(\cdot, z, q) \in L_t^2(\Omega)$ *of the inequality*

$$\|Fg_{\delta,\epsilon}(\cdot, z) - E_{e,\infty}(\cdot, z, q)\|_{L_t^2(\Omega)} < \epsilon + \delta,$$

such that

$$\lim_{\delta \to 0} \|E_{g_{\delta,\epsilon}(\cdot, z)}\|_X = \infty, \quad \text{and} \quad \lim_{\delta \to 0} \|g_{\delta,\epsilon}(\cdot, z)\|_{L_t^2(\Omega)} = \infty.$$

Here

$$\|u\|_X^2 = \|u\|_{H(curl, D)}^2 + \|\nu \times u\|_{L_t^2(\Gamma_I)}^2.$$

The proof of the theorem is based on two key results stated in the following lemmas which are proved in [4]. First we note that by superposition it is easy to see that

$$Fg = -ik\mathcal{B}(E_g)$$

where the operator $\mathcal{B} : H_{loc}(\mathrm{curl}, \mathbb{R}^3)) \cap \{u : \nu \times u|_{\Gamma_I} \in L_t^2(\Gamma_I)\} \to L_t^2(\Omega)$ maps the electric incident field E^i to the electric far field pattern of the corresponding scattered solution to (1)-(5) and E_g is the electric field of the electromagnetic Herglotz pair with kernel g given by (13).

Lemma 2.3. *The electric field of the interior mixed boundary value problem (6)-(7) can be approximated arbitrarily close by the electric field of an electromagnetic Herglotz pair in $H(curl, D) \cap \{u : \nu \times u|_{\Gamma_I} \in L_t^2(\Gamma_I)\}$.*

Lemma 2.4. *The operator $\mathcal{B}$ is continuous, compact and has dense range. Moreover $E_{e,\infty}(\cdot, z, q)$ is in the range of $\mathcal{B}$ if and only if $z \in D$.*

The function g of Theorem 2.2 is sought for by applying Tikhonov regularization and the Morozov discrepancy principle to the far field equation. The relationship between such a regularized solution and the function g has recently been examined by Arens [1] and a numerical study of this approach for determining D has recently been given by Collino, Fares and Haddar [6].

Having determined D, the function g of the above theorem can also be used to determine the essential supremum of the surface impedance $\lambda = \lambda(x)$ from a knowledge of the electric far field pattern [2]. In particular, let E_z denote the solution of the interior mixed boundary value problem (6)-(7) with boundary data

$$f = -\nu \times E_e(\cdot, z, q)$$

$$h = -\nu \times \mathrm{curl}\, E_e(\cdot, z, q) + i\lambda(x)(\nu \times E_e(\cdot, z, q)) \times \nu$$

(14)

for $z \in B_r \subset D$ where B_r is a ball of radius r. In [4] as part of the proof of Theorem 2.2 it is shown that E_z can be approximated in the norm $\|\cdot\|_X$ of the above theorem by the electric field of an electromagnetic Herglotz pair with kernel ikg where g is the above approximate (regularized) solution of the far field equation. Noting that if we define

$$W_z := E_z + E_e(\cdot, z, q) \tag{15}$$

then $\nu \times W_z = 0$ on Γ_D, the following theorem provides a variational method for determine $\|\lambda\|_{L_\infty(\Gamma_I)}$. Note that Γ_I is *not* determined by this method.

Theorem 2.5. *Let $\lambda \in L_\infty(\Gamma_I)$ be the surface impedance of the scattering problem (1)-(5). Then*

$$\|\lambda\|_{L_\infty(\Gamma_I)} =$$

$$\sup_{\substack{z_i \in B_r, q \in \mathbb{R}^3 \\ \alpha_i \in \mathbb{C}}} \frac{\sum\limits_{i,j} \alpha_i \overline{\alpha_j} \left[-\|q\|^2 A(z_i, z_j, k, q) + k \left(q \cdot E_{z_i}(z_j) + q \cdot \overline{E}_{z_j}(z_i) \right) \right]}{2 \| \sum\limits_i \alpha_i (W_{z_i})_T \|^2_{L_t^2(\Gamma)}}$$

where $\hfill (16)$

$$A(z_i, z_j, k, q) = \frac{k^3}{6\pi} \left[2j_0(k|z_i - z_j|) + j_2(k|z_i - z_j|)(3\cos^2\phi - 1) \right]$$

with j_0 and j_2 being spherical Bessel functions of order 0 and 2, respectively, and ϕ is the angle between $(z_i - z_j)$ and q.

The proof of this theorem follows from the following lemmas (for the proofs see [2]).

Lemma 2.6. *For every two points z_1 and z_2 in D and polarization $q \in \mathbb{R}^3$ we have that*

$$2 \int\limits_{\Gamma_I} (W_{z_1})_T \cdot \lambda (\overline{W}_{z_2})_T \, ds = -\|q\|^2 A(z_1, z_2, k, q) + k \left(q \cdot E_{z_1}(z_2) + q \cdot \overline{E}_{z_2}(z_1) \right)$$

Lemma 2.7. *Let*

$$\mathcal{E} := \left\{ f \in L_t^2(\Gamma_I) : \begin{array}{l} f = (W_z)_T|_{\Gamma_I} \text{ with } W_z \text{ define by (15)}, \\ z \in B_r, \text{ and } q \in \mathbb{R}^3 \end{array} \right\}.$$

Then $\mathcal{E}$ is complete in $L_t^2(\Gamma_I)$.

In the particular case where λ is a positive constant the formula (16) simplifies to

$$\lambda = \frac{-\frac{k^2}{6\pi}\|q\|^2 + k\mathrm{Re}\,(q \cdot E_{z_0})}{\|(W_{z_0})_T\|^2_{L^2_t(\Gamma)}} \qquad (17)$$

where $z_0 \in B_r$.

3. The inverse scattering problem for screens

Let S be a bounded, simply connected, orientated, smooth open surface in $\mathbb{R}^3$ with piecewise smooth boundary which does not intersect itself. We consider S as part of the smooth boundary ∂D of some bounded connected open set $D \subset \mathbb{R}^3$ and denote by ν the normal vector to Γ that coincides with the outward normal vector to ∂D. We denote by $\nu \times E^+$, $\gamma_T E^+$ and $\nu \cdot E^+|_\Gamma$, $(\nu \times E^-$, $\gamma_T E^-$ and $\nu \cdot E^-|_\Gamma)$ the restriction to S of the traces $\nu \times E|_{\partial D}$, $\gamma_T E|_{\partial D}$ and $\nu \cdot E|_{\partial D}$ respectively, from the outside (from the inside) of ∂D where $\gamma_T u := \nu \times (u \times \nu)$ is the tangential component of u. The scattering of electromagnetic waves by the screen S, under the assumption that S is perfectly conducting, leads to the boundary value problem

$$\mathrm{curl}\,E - ikH = 0$$
$$\mathrm{curl}\,H + ikE = 0 \qquad (18)$$

in $\mathbb{R}^3 \setminus \overline{S}$ such that

$$\gamma_T E^\pm = 0 \qquad \text{on} \quad S \qquad (19)$$

where

$$E = E^i + E^s, \qquad H = H^i + H^s, \qquad (20)$$

E^i, H^i are given by (5) and E^s satisfies the Silver-Müller radiation condition (4). We refer to this problem as the *screen problem*. The following theorem is proved in [3]:

Theorem 3.1. *The screen problem has a unique solution* $E, H \in \dot{H}_{loc}(\mathrm{curl}, \mathbb{R}^3 \setminus \overline{S})$.

We note that the solution of the screen problem always has a singularity near the edge! In particular, even if S is smooth and has smooth boundary, the solution E, H to the screen problem is such that $E, H \in H_{loc}^{\frac{1}{2}-\epsilon}(\mathbb{R}^3 \setminus \overline{S})$

for every $\epsilon > 0$ but $E, H \notin H_{loc}^{\frac{1}{2}}(\mathbb{R}^3 \setminus \overline{S})$! Hence, in order to develop a numerical method for solving the screen problem it is necessary to take into account edge singularities.

In order to formulate the inverse scattering problem associated with (18)-(20), we note that the electric field E again has the asymptotic behavior (8). Hence our inverse problem is to determine S from a knowledge of $E_\infty(\hat{x}, d, p)$ for $\hat{x}, d$ on the unit sphere Ω, three linearly independent polarizations $p \in \mathbb{R}^3$ and k fixed. The following theorem can be established in the same way as the corresponding uniqueness theorem for domains with non-empty interior [9].

Theorem 3.2. *The screen S is uniquely determined from a knowledge of $E_\infty(\hat{x}, d, p)$ for all $\hat{x}, d \in \Omega$ and three linearly independent polarizations $p \in \mathbb{R}^3$.*

A reasonable question to ask is if only a finite number of incident waves d suffice to determine uniqueness. In the case of scattering of TM-polarized incident waves by an infinite cylinder (i.e. the Dirichlet crack problem for the two dimensional Helmholtz equation) partial results in this direction have been established by Rondi [10]. However, the case of a perfectly conducting screen in $\mathbb{R}^3$ remains an open problem.

Having established uniqueness, the next step is to reconstruct S from the far field data. We will again do this by using the linear sampling method. To this end we define

$$H_{div}^{-\frac{1}{2}}(S) := \left\{ u \in H^{-\frac{1}{2}}(S) : \nu \cdot u = 0 \quad \text{div}_S u \in H^{-\frac{1}{2}}(S) \right\}$$

and let $\tilde{H}_{div}^{-\frac{1}{2}}(S)$ denote those functions in $H_{div}^{-\frac{1}{2}}(S)$ that can be extended by zero to a function in $H_{div}^{-\frac{1}{2}}(\partial D)$. Note that for $u \in H_{loc}(\text{curl}, \mathbb{R}^3 \setminus \overline{S})$ the trace $\nu \times u|_S$ is in $\in H_{div}^{-\frac{1}{2}}(S)$. We again define the far field operator by (9), where now E_∞ is the electric far field pattern for the screen problem, and define the far field equation by

$$(Fg)(\hat{x}) = \Phi_\infty^L(\hat{x}) \tag{21}$$

where

$$\Phi_\infty^L(\hat{x}) := \hat{x} \times \left(\int_L \alpha_L(y) e^{-ik\hat{x} \cdot y} ds_y \right) \times x \tag{22}$$

for $L \subset S$ and $\alpha_L \in \tilde{H}_{div}^{-\frac{1}{2}}(L)$. The far field operator corresponding to the screen problem can be factored as

$$(Fg)(\hat{x}) = -ik\mathcal{G}(\nu \times E_g) \tag{23}$$

where the operator $\mathcal{G} : H_{div}^{-\frac{1}{2}}(S) \to L_t^2(\Omega)$ maps $\nu \times E^i|_S$ to the electric field of the far field pattern of the scattered field corresponding to an arbitrary incident field E^i, H^i such that $E^i \in H_{loc}(\text{curl}, \mathbb{R}^3 \setminus \overline{S})$. Here E_g is the electric field of the electromagnetic Herglotz pair with kernel g given by (13). The following lemma was proved in [3]:

Lemma 3.3. *The operator $\mathcal{G} : H_{div}^{-\frac{1}{2}}(S) \to L_t^2(\Omega)$ is compact, injective and has dense range. Moreover the range of $\mathcal{B}$ consists of functions Φ_∞^L defined by (22) for $L \subset S$ and $\alpha_L \in \tilde{H}_{div}^{-\frac{1}{2}}(L)$.*

Combining this lemma with the fact that any function in $H_{div}^{-\frac{1}{2}}(S)$ can be approximated by $\nu \times E_g$ with respect to the $H_{div}^{-\frac{1}{2}}(S)$-norm we can prove the following theorem.

Theorem 3.4. *Let S be a perfectly conducting screen having electric far field pattern E_∞. Then*

(1) *if $L \subset S$ then for every $\epsilon > 0$ there exists a solution $g_\epsilon^L \in L^2(\Omega)$ of the inequality*

$$\|Fg_\epsilon^L - \Phi_\infty^L\|_{L^2(\Omega)} \le \epsilon.$$

(2) *if $L \not\subset S$ then for every $\epsilon > 0$ and $\delta > 0$ there exists a solution $g_{\epsilon,\delta}^L \in L^2(\Omega)$ of the inequality*

$$\|Fg_{\epsilon,\delta}^L - E_\infty^L\|_{L^2(\Omega)} \le \epsilon + \delta$$

such that

$$\lim_{\delta \to 0} \|g_{\epsilon,\delta}^L\|_{L^2(\Omega)} = \infty \quad \text{and} \quad \lim_{\delta \to 0} \|E_{g_{\epsilon,\delta}^L}\|_{H(\text{curl},B_R)} = \infty$$

where $E_{g_{\epsilon,\delta}^L}$ is the electric field of the electromagnetic Herglotz pair with kernel $g_{\epsilon,\delta}^L$.

In order for the above theorem to be of practical value, we need to have the right hand side of the far field equation to be independent of L! To this end, we let L degenerate to a point with α_L being an appropriate delta sequence in order to replace Φ_∞^L by $E_{e\infty}(\hat{x}, z, q)$, i.e. we now have the same far field equation as we did for a partially coated obstacle with non-empty interior (c.f. (12))! In particular this is the far field equation we use to reconstruct S from E_∞. For numerical examples using this approach, see [3].

The above method for reconstructing the shape of a perfect conducting screen S can also be used to reconstruct a partially coated screen, i.e. one in which one side is coated by a dielectric and the other side is a perfect conductor [5]. In this case the total electric field $E = E^s + E^i$ satisfies the following boundary conditions

$$\gamma_T E^- = 0 \qquad \text{on} \qquad S^-$$

$$\nu \times \operatorname{curl} E^+ - i\lambda(x)\gamma_T E^+ = 0 \qquad \text{on} \qquad S^+$$

(24)

where $\lambda(x) \geq \lambda_0$ and $\lambda \in L_\infty(S)$ and S^+ and S^- denote the positive and negative sides of the screen respectively. One can prove [5] that there exists a unique solution of this problem such that $E \in H_{loc}(\operatorname{curl}, \mathbb{R}^3 \setminus \overline{S}) \cap \{u : \nu \times u^+ \in L_t^2(S)\}$. But in this case the singularity of the solution near the scree edge is much stronger than in the case of a perfect conductor because roughly speaking the change of the boundary conditions contributes to the edge singularity. In general, even if S is smooth with smooth boundary, the electric field cannot be more regular than $H_{loc}^{\frac{1}{4}-\epsilon}(\mathbb{R}^3 \setminus \overline{S})$ for every $\epsilon > 0$. A detailed analysis of the associated inverse problem shows that Theorem 3.4 is also valid in the case of mixed boundary value problem for screens. Examples of reconstructions based on this theorem for partially coated screens are presented in [5]. However in this case we do not know how to use g to determine the essential supremum of the surface impedance $\lambda(x)$ as we did for partially coated obstacles with non-empty interior.

References

1. T. Arens, Why linear sampling works!, *Inverse Problems* (to appear).
2. F. Cakoni and D. Colton, The determination of the surface impedance of a partially coated obstacle from the far field data, *SIAM J. Applied Math.*, (to appear).
3. F. Cakoni, D. Colton and E. Darrigrad, The inverse electromagnetic scattering problem for screens, *Inverse Problems* **19**, 627 (2003).
4. F. Cakoni, D. Colton and P. Monk, The electromagnetic inverse scattering problem for partially coated Lipschitz domains, (to appear).
5. F. Cakoni and E. Darrigrad, The inverse electromagnetic scattering problem for mixed boundary value problem for screens, (to appear).

6. F. Collino, M. Fares and H. Haddar, Numerical and analytical studies of the linear sampling method in electromagnetic scattering problems, *Inverse Problems,* **19**, (6) (2003).

7. D. Colton, H. Haddar and M. Piana, The linear sampling method in inverse electromagnetic scattering theory, *Inverse Problems,* **19**, (6) (2003).

8. D. Colton and R. Kress, Inverse Acoustic and Electromagnetic Scattering Theory 2nd Edition, *Springer Verlag,* Berlin, (1998)

9. R. Kress, Uniqueness in inverse obstacle scattering for electromagnetic waves, *Proceedings of the URSI General Assembly 2002, Maastricht;* Dounloadable from the web site http://www.num.math.uni-goettingen.de/kress/researchlist.html.

10. L. Rondi, Unique determination of non-smooth sound-soft scatterers by finitely many far-field measurements, *Indiana Uni. Math. Jour.* **52** (6) (2003).

SCATTERING IN ANISOTROPIC MEDIA

G. DASSIOS and K.S. KARADIMA

*Division of Applied Mathematics, Department of Chemical Engineering,
University of Patras and ICEHT-FORTH, Patras, GR-265 04, Greece*

As it is well know, there is an enormous literature on scattering theory that has been developed during the 130 years since Rayleigh gave birth to this fascinating subject. Almost all this literature concerns work on isotropic media. Relatively few recent papers deal with scattering in anisotropic media and most of the times the anisotropy is associated with the scattering region. Although this is a more practical scattering problem, it is of interest to investigate what actually happens when the exterior medium of propagation is also anisotropic. This is the point that the present report wants to focus on. So, we look at anisotropic acoustic scattering where the anisotropic characteristics of the exterior space do not necessarily coincide with the corresponding characteristics of the scattering region. A particular redirected gradient operator is introduced, which carries all directional characteristics of the anisotropic media. Once the fundamental solution is obtained integral representations for the scattered as well as for the interior and the total fields are generated. This is done through a modified version of Green's second integral theorem. For such media even the handling of the singularities, in generating integral representations, depends on the characteristics of the particular medium. In fact, the standard way to isolate a singularity, via the elimination of a centrally located sphere in isotropic media has to be replaced by the elimination of a centrally located and medium dependent ellipsoid when the space is anisotropic. Furthermore, a modified, also medium dependent, radiation condition has to be imposed. Detailed asymptotic analysis leads to an integral representation for the scattering amplitude. The associated energy functionals are presented and the effects of low frequencies are further discussed.

1. Postulation of the problem

Let us assume that a scalar incident field u^i, which propagates in an anisotropic space characterized by a real symmetric and positive definite constant dyadic $\tilde{\mathbf{A}}_+$, is disturbed by another real, symmetric and positive definite constant dyadic $\tilde{\mathbf{A}}_-$, which occupies the closure of a bounded domain V^-. The boundary $S = \partial V^-$ of the obstacle is considered to be $C^{(1)}$ smooth. A scattered field u, which lives in the exterior open domain $V = \left(V^- \cup S\right)^C$, is generated as a result of the interaction between the incident field u^i and the obstacle. In addition, an interior field u^- is generated in V^- caused by the vibrations in the obstacle's interior. If the time dependence is introduced via the spectral component $\exp\{-i\omega t\}$ of angular frequency ω, then the spatial form of the above scattering problem is postulated in mathematical terms as follows [4]: Find the total exterior and the interior field u^+ and u^- respectively, which solve the Helmholtz's equations

$$\left(\tilde{\mathbf{A}}_{+} : \nabla\nabla + k^{2} \right) u^{+}\left(\mathbf{r} \right) = 0 \qquad \text{in } V^{+} \tag{1}$$

$$\left(\tilde{\mathbf{A}}_{-} : \nabla\nabla + k^{2}\eta^{2} \right) u^{-}\left(\mathbf{r} \right) = 0 \qquad \text{in } V^{-} \tag{2}$$

and satisfy the transmission conditions

$$u^{+}\left(\mathbf{r} \right) = u^{-}\left(\mathbf{r} \right) \qquad \text{on } S \tag{3}$$

$$\hat{\mathbf{n}} \cdot \tilde{\mathbf{A}}_{+} \cdot \nabla u^{+}\left(\mathbf{r} \right) = \hat{\mathbf{n}} \cdot \tilde{\mathbf{A}}_{-} \cdot \nabla u^{-}\left(\mathbf{r} \right) \qquad \text{on } S \tag{4}$$

where u^{+} denotes the total field

$$u^{+}\left(\mathbf{r} \right) = u\left(\mathbf{r} \right) + u^{i}\left(\mathbf{r} \right). \tag{5}$$

Furthermore, u has to satisfy a radiation condition at infinity. The wave vector $\mathbf{k}$ is connected to the angular frequency ω by the dispersion condition

$$\omega^{2} = c^{2}\tilde{\mathbf{A}}_{+} : \mathbf{kk}, \tag{6}$$

where c is the phase velocity of the medium occupying the region of propagation V. The constant η in (2) stands for the relative index of refraction. References [2,3,5,6] deal with related formulations.

2. The integral representations and radiation condition

In order to analyze scattering problems an integral representation of the solution on the boundary of the scatterer is needed. In our case the operator is the modified Laplace operator $\left(\Delta_{A} = \tilde{\mathbf{A}} : \nabla\nabla \right)$ and the form of the normal derivative is described in the transmission condition (4). We need to find the appropriate form of Green's identity, needed to arrive at the desired integral representation.
Proposition: For $u, \upsilon \in C^{(2)}\left(\Omega \right)$ and the symmetric and constant dyadic $\tilde{\mathbf{A}}$ the following identity holds true:

$$\int_{\Omega} \left[u\left(\nabla \cdot \tilde{\mathbf{A}} \cdot \nabla \upsilon \right) - \upsilon\left(\nabla \cdot \tilde{\mathbf{A}} \cdot \nabla u \right) \right] dv = \int_{\partial\Omega} \left[u\left(\hat{\mathbf{n}} \cdot \tilde{\mathbf{A}} \cdot \nabla \upsilon \right) - \upsilon\left(\hat{\mathbf{n}} \cdot \tilde{\mathbf{A}} \cdot \nabla u \right) \right] ds. \tag{7}$$

The above proposition is applied in the region V_{R}^{+} which involves the space exterior to S and interior to a large sphere S_{R} centered at the origin with radius R so large as to include in its interior both the scatterer V^{-} and the observation point $\mathbf{r} \in V^{+}$. Replacing u by the scattered field and υ by the fundamental solution of the modified Helmholtz equation (1) $G^{A_{+}}\left(\mathbf{r}, \mathbf{r}' \right)$ [1], where

$$G^{A_+}\left(\mathbf{r},\mathbf{r}'\right)=\frac{1}{4\pi\sqrt{\det\tilde{\mathbf{A}}_+}}\frac{e^{ik\sqrt{\tilde{\mathbf{A}}_+^{-1}:(\mathbf{r}-\mathbf{r}')(\mathbf{r}-\mathbf{r}')}}}{\sqrt{\tilde{\mathbf{A}}_+^{-1}:(\mathbf{r}-\mathbf{r}')(\mathbf{r}-\mathbf{r}')}} \tag{8}$$

is a solution of the equation

$$\left(\tilde{\mathbf{A}}_+:\nabla\nabla+k^2\right)G^A\left(\mathbf{r},\mathbf{r}'\right)=-\delta\left(\mathbf{r},\mathbf{r}'\right)\text{ in }V^+ \tag{9}$$

and applying identity (7) for u and G^{A_+} in V_R^+ we obtain

$$-\int_S\left[u\left(\mathbf{r}'\right)\left(\hat{\mathbf{n}}'\cdot\tilde{\mathbf{A}}_+\cdot\nabla_{r'}G^{A_+}\left(\mathbf{r},\mathbf{r}'\right)\right)-G^{A_+}\left(\mathbf{r},\mathbf{r}'\right)\left(\hat{\mathbf{n}}'\cdot\tilde{\mathbf{A}}_+\cdot\nabla_{r'}u\left(\mathbf{r}'\right)\right)\right]ds\left(\mathbf{r}'\right)$$

$$+\int_{S_R}\left[u\left(\mathbf{r}'\right)\left(\hat{\mathbf{r}}'\cdot\tilde{\mathbf{A}}_+\cdot\nabla_{r'}G^{A_+}\left(\mathbf{r},\mathbf{r}'\right)\right)-G^{A_+}\left(\mathbf{r},\mathbf{r}'\right)\left(\hat{\mathbf{r}}'\cdot\tilde{\mathbf{A}}_+\cdot\nabla_{r'}u\left(\mathbf{r}'\right)\right)\right]ds\left(\mathbf{r}'\right) \tag{10}$$

$$=\int_{V_R^+}\left[u\left(\mathbf{r}'\right)\left(\tilde{\mathbf{A}}_+:\nabla_{r'}\nabla_{r'}G^{A_+}\left(\mathbf{r},\mathbf{r}'\right)\right)-G^{A_+}\left(\mathbf{r},\mathbf{r}'\right)\left(\tilde{\mathbf{A}}_+:\nabla_{r'}\nabla_{r'}u\left(\mathbf{r}'\right)\right)\right]dv\left(\mathbf{r}'\right)$$

where $\hat{\mathbf{n}}$ on S is taken in the outward direction. Expression (10) involves the integral over S_R which we want to vanish as $R\to\infty$. In fact, it is the condition for the vanishing of this limit that will provide us with the appropriate radiation condition.

Using (1) in (10) and the asymptotic analysis of d_{A_+} as $R=|\mathbf{r}'|\to\infty$, where

$$d_{A_+}=\sqrt{\tilde{\mathbf{A}}_+^{-1}:(\mathbf{r}-\mathbf{r}')(\mathbf{r}-\mathbf{r}')} \tag{11}$$

we see that the vanishing of the integral over S_R, as $R\to\infty$, demands the radiation condition

$$\sqrt{\tilde{\mathbf{A}}_+^{-1}:\hat{\mathbf{r}}'\hat{\mathbf{r}}'}\,\hat{\mathbf{r}}\cdot\tilde{\mathbf{A}}_+\cdot\nabla u\left(\mathbf{r}\right)-iku\left(\mathbf{r}\right)=O\left(\frac{1}{r^2}\right),\ r\to\infty \tag{12}$$

which should hold true uniformly over directions. Hence, the effects of anisotropy is the replacement of the operator ∂_r by the operator

$$\partial_r^A=\sqrt{\tilde{\mathbf{A}}_+^{-1}:\hat{\mathbf{r}}\hat{\mathbf{r}}}\,\hat{\mathbf{r}}\cdot\tilde{\mathbf{A}}\cdot\nabla\ . \tag{13}$$

So u has to satisfy (12) as well.

By virtue of the condition (12) the scattered field u satisfies the following integral representation in V^+

$$u\left(\mathbf{r}\right)=\int_S\left[u\left(\mathbf{r}'\right)\hat{\mathbf{n}}'\cdot\tilde{\mathbf{A}}_+\cdot\nabla_{r'}G^{A_+}\left(\mathbf{r},\mathbf{r}'\right)-G^{A_+}\left(\mathbf{r},\mathbf{r}'\right)\hat{\mathbf{n}}'\cdot\tilde{\mathbf{A}}_+\cdot\nabla_{r'}u\left(\mathbf{r}'\right)\right]ds\left(\mathbf{r}'\right). \tag{14}$$

If the standard technique is used to add the incident field then we obtain the following representation for the total field u^+ in V^+

$$
\begin{aligned}
u^+(\mathbf{r}) = u^i(\mathbf{r}) + \int_S \Big[& u^+(\mathbf{r}')\big(\hat{\mathbf{n}}' \cdot \tilde{\mathbf{A}}_+ \cdot \nabla_{r'} G^{A_+}(\mathbf{r},\mathbf{r}')\big) \\
& -G^{A_+}(\mathbf{r},\mathbf{r}')\big(\hat{\mathbf{n}}' \cdot \tilde{\mathbf{A}}_+ \cdot \nabla_{r'} u^+(\mathbf{r}')\big) \Big] ds(\mathbf{r}') ,
\end{aligned}
\tag{15}
$$

where $G^{A_+}(\mathbf{r},\mathbf{r}')$ is given by (8).

In a similar way, formula (7) applied in the region V_ε^- for $u = u^-$ and $\upsilon = G^{A_-}$, where

$$
G^{A_-} = \frac{1}{4\pi\sqrt{\det \tilde{\mathbf{A}}_-}} \frac{e^{ik\eta\sqrt{\tilde{\mathbf{A}}_-^{-1}:(\mathbf{r}-\mathbf{r}')(\mathbf{r}-\mathbf{r}')}}}{\sqrt{\tilde{\mathbf{A}}_-^{-1}:(\mathbf{r}-\mathbf{r}')(\mathbf{r}-\mathbf{r}')}}
\tag{16}
$$

is a solution of

$$
\big(\tilde{\mathbf{A}}_- : \nabla\nabla + k^2 n^2\big) G^{A_-}(\mathbf{r},\mathbf{r}') = -\delta(\mathbf{r}-\mathbf{r}') \quad \text{in } V^-,
\tag{17}
$$

implies the relation

$$
\begin{aligned}
\int_{V_\varepsilon^-} & \Big[u^-(\mathbf{r}')\tilde{\mathbf{A}}_- : \nabla_{r'}\nabla_{r'} G^{A_-}(\mathbf{r},\mathbf{r}') - G^{A_-}(\mathbf{r},\mathbf{r}')\tilde{\mathbf{A}}_- : \nabla_{r'}\nabla_{r'} u^-(\mathbf{r}') \Big] dv(\mathbf{r}') \\
= & \int_S \hat{\mathbf{n}}' \cdot \tilde{\mathbf{A}}_- \cdot \Big[u^-(\mathbf{r}')\nabla_{r'} G^{A_-}(\mathbf{r},\mathbf{r}') - G^{A_-}(\mathbf{r},\mathbf{r}')\nabla_{r'} u^-(\mathbf{r}') \Big] ds(\mathbf{r}') \\
& + \int_{S_\varepsilon} \frac{\mathbf{r}-\mathbf{r}'}{|\mathbf{r}-\mathbf{r}'|} \cdot \tilde{\mathbf{A}}_- \cdot \Big[u^-(\mathbf{r}')\nabla_{r'} G^{A_-}(\mathbf{r},\mathbf{r}') - G^{A_-}(\mathbf{r},\mathbf{r}')\nabla_{r'} u^-(\mathbf{r}') \Big] ds(\mathbf{r}')
\end{aligned}
\tag{18}
$$

where S_ε is the surface of a sphere of radius ε centered at $\mathbf{r}$ and V_ε^- is what is left from V^- after we remove the interior of S_ε.

Taking the limit of (18) as $\varepsilon \to 0$ and using (2) and (17), we arrive at the representation

$$
u^-(\mathbf{r}) = \int_S \hat{\mathbf{n}}' \cdot \tilde{\mathbf{A}}_- \cdot \Big[G^{A_-}(\mathbf{r},\mathbf{r}')\nabla_{r'} u^-(\mathbf{r}') - u^-(\mathbf{r}')\nabla_{r'} G^{A_-}(\mathbf{r},\mathbf{r}') \Big] ds(\mathbf{r}'), \mathbf{r}\in V^-.
\tag{19}
$$

3. The scattering amplitude

In order to analyze the behavior of the scattered field in the radiation zone we have to perform an asymptotic expansion of (15) as $r \to \infty$. Therefore some delicate but straight-forward calculations lead to the far-field representation:

$$u(\mathbf{r}) = g^A\left(\hat{\mathbf{r}};\hat{\mathbf{k}}\right)h\left(a\left(\hat{\mathbf{r}}\right)kr\right)+O\left(\frac{1}{r^2}\right), \tag{20}$$

where the anisotropic scattering amplitude enjoys the representation

$$g^{A_+}\left(\hat{\mathbf{r}};\hat{\mathbf{k}}\right) = -\frac{ik}{4\pi\sqrt{\det \tilde{\mathbf{A}}_+}}\int_S \hat{\mathbf{n}}'\cdot\left[\hat{\mathbf{r}}u^+\left(\mathbf{r}'\right)\frac{ik}{a\left(\hat{\mathbf{r}}\right)}\right.$$

$$\left. +\tilde{\mathbf{A}}_+\cdot\nabla_{\mathbf{r}'}u^+\left(\mathbf{r}'\right)\right]\times\exp\left\{-\frac{ik\left(\tilde{\mathbf{A}}_+^{-1}:\hat{\mathbf{r}}\mathbf{r}'\right)}{a\left(\hat{\mathbf{r}}\right)}\right\}ds\left(\mathbf{r}'\right), \tag{21}$$

with

$$h\left(a\left(\hat{\mathbf{r}}\right)kr\right) = \frac{e^{ia(\hat{\mathbf{r}})kr}}{ia\left(\hat{\mathbf{r}}\right)kr} \tag{22}$$

denoting the radiative fundamental solution for a point source at the origin and

$$a\left(\hat{\mathbf{r}}\right) = \sqrt{\tilde{\mathbf{A}}_+^{-1}:\hat{\mathbf{r}}\hat{\mathbf{r}}} \tag{23}$$

denotes a factor which establishes the anisotropic directivity of the medium.

4. Energy functionals

We develop expressions for energy and power flux in case of anisotropic medium with mean compressibility $\tilde{\mathbf{A}}_+^{-1}$, compressional viscosity δ and mass density ρ from the basic equations

$$\dot{U} = -\nabla_{A_+}\cdot\mathbf{V} = -\tilde{\mathbf{A}}_+\cdot\nabla\cdot\mathbf{V} \tag{24}$$

$$\nabla U = \delta\nabla\nabla\cdot\mathbf{V}-\rho\dot{\mathbf{V}}, \tag{25}$$

where the anisotropy is introduced via (24).Using standard energy arguments we obtain

$$\frac{\partial}{\partial t}\left(\frac{\rho}{2}\left|\tilde{\mathbf{B}}_+\cdot\mathbf{V}\right|^2+\frac{1}{2}|U|^2\right)+\nabla_{A_+}\cdot\text{Re}\left\{UV^*\right\} = \delta\,\text{Re}\left\{\mathbf{V}\cdot\nabla_{A_+}\left(\nabla\cdot\mathbf{V}^*\right)\right\}, \tag{26}$$

where

$$\tilde{\mathbf{B}}_+^2 = \tilde{\mathbf{A}}_+. \tag{27}$$

Recognizing the terms $\frac{\rho}{2}\left|\tilde{\mathbf{B}}_+\cdot\mathbf{V}\right|^2$ as kinetic energy and $\frac{1}{2}|U|^2$ as potential energy, we define the energy density function

$$W_{A_+}(\mathbf{r},t) = \frac{\rho}{2}\left|\tilde{\mathbf{B}}_+ \cdot \mathbf{V}(\mathbf{r},t)\right|^2 + \frac{1}{2}\left|U(\mathbf{r},t)\right|^2 \tag{28}$$

and the power flux vector

$$\mathbf{I}_{A_+}(\mathbf{r},t) = \mathrm{Re}\left\{U(\mathbf{r},t)\mathbf{V}^*(\mathbf{r},t)\right\}. \tag{29}$$

Then (26), now written as

$$\frac{\partial}{\partial t}W_{A_+}(\mathbf{r},t) + \nabla_{A_+}\cdot\mathbf{I}_{A_+}(\mathbf{r},t) = \delta\,\mathrm{Re}\left\{\mathbf{V}(\mathbf{r},t)\cdot\nabla_{A_+}\left(\nabla\cdot\mathbf{V}^*(\mathbf{r},t)\right)\right\}, \tag{30}$$

is the energy conservation law which states that the time rate of change of energy density plus the spatial power flux is equal to the energy dissipated in the medium represented by the term $\delta\,\mathrm{Re}\left\{\mathbf{V}(\mathbf{r},t)\cdot\nabla_{A_+}\left(\nabla\cdot\mathbf{V}^*(\mathbf{r},t)\right)\right\}$.

Introducing the harmonic time dependence $e^{-i\omega t}$ the energy density function and the power flux vector become

$$W_{A_+}(\mathbf{r}) = \frac{\rho}{2}\left|\tilde{\mathbf{B}}_+ \cdot \mathbf{v}(\mathbf{r})\right|^2 + \frac{1}{2}\left|u(\mathbf{r})\right|^2 \tag{31}$$

$$\mathbf{I}_{A_+}(\mathbf{r}) = \mathrm{Re}\left\{u(\mathbf{r})\mathbf{v}^*(\mathbf{r})\right\}. \tag{32}$$

Note that the dyadic $\tilde{\mathbf{A}}_+$, which carries the characteristics of the medium, appears both in the derivative as well as in the norm of the velocity.

5. Low frequency expansions

Here we assume that the anisotropic scattering medium, characterized by the dyadic $\tilde{\mathbf{A}}$, is located in an isotropic space. We are restricting attention to incident fields which are plane waves. Furthermore, all incident fields admit power series expansions [4] and therefore for a plane wave the expansion is given by

$$u^i(\mathbf{r}) = e^{ik\hat{\mathbf{k}}\cdot\mathbf{r}} = \sum_{n=0}^{\infty}\frac{(ik)^n}{n!}\left(\hat{\mathbf{k}}\cdot\mathbf{r}\right)^n. \tag{33}$$

We also assume that

$$u^+(\mathbf{r}) = \sum_{n=0}^{\infty}\frac{(ik)^n}{n!}u_n^+(\mathbf{r}), \quad \mathbf{r}\in V^+ \tag{34}$$

and

$$u^-(\mathbf{r}) = \sum_{n=0}^{\infty} \frac{(ik)^n}{n!} \eta^n u_n^-(\mathbf{r}), \quad \mathbf{r} \in V^-. \tag{35}$$

The total field $u^+(\mathbf{r})$ satisfies (5) while $u(\mathbf{r})$ now satisfies the equation

$$(\Delta + k^2) u(\mathbf{r}) = 0, \quad \mathbf{r} \in V^+ \tag{36}$$

and the Sommerfeld's radiation condition

$$\lim_{r \to \infty} r \left(\frac{\partial}{\partial r} u(\mathbf{r}) - iku(\mathbf{r}) \right) = 0 \tag{37}$$

uniformly over all directions.

The interior field satisfies (2) while the transmission conditions now are written as

$$\left. \begin{array}{c} u^+(\mathbf{r}) = u^-(\mathbf{r}) \\ \hat{\mathbf{n}} \cdot \tilde{\mathbf{A}} \cdot \nabla u^-(\mathbf{r}) = \hat{\mathbf{n}} \cdot \nabla \cdot u^+(\mathbf{r}) \end{array} \right\} \mathbf{r} \in S. \tag{38}$$

Furthermore, the fundamental solution for the exterior space is

$$G(\mathbf{r}, \mathbf{r}') = \frac{e^{ik|\mathbf{r} - \mathbf{r}'|}}{ik|\mathbf{r} - \mathbf{r}'|} = h(k|\mathbf{r} - \mathbf{r}'|), \tag{39}$$

which is a solution of the equation

$$(\Delta + k^2) G(\mathbf{r}, \mathbf{r}') = -\frac{4\pi}{ik} \delta(k|\mathbf{r} - \mathbf{r}'|) \tag{40}$$

and the integral representation for u^+ assumes the standard form

$$u^+(\mathbf{r}) = u^i(\mathbf{r}) + \frac{ik}{4\pi} \int_S \left[u^+(\mathbf{r}') \frac{\partial G(\mathbf{r}, \mathbf{r}')}{\partial n'} - G(\mathbf{r}, \mathbf{r}') \frac{\partial u^+(\mathbf{r}')}{\partial n'} \right] ds(\mathbf{r}'), \mathbf{r} \in V^+. \tag{41}$$

If we employ the low frequency expansions of the incident field (33), the total field (34) and the expansion of the fundamental solution

$$ikG(\mathbf{r}, \mathbf{r}') = \sum_{n=0}^{\infty} \frac{(ik)^n}{n!} |\mathbf{r} - \mathbf{r}'|^{n-1} \tag{42}$$

in the integral representation (41), we rearrange the terms in the product of two series using the Cauchy product and equating like powers of ik we recover the known results [4] for all $\mathbf{r}$.

Similarly, the low frequency expansion of $u^-(\mathbf{r})$ (35) together with the expansion of the fundamental solution

$$G^A(\mathbf{r},\mathbf{r}') = \frac{1}{4\pi\sqrt{\det\tilde{\mathbf{A}}}} \sum_{n=0}^{\infty} \frac{(ik)^n}{n!} \eta^n d_A^{n-1}(\mathbf{r},\mathbf{r}'), \qquad (43)$$

where $d_A(\mathbf{r},\mathbf{r}')$ given by (11), with $\tilde{\mathbf{A}}_+$ replaced by $\tilde{\mathbf{A}}$, may be substituted in (19) for $\tilde{\mathbf{A}}_- = \tilde{\mathbf{A}}$ and after rearranging terms and equating equal powers of ik, we obtain

$$u_n^-(\mathbf{r}) = f_n^-(\mathbf{r}) + \frac{1}{4\pi\sqrt{\det\tilde{\mathbf{A}}}} \int_S \Big[\hat{\mathbf{n}}' \cdot \tilde{\mathbf{A}} \cdot \nabla_{\mathbf{r}'} u_n^-(\mathbf{r}') \\ - u_n^-(\mathbf{r}')\hat{\mathbf{n}}' \cdot \tilde{\mathbf{A}} \cdot \nabla_{\mathbf{r}'} d_A^{-1}(\mathbf{r},\mathbf{r}') \Big] ds(\mathbf{r}'), \qquad (44)$$

where $f_0^-(\mathbf{r}) = 0$ and

$$f_n^-(\mathbf{r}) = \frac{1}{4\pi\sqrt{\det\tilde{\mathbf{A}}}} \sum_{m=1}^{n} \binom{n}{m} \eta^m \int_S \Big[d_A^{m-1}(\mathbf{r},\mathbf{r}')\hat{\mathbf{n}}' \cdot \tilde{\mathbf{A}} \cdot \nabla_{\mathbf{r}'} u_{n-m}^-(\mathbf{r}') \\ - u_{n-m}^-(\mathbf{r}')\hat{\mathbf{n}}' \cdot \tilde{\mathbf{A}} \cdot \nabla_{\mathbf{r}'} d_A^{m-1}(\mathbf{r},\mathbf{r}') \Big] ds(\mathbf{r}'), n \in S^*. \qquad (45)$$

It is useful to note that not only $f_0^- = 0$, but also $f_1^- = 0$.

6. Reduction to isotropic media

At this point it should be mentioned that every result, which contains the dyadic $\tilde{\mathbf{A}}$ ($\tilde{\mathbf{A}}_+$ or $\tilde{\mathbf{A}}_-$) reduces to the corresponding known expression for isotropic media whenever $\tilde{\mathbf{A}}$ is replaced by $\tilde{\mathbf{I}}$.

As far as the new functions $d_A(\mathbf{r},\mathbf{r}')$ and $a(\hat{\mathbf{r}})$ are concerned, they are reduced to the Euclidian distance

$$d_I(\mathbf{r},\mathbf{r}') = |\mathbf{r}-\mathbf{r}'| \qquad (46)$$

and to unit, respectively.

References

1. I.V. Lindell, Methods for Electromagnetic Field Analysis, 2nd ed, Oxford University Press, Oxford, (1995).
2. R.C. Stevenson, *SIAM J. Applied Mathematics*, **50**, 199 (1990).
3. A. Charalambopoulos, *Journal of Elasticity*, **67**, 149 (2000)

4. G. Dassios and R. Kleinman, Low Frequency Scattering, Oxford University Press, Oxford, (2000).
5. P. Hähner, *J. of Computational and Applied Mathematics*, **116**, 167 (2000).
6. F. Cakoni, D. Calton and H. Haddar, *Journal of Computational and Applied Mathematics*, **146**, 285 (2002).

THE FACTORIZATION METHOD FOR MAXWELL'S EQUATIONS

ANDREAS KIRSCH

University of Karlsruhe
Department of Mathematics
76128 Karlsruhe
Germany

1. Introduction

The factorization method is a relatively new method for solving certain kinds of inverse scattering problems in the frequency domain or problems in impedance tomography. It is designed to determine the shape of the support of the unknown contrast. So far, this method has been successfully applied to many scalar problems, i.e. problems for the scalar Helmholtz equation. We refer to [12,13,17,18,19,20,23] for some of the results. Also, inverse scattering problems for elastic waves have been treated by this method see, e.g., [15,25]. So far, no corresponding theoretical result is available when the wave propagation is described by the full Maxwell system. This is because of some, probably only technical, assumptions of a crucial functional analytic result (see Theorem 4.1 below. However, practical experiments with this method have been reportet in [9]. In this paper we will show first theoretical results for the Maxwell system for two special cases: In the first case we will assume that the scattering medium Ω is absorbing, i.e. the index of refraction is complex and the imaginary part can be bounded below on Ω by some positive constant. In the second case we will assume the index of refraction to be real-valued but, unfortunately, it has to decay smoothly to one (the background value) at the boundary of Ω. The interesting case when the index "jumps to zero" is still open.

The paper is organized as follows. In Section 2 we formulate the direct scattering, introduce the far field pattern, and formulate the inverse scattering problem. In Section 3 we prove a factorization of the far field operator F in the form $F = H^*TH$ with some operators H and T. Here, F is the integral operator with the far field pattern as its kernel, and H^* denotes the adjoint of H. This factorization is central for the following and holds under weak assumptions on the index of refraction. The operators H and H^* depend explicitly on the domain Ω. Moreover, Ω can be characterized by those points $z \in \mathbb{R}^3$ for which an explicitly given function φ_z belongs to the range of H^*. The factorization $F = H^*TH$ has now to be used to describe the range of H^* by F alone (since F is known by the data of the inverse problem). It is this step where functional analytic tools are used. In Section 4 and 5 we consider the cases of absorbing and non-absorbing media, respectively.

2. The Direct Scattering Problem

In this section we study the direct scattering problem. Let $k = \omega\sqrt{\varepsilon_0\mu_0} > 0$ be the wave number with frequency ω, dielectricity ε_0, and permeability μ_0 in vacuum. Furthermore, let $\hat{\theta} \in S^2 = \{x \in \mathbb{R}^3 : |x| = 1\}$ be a unit vector in $\mathbb{R}^3$ and $p \in \mathbb{C}^3$ such that $p \cdot \hat{\theta} := \sum_{j=1}^{3} p_j\hat{\theta}_j = 0$. The plane electromagnetic waves of direction $\hat{\theta}$ and polarization p are given by

$$E^{inc}(x) = p\,e^{ik\hat{\theta}\cdot x} \quad \text{and} \quad H^{inc}(x) = \sqrt{\frac{\varepsilon_0}{\mu_0}}\,(\hat{\theta} \times p)\,e^{ik\hat{\theta}\cdot x}, \quad x \in \mathbb{R}^3.$$

This plane wave is scattered by a medium with (space dependent) dielectricity $\varepsilon = \varepsilon(x)$, permeability $\mu = \mu(x)$, and conductivity $\sigma = \sigma(x)$. We assume that $\varepsilon \equiv \varepsilon_0$, $\mu \equiv \mu_0$ and $\sigma \equiv 0$ outside of some bounded domain Ω. The total fields are superpositions of the incident and scattered fields, i.e. $E = E^{inc} + E^s$ and $H = H^{inc} + H^s$ and satify the Maxwell system

$$\operatorname{curl} E - i\omega\mu H = 0 \quad \text{in } \mathbb{R}^3, \tag{1a}$$

$$\operatorname{curl} H + i\omega\varepsilon E = \sigma E \quad \text{in } \mathbb{R}^3. \tag{1b}$$

We define the contrast

$$q = \frac{\mu\varepsilon}{\mu_0\varepsilon_0} + i\frac{\mu\sigma}{\omega\mu_0\varepsilon_0} - 1$$

and observe that q vanishes outside of Ω. Furthermore, the scattered field has to satisfy the Silver-Müller radiation condition

$$\sqrt{\frac{\mu_0}{\varepsilon_0}}\,H^s(x) \times x - |x|E^s(x) = O(1/|x|) \quad \text{as } |x| \to \infty$$

uniformly w.r.t. $\hat{x} = x/|x|$.

In this paper we will always work with the electrical field E only. Eliminating the magnetic field H from the system (1a), (1b) and using the definition of q leads to

$$\operatorname{curl}^2 E - k^2(1+q)E = 0 \quad \text{in } \mathbb{R}^3 \tag{2}$$

for the total field while the incident field satisfies

$$\operatorname{curl}^2 E^{inc} - k^2 E^{inc} = 0 \quad \text{in } \mathbb{R}^3. \tag{3}$$

Subtracting both equations yields

$$\operatorname{curl}^2 E^s - k^2(1+q)E^s = k^2 q\, E^{inc} \quad \text{in } \mathbb{R}^3. \tag{4}$$

The Silver-Müller radiation condition turns into

$$\operatorname{curl} E^s(x) \times \hat{x} - ikE^s(x) = O\left(\frac{1}{|x|^2}\right), \quad |x| \to \infty. \tag{5}$$

It will be necessary to allow more general source terms on the right-hand side of (4). In particular, we will consider the following problem:

$$\operatorname{curl}^2 v - k^2(1+q)v = k^2 qf \quad \text{in } \mathbb{R}^3 \tag{6}$$

and v satisfies the Silver-Müller radiation condition (5). Here $f \in L^2(\mathbb{R}^3, \mathbb{C}^3)$ is an arbitrary vector function with compact support. The solutions v of (6) as well as of (2) and (4) have to be understood in the variational sense, i.e. are sought in the space

$$H_{loc}(\operatorname{curl}, \mathbb{R}^3) = \left\{ v : \mathbb{R}^3 \to \mathbb{C}^3 : v|_B \in H(\operatorname{curl}, B) \text{ for all balls } B \subseteq \mathbb{R}^3 \right\}$$

where $H(\operatorname{curl}, B)$ is defined as the completion of $C^1(\overline{B}, \mathbb{C}^3)$ with respect to the norm

$$\|v\|_{H(\operatorname{curl}, B)} := \left(\|v\|_{L^2(B, \mathbb{C}^3)}^2 + \|\operatorname{curl} B\|_{L^2(B, \mathbb{C}^3)}^2 \right)^{1/2}.$$

Existence of a variational solution of (6) can be shown under mild conditions on q:

Theorem 2.1. *Let $q \in L^\infty(\mathbb{R}^3)$ with compact support in B such that $\Re q \geq 0$ and $\Im q \geq 0$ on $\mathbb{R}^3$. Then (6) has a (unique) solution $v \in H_{loc}(\operatorname{curl}, \mathbb{R}^3)$ for every $f \in L^2(\mathbb{R}^3, \mathbb{C}^3)$ with compact support in B provided the homogeneous equation (i.e. for $f \equiv 0$) admits only the trivial solution $v \equiv 0$. Furthermore, there exists a constant $c > 0$ (depending only on B, k, and q) with $\|v\|_{H(\operatorname{curl}, B)} \leq c\|f\|_{L^2(B, \mathbb{C}^3)}$ for all $f \in L^2(B, \mathbb{C}^3)$.*

For a proof we refer to, e.g., Monk [24]. The assumption of uniquenss is satisfied if, e.g. $\Im q > 0$ on some open and bounded set $\Omega \subset \mathbb{R}^3$ and $q \equiv 0$ on $\mathbb{R}^3 \setminus \Omega$. ($\overline{\Omega}$ is then the support of q.) We note that the solution v lies in the space

$$H(\mathrm{curl}^2, B) = \{v \in H(\mathrm{curl}, B) : \mathrm{curl}\, v \in H(\mathrm{curl}, B)\} \qquad (7)$$

for every ball B and $\|v\|_{H(\mathrm{curl}^2, B)} \le c\|f\|_{L^2(B,\mathbb{C}^3)}$ for all $f \in L^2(B, \mathbb{C}^3)$.

Furthermore, from [7] or [24] it is well known that every radiating solution v of (6) has an asymptotic behaviour of the form

$$v(x) = \frac{\exp(ik|x|)}{4\pi|x|} v^\infty(\hat{x}) + O\left(\frac{1}{|x|^2}\right), \quad |x| \to \infty,$$

uniformly w.r.t. $\hat{x} = x/|x|$. The vector field v^∞ is called the far field pattern of v. It is an analytic function on S^2 with respect to $\hat{x}$ and is a tangential field, i.e. satisfied $v^\infty(\hat{x}) \cdot \hat{x} = 0$ for all $\hat{x} \in S^2$.

In particular, the far field pattern E^∞ of E^s is defined. Since it depends on $\hat{\theta}$ and p as well, we will also write $E^\infty = E^\infty(\hat{x}, \hat{\theta}; p)$. Note again, that it is tangential, i.e. $E^\infty(\hat{x}, \hat{\theta}; p) \cdot \hat{x} = 0$ for all $\hat{x} \in S^2$ and all $\hat{\theta} \in S^2$ and $p \in \mathbb{C}^3$ with $p \cdot \hat{\theta} = 0$.

We are now able to define the *inverse problem*. Given $E^\infty(\hat{x}, \hat{\theta}; p)$ for all $\hat{x}, \hat{\theta} \in S^2$ and $p \in \mathbb{C}^3$ with $p \cdot \hat{\theta} = 0$, find the support Ω of q. Because of the linear dependence of E^∞ on p it is sufficient to know E^∞ only for a basis of three vectors for p. The task of determining only Ω is rather modest since it is well known that one can even reconstruct q uniquely from this set of data, see [8] – at least if q is smooth enough. However, the proof of uniqueness is non-constructive while we will give an explicit characterization of the characteristic function of Ω which holds for less smooth q and can, e.g., be used for numerical purposes.

3. Factorization of the Far Field Operator

For this section we make the general assumption that the problem (6) has a unique solution for every $f \in L^2(B, \mathbb{C}^3)$. In particular, the scattering problem (1a), 1b), (5) has a unique solution for every incident plane wave. Let $\Omega \subset \mathbb{R}^3$ be open and bounded such that $\overline{\Omega}$ is the support of q. We introduce the subspace $L^2_t(S^2)$ of $L^2(S^2, \mathbb{C}^3)$ consisting of all tangential fields, i.e.

$$L^2_t(S^2) := \{v \in L^2(S^2, \mathbb{C}^3) : v(\hat{x}) \cdot \hat{x} = 0, \ \hat{x} \in S^2\}.$$

The *far field operator* $F : L_t^2(S^2) \to L_t^2(S^2)$ is defined as

$$(Fp)(\hat{x}) := \int\limits_{S^2} E^\infty(\hat{x}, \hat{\theta}; p(\hat{\theta})) \, ds(\hat{\theta}), \quad \hat{x} \in S^2. \tag{8}$$

It is linear since E^∞ depends linearly on the polarization p. We note that Fp is the far field pattern of the electrical field which corresponds to the incident field

$$v_p(x) = \int\limits_{S^2} E^{inc}(x, \hat{\theta}; p(\hat{\theta})) \, ds(\hat{\theta}) = \int\limits_{S^2} p(\hat{\theta}) \, e^{ikx \cdot \hat{\theta}} \, ds(\hat{\theta}), \quad x \in \mathbb{R}^3.$$

These entire solutions of $\text{curl}^2 v - k^2 v = 0$ are called *Herglotz wave functions*. We define the operator $H : L_t^2(S^2) \to L^2(\Omega, \mathbb{C}^3)$ by

$$(Hp)(x) := \int\limits_{S^2} p(\hat{\theta}) \, e^{ikx \cdot \hat{\theta}} \, ds(\hat{\theta}), \quad x \in \Omega, \tag{9}$$

and call H the *Herglotz operator*. H is one-to-one as it is easily seen. The following result is one of the main ingredients of the method. For a proof we refer to [22].

Theorem 3.1. *Let F and H be defined by (8) and (9), respectively. Then*

$$F = H^* T H \tag{10}$$

where $H^* : L^2(\Omega, \mathbb{C}^3) \to L_t^2(S^2)$ denotes the adjoint of H and $T : L^2(\Omega, \mathbb{C}^3) \to L^2(\Omega, \mathbb{C}^3)$ is given by $Tf = k^2 q (f + v)$ and $v \in H_{loc}(\text{curl}, \mathbb{R}^3)$ solves

$$\text{curl}^2 v - k^2(1 + q)v = k^2 q f \quad in \quad \mathbb{R}^3 \tag{11}$$

and the radiation condition (5).

We note that the operator T is one-to-one. Indeed, $Tf = 0$ implies $v + f \equiv 0$ in Ω, i.e. $\text{curl}^2 v + k^2 v = 0$ in $\mathbb{R}^3$. This yields $v \equiv 0$ in $\mathbb{R}^3$ since $\nu \times v$ and $\nu \times \text{curl}\, v$ are continuous through $\partial \Omega$ and v satisfies the radiation condition. Therefore, also $f \equiv 0$.

4. The Factorization Method for Absorbing Media

In this section we make the general assumption that there exists $q_0 > 0$ with $\Im q(x) \geq q_0$ for almost all $x \in \Omega$. Recall that $\overline{\Omega}$ is the support of q, i.e. $q = 0$ outside of (the open and bounded set) Ω. As we mentioned above (following Theorem 2.1), problem (6) is then uniquely solvable for all $f \in L^2(\Omega, \mathbb{C}^3)$.

We recall the factorization $F = H^*TH$ and introduce the operators $\Im F := \frac{1}{2i}(F - F^*)$ and, analogously, $\Im T := \frac{1}{2i}(T - T^*)$. Then $\Im F$ and $\Im T$ are selfadjoint and

$$\Im F = H^*(\Im T)H. \tag{12}$$

The selfadjoint operator $\Im T$ is an isomorphism from $L^2(\Omega, \mathbb{C}^3)$ onto itself and, even more, it is coercive (see [22]):

Lemma 4.1. *Under the assumptions of this section there exists $\gamma > 0$ with*

$$\langle (\Im T)f, f \rangle_{L^2(\Omega,\mathbb{C}^3)} = \Im \langle Tf, f \rangle_{L^2(\Omega,\mathbb{C}^3)} \geq \gamma \|f\|^2_{L^2(\Omega,\mathbb{C}^3)}$$

for all $f \in L^2(\Omega, \mathbb{C}^3)$.

The following functional analytic theorem taken from [21] is more general than needed for the present situation. We will, however, apply it a second time for real valued indices in the next section.

Theorem 4.1. *Let X and Y be Hilbert spaces, $B : Y \to Y$, $H : Y \to X$, and $A : X \to X$ be linear and bounded operators with*

$$B = H^* A H. \tag{13}$$

We make the following assumptions:

(a) H is one-to-one and compact.
(b) $\Re A$ has the form $\Re A = C + K$ with some coercive operator $C : X \to X$ and compact operator K.
(c) $\Im A$ is non-negative on X, i.e. $\Im \langle A\varphi, \varphi \rangle \geq 0$ for all $\varphi \in X$.
(d) $\Re A$ is one-to-one (and thus an isomorphism onto X) or $\Im A$ is positive on X, i.e. $\Im \langle A\varphi, \varphi \rangle > 0$ for all $\varphi \in X$, $\varphi \neq 0$.

Then the operator $B_{\#} := |\Re B| + \Im B$ is also positive, and the ranges of $H^* : X \to Y$ and $B_{\#}^{1/2} : Y \to Y$ coincide. Furthermore, the operator $B_{\#}^{-1/2} H^*$ is a bounded isomorphism from X onto Y.

We apply this theorem to the factorization (12), i.e., we set $B = \Im F$ and $A = \Im T$. Then A and B themselves are selfadjoint and A is coercive. Therefore, we set $C = A$ and $K = 0$. All of the other assumptions are also satisfied. Furthermore, $\Re B = B = \Im F$, $\Im B = 0$, and $B_\# = \Im F$. Therefore, we conclude that the ranges of $(\Im F)^{1/2}$ and H^* coincide, i.e.

$$R\big((\Im F)^{1/2}\big) = R(H^*). \tag{14}$$

The advantage of this relation lies in the fact that we we can explicitly characterize Ω by the range of H^*. Indeed, recall, that $H^* : L^2(\Omega, \mathbb{C}^3) \to L^2(S^2)$ is given by

$$(H^*g)(\hat{x}) = \hat{x} \times \iint_\Omega g(y)\, e^{-ik\hat{x}\cdot y}\, dy \times \hat{x} \quad \text{for } \hat{x} \in S^2. \tag{15}$$

For any $z \in \mathbb{R}^3$ and fixed $p \in \mathbb{C}^3$ we define $\phi_z \in L^2_t(S^2)$ by

$$\phi_z(\hat{x}) = \frac{ik}{4\pi}(\hat{x} \times p) \times \hat{x}\, e^{-ik\hat{x}\cdot z}, \quad \hat{x} \in S^2. \tag{16}$$

Then it can be shown (see [22]) that $z \in \Omega$ if, and only if, $\phi_z \in R(H^*)$.

The combination of this result with the expression (14) yields the main result of this section:

Theorem 4.2. *Let the assumption of the beginning of the section be satisfied. Furthermore, let ϕ_z be defined in (16) for $z \in \mathbb{R}^3$ (where again $p \in \mathbb{C}^3$ is kept fixed). Then $z \in \Omega$ if, and only if, $\phi_z \in R\big((\Im F)^{1/2}\big)$.*

This result characterizes the support $\overline{\Omega}$ of q by the data of the inverse scattering problem which are collected in the operator $\Im F$. In particular, it gives a new "direct" proof of uniqueness of the inverse scattering problem to determine Ω from $E^\infty(\hat{x}, \hat{\theta}, p)$ for all $\hat{x}, \hat{\theta} \in S^2$ and $p \in \mathbb{C}^3$ with $p \cdot \hat{\theta} = 0$.

5. The Case of Non-absorbing Media

For this section we make the general assumption that $v \equiv 0$ is the only solution of

$$\operatorname{curl}^2 v - k^2(1 + q)v = 0 \quad \text{in } \mathbb{R}^3$$

satisfying the radiation condition (5). Furthermore, we assume that $q \in L^\infty(\mathbb{R}^3)$ has compact support (which we again denote by $\overline{\Omega}$) and is *real-valued* and *positive*. Unfortunately, below we will have to make a further strong smoothness assumption on q.

We recall the definition of the operator T as $Tf = k^2q(f + v)$ and v solves (11). Up to the factor k^2q the operator T is the sum of the identity operator and a second one which we will show to be compact on a suitable subspace under the before-mentioned smoothness assumption on q.

The factor q motivates us to work in the weighted L^2-space $L_q^2(\Omega, \mathbb{C}^3)$ defined as

$$L_q^2(\Omega, \mathbb{C}^3) = \{\psi : \Omega \to \mathbb{C}^3 : \sqrt{q}\,\psi \in L^2(\Omega, \mathbb{C}^3)\}$$

and equipped with the inner product

$$\langle \psi, \varphi \rangle_{L_q^2(\Omega, \mathbb{C}^3)} = \iint\limits_{\Omega} q(x)\,\psi(x)\,\overline{\varphi(x)}\,dx\,.$$

First, we note that $L^2(\Omega, \mathbb{C}^3)$ is imbedded in $L_q^2(\Omega, \mathbb{C}^3)$ with bounded imbedding. This follows from

$$\|\psi\|_{L_q^2(\Omega, \mathbb{C}^3)}^2 = \iint\limits_{\Omega} q(x)\,|\psi(x)|^2 dx \leq \|q\|_\infty \iint\limits_{\Omega} |\psi(x)|^2 dx$$

$$= \|q\|_\infty \|\psi\|_{L^2(\Omega, \mathbb{C}^3)}^2\,.$$

We denote the adjoint of H considered as an operator into $L_q^2(\Omega, \mathbb{C}^3)$ by $H^\dagger$. It is given by $H^\dagger \varphi = H^*(q\varphi)$, $\varphi \in L_q^2(\Omega, \mathbb{C}^3)$. Therefore, the factorization (10) takes the form

$$F = H^\dagger q^{-1} T H = H^\dagger \tilde{T} H \tag{17}$$

with $\tilde{T}f = k^2(f + v)$ and v solves (11). The operator $\tilde{T}$ is well defined and bounded on $L_q^2(\Omega, \mathbb{C}^3)$ since, if $f \in L_q^2(\Omega, \mathbb{C}^3)$ then the right hand side of (11) is still in $L^2(\Omega, \mathbb{C}^3)$.

It is our aim to apply Theorem 4.1 to the factorization (17). Setting $A = \tilde{T}$ and $Cf = k^2 f$ and K the real part of the operator $f \mapsto \frac{k^2}{4\pi} v$ we have to show that the latter operator is compact. However, this fails to be the case due to the fact that $H(\mathrm{curl}, \Omega)$ is not compactly imbedded in $L^2(\Omega, \mathbb{C}^3)$. To overcome this difficulty we note that it is sufficient to consider $\tilde{T}$ on the closure of the range $R(H)$ of H only. To formulate this space weakly we define

$$H_{00}(\mathrm{curl}^2, \Omega) = \{v \in H_0(\mathrm{curl}, \Omega) : \mathrm{curl}\,v \in H_0(\mathrm{curl}, \Omega)\}$$

where $H_0(\mathrm{curl}, \Omega)$ denotes the closed subspace of $H(\mathrm{curl}, \Omega)$ with vanishing traces $\nu \times v$ and

$$X := \left\{ f \in L_q^2(\Omega, \mathbb{C}^3) : \begin{array}{c} \iint\limits_{\Omega} f \cdot (\mathrm{curl}^2 w - k^2 w)\,dx = 0 \\ \text{for all } w \in H_{00}(\mathrm{curl}^2, \Omega) \text{ with} \\ q^{-1/2}(\mathrm{curl}^2 w - k^2 w) \in L^2(\Omega, \mathbb{C}^3) \end{array} \right\} \tag{18}$$

Then X is a closed subspace of $L_q^2(\Omega, \mathbb{C}^3)$ which contains the range of H. Then the following theorem holds $(^{22})$.

Theorem 5.1. *In addition to the assumption at the beginning of this section let $q \in C^1(\mathbb{R}^3)$ with compact support $\overline{\Omega}$ such that there exists $c > 0$ with*

$$|\nabla q(x)|^2 \leq c\, q(x) \quad \text{for all } x \in \Omega. \tag{19}$$

Then $f \mapsto v|_\Omega$ is compact from X into $L_q^2(\Omega, \mathbb{C}^3)$. Here, X is defined in (18) and $v \in H_{loc}(curl, \mathbb{R}^3)$ denotes the radiating solution of (11).

Remark: In the radially symmetric case, i.e. if Ω is the unit ball and $q = q(r)$, $r \geq 0$, the condition reduces to

$$q'(r)^2 \leq c\, q(r) \quad \text{for all } r \geq 0,$$

which is, e.g., satisfied if q is of polynomial form $q(r) = \alpha(1-r)^\beta$, $0 \leq r \leq 1$, for $\alpha > 0$ and $\beta \geq 2$. Therefore, q has to decay to zero sufficiently fast (compare the Example given in ^{22}).

As a next step we have to formulate the factorization in the space X rather than in $L_q^2(\Omega, \mathbb{C}^3)$. Therefore, let $P : L_q^2(\Omega, \mathbb{C}^3) \longrightarrow X$ be the orthogonal projection operator onto X. Then $L^\dagger = L^\dagger P$ since the range of L is contained in X and thus the orthogonal complement $X^\perp$ of X is contained in the complement of the range of L which is the nullspace of $L^\dagger$. Therefore, we an write the factorization (17) in the form

$$F = L^\dagger P \tilde{T} L. \tag{20}$$

It can now easily be shown $(^{22})$ that assumptions (a), (b), and (c) of Theorem 4.1 are satisfies for the case $B = F$ and $A = P\tilde{T}\big|_X$ and $Y = L_t^2(S^2)$. For assumption (d) one has to study under which conditions $\Im\langle P\tilde{T}f, f\rangle_{L_q^2(\Omega, \mathbb{C}^3)}$ is strictly positive. By Green's theorem it can be shown that $\Im\langle P\tilde{T}f, f\rangle_{L_q^2(\Omega, \mathbb{C}^3)}$ vanishes if, and only if, v^∞ vanishes identically, where v is the radiating solution of $curl^2 v - k^2(1+q)v = k^2 q f$ in $\mathbb{R}^3$. From this one concludes that f and $\tilde{f} := f + v$ satisfy the following equations in the weak sense:

$$\tilde{f}, f \in L_q^2(\Omega, \mathbb{C}^3), \quad \tilde{f} - f \in H_{00}(curl^2, \Omega), \tag{21a}$$

$$curl^2 \tilde{f} - k^2(1+q)\tilde{f} = 0 \quad \text{in } \Omega, \tag{21b}$$

$$curl^2 f - k^2 f = 0 \quad \text{in } \Omega. \tag{21c}$$

(see [22] for details.) This is an eigenvalue problem with respect to the parameter k. Similar "transmission eigenvalue problems" have been studied before in connection with the Linar Sampling Method and the factorization method (see [7,4,14]). At the moment it is not known to the author whether the set of eigenvales is discrete or not.

Application of Theorem 4.1 yields the main theorem of this section.

Theorem 5.2. *Let the assumption of the beginning of the section be satisfied. Furthermore, assume that k is not an eigenvalue of the problem (21). Let ϕ_z be defined in (16) for $z \in \mathbb{R}^3$ (where again $p \in \mathbb{C}^3$ is kept fixed). Then $z \in \Omega$ if, and only if, $\phi_z \in R\left(F_\#^{1/2}\right)$ where $F_\# := |\Re F| + \Im F$.*

References

1. BRÜHL, M.: Gebieterkennung in der elektrischen Impedanztomographie. Dissertation thesis, University of Karlsruhe, 1999.
2. BRÜHL, M: Explicit Characterization of Inclusions in Electrical Impedance Tomography. SIAM J. Math. Anal. **32** (2001), 1327–1341.
3. BRÜHL, M., and HANKE, M.: Numerical Implementation of Two Noniterative Methods for Locating Inclusions by Impedance Tomography. Inverse Problems **16** (2000), 1029–1042.
4. CAKONI, F., COLTON, D., and HADDAR, H.: The Linear Sampling Method for Anisotropic Media. Preprint, 2002.
5. CAKONI, F., and HADDAR, H.: The Linear Sampling Method for Anisotropic Media: Part II. Preprint, 2002.
6. COLTON, D., and KIRSCH, A.: A Simple Method for Solving Inverse Scattering Problems in the Resonance Region. Inverse Problems **12** (1996), 383–393.
7. COLTON, D., and KRESS, R.: Inverse Acoustic and Electromagnetic Scattering Problems. 2nd edition, Springer-Verlag, New-York, 1998.
8. COLTON, D., and PÄIVÄRINTA, L.: The Uniqueness of a Solution to an Inverse Scattering Problem for Electromagnetic Waves. Arch. Rational Mech. Anal. **119** (1992), 59-70.
9. FISCHER, C.: Multistatisches Radar zur Lokalisierung von Objekten im Boden. PhD-Thesis, University of Karlsruhe, 2003.
10. GIBARG, D., and TRUDINGER, N. S.: Elliptic Partial Differential Equations. 2nd edition, Springer, New-York, 1983.
11. GIRAULT, V., and RAVIART, P.A.: Finite Element Methods for Navier-Stokes Equations. Springer, New-York, 1986.
12. GRINBERG, N.: On the Inverse Obstacle Scattering Probblem with Robin or Mixed Boundary Condition: Application of the Modified Kirsch Factorization Method. University of Karlsruhe, Department of Mathematics, Preprint 02/4.

13. GRINBERG, N., and KIRSCH, A.: The Linear Sampling Method in Inverse Obstacle Scattering for Impedance Boundary Conditions. Journal of Inverse and Ill-Posed Problems, to appear.

14. HADDAR, H.: The Interior Transmission Problem for Anisotropic Maxwell's Equations and its Applications to the Inverse Problem. Preprint, 2002.

15. HÄHNER, P.: An Inverse Problem in Elastostatics. Inverse Problems **15** (1999), 961–975.

16. KIRSCH, A.: Introduction to the Mathematical Theory of Inverse Problems. Springer-Verlag, New-York, 1998.

17. KIRSCH, A.: Characterization of the shape of a scattering obstacle using the spectral data of the far field operator. Inverse Problems **14** (1998), 1489–1512.

18. KIRSCH, A.: Factorization of the Far Field Operator for the Inhomogeneous Medium Case and an Application in Inverse Scattering Theory. Inverse Problems **15** (1999), 413–429.

19. KIRSCH, A.: New Characterizations of Solutions in Inverse Scattering Theory. Applicable Analysis **76** (2000), 319–350.

20. KIRSCH, A.: The MUSIC-Algorithm and the Factorization Method in Inverse Scattering Theory for Inhomogeneous Media. Inverse Problems **18** (2002), 1025–1040.

21. KIRSCH, A.: The Factorization Method for a Class of Inverse Elliptic Problems. To appear in Math. Nachr.

22. KIRSCH, A.: Factorization Results for Electromagnetic Inverse Scattering Problems. Submitted to the 2004 Inverse Problems' Special Section "Electromagnetic Characterization of Buried Obstacles".

23. KIRSCH, A., and RITTER, S.: A Linear Sampling Method for Inverse Scattering from an Open Arc. Inverse Problems **16** (2000), 89–105.

24. MONK, P.: Finite Element Methods for Maxwell's Equations. Oxford Science Publications. Oxford, 2003.

25. PELOKANOS, G., and SEVROGLOU, V.: The $(F^*F)^{1/4}$–Method for 2D Penetrable Elastic Bodies. Talk presented at the 6th International Workshop on Mathematical Methods in Scattering Theory and Biomedical Engineering. Tsepelovo, 2003.

26. RYNNE, B. P., and SLEEMAN, B. D.: The Interior Transmission Problem and Inverse Scattering from Inhomogeneous Media. SIAM J. Math. Anal. **22** (1991), 1755–1762.

DECOMPOSITION METHODS IN INVERSE OBSTACLE SCATTERING VIA NEWTON ITERATIONS

R. KRESS

Institut für Numerische und Angewandte Mathematik,
Universität Göttingen,
Lotzestr. 16–18
37083 Göttingen, Germany
E-mail: kress@math.uni-goettingen.de

We describe a new method for solving the inverse scattering problem to determine the shape of a scatterer D from the far field pattern u_∞ for scattering of time-harmonic waves. This method may be interpreted as a hybrid of the decomposition method due to Kirsch and Kress and regularized Newton iteration methods for solving the nonlinear ill-posed operator equation $F(\Gamma) = u_\infty$ with the operator F mapping the boundary Γ of the scatterer D onto the far field. The new method does not require a forward solver and its feasibility is demonstrated through a numerical example.

1. Introduction

The mathematical modelling of the application of scattering phenomena in technologies such as radar and sonar, nondestructive evaluation, geophysical exploration and medical imaging leads to inverse scattering problems for time-harmonic acoustic, electromagnetic and elastic waves. In principle, in these approaches the effects of scattering objects on the propagation of waves are exploited to obtain images of the nature of the scattering objects. In inverse obstacle scattering problems the scattering object is a homogeneous obstacle and the inverse problem is to obtain an image of the shape of the obstacle from a knowledge of the scattered wave at large distances, i.e., from the far field pattern. For the sake of simplicity and in order to present basic ideas we confine ourselves to the case of time-harmonic acoustic waves and the inverse scattering problem for sound-soft obstacles. However, the following concepts, in principle, extend to other types of inverse obstacle scattering problems for time-harmonic waves.

Given an obstacle D, i.e., a bounded domain $D \subset \mathbb{R}^3$ with a connected C^2 boundary Γ, and an incident field u^i, the direct obstacle scattering problem consists of finding the total field $u = u^i + u^s$ as a solution to the Helmholtz equation

$$\Delta u + k^2 u = 0 \quad \text{in } \mathbb{R}^3 \setminus \bar{D} \tag{1}$$

with positive wave number k satisfying the Dirichlet boundary condition

$$u = 0 \quad \text{on } \Gamma \tag{2}$$

such that the scattered wave u^s fulfills the Sommerfeld radiation condition

$$\lim_{r \to \infty} r \left(\frac{\partial u^s}{\partial r} - iku^s \right) = 0, \quad r = |x|, \tag{3}$$

uniformly with respect to all directions. The latter is equivalent to an asymptotic behavior of the form

$$u^s(x) = \frac{e^{ik|x|}}{|x|} \left\{ u_\infty \left(\frac{x}{|x|} \right) + O \left(\frac{1}{|x|} \right) \right\}, \quad |x| \to \infty, \tag{4}$$

uniformly in all directions, with the far field pattern u_∞ defined on the unit sphere Ω in $\mathbb{R}^3$ (see Theorem 2.5 in Colton and Kress[4]). The inverse obstacle scattering problem we are considering is, given the far field pattern u_∞ for one incident plane wave $u^i(x) = e^{ik\,x \cdot d}$ with incident direction $d \in \Omega$ to determine the boundary Γ of the scatterer D. From the numerous methods that have been developed for the solution of this inverse obstacle scattering problem, in this survey we will confine ourselves to two groups of approaches, namely regularized Newton iterations and decomposition methods.

The solution to the direct scattering problem with a fixed incident wave u^i defines an operator

$$F : \Gamma \mapsto u_\infty$$

that maps the boundary Γ of the sound-soft scatterer D onto the far field pattern u_∞ of the scattered wave. In terms of this operator, given a far field pattern u_∞, the inverse problem consists of solving the nonlinear and ill-posed operator equation

$$F(\Gamma) = u_\infty \tag{5}$$

for the unknown boundary surface Γ. The equation (5) is nonlinear, since the solution to the direct scattering problem depends nonlinearly on the boundary, and it is ill-posed, because the far field mapping is extremely smoothing due to the fact that the far field pattern is an analytic function on the unit sphere (see Theorem 2.5 in Colton and Kress[4]). It seems quite natural to apply Newton iterations for the solution of the nonlinear operator equation (5). Because of the ill-posedness these iterations need to be regularized.

This approach has the advantages that, in principle, it is conceptually simple and that it leads to highly accurate reconstructions (see for example Farhat et al[6], Hohage[8], Kirsch[10], and Kress[6]). However, as disadvantages we note that the numerical implementation requires the forward solution of the direct scattering problem (1)–(3) in each step of the Newton iteration and a good a priori information for the initial approximation. Furthermore, in the theoretical foundation both the linearization, i.e., the computation of the derivative of F with respect to Γ (see Section 5.3 in Colton and Kress[4]) is rather involved and the convergence analysis for the regularized iterations (see Hohage[9] and Potthast[18]), at the time this is written, is not completely settled.

In a second group of methods the ill-posedness and the nonlinearity of the inverse obstacle scattering problem are decomposed. In a first step the scattered wave u^s is reconstructed from the given far field pattern u_∞. For example in the method due to Kirsch and Kress[11] it is assumed that a priori enough information is known to place an auxiliary closed surface Γ_0 in the interior of the unknown scatterer D. Then trying to represent u^s as a single-layer potential with unknown densities φ on Γ_0 and matching the far field patterns leads to an ill-posed integral equation of the first kind for φ that can be solved, for example, by Tikhonov regularization. Then in a second step the unknown boundary Γ is determined as the location where the boundary condition (2) is satisfied in some least squares sense. For a detailed description of this method we refer to Section 5.4 in Colton and Kress[4].

This method enjoys as advantages again a conceptual simplicity and the fact that its numerical implementation does not require a forward solver. As disadvantages we note that this method does not result in high accuracy for the reconstructions and that in the first step the domain where the scattered wave needs to be reconstructed is unknown. Furthermore there is a gap between the theoretical foundation of the method and the numerical implementation: Whereas the latter usually is done in two steps as described above, for a satisfactory mathematical analysis the two steps must be combined into one optimization problem.

It is the purpose of the first part of this survey to describe a particular variant of the implementation of the Newton iterations as independently suggested by Kress and Rundell[15] and Potthast[18]. Then in the second part we modify this approach into an iterative method that does not require the solution of the forward problem in each iteration step and that may be considered as a hybrid of decomposition and Newton methods. We will conclude with a numerical example illustrating the feasibility of the hybrid method.

Before we proceed, however, we would like to emphasize that, of course, besides regularized Newton iterations for nonlinear ill-posed operator equations and decomposition methods a variety of other methods have been developed and successfully applied for the solution of inverse obstacle scattering problems. For a description of other approaches the reader is referred to the monographs[4,19] and the survey[2]. Some of these methods such as the linear sampling method or the factorization method are also presented here at this workshop.

2. Newton iterations

In order to define the operator F properly a parameterization of the boundary surface is required. For simplicity and to avoid lengthy notations we consider only starlike domains, i.e., we choose a parametrization in polar coordinates of the form

$$\Gamma_r = \{\gamma_r(\hat{x}) : \hat{x} \in \Omega\}, \tag{6}$$

where

$$\gamma_r(\hat{x}) := r(\hat{x})\,\hat{x}, \quad \hat{x} \in \Omega,$$

is defined in terms of a function $r : \Omega \rightarrow (0, \infty)$ representing the radial distance from the origin. After introducing $C_+^2(\Omega) := \{r \in C^2(\Omega) : r > 0\}$ we can interpret the far field mapping as an operator

$$F : C_+^2(\Omega) \rightarrow L^2(\Omega)$$

that maps the radial function r onto the far field pattern u_∞ for the scattering of the plane wave u^i from the obstacle bounded by Γ_r. However, we wish to emphasize that the concepts described below, in principle, are not confined to starlike scatterers only.

In this setting, given a far field pattern u_∞, the inverse scattering problem consists of solving the nonlinear and ill-posed operator equation

$$F(r) = u_\infty \tag{7}$$

for the radial function r. Uniqueness for this equation is settled through the following theorem that is a reformulation of the uniqueness result for inverse obstacle scattering due to Colton and Sleeman[5] (see also Corollary 5.3 in Colton and Kress[4]).

Theorem 2.1. *On the subset* $C^2(\Omega_{+,R}) := \{r \in C^2(\Omega_+) : \|r\|_\infty \leq R\}$ *with* $kR < \pi$ *the boundary to far field mapping* $F : r \mapsto u_\infty$ *is injective.*

For differentiability of F we have the following theorem.

Theorem 2.2. *The boundary to far field mapping* $F : r \mapsto u_\infty$ *is Fréchet differentiable and the derivative is given by*

$$F'(r) : q \to v_{q,\infty},$$

where $v_{q,\infty}$ *is the far field pattern of the solution* v_q *to the Dirichlet problem for the Helmholtz equation in the exterior of* Γ_r *satisfying the Sommerfeld radiation condition and the Dirichlet boundary condition*

$$v_q = -\nu \cdot (\gamma_q \circ \gamma_r^{-1}) \, \frac{\partial u}{\partial \nu} \quad on \; \Gamma_r, \tag{8}$$

where u *is the total field for scattering from* Γ_r *and* ν *the outward unit normal.*

The boundary condition (8) for the derivative can be obtained formally by differentiating the boundary condition $u = 0$ on Γ_r with respect to Γ_r by the chain rule. It was obtained by Roger[20] who first employed Newton type iterations for the approximate solution of inverse obstacle scattering problems. Rigorous foundations for the Fréchet differentiability were given by Kirsch[10] in the sense of a domain derivative via variational methods and by Potthast[17] via boundary integral equation techniques. Alternative proofs were contributed by Kress and Päivärinta[14] based on Green's theorems and a factorization of the difference of the far field patterns for two scatterers and by Hohage[9] and Schormann[21] via the implicit function theorem. We note that

$$(\nu \circ \gamma_r) \cdot \gamma_q = \frac{r \, q}{\sqrt{r^2 + |\operatorname{Grad} r|^2}} \quad on \; \Omega, \tag{9}$$

where Grad denotes the surface gradient.

In Newton's method one linearizes

$$F(r + q) = F(r) + F'(r)q + O(q^2)$$

and solves the linear equation

$$F(r) + F'(r)q = u_\infty \tag{10}$$

for q in order to improve an approximate boundary given by the radial function r into a new approximation with radial function $r + q$. For uniqueness we have the following theorem which follows easily from Holmgren's theorem (see Theorem 5.15 in Colton and Kress[4]).

Theorem 2.3. *The linear operator $F'(r)$ is injective.*

In the usual fashion, Newton's method consists in iterating the above procedure, i.e., in solving

$$F'(r_n)q_{n+1} = u_\infty - F(r_n), \quad n = 0, 1, 2, \ldots, \tag{11}$$

to obtain a sequence of approximations through $r_{n+1} = r_n + q_{n+1}$ starting with some initial approximation r_0. Since the derivative $F'(r)$ inherits the ill-posedness of F, it is compact. This also can be seen from the explicit representation of the derivative in Theorem 2.2. Hence, regularization techniques have to be employed on the basis that only inexact data u_∞^δ satisfying $\|u_\infty^\delta - u_\infty\|_{L^2(\Omega)} \leq \delta$ with some error level $\delta > 0$ are available for the right-hand side, i.e., we need to solve (10) with u_∞ replaced by a perturbed u_∞^δ. In principle, any regularization scheme for ill-posed linear operator equations such as Tikhonov regularization, Landweber iteration or singular value cut-off can be used.

We proceed by describing an approach for the implementation of the Newton iteration that is based on pulling back the equation (10) on the boundary via (8) (see Kress and Rundell[15] and Potthast[18]). For this, with the aid of the fundamental solution

$$\Phi(x, y) := \frac{1}{4\pi} \frac{e^{ik|x-y|}}{|x - y|}, \quad x \neq y,$$

we introduce the single-layer potential

$$(S_r\varphi)(x) := \int_{\Gamma_r} \Phi(x, y)\varphi(y) \, ds(y), \quad x \in \mathbb{R}^3,$$

with corresponding far field pattern

$$(S_{r,\infty}\varphi)(\hat{x}) := \frac{1}{4\pi} \int_{\Gamma_r} e^{-ik\,\hat{x}\cdot y}\varphi(y) \, ds(y), \quad \hat{x} \in \Omega.$$

In the sequel we assume that the assumptions of Theorem 2.1 are satisfied, that is, $\|r\|_\infty < \pi/k$. By the strong monotonicity properties for eigenvalues of the negative Laplacian under Dirichlet boundary conditions this implies that k^2 is not an eigenvalue for the interior of Γ_r (compare the proof of Theorem 5.2 in Colton and Kress[4]). Hence, the single-layer operator S_r is a homeomorphism in a Hölder space setting from $C^{0,\alpha}(\Gamma_r)$ onto $C^{1,\alpha}(\Gamma_r)$ (see Colton and Kress[3]) and in a Sobolev space setting from $H^{-1/2}(\Gamma_r)$ onto $H^{1/2}(\Gamma_r)$ (see McLean[16]), since the above assumption implies injectivity. Furthermore, the compact operator

$$S_{r,\infty} : L^2(\Gamma_r) \to L^2(\Omega)$$

is injective and has dense range (see Theorem 5.17 in Colton and Kress[4]) which is of importance in connection with the regularization of (12). In addition, we introduce the restriction operator

$$R_r v := v|_{\Gamma_r}$$

for functions defined in the closed exterior of Γ_r.

Given a (perturbed) far field u_∞^δ and a current approximation Γ_r we approximately solve the ill-posed linear integral equation

$$S_{r,\infty}\varphi^\delta = u_\infty^\delta, \tag{12}$$

for example, via Tikhonov regularization and set

$$u_\delta^s = S_r\varphi^\delta. \tag{13}$$

Then we solve the boundary integral equation

$$R_r S_r \varphi = -R_r u^i \tag{14}$$

to obtain the solution of the forward problem for the domain with boundary Γ_r via the single-layer potential

$$u^s = S_r\varphi \tag{15}$$

Through the jump relations, we can also evaluate the normal derivative of u^s on the boundary Γ_r. Via (8) and (9) the far field formulation of the Newton iteration (10) corresponds to the boundary relation

$$\frac{r\,q}{\sqrt{r^2 + |\operatorname{Grad} r|^2}} \frac{\partial}{\partial \nu} (u^i + u^s) \circ \gamma_r - [u^s - u_\delta^s] \circ \gamma_r = 0 \quad \text{on } \Omega. \tag{16}$$

To take care of possible zeros of the normal derivative $\partial u/\partial \nu$ and to achieve a real-valued q, the latter equation now can be collocated at appropriate collocation points on Ω and then solved in a least squares sense with respect to a finite dimensional subspace to yield the update q. To acknowledge the ill-posedness involved in (16) the least square approach requires a penalty term for regularization, i.e., abbreviating

$$R(q) =: \frac{r\,q}{\sqrt{r^2 + |\operatorname{Grad} r|^2}} \frac{\partial}{\partial \nu} (u^i + u^s) \circ \gamma_r - [u^s - u_\delta^s] \circ \gamma_r$$

one needs to minimize the penalized residual

$$\lambda \|q\|^2_{H^\ell(\Omega)} + \|R(q)\|^2_{L^2(\Omega)} \tag{17}$$

with some positive regularization parameter λ and some moderately large Sobolev index ℓ. For details in the two-dimensional case we refer to Kress and Rundell[15].

3. A hybrid method

Motivated by the above it seems reasonable to try the following procedure that might be interpreted as Newton iteration for retrieving the unknown boundary from the boundary condition $u = 0$ on the boundary of the scatterer. As above, given a (perturbed) far field u^δ_∞ and a current approximation Γ_r, we approximately solve the ill-posed linear integral equation (12) and define u^s_δ by (13). However, then we proceed differently and, in principle, view u^s_δ as an approximation for the solution to the scattering problem for the scatterer with boundary Γ_r. We update r by linearizing the boundary condition

$$R_{r+q}(u^i + u^s_\delta) = 0$$

through Taylor's formula. This leads to

$$R_r[u^i + u^s_\delta] + R_r \operatorname{grad}(u^i + u^s_\delta) \cdot (\gamma_q \circ \gamma_r^{-1}) = 0 \quad \text{on } \Gamma_r,$$

that is,

$$[\operatorname{grad}(u^i + u^s_\delta)] \circ \gamma_r \cdot \hat{x}\, q + (u^i + u^s_\delta) \circ \gamma_r = 0 \quad \text{on } \Omega, \tag{18}$$

and this can be solved analogously to (16) in a least squares sense. Again a penalty term as in (17) is required to take care of the ill-posedness involved in the inverse scattering problem. The derivative in the direction $\hat{x}$ as coefficient of the update q in (18) can be obtained from the value of u^s_δ and the normal derivative on Γ_r that are available through the jump relations for the single-layer potential.

Obviously, iterating this procedure does not need the solution of the forward scattering problem as necessary for the Newton iterations for solving (7). This method can also be viewed as a modification of the decomposition method of Kirsch and Kress[11]. Using the above notations, in this method one assumes that a surface Γ_{r_0} with radial function r_0 contained in the interior of the unknown scatterer D is a priori known. Then, in a first step, for $r = r_0$, the ill-posed linear integral equation (12) is solved by Tikhonov regularization and u^s_δ defined via (13). In the second step the unknown scatterer is found by determining a surface that satisfies $u^i + u^s_\delta = 0$ in some least squares sense. Note, that in contrast to the above new method in the original version of the method of Kirsch and Kress the auxiliary interior surface is kept fixed during the iteration procedure for finding the unknown scatterer as the location where the boundary condition is satisfied. However, we note that adaptive versions of the method of Kirsch and Kress with updates of the auxiliary surface during the least squares iterations have also been reported by Haas and Lehner[7]. With this point of view, the iterative version of (18) can be considered as a bridge connecting decomposition and Newton type methods in inverse obstacle scattering, i.e., as a hybrid of both methods.

Revisiting the theoretical foundation of the Kirsch–Kress decomposition method as presented in Section 5.4 of Colton and Kress[4] suggests to associate an optimization problem with the hybrid method. For this choose a bounded subset $V \subset H^\ell(\Omega)$ representing the admissible boundary surfaces. Then, given an incident field u^i and a (perturbed) far field pattern u^δ_∞, for a regularization parameter $\alpha > 0$ consider the cost function

$$\mu(\cdot,\cdot;\alpha) : L^2(\Omega) \times V(\Omega) \to \mathbb{R}$$

given by

$$\mu(\varphi, r; \alpha) := \mu_1(\varphi, r; \alpha) + \mu_2(\varphi, r),$$

where

$$\mu_1(\varphi, r; \alpha) := \alpha \|\varphi\|^2_{L^2(\Omega)} + \|S_{r,\infty}(\varphi \circ \gamma_r^{-1}) - u^\delta_\infty\|^2_{L^2(\Omega)}$$

and

$$\mu_2(\varphi, r) := \|u^i + S_r(\varphi \circ \gamma_r^{-1})\|^2_{L^2(\Gamma_r)}.$$

Then a surface Γ_{r_0} with radial function $r_0 > 0$ is called optimal if there exists $\varphi_0 \in L^2(\Omega)$ such that the pair φ_0 and r_0 minimizes the cost function μ over $L^2(\Omega) \times V$. Proceeding as in the proofs of Theorems 5.20 and 5.22 in Colton and Kress[4] it can be shown that for a positive regularization parameter optimal surfaces exist and that the optimization formulation is a regularization scheme for the inverse scattering problem in the sense that we have convergence of the optimal surfaces if the error level δ and the regularization parameter α tend to zero. Precise formulations and detailed proofs for these statements will be given in a forthcoming paper[1].

We note that the actual implementation of the hybrid method as given above separates the minimization the cost function μ into two parts. In each iteration step in the first part, for fixed r, the quadratic function $\mu_1(\cdot, r; \alpha)$ is minimized with respect to φ through solving (12) by Tikhonov regularization. Then in the second part, for fixed φ, the function $\mu_2(\varphi, \cdot)$ is decreased with respect to r by a Newton step via penalized least squares for (18). Hence, the above analysis can be viewed only as a preliminary mathematical foundation of our hybrid method and more research is required.

The analysis, of course, extends to the two-dimensional case after the appropriate changes in the radiation condition, the far field pattern and the fundamental solution with the latter given by

$$\Phi(x, y) = \frac{i}{4} H_0^{(1)}(k|x - y|)$$

in terms of the Hankel function $H_0^{(1)}$ of order zero and of the first kind. For the implementation in our concluding numerical example q is taken from a finite dimensional subspace with approximations of the form

$$\Gamma_r \approx \{p(t) (\cos t, \sin t) : t \in [0, 2\pi]\}$$

with a trigometric polynomial p of degree N. The update is obtained after collocating (18) at L equidistantly spaced points on the parameter interval $[0, 2\pi]$ and solving the resulting linear system for the coefficients of the trigonometric polynomial in the least squares sense with an H^2 penalty term and with L considerably larger than the number $2N + 1$ of unknowns. The severely ill-posed linear integral equation (12) is solved by Tikhonov regularization with L^2 penalty term and with the trapezoidal rule applied for the smooth periodic integrands on the unit circle. The regularization parameters α and λ and the number of iteration steps where chosen by trial and error. For details on the evaluation of u_δ^s and its normal derivative and for further examples we refer to Kress[13].

We used $L = 32$ collocation points for the update of q and the same number of equidistant quadrature points for the Nyström solution of the Tikhonov regularization for (12). To avoid a blatant inverse crime, the synthetic data were obtained via using the single-layer boundary integral equation of the first kind with 64 collocation points. For the noisy data random errors were added pointwise to the values of the far field pattern at the 32 Nyström points with the percentage given in terms of the L^2 norm.

In the example, we consider the identification of a kite shaped boundary curve given by the parametrization

$$\Gamma = \{(\cos t + 0.65 \cos 2t - 0.65, 1.5 \sin t) : t \in [0, 2\pi]\}.$$

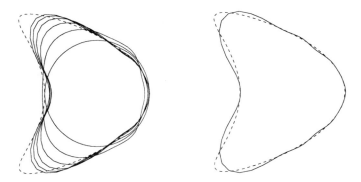

Figure 1.   Reconstructions with exact data.

The parameters in the reconstruction algorithm for the wave number $k = 1$ were chosen as $d = (-1, 0)$, $N = 8$, $\alpha = 0.0001$ and $\lambda = 0.1 \cdot [0.5]^j$ for the j th step and the iterations were started with the circle of radius one centered at the origin. The left part of Figure 1 shows the progress of the iteration through the first five steps and the right part shows the reconstruction after eight iteration steps. The exact boundary curve is given through the dashed line. Figure 2 shows reconstructions with noisy data, again with eight iteration. Here, the regularization parameter λ was changed into $\lambda = 0.1 \cdot [0.8]^j$. Clearly, the results indicate satisfactory boundary reconstructions with reasonable stability against noisy data.

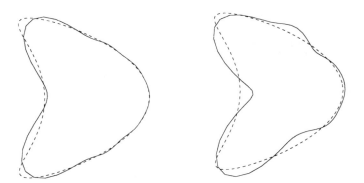

Figure 2.   Reconstructions with 2% (left) and 5% (right) noise.

References

1. R. Chapko and R. Kress, A hybrid method for inverse boundary value problems in potential theory (to appear).
2. D. Colton, J. Coyle, and P. Monk, SIAM Review **42**, 369 (2000).
3. D. Colton and R. Kress, *Integral Equation Methods in Scattering Theory*. Wiley-Interscience Publication, New York 1983.
4. D. Colton and R. Kress, *Inverse Acoustic and Electromagnetic Scattering Theory*. 2nd. ed. Springer, Berlin 1998.
5. D. Colton and B.D. Sleeman, IMA J. Appl. Math. **31**, 253 (1983).
6. C. Farhat, R. Tezaur, and R. Djellouli: Inverse Problems **18**, 1229 (2002).
7. M. Haas and G. Lehner, IEEE Transactions on Magnetics **33**, 1958 (1997).
8. T. Hohage, Inverse Problems **13**, 1279 (1997).
9. T. Hohage, *Iterative Methods in Inverse Obstacle Scattering: Regularization Theory of Linear and Nonlinear Exponentially Ill-Posed Problems*. Dissertation, Linz 1999.
10. A. Kirsch, Inverse Problems **9**, 81(1993).
11. A. Kirsch and R. Kress, In: *Boundary elements IX, Vol 3. Fluid Flow and Potential Applications*, (Brebbia et al, eds.), pp 3–18, Springer-Verlag, Berlin 1987.
12. R. Kress, In: *Boundary Integral Formulations for Inverse Analysis*, (Ingham, Wrobel, eds.), pp 67–92, Computational Mechanics Publications, Southampton 1997.
13. R. Kress, Inverse Problems **19**, (2003).
14. R. Kress and L. Päivärinta, SIAM J. Appl. Math. **59**, 1413 (1999).
15. R. Kress and W. Rundell, Inverse Problems **17**, 1075 (2001).
16. W. McLean, *Strongly Elliptic Systems and Boundary Integral Equations*. Cambridge University Press, 2000.
17. R. Potthast, Inverse Problems **10**, 431 (1994).
18. R. Potthast, Inverse Problems **17**, 1419 (2001).
19. R. Potthast, *Point-sources and multipoles in inverse scattering theory*. Chapman & Hall, London 2001.
20. A. Roger, IEEE Trans. Ant. Prop. **AP-29**, 232 (1981).
21. C. Schormann, *Analytische und numerische Untersuchungen bei inversen Transmissionsproblemen zur zeitharmonischen Wellengleichung*. Dissertation, Göttingen 2000.

ELECTROMAGNETIC SCATTERING FROM REAL SCATTERERS: IMPACT OF MATERIAL ELECTRICAL CHARACTERISTICS TO THE SCATTERED FIELD

PANAYIOTIS P. LEMOS

A.N. MAGOULAS AND I.K. HATZILAU

Chair of Electrical Engineering

Hellenic Naval Academy

18503 Piraeus, GREECE

E-mail: lemospan@ieee.org - aris@snd.edu.gr - ikx@snd.edu.gr

The problem of scattering of electromagnetic waves is one of the most essential problems of electromagnetic theory and presents a significant interest in both theoretical and practical applications (e.g. telecommunications, electromagnetic interaction, biomedicine, remote sensing, etc.) The solution to such a problem involves the computation of surface or spatial current density on scatterers followed by the computation of scattering field (near or far) as well as various other parameters (i.e. scattering cross section). For real scatterers the scattered field is also a function of their electrical characteristics beyond all other parameters that are involved in such problems. This paper investigates the impact of the electrical characteristics of several materials to the scattered electromagnetic field.

Keywords: Scattering, Radar Cross-Section, Method of Moments, Real Scatterers, Aluminum Alloys, Carbon Steel, Composite Materials, Fiber Glass.

1. Introduction

When an incident plane wave arrives on the surface of a real scatterer, it induces currents on both the scatterer's surface and interior. For source-free interior regions in homogeneous dielectric bodies, if we want to have an approximate calculation of the scattered field, it is important that we do not need to know the interior distributions of currents [1], [2]. In the present paper, a comparative presentation of the scattered electromagnetic fields on several finite conducting material (real) scatterers is attempted. The initial investigation is based on electrical characteristics of Aluminium. Then, the investigation is extended to some kind of carbon steels and it concludes with the examination of some kinds of dielectric composite materials (fiberglass). Table 1 presents some of the physical and electrical characteristics of the materials that have been selected for the current investigation [3], [4]. The Method of Moments (MoM) is used for this investigation based on FEKO that has been evaluated to produce accurate Radar Cross Section predictions for real axisymmetric scatterers. The materials selected present a very big interest in the naval ship construction industry, since most of the superstructure of the naval vessels is constructed with one of these or similar to these materials.

Table 1. Physical and Electrical Characteristics of the materials used for FEKO modeling.

No	Material	Components Wt (%)		Density $\left(\frac{g}{cc}\right)$	Electrical Resistivity $(Ohm \cdot cm)$	Conductivity $\times \dfrac{10^7}{\Omega \cdot m}$	Relative Permitivity (24°C) (e_r)	Relative Permeability (μ_r)
1.	Aluminum Alloys, General	Al	87 – 100	2.7	5e-006	3.77	~1	1.000021
2.	AISI 1005 Low Carbon Steel	C Fe Mn P S	max 0.06 99.5-100 max 0.35 max 0.04 max 0.05	7.872	1.74e-005	~1		~2000
3.	AISI 1030 Medium Carbon Steel	C Fe Mn P S	0.27 – 0.34 98.7 – 99.1 0.6 – 0.9 max 0.04 max 0.05	7.85	1.66e-005	~1		~2000
4.	AISI 1059 High Carbon Steel	C Fe Mn P S	0.55 – 0.65 98.5 – 98.9 0.5 – 0.8 max 0.04 max 0.05	7.87	1.74e-005	~1		~2000
5.	FiberGlass BK-174			2.1-2.8	9.0e+010		5.9+j0.52 at 10^5 Hz / 5.3+j0.24 at 10^6 Hz / 5.0+j0.17 at 10^7 Hz	

2. Geometry Setup

In order to proceed to Radar Cross Section computations, the MoM-based FEKO software package has been used. For these series of computations, triangular surface patches have been selected so that the length of the triangle's side is in the order of the tenth of the free space wavelength. This restriction introduces a significant increase in the number of surface patches as the electrical size of the sphere increases. The package has been extensively evaluated by computing the bistatic Radar Cross Section of a perfectly conducting sphere and then the results were compared to those provided by Mie series [5]. The mean square error, as defined in [6], computed for various electric sizes of the perfect sphere, is constantly below 7%. These results, obtained by FEKO and listed in Table 2, are far better than all other computations and techniques performed in [6] and [7] for similar number of surface patches. The fact that the length of the side of the surface triangles has been always kept equal to the tenth of the wavelength assisted in obtaining such accuracy. The actual setup of the geometry computer model is shown in Figure 1.

Table 2. Accuracy of FEKO computation of the bistatic RCS for various electric sizes of the perfect sphere compared to Mie series

Electrical size of the perfect sphere (kα in free space)	Number of surface triangles N	Mean Square Error (%) $s = \sqrt{\dfrac{1}{N_s} \sum_{i=0}^{N_s} \left(\dfrac{\sigma_{r,i} - \sigma_i}{\sigma_{r,i}} \right)^2}$
3	672	0.7723
5	1768	4.4941
7	3360	6.8624
10	6568	6.7587

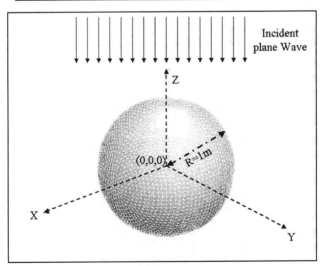

Figure 1. Geometry Setup

3. Results

3.1. *Aluminum Alloy*

Following the evaluation of the package accuracy, an aluminium alloy (87-100% Al) sphere has been designed and the electrical characteristics of aluminium have been introduced to the computations. Several frequencies have been attempted (keeping the problem dimensionless, i.e. ka=3, 5, 7 and 10). The length of the side of the surface triangles has been kept constantly equal to the tenth of the wavelength, which is still equal to the free space wavelength, since the wave number k of the aluminum is almost equal to the wave number of the free space according to equation 1 due to the fact that the relative permittivity and permeability of aluminium alloy are almost equal to 1.

$$k_1 = \omega \sqrt{\varepsilon_1 \cdot \mu_1} = k_0 \cdot m_1 \tag{1}$$

where: $\varepsilon_1 = \varepsilon_0 \cdot \varepsilon_{r\_AL}$, and $\mu_1 = \mu_0 \cdot \mu_{r\_AL}$

are the effective permittivity and permeability of the aluminium sphere, and m_1, as defined above, is the refractive index of the aluminium [1]. For this aluminium alloy sphere, the bistatic Radar Cross Section is computed and is compared to the RCS of the perfect sphere for various frequencies (ka). The deviation for these computations relatively to the analytical solution and the computations for the perfect sphere for various ka's is constantly less than 1% thus it is concluded that the aluminum alloy sphere behaves almost like a perfect conducting sphere. These results are presented in Table 3 and an indicative graph for ka=10 is included in Figure 2.

Table 3. Mean Square error relative to the Analytical solution and the perfect sphere computations for aluminum alloy sphere.

Electrical size of the Al sphere (almost equivalent to perfect sphere)	Mean Square Error (%) $$s = \sqrt{\frac{1}{N_s} \sum_{i=0}^{N_s} \left(\frac{\sigma_{r,i} - \sigma_i}{\sigma_{r,i}} \right)^2}$$	
	vs. Mie series	Deviation vs. FEKO computation
3	0.3113	0.2871
5	0.1930	0.5295
7	0.7004	0.7371
10	0.6336	0.2768

Bistatic RCS of Aluminium Alloy compared to Analytical Solution and FEKO approximation for Perfect Sphere (ka=10)

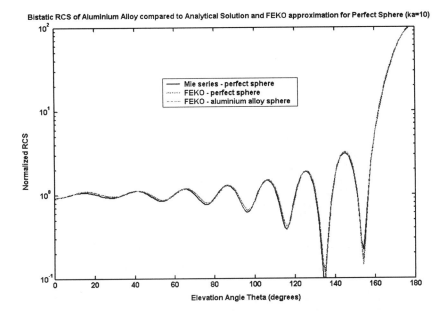

Figure 2. Bistatic RCS of Aluminium Alloy compared to Analytical solution and FEKO approximation for perfect sphere (ka=10)

3.2. Carbon Steel

The next step of computations included a Carbon Steel sphere with electrical characteristics listed in Table 1. As shown, the electrical characteristics do not vary significantly in relation to the composition of the carbon steel (especially in relation to the percentage of the carbon included), thus they do not affect the resulting computations. Similar to the aluminium sphere computations have been performed and the deviation for these computations relatively to the analytical solution and the computations for the perfect sphere for various kα's is presented in Table 4 and an indicative graph for kα=5 is included in Figure 3. The results present behavior similar to the aluminum sphere and the mean square error is kept constantly below 1% for a significant range of the frequency of the incident plane wave (kα).

3.3. FiberGlass

The case of the computations of the RCS for a dielectric material as the fiberglass selected, presents some peculiarities since the difference between the physical and the electrical size of the solid sphere under examination is significant, due to the values of the dielectric constant, which in fact is not constant anymore. Now, the dielectric constant is a complex number that varies with frequency. Therefore, modeling cannot anymore be dimensionless, since the dielectric constant varies with frequency. Figure 4 presents the bistatic RCS for a fiberglass unit sphere at various frequencies (0.1, 1 and 10 MHz).

Table 4. Mean Square error relative to the Analytical solution and the perfect sphere computations for carbon steel sphere

Electrical size of the Carbon Steel sphere (almost equivalent to perfect sphere)	Mean Square Error (%) $$s = \sqrt{\frac{1}{N_s} \sum_{i=0}^{N_s} \left(\frac{\sigma_{r,i} - \sigma_i}{\sigma_{r,i}} \right)^2}$$	
	vs.Mie series	deviation from FEKO computation
3	0.3112	0.2871
5	0.1675	0.5331
7	0.5742	0.7039
10	0.3386	0.4427

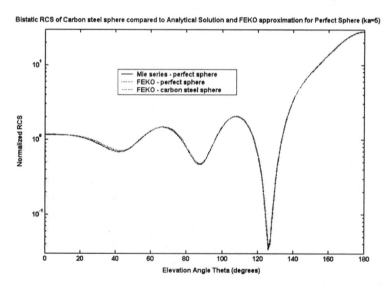

Figure 3. Bistatic RCS of carbon steel sphere compared to Analytical solution and FEKO approximation for perfect sphere (ka=5)

4. Conclusions

A comparative investigation of the impact of the electrical characteristics of materials that present a big interest in the Naval ship construction industry to the

Radar Cross Section has been performed. Aluminum, Low-Medium-High Carbon Steel and Fiberglass have been selected to be modeled and FEKO has been used for the modeling, based on MoM after it has been successfully evaluated. Figure 5 contains a comparative presentation of an aluminum vs. a Fiberglass sphere for various $k\alpha$'s (3, 5 and 7).

In the frequency range of interest, both Aluminum and Carbon steel approximate the behavior of perfect conductors, thus the problem can be

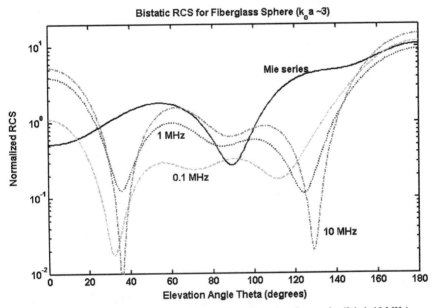

Figure 4. Bistatic RCS for Fiberglass Sphere for various frequencies (0.1, 1, 10 MHz)

considered dimensionless. Fiberglass behaves differently depending on the frequency of the incident electromagnetic plane wave, due to its variable dielectric constant vs frequency, thus the problem cannot be considered dimensionless.

Furthermore, fiberglass presents, in general, different elevation angles (theta) that the RCS is minimum vs perfect conducting materials (aluminum, etc.). It is also noticeable that Back-scattering (theta=0°) RCS is getting bigger in Fiberglass than perfect conductors (aluminum, etc.) as frequency increases.

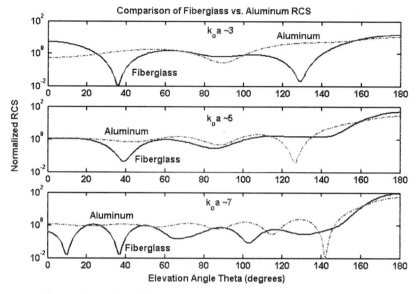

Figure 5. Comparison of Fiberglass vs. Aluminium Sphere Radar Cross Section.

Acknowledgments

The authors would like to thank Niels Berger, EM Software & Systems GmbH, Germany, for his support in effective modeling in calculations with FEKO.

References

[1]   Knott, Eugene, F., Shaeffer, John, F., and Tuley, Michael, T., "*Radar Cross Section*", 2nd edition, Artech House, ISBN: 0-89006-618-3, 1993.

[2]   Ruck, G. T., D.E. Barrick, W.D. Stuart, C.K. Krichbaum, "*Radar Cross Section Handbook*", Plenum Press, New York (1970).

[3]   Harrington, Roger, F., "*Time-Harmonic Electromagnetic Fields*", Wley-IEEE Press, ISBN: 0-471-20806-X, 2001 .

[4]   Cheng, K., David, "*Field and Wave Electromagnetics, 2nd edition*", Addison-Wesley Publishing Company, Inc., 1989.

[5]   Mie, G., "*Beitraege zur Optik trueber Medien, speziell kolloidaler Metalloesungen*", Ann. Phys, 25:377 (1908).

[6]   P.P. Lemos, A.N. Magoulas, I.K. Hatzilau, *"Scattering from Perfect Conducting Scatterers: Comparison of two Methods for the Solution of the Problem"*, Proceedings of 5th International Workshop on Mathematical Methods in Scattering Theory and Biomedical Engineering, World Scientific, ISBN: 981-238-054-X (2002).

[7]   P. Lemos, A. Magoulas, I.K. Hatzilau, N. Mastorakis *"Mesh Variations for the Computation of Electromagnetic Scattering on Perfect Conductors"*, International Journal of Computer Research, Vol. 11, No. 4, 2002.

[8]   R.F. Harrington, *"Field Computation by Moment Methods"*, The Macmillan Company, New York (1968).

[9]   Jakobus, U., *"Comparison of Different Techniques for the Treatment of Lossy Dielectric/Magnetic Bodies within the Method of Moments Formulation"*, Int. J. Electron. Commun (AEU) 54, No. 1, 1 1, (2000).

[10]   Rao, M., Sadasiva, Sarkar, K., Tapan, Midya, Pallab, and Djordevic, R., Antonije, *"Electromagnetic Radiation and Scattering from Finite Conducting and Dielectric Structures: Surface / Surface Formulation"*, IEEE Transactions on Antennas and Propagation, Vol. 39, No. 7, July 1991.

SHAPE DEPENDENCE OF THE SCATTERED ELECTROMAGNETIC FIELD OF SPHEROIDAL REAL SCATTERERS

PANAYIOTIS P. LEMOS

A.N. MAGOULAS AND I.K. HATZILAU

Chair of Electrical Engineering

Hellenic Naval Academy

18503 Piraeus, GREECE

E-mail: lemospan@ieee.org - aris@snd.edu.gr - ikx@snd.edu.gr

In the present paper the approach to investigate the shape dependence of a real axissymetric scatterer and especially a spheroid, with relation to its scattered electromagnetic field is continued. For this investigation a spheroid has been selected with varying ratio of its dimensions. The first approach, in the first of this series' papers, has been based on the principle of maintaining the scatterer's area constant while changing its shape from oblate to spherical and then to prolate, allowing the geometric cross section on the plane of the incident electromagnetic wave to vary in order to maintain the total surface area constant. For this attempt, the geometric cross section on the horizontal (x-y) plane, assuming incidence of the electromagnetic plane wave towards the –z axis, is kept constant while the length of the semi-axis on the direction of the incident plane wave is changing. resulting to change of the size of the spheroid and consequently to its total surface area. For this purpose, two different kinds of real scatterers are modeled, an aluminum and a fiberglass spheroids, and the Method of Moments with the use of FEKO is applied and the results of the Bistatic Radar Cross Section computations are analyzed and presented.

Keywords: Scattering, Radar Cross Section, Method of Moments, Real Spheroidal Scatterer, Composite materials.

1. Introduction

A plane electromagnetic wave impinging on a structure, results to inductive charges and currents in both its surface and interior [1]. When this structure is a perfect conducting one, then the induced charges and currents are confined in its surface, whereas in the case of a dielectric the induced charges and currents are extended to its interior. Investigation on the shape dependence of the Scattered Electromagnetic field for the case of perfect conducting spheroids as well as some interesting points of the computation of the solution and the angles that do not contribute significantly on the RCS contribution of the spheroids is presented in [2]. For that investigation, perfect spheroids have been modelled with the length of their semi-axes changing but with the overall surface area remaining constant. In this paper, the investigation is extended to real scatterers. Moreover, in this attempt, the constant parameter is the geometric cross section of the spheroid along the direction of the incident plane wave (x-y plane, i.e. semi-axis α on the horizontal plane is constant) resulting to oblate and prolate spheroids with various surface areas. Another parameter inserted in the current investigation is the electrical characteristics of the spheroids under investigation, resulting to two major categories of materials (real metals – aluminum and dielectric composite materials – fiberglass). At microwave frequencies, even a poorly conducting metal like steel approximates the behavior of a perfect conductor.

2. Geometry

2.1. *Model Generation*

This investigation is considered essential as the geometric cross section that is "seen" by the incident plane wave is always constant and equal to $\pi\alpha^2$ (the normalization factor for the computed RCS). Table 1 presents the dimensions of the spheroids as well as the number of surface triangular patches used for the generation of the spheroids, which are depicted in Figure 1. In order to achieve an adequate accuracy for the computations, the length of the side of the triangular surface patches has been selected to be in the order of the tenth of the incident wavelength whilst it remains constantly bigger than the fifth of the radius of the circular geometric cross section on the xy plane. Modifying the spheroids semi-axis on the zz' direction, thus varying the semi-axes ratio from 0.1 to 2.9, we produce spheroids with total surface area varying from 6.4722 to 29.9412 m^2 providing a range of approximately 50% to 240% compared to the surface of the unity sphere. The medium of the spheroidal models behaves as a material with a wave number

$$k_1 = \omega\sqrt{\varepsilon_1 \cdot \mu_1} = k_0 \cdot m_1 \tag{1}$$

where: $\varepsilon_1 = \varepsilon_0 \cdot \varepsilon_r$, and $\mu_1 = \mu_0 \cdot \mu_r$ are the effective permittivity and permeability of the aluminium sphere, and m_1, as defined above, is the refractive index of the material of the spheroid [11].

2.2. *Model Evaluation*

In order to proceed to Radar Cross Section computations, the MoM-based FEKO software package has been used. For these series of computations, triangular surface patched have been selected so that the length of the triangle's side is in the order of the tenth of the free space wavelength. This restriction introduces a significant increase in the number of surface patches as the electrical size of the sphere increases, especially in the case of the fibreglass where the dielectric constant is significantly greater than 1. The package has been extensively evaluated by computing the bistatic Radar Cross Section of a perfectly conducting sphere and then the results were compared to those provided by Mie series [5]. The mean square error, as defined in [6], computed for various electric sizes of the perfect sphere, is constantly below 7%. These results, obtained by FEKO, are far better than all other computations and techniques performed in [6] and [7] for similar number of surface patches.

Table 1. Geometric Data of Spheroidal models

Semi-axis a (m)	Semi-axis c (m)	Semi-axis ratio $\left(\frac{c}{a}\right)$	Spheroid area (m^2)	Norma-lized area (compa-red to area of unity sphere)	Number of triangular surface patches			
					Aluminum Alloy		Fiberglass	
					ka=3	ka=7	ka=3	ka=5
1.0	0.1		6.4722	0.5150	872	2080	3064	8160
	0.3		7.3940	0.5884	872	2080	3064	8160
	0.5		8.6719	0.6901	872	2080	3064	8160
	0.7		10.1442	0.8072	872	2080	3064	8160
	0.9		11.7375	0.9340	872	2080	3064	8160
	1.1		13.4121	1.0673	1056	2608	3720	9976
	1.3		15.1440	1.2051	1520	3656	5072	14040
	1.5		16.9182	1.3463	2032	4664	6848	18240
	1.7		18.7244	1.4900	2592	5968	8648	23600
	1.9		20.5553	1.6357	3280	7544	10624	29648
	2.1		22.4059	1.7830	3768	9216	13184	35768
	2.3		24.2724	1.9315	4600	11112	15624	43112
	2.5		26.1518	2.0811	5464	12760	18240	50368
	2.7		28.0420	2.2315	6392	14880	21616	59040
	2.9		29.9412	2.3826	7336	17336	24680	68320

The fact that the length of the side of the surface triangles has been always kept equal to the tenth of the wavelength assisted in obtaining such accuracy. Similar evaluation has been performed for the case of non-perfectly conducting homogeneous spheres with adequate accuracy.

The materials selected for this study (aluminum alloy and fiberglass) present a very big interest in the naval ship construction industry, since most of the superstructure of the naval vessels is constructed with one of these or similar to these materials.

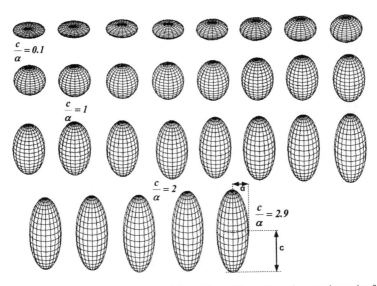

Figure 1. Various Spheroids used for the modeling of the problem with semi-axes ratio varying from 0.1 (oblate) to 2.9 (prolate).

3. Results

3.1 Aluminum Alloy

The computation of the scattered field of real conducting spheroids was initiated using an aluminium spheroidal model with finite relative electrical permittivity ($\sim$1) and conductivity 3.77×10^7 / Ωm. The results obtained by FEKO are presented in Figure 3 for oblate and in Figure 4 for prolate aluminium spheroids. For this model, all the necessary electrical characteristics have been introduced into the FEKO model and computations have been performed for frequencies that would produce $k_o a = 3$ (Figures 2 and 3) and $k_o a = 7$ (Figures 4 and 5).

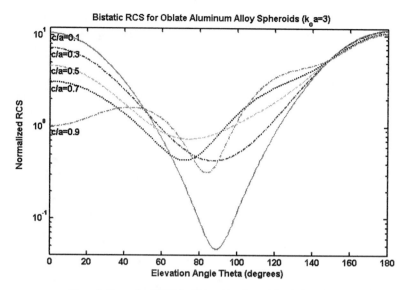

Figure 2. Normalized RCS for Oblate Aluminum Alloy Spheroids ($k_0\alpha$=3).

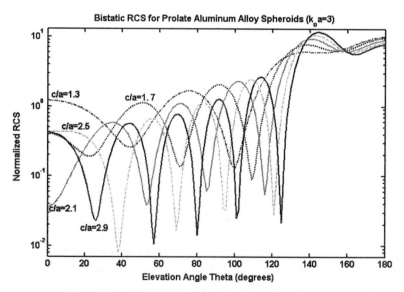

Figure 3. Normalized RCS for Prolate Aluminum Alloy Spheroids ($k_0\alpha$=3).

For these computations, the aluminum alloy sphere behaves as a perfect conductor, since its electrical characteristics do not allow the induction of currents in the interior of the aluminum scatterer and the current density $\vec{J}_S$ is distributed on its surface.

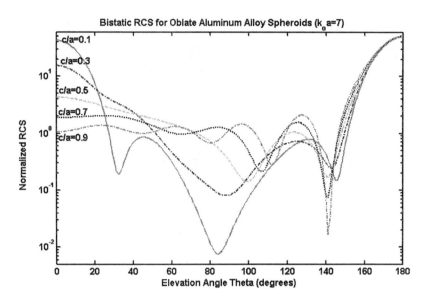

Figure 4. Normalized RCS for Oblate Aluminum Alloy Spheroids ($k_0\alpha$=7).

3.2 Fiberglass

For this model generation, the dielectric constant is not a "constant" anymore as it varies with frequency changes ($e_r = 5.9 + j0.52$ for 10^5 Hz, $e_r = 5.3 + j0.24$ for 10^6 Hz and $e_r = 5 + j0.17$ for 10^7 Hz). These values of dielectric constant do not restrict anymore the induced currents to the scatterers' surface but allow the induction of spatial current density within the interior of the scatterer, which contribute to the scattered electromagnetic field. Some indicative results for these computations are presented in Figures 6 and 7.

118

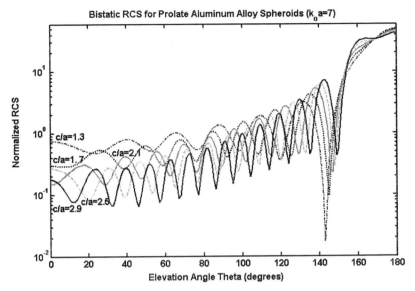

Figure 5. Normalized RCS for Prolate Aluminum Alloy Spheroids ($k_0\alpha$=7).

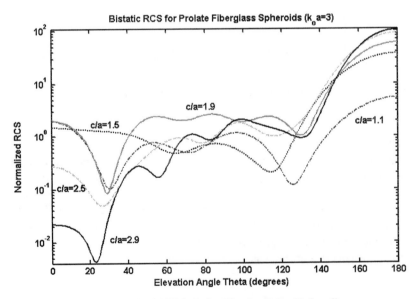

Figure 6. Normalized RCS for Prolate Fiberglass Spheroids ($k_0\alpha$=3).

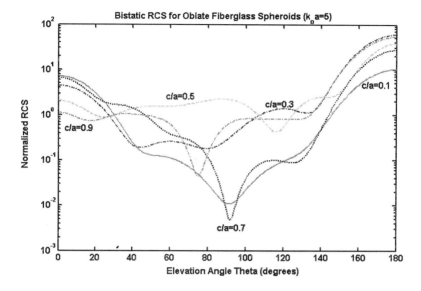

Figure 7. Normalized RCS for Oblate Fiberglass Spheroids ($k_0 \alpha = 7$).

4. Conclusions

The bistatic Radar Cross Section of several axissymetric finite conducting aluminium spheroids for various values of elevation angles theta $(\theta°)$, has been calculated and investigated in relation to the ratio (c/a). The ratio (c/a) of the spheroids has been used as the main parameter for this comparison, keeping the geometric cross section constant, thus resulting to changes in the spheroids' surface area. Real finite materials, with high interest in the naval ship construction industry (aluminum and fiberglass), have been modeled. The electrical characteristics of these materials "grow" significantly the number of surface triangular patches required for effective modeling using MoM, making the task very computer resources' demanding. Another important observation is that as aluminium spheroids transforms from oblate to prolate, several side-lobes appear on the scattered field, even in low frequencies. Moreover, it can be clearly shown that forward scattering region (150°-180°) appears to reach a certain range of values (30-60db) independently of the shape and the material of the spheroid.

120

Acknowledgments
The authors would like to thank Niels Berger, EM Software & Systems GmbH, Germany, for his support in effective modeling in calculations with FEKO.

References

[1]   Knott, Eugene, F., Shaeffer, John, F., and Tuley, Michael, T., *"Radar Cross Section"*, 2nd edition, Artech House, ISBN: 0-89006-618-3, 1993.

[2]   Lemos, Magoulas, Peppas, Hatzilau, "Shape Dependence of the Scattered Electromagnetic Field of Spheroidal Perfect Scatterers", Records of the 2003 IEEE International Symposium on Electromagnetic Compatibility (EMC), Istanbul, Turkey, May 2003, *ISBN: 0-7803-7780-X.*

[3]   Harrington, Roger, F., *"Time-Harmonic Electromagnetic Fields"*, Wley-IEEE Press, ISBN: 0-471-20806-X, 2001 .

[4]   Cheng, K., David, *"Field and Wave Electromagnetics, 2nd edition"*, Addison-Wesley Publishing Company, Inc., 1989.

[5]   Mie, G., *"Beitraege zur Optik trueber Medien, speziell kolloidaler Metalloesungen"*, Ann. Phys, 25:377 (1908).

[6]   P.P. Lemos, A.N. Magoulas, I.K. Hatzilau, *"Scattering from Perfect Conducting Scatterers: Comparison of two Methods for the Solution of the Problem"*, Proceedings of 5th International Workshop on Mathematical Methods in Scattering Theory and Biomedical Engineering, World Scientific, ISBN: 981-238-054-X (2002).

[7]   P. Lemos, A. Magoulas, I.K. Hatzilau, N. Mastorakis *"Mesh Variations for the Computation of Electromagnetic Scattering on Perfect Conductors"*, International Journal of Computer Research, Vol. 11, No. 4, 2002.

[8]   R.F. Harrington, *"Field Computation by Moment Methods"*, The Macmillan Company, New York (1968).

[9]   Jakobus, U., *"Comparison of Different Techniques for the Treatment of Lossy Dielectric/Magnetic Bodies within the Method of Moments Formulation"*, Int. J. Electron. Commun (AEU) 54, No. 1, 1 1, (2000).

[10]   Rao, M., Sadasiva, Sarkar, K., Tapan, Midya, Pallab, and Djordevic, R., Antonije, *"Electromagnetic Radiation and Scattering from Finite Conducting and Dielectric Structures: Surface / Surface Formulation"*, IEEE Transactions on Antennas and Propagation, Vol. 39, No. 7, July 1991.

[11]   Ruck, G. T., D.E. Barrick, W.D. Stuart, C.K. Krichbaum, *"Radar Cross Section Handbook"*, Plenum Press, New York (1970).

ACOUSTIC SCATTERING BY AN IMPENETRABLE SPHEROID

JOHN A.ROUMELIOTIS, ARISTIDES D. KOTSIS AND GEORGE KOLEZAS

School of Electrical and Computer Engineering National Technical University of AthensAthens 15773, Greece

The scattering of a plane acoustic wave from an impenetrable, soft or hard prolate or oblate spheroid is considered. Two different methods are used for the evaluation. In the first, the pressure field is expressed in terms of spheroidal wave functions. In the second, a shape perturbation method, the field is expressed in terms of spheroidal wave functions only, while the equation of the spheroidal boundary is given in spherical coordinates. Analytical expressions are obtained for the scattered pressure field and the scattering cross-sections, when the solution is specialized to small values of $h=d/(2a)$, ($h\ll1$), with d the interfocal distance of the spheroid and 2a the length of its rotation axis. In this case exact, closed-form expressions, valid for each small h, are obtained for the expansion coefficients $g^{(2)}$ and $g^{(4)}$ in the relation $S(h)=S(0)[1+g^{(2)}h^2+g^{(4)}h^4+O(h^6)]$ expressing the scattered field and the various scattering cross-sections.

1. Introduction

Study of acoustic scattering by spheroids is an old problem with numerous applications. Many researchers have been involved with its solution in the past. Among a great number of papers, treating the problem by various methods, we refer only some recent ones here [1-3].

In this paper we consider the scattering of a plane acoustic wave from an impenetrable, soft or hard, prolate or oblate spheroid. In Fig. 1 we give the geometry of the prolate spheroid. Its interfocal distance is d, while a and b are the lengths of its major and minor semiaxes, respectively. The prolate spheroid is the only one to be considered explicitly, but corresponding formulas for the oblate one are obtained immediately.

We use two different methods for the solution. In the first the pressure field is expressed in terms of spherical wave functions, while in the second, a shape perturbation method, the field is expressed in terms of spherical wave functions only and the equation of the spheroidal boundary is given in spherical coordinates. When the solution is specialized to small values of of the parameter $h=d/(2a),(h\ll1)$ analytical expressions of the form$S(h)=S(0)[1+g^{(2)}h^2+g^{(4)}h^4+O(h^6)]$ are obtained for the scattered pressure field and the various scattering cross-sections. The expansion coefficients $g^{(2)}$ and $g^{(4)}$ are given by exact, closed-form expressions, independent of h, while S(0) corresponds to a sphere with radius a (h=0).

122

The main advantage of such an analytical solution is its general validity for all small values of h, while all numerical techniques require repetition of the evaluation for each different h.

The scattering by a soft spheroid is examined in section 2, while in section 3 is examined the scattering by a hard one. Finally, in section 4 some numerical results are given.

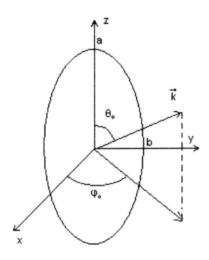

Figure 1 Geometry of the scatterrer .

2. Scattering by a soft spheroid

We apply first the method using spheroidal wave functions. The incident plane pressure wave impinging on the scatterer of Fig.1is expressed as [4],(the time factor exp(-jωt) is suppressed)

$$p_i = 2\sum_{n=0}^{\infty}\sum_{m=0}^{n} j^n \frac{\epsilon_m}{N_{mn}} R_{mn}^{(1)}(c, \cosh\mu) S_{mn}(c, \eta) S_{mn}(c, \cos\theta_0) \cos\left[m(\phi - \phi_0)\right] (1)$$

where the angles θ_0 and ϕ_0 define the direction of the incidence, μ, η, ϕ are the spheroidal coordinates, $R_{mn}^{(1)}$ and S_{mn} are the radial and the angular respectively, spheroidal functions of the first kind , $\epsilon_0 = 1$ and $\epsilon_m = 2, m \geq 1$ is the Neumann factor, c=kd/2 (k is the wavenumber), while the normalization constant is [5]

$$N_{mn} = 2 \sum_{r=0,1}^{\infty} \frac{\cdot (d_r^{mn})^2 (r+2m)!}{(2r+2m+1)r!} \tag{2}$$

The prime over the summation symbol in (2) indicates that when n-m is even/odd this summation starts with the first/second value of r and continues only with values of the same parity with it. The expansion coefficients d_r^{mn} are defined in [5,6].

The scattered pressure field is

$$p_s = \sum_{n=0}^{\infty} \sum_{m=0}^{n} R_{mn}^{(3)}(c,\cosh\mu) S_{mn}(c,\eta) \left(A_{mn} \cos m\phi + B_{mn} \sin m\phi \right) \tag{3}$$

where $R_{mn}^{(3)}$ is the radial spheroidal function of the third kind. The unknown expansion coefficients A_{mn} and B_{mn} are calculated by satisfying the Dirichlet boundary condition $p_i + p_s = 0$ at $\mu = \mu_0$ and using next the orthogonal properties of the angular spheroidal [5] and the trigonometric functions. So we obtain

$$\begin{pmatrix} A_{mn} \\ B_{mn} \end{pmatrix} = C_{mn} \begin{pmatrix} \cos m\phi_0 \\ \sin m\phi_0 \end{pmatrix} \tag{4}$$

where
$$C_{mn} = -2j^n \frac{\in_m}{N_{mn}} S_{mn}(c,\cos\theta_0) \frac{R_{mn}^{(1)}(c,\cosh\mu_0)}{R_{mn}^{(3)}(c,\cosh\mu_0)} . \tag{5}$$

We next express the spheroidal functions in (3) into spherical functions by using the expansion formula [5]

$$R_{mn}^{(3)}(c,\cosh\mu) S_{mn}(c,\eta) = \sum_{s=0,1}^{\infty} {}' j^{m-n+s} d_s^{mn}(c) P_{m+s}^m(\cos\theta) h_{m+s}(kr) \tag{6}$$

where r,θ are the spherical coordinates, P_{m+s}^m are the associated Legendre functions and h_{m+s} the spherical Hankel functions of the first kind. For the prime over Σ the remarks as in (2) are valid. Using next the asymptotic expansion for $h_{m+s}(kr)$ we find the scattered far pressure field expression

$$p_s = \frac{e^{jkr}}{kr} G(\theta,\phi) \tag{7}$$

where

$$G(\theta,\phi) = \sum_{n=0}^{\infty} \sum_{m=0}^{n} \sum_{s=0,1}^{\infty} {}' j^{-n-1} d_s^{mn} P_{m+s}^m(\cos\theta) C_{mn} \cos\left[m(\phi-\phi_0) \right] \tag{8}$$

is the scattering amplitude.

The backscattering or sonar (σ_b), the forward (σ_f) and the total (Q_t) scattering cross-sections are defined as follows:

$$\frac{\sigma_b}{\lambda^2} = \frac{1}{\pi}\left|G\left(\pi - \theta_0, \pi + \phi_0\right)\right|^2, \quad \frac{\sigma_f}{\lambda^2} = \frac{1}{\pi}\left|G\left(\theta_0, \phi_0\right)\right|^2,$$

$$\frac{Q_t}{\lambda^2} = \frac{1}{4\pi^2}\int_{\theta=0}^{\pi}\int_{\phi=0}^{2\pi}\left|G\left(\theta, \phi\right)\right|^2 \sin\theta \, d\theta \, d\phi \tag{9}$$

with λ the wavelength of the acoustic waves.

For large values of c one can proceed only numerically, but for c small analytical, closed-form expressions can be found for $G(\theta,\varphi)$ as well as σ_b, σ_f and Q_t. Using the parameter h=d/2a (instead of c=kah) we obtain, after lengthy but straightforward calculations, expressions of the form

$$S(h)=S(0)[1+g^{(2)}h^2+g^{(4)}h^4+O(h^6)] \tag{10}$$

for the scattered field and the scattering cross-sections, for h<<1. S(0) corresponds to a sphere with radius a (h=0).

We next apply the second method using spherical wave functions only. The incident plane wave is now expressed as [4]

$$p_i = \sum_{n=0}^{\infty}\sum_{m=0}^{n} j^n(2n+1)\,\epsilon_m\,\frac{(n-m)!}{(n+m)!}\,j_n(kr)P_n^m(\cos\theta)P_n^m(\cos\theta_0)\cos\left[m(\phi-\phi_0)\right] \tag{11}$$

where j_n are the spherical Bessel functions of the first kind, while the angle φ is common in both spherical and spheroidal coordinates.

The scattered field is

$$p_s = \sum_{n=0}^{\infty}\sum_{m=0}^{n} h_n(kr)P_n^m(\cos\theta)\left(A_{mn}\cos m\phi + B_{mn}\sin m\phi\right) \tag{12}$$

where certainly A's And B's are different of those in the first method.

In order to satisfy the boundary condition $p_i+p_s=0$ at the spheroidal surface, we express the equation of this surface in terms of r and θ and use its expansion into power series in h for h<<1 [7]

$$r = \frac{a}{\sqrt{1-v\sin^2\theta}} = a\left[1-\frac{h^2}{2}\sin^2\theta - \frac{h^4}{2}\left(\sin^2\theta - \frac{3}{4}\sin^4\theta\right) + O(h^6)\right] \tag{13}$$

where

$$v = 1 - \frac{a^2}{b^2} = -h^2 - h^4 + O(h^6) \tag{14}$$

By using (13) we obtain the expression

$$j_n(kr) = j_n(x) - \frac{h^2}{2} x j_n'(x) \sin^2 \theta +$$

$$h^4 \left\{ -\frac{x}{2} j_n'(x) \sin^2 \theta + \frac{1}{8} \left[3 x j_n'(x) + x^2 j_n''(x) \right] \sin^4 \theta \right\} + O(h^6), x = ka \tag{15}$$

and a similar one for $h_n(kr)$, where the primes denote derivatives with respect to the argument.

We next substitute these expansions into (11) and (12), satisfying the boundary condition $p_i + p_s = 0$ and use the orthogonal properties of the associated Legendre and the trigonometric functions, concluding finally to an infinite set of linear inhomogeneous equations for the expansion coefficients A_{mn} (or B_{mn}) of the following form (up to the order h^4):

$$a_{n,n-4} A_{n-4} + a_{n,n-2} A_{n-2} + a_{n,n} A_n + a_{n,n+2} A_{n+2} + a_{n,n+4} A_{n+4} = K_n \tag{16}$$

The subscript m is ommited from the various α's and A's in eq. (16) for simplicity.

For small values of h we can set, up to the order h^4

$$a_{n,n} = D_{n,n}^{(0)} + h^2 D_{n,n}^{(2)} + h^4 D_{n,n}^{(4)} + O(h^6), a_{n,n\pm2} = h^2 D_{n,n\pm2}^{(2)} + h^4 D_{n,n\pm2}^{(4)} + O(h^6),$$

$$a_{n,n\pm4} = h^4 D_{n,n\pm4}^{(4)} + O(h^6) \tag{17}$$

$$K_{n,n} = K_n^{(0)} + h^2 K_n^{(2)} + h^4 K_n^{(4)} + O(h^6) \tag{18}$$

where D's and K's are known. A's (or B's) are obtained from the solution of (16) by Cramer's rule, following steps similar with the ones in [8].

Using the asymptotic expansion for h_n in (12) we obtain again (7) where in this case

$$G(\theta,\phi) = \sum_{n=0}^{\infty} \sum_{m=0}^{n} (-j)^{n+1} P_n^m(\cos\theta) \left(A_{mn} \cos m\phi + B_{mn} \sin m\phi \right) \tag{19}$$

The next steps are analogous with the ones in the first method, so they will not be repeated here.

For the oblate spheroid we simply replace h^2 with $-h^2$ in each case.

3. Scattering by a hard spheroid

We first start with the method using spheroidal functions. The incident and the scattered waves are again given by (1) and (3), respectively. The new A's and B's are calculated by satisfying the homogenous Neumann boundary condition $\partial(p_i + p_s)/\partial\mu = 0$ at $\mu = \mu_0$ and using next the same steps as for the soft spheroid. Eq.(4) is again valid but in eq.(5) $R_{mn}^{(1)}$ and $R_{mn}^{(3)}$ are replaced by $\partial R_{mn}^{(1)}/\partial\mu$ and $\partial R_{mn}^{(3)}/\partial\mu$ respectively. The next steps are exactly the same as in section 2 and will not be repeated here.

We continue with the second method. The incident and the scattered waves are again given by (11),(12), with new A's and B's. The Neumann boundary condition at the spheroidal surface is[7]

$$\frac{\partial(p_i + p_s)}{\partial r} + \frac{h^2}{2}\sin 2\theta\left(1 + h^2\cos^2\theta\right)\frac{1}{r}\frac{\partial(p_i + p_s)}{\partial\theta} = 0. \quad (20)$$

By using (13) we obtain expansions similar to (15) for $j_n'(kr)$ and $h_n'(kr)$, respectively, with the only difference that one more prime is added in each of the Bessel functions met there. Also we obtain the expansion

$$\frac{j_n(kr)}{kr} = \frac{j_n(x)}{x} - \frac{h^2}{2}\left[-\frac{j_n(x)}{x} + j_n'(x)\right]\sin^2\theta + \frac{h^4}{2}\left\{\left[\frac{j_n(x)}{x} - j_n'(x)\right]\sin^2\theta\right.$$

$$\left. + \frac{1}{4}\left[-\frac{j_n(x)}{x} + j_n'(x) + xj_n''(x)\right]\sin^4\theta\right\} + O(h^6), x = ka \quad (21)$$

and a similar one for $\dfrac{h_n(kr)}{kr}$.

We next substitute these expansions into(11),(12), satisfying the boundary condition (20) and follow the same steps as in section 2. For the oblate spheroid we replace h^2 with $-h^2$.

4. Numerical results

In Fig. 2 we plot σ_b for a hard prolate spheroid, while in Fig.3 we plot σ_f for a soft oblate spheroid, both versus θ_0, for $a/\lambda = 0.5$ and h=0.1. The results are independent of φ_0 and symmetric about $\theta_0=90^0$, as it is imposed by the geometry of the scatterer. Another check for their correctness, moreover to the use of two different methods for the solution, is the validity of the forward scattering theorem, which in the present case has the form $Q_t / \lambda^2 = I_m[G(\theta_0, \varphi_0)]/\pi$.

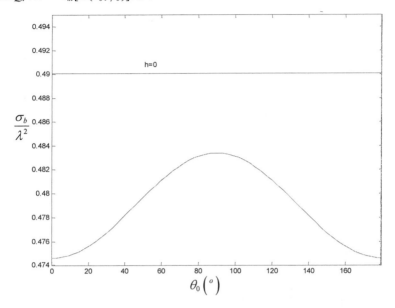

Figure 2 Backscattering cross section for a/λ=0.5 and h=0.1 for a hard prolate spheroid

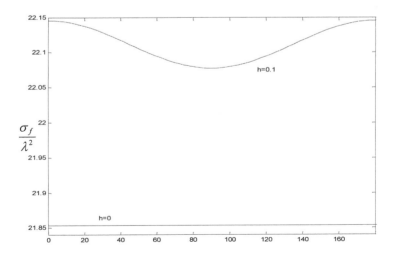

Figure 3 Forward scattering cross section for a/λ=0.5 and h=0.1 for a soft oblate spheroid

128

References

1. J.P.Barton, N.L.Wolff, H.Zhang and C.Tarawneh, "Near-Field calculations for a rigid spheroid with an arbitrary incident acoustic field", J.Acoust.Soc.Am.,vol 113(3)pp. 1216-1222, March 2003.
2. T.M.Acho, "Scalar wave scattering of a prolate spheroid as a parameter expansion of that of a sphere", Quart. Appl. Math, vol.L(3), pp.451-468, Sept.1992.
3. G.S.Sammelmann, D.H.Trivett and R.H.Hackman, "High-frequency scattering from rigid prolate spheroids", J.Acoust.Soc. Am. vol.83 (1), pp 46-54, Jan.1988.
4. P.M.Morse and H.Feshbach, Method of theoretical Physics, McGraw-Hill,New York, 1953. $\theta_0\left(^o\right)$
5. C.Flammer, Spheroidal Wave Functions, Stanford CA, Stanford Univ. Press, 1957.
6. G.C. Kokkorakis and J.A.Roumeliotis, "Acoustic eigenfrequencies in concentric spheroidal-spherical cavities", J.Sound Vibr. Vol.206 (3), pp.287-308, 1997.
7. G.C.Kokkorakis and J.A.Roumeliotis, "Acoustic eigenfrequencies in concentric spheroidal-spherical cavities: calculation by shape perturbation", J.Sound Vibr., vol.212 (2), pp337-355, 1998.
8. J.A.Roumeliotis and J.G.Fikioris "Scattering of plane waves from an eccentrically coated metallic sphere", J.Franklin Inst., vol.312 (1), pp.41-59, July 1981.

SCATTERING BY AN INFINITE CIRCULAR DIELECTRIC CYLINDER COATING ECCENTRICALLY AN ELLIPTIC DIELECTRIC ONE

STYLIANOS P. SAVAIDIS

Department of Electrical Engineering, Technological Institute of Piraeus, 12244 Athens, Greece

JOHN A. ROUMELIOTIS

Department of Electrical and Computer Engineering, National Technical University of Athens, 15773 Athens, Greece

The scattering of a plane electromagnetic wave from an infinite circular dielectric cylinder, coating eccentrically an elliptic dielectric one, is examined. Both E and H polarizations are treated for normal incidence. The electromagnetic field is expressed in terms of both circular and elliptical cylindrical wave functions. Using transformation theorems between the field expressions in different coordinate systems, for the satisfaction of the boundary conditions, we obtain two infinite sets of linear nonhomogeneous equations for the expansion coefficients of the field. For small values of $h=k_2c/2$, where k_2 is the wavenumber of the dielectric coating and c is the interfocal distance of the elliptic cylinder, we conclude with semianalytical expressions of the form $S(h)=S(0)[1+gh^2+O(h^4)]$ for the scattered field and the scattering cross sections. The coefficients g are independent of h while $S(0)$ corresponds to the eccentric circular problem. Numerical results are given for various values of the parameters.

1. Introduction

Scattering from composite bodies is often used for detecting their possible internal structure. On the other hand, construction of such scatterers may also modify the various scattering cross sections, resulting to suitable forms for the scattered field.

The shape of the boundaries in such problems imposes limitations to analytical scattering theories and enforces the use of pure numerical methods. The present paper investigates the potential of analytical solutions in scattering problems involving cylindrical elliptic geometries. The scattering of a plane electromagnetic wave by an infinite circular dielectric cylinder, coating an off-axis dielectric elliptic cylinder, is examined. The geometry of the scatterer, shown in Figure 1., is a perturbation of the eccentric circular one, with radii α and b. All materials are lossless, while both polarizations are considered for normal incidence.

Using translational addition theorems for circular cylindrical wave functions [1], and expansion formulas between circular and elliptical wave functions [2], we obtain, after the satisfaction of the boundary conditions, two infinite sets of linear nonhomogeneous equations for the expansion coefficients of the field inside the elliptic dielectric cylinder.

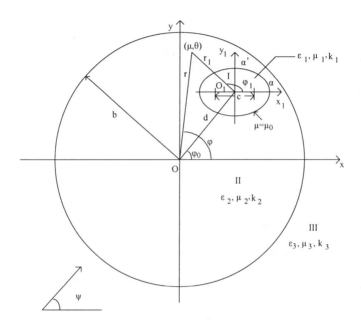

Figure 1. Geometry of the scatterer.

For general values of $h=k_2c/2$ these sets can be solved only numerically, by truncation, while for $h<<1$ a semianalytical solution is possible. In this latter case, we obtain expressions of the form $S(h)=S(0)[1+gh^2+O(h^4)]$ for the scattered field and the scattering cross sections. The coefficients g are independent of h, while $S(0)$ corresponds to the eccentric circular problem. Therefore, our expressions are valid for each small h, while purely numerical techniques [3, 4] involve rather complicated calculations of Mathieu functions for each different h.

2. Theoretical formulation

We start first with E polarization. The incident plane wave, impinging normally on the z-axis, is expressed as [1],

$$E_z^{inc} = exp\left[-jk_3 r \cos(\varphi-\psi)\right] = \sum_{n=0}^{\infty} \varepsilon_n j^{-n} J_n(k_3 r) \cos\left[n(\varphi-\psi)\right] \qquad (1)$$

where r, φ are the polar coordinates with respect to xOy, J_n is the cylindrical Bessel function of the first kind and $\varepsilon_0 = 1$, $\varepsilon_n = 2$ ($n \geq 1$) is the Neuman factor. Angle ψ defines the direction of incidence with respect to x-axis. The time factor $exp(j\omega t)$ is suppressed throughout.

The scattered field has the form

$$E_z^{sc} = \sum_{n=0}^{\infty} H_n(k_3 r)\left[P_n \cos n\varphi + Q_n \sin n\varphi\right] \qquad (2)$$

where $H_n(k_3 r)$ is the Hankel function of the second kind.

The field in region I with respect to $x_1 O_1 y_1$, is

$$E_z^I = \sum_{\ell=0}^{\infty} \left[F_\ell Je_\ell(h_1, \cosh\mu) Se_\ell(h_1, \cos\theta) + G_\ell Jo_\ell(h_1, \cosh\mu) So_\ell(h_1, \cos\theta)\right], h_1 = k_1 \frac{c}{2}$$
$$(3)$$

where μ, θ are the transverse elliptical cylindrical coordinates with respect to $x_1 O_1 y_1$, Je_ℓ (Jo_ℓ) are the even (odd) radial Mathieu functions of the first kind and Se_ℓ (So_ℓ) are the even (odd) angular Mathieu functions [2].

The field expression in region II with respect to $x_1 O_1 y_1$, has the form

$$E_z^{II} = \sum_{\ell=0}^{\infty}\sum_{i=0}^{\infty} \left\{\left[R_{\ell i}^e Je_i(h, \cosh\mu) - S_{\ell i}^e Ne_i(h, \cosh\mu)\right] F_\ell Se_i(h, \cos\theta)\right.$$

$$\left. + \left[R_{\ell i}^o Jo_i(h, \cosh\mu) - S_{\ell i}^o No_i(h, \cosh\mu)\right] G_\ell So_i(h, \cos\theta)\right\} h = k_2 \frac{c}{2} \qquad (4)$$

after the satisfaction of the boundary conditions at the elliptical boundary $\mu = \mu_0$ ($E_z^I = E_z^{II}, \mu_1^{-1} \partial E_z^I / \partial\mu = \mu_2^{-1} \partial E_z^{II} / \partial\mu$). In Eq. (4) Ne_i (No_i) are the even (odd) radial Mathieu functions of the second kind, while

$$R_{\ell i}^{\substack{e\\o}} = \frac{M_{\ell i}^{\substack{e\\o}}(h_1, h)}{M_i^{\substack{e\\o}}(h)} \times$$

$$
\times \frac{J_{e_{o}}_{\ell}(h_1,\cosh\mu o)\left.\dfrac{dN_{e_{o}}_{i}(h,\cosh\mu)}{d\mu}\right|_{\mu=\mu_0} -\dfrac{\mu_2}{\mu_1}N_{e_{o}}_{i}(h,\cosh\mu o)\left.\dfrac{dJ_{e_{o}}_{\ell}(h_1,\cosh\mu)}{d\mu}\right|_{\mu=\mu_0}}{J_{e_{o}}_{i}(h,\cosh\mu o)\left.\dfrac{dN_{e_{o}}_{i}(h,\cosh\mu)}{d\mu}\right|_{\mu=\mu_0} -N_{e_{o}}_{i}(h,\cosh\mu o)\left.\dfrac{dJ_{e_{o}}_{i}(h,\cosh\mu)}{d\mu}\right|_{\mu=\mu_0}} \tag{5}
$$

and $S^o_{\ell i}$ is given by Eq. (5) if we replace $N_{e_{o}}_{i}$ and $dN_{e_{o}}_{i}/d\mu$ with $J_{e_{o}}_{i}$ and $dJ_{e_{o}}_{i}/d\mu$, respectively, in the numerator of the above fraction. Also,

$$
M^o_{\ell i}(h_1,h)=\int_0^{2\pi}S_{e_{o}}_{\ell}(h_1,\cos\theta)S_{e_{o}}_{i}(h,\cos\theta)d\theta=\pi\sum_{v=0}^{\infty}\binom{2/\varepsilon_v}{1}B^o_v(h_1,\ell)B^o_v(h,i) \tag{6}
$$

and $M^o_i(h)=M^o_{ii}(h,h)$, where $B^o_v(h,m)$ are the expansion coefficients of the Mathieu functions [2], with n and m both even or odd.

In order to satisfy the boundary conditions at the circular boundary $r=b$, we first use the expansion formulas connecting elliptical cylindrical wave functions with the axial circular ones with respect to $x_1O_1y_1$ [2], and next the translational addition theorem for the circular cylindrical wave functions [1]. The procedure, through which we obtain an expression for E_z^{II} in terms of circular cylindrical wave functions with respect to xOy, is identical with that in [5]. Substituting the latter expressions into the boundary condition $E_z^{II}=E_z^{III}(=E_z^{inc}+E_z^{sc})$ at $r=b$, we express P_n and Q_n in terms of F_ℓ's and G_ℓ's

$$
\binom{P_n}{Q_n}=\binom{\varepsilon_n/2}{1}\frac{\sqrt{\pi/2}}{H_n(x_4)}\sum_{\ell=0}^{\infty}\sum_{i=0}^{\infty}\sum_{m=0}^{\infty}j^{m-i}\Bigg\{F_\ell B_m^e(h,i)\Big[R^e_{\ell i}J_n(x_2)-S^e_{\ell i}N_n(x_2)\Big]J_{nm}^{c+}\mp
$$

$$
\mp G_\ell B_m^o(h,i)\Big[R^o_{\ell i}J_n(x_2)-S^o_{\ell i}N_n(x_2)\Big]J_{nm}^{c-}\Bigg\}-\binom{\varepsilon_n}{2}j^{-n}J_n(x_4)\frac{\cos}{\sin}n\psi/H_n(x_4),(n\gtreqless^0_1) \tag{7}
$$

In Eq. (7) $x_2=k_2R_2, x_4=k_3R_2, N_n$ is the cylindrical Neumann function, while

$$
J_{nm}^{c\pm}=J_{n-m}(k_2d)\cos(n-m)\varphi_o \pm(-1)^m J_{n+m}(k_2d)\cos(n+m)\varphi_o \tag{8}
$$

and $J_{nm}^{s\pm}$ is given by Eq. (8) if we replace cos with sin.

The combination of Eq. (7) with the boundary condition at $r=b$ ($\mu_2^{-1}\partial E_z^{II}/\partial r=\mu_3^{-1}\partial(E_z^{III})/\partial r$), provides two infinite sets of linear nonhomogeneous equations for the expansion coefficients F_ℓ and G_ℓ:

$$\sum_{\ell=0}^{\infty} a_{n\ell}^{ec} F_\ell + \sum_{\ell=1}^{\infty} b_{n\ell}^{os} G_\ell = d_n^c , (n \geq 0), \quad \sum_{\ell=0}^{\infty} a_{n\ell}^{es} F_\ell + \sum_{\ell=1}^{\infty} b_{n\ell}^{oc} G_\ell = d_n^s , (n \geq 1) \quad (9)$$

In Eq. (9) we have made the substitutions

$$a_{n\ell}^{c} = \sqrt{\frac{\pi}{2}} \sum_{i=0}^{\infty} \sum_{m=0}^{\infty} j^{m-i} B_m^e(h,i) U_{n\ell i}^e J_{nm}^{s+}, \quad b_{n\ell}^{c} = \pm \sqrt{\frac{\pi}{2}} \sum_{i=1}^{\infty} \sum_{m=1}^{\infty} j^{m-i} B_m^o(h,i) U_{n\ell i}^o J_{nm}^{s-} \quad (10)$$

where

$$U_{n\ell i}^{e} = \left[\frac{k_2}{\mu_2} J_n'(x_2) - \frac{k_3}{\mu_3} \frac{H_n'(x_4)}{H_n(x_4)} J_n(x_2) \right] R_{\ell i}^{e} - \left[\frac{k_2}{\mu_2} N_n'(x_2) - \frac{k_3}{\mu_3} \frac{H_n'(x_4)}{H_n(x_4)} N_n(x_2) \right] S_{\ell i}^{e} \quad (11)$$

and

$$d_n^{c} = 2 \frac{k_3}{\mu_3} j^{-n} \left[J_n'(x_4) - \frac{H_n'(x_4)}{H_n(x_4)} J_n(x_4) \right] \frac{\cos}{\sin} n\psi \quad (12)$$

with ℓ, i, m all even or odd, and the prime denoting derivative with respect to the argument.

For general values of h ($h_1 = hk_1/k_2$) the sets of Eq. (9) can be solved only numerically. This approach leads to rather complex calculations, due to the evaluation of the Mathieu functions for each different h. However, for small h ($\ll 1$) a semianalytical solution is possible. After lengthy but straightforward calculations, we obtain expansions of the form

$$a_{n\ell}^{es}(h) = C_{n\ell}^{es} + C_{n\ell}^{(2)es} h^2 + O(h^4), \quad b_{n\ell}^{os}(h) = T_{n\ell}^{os} + T_{n\ell}^{(2)os} h^2 + O(h^4) \quad (13)$$

The evaluation of F_ℓ's and G_ℓ's is possible, first by truncating the infinite sets of Eq. (9) and then by using Cramer's rule and the expansions of Eq. (13). The set structure and the calculation procedure are identical with those described in [6].

By using the asymptotic expansion for the Hankel function in Eq. (2), we obtain the scattered far-field expression and next the differential [σ(φ)], backscattering or radar (σ$_b$), forward (σ$_f$), and total scattering (Q$_t$) cross sections[7]:

$$\sigma(\varphi) = 2|G(\varphi)|^2 /(\pi k_3), \sigma_b = 4|G(\psi+\pi)|^2 / k_3, \sigma_f = 4|G(\psi)|^2 / k_3 \quad (14)$$

$$Q_t = \left[4|P_0|^2 + 2\sum_{n=1}^{\infty} (|P_n|^2 + |Q_n|^2) \right] \Big/ k_3 \quad (15)$$

where

$$G(\varphi)=\sum_{n=0}^{\infty} j^n \left[P_n \cos n\varphi + Q_n \sin n\varphi\right] \tag{16}$$

Substituting in Eq. (7) the expansions of the various quantities for small h, P_n and Q_n can be set in the form

$$P_n(h)=P_n^0 + P_n^{(2)} h^2 + O(h^4), \quad Q_n(h)=Q_n^0 + Q_n^{(2)} h^2 + O(h^4) \tag{17}$$

Similar expansions are obtained for $G(\varphi)$, $\sigma(\varphi)$, and Q_t. So, $G^0(\varphi)$ and $G^{(2)}(\varphi)$ are given by (16), with upperscripts 0 and (2), respectively, in P_n and Q_n, $\sigma^0(\varphi)$ and Q_t^0 are given by Eq. (14) and Eq. (15), with upperscripts 0 in $G(\varphi)$ and in P_0, P_n, Q_n, respectively, while $\sigma^{(2)}(\varphi)$ and $Q_t^{(2)}$ are

$$\sigma^{(2)}(\varphi)=4 Re\left[G^0(\varphi)\overline{G^{(2)}(\varphi)}\right]\Big/(\pi k_3) \tag{18}$$

$$Q_t^{(2)} = Re\left[8 P_0^0 \overline{P_0^{(2)}} + 4\sum_{n=1}^{\infty}(P_n^0 \overline{P_n^{(2)}} + Q_n^0 \overline{Q_n^{(2)}})\right]\Big/k_3 \tag{19}$$

with Re the real part and (—) the conjugate quantity.

We write $\sigma(\varphi)$ and Q_t as

$$\sigma(\varphi)=\sigma^0(\varphi)\left[1+g_\sigma(\varphi)h^2 + O(h^4)\right], Q_t=Q_t^0\left[1+g_{Q_t}h^2 + O(h^4)\right] \tag{20}$$

where $\sigma^0(\varphi)$ and Q_t^0 correspond to the eccentric circular geometry and

$$g_\sigma(\varphi)=\sigma^{(2)}(\varphi)\Big/\sigma^0(\varphi), g_{Q_t}=Q_t^{(2)}\Big/Q_t^0 \tag{21}$$

are independent of h.

For the H wave the procedure and the steps are identical with the above mentioned ones. Actually, the field expressions, the boundary conditions and all the resulting formulas are still valid if we replace E, μ_1, μ_2, μ_3 with H, ε_1, ε_2, ε_3, respectively.

3. Numerical results and discussion

In Figures 2 -4 the g's for the scattering cross sections of the configuration of Figure 1. are plotted for both polarizations. In each figure, we plot also the corresponding scattering cross section for $h=0$ (eccentric circular geometry), which depends only on $\psi-\varphi_0$ and not on their distinct values, as for $h\neq 0$. In all figures $\mu_1=\mu_2=\mu_3$, while $\lambda_3=2\pi/k_3$.

For $g>0$ (<0), $S(h)$ is increased (decreased) with respect to $S(0)$. Therefore, by making the inner cylinder elliptic, one can increase or decrease the scattering cross sections, as compared to the corresponding circular inner cylinder. Inversely, this may be useful for detecting a shape perturbation of the inner cylinder.

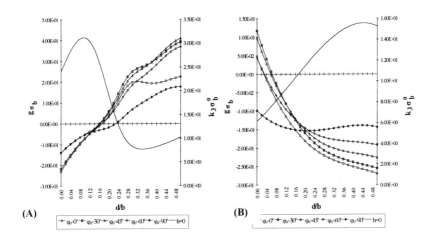

Figure 2. Backscattering cross section for (A) α/b=0.5, b/λ_3=0.7, $\psi-\varphi_0$=45°, $\varepsilon_1/\varepsilon_3$=5.5, $\varepsilon_2/\varepsilon_3$=2.54 (E wave) (B) α/b=0.5, b/λ_3=0.5, $\psi-\varphi_0$=60°, $\varepsilon_1/\varepsilon_3$=5.5, $\varepsilon_2/\varepsilon_3$=2.54 (H wave).

136

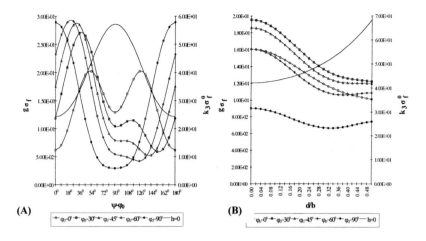

Figure 3. Forward scattering cross section for (A) $a/b=0.5$, $d/b=0.3$, $b/\lambda_3=0.5$, $\varepsilon_1/\varepsilon_3=5.5$, $\varepsilon_2/\varepsilon_3=2.54$ (E wave) (B) $a/b=0.5$, $b/\lambda_3=0.5$, $\psi-\varphi_0=60^0$, $\varepsilon_1/\varepsilon_3=5.5$, $\varepsilon_2/\varepsilon_3=2.54$ (H wave).

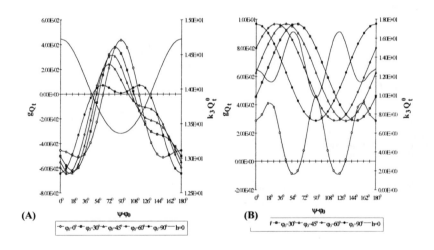

Figure 4. Total scattering cross section for (A) $a/b=0.5$, $d/b=0.4$, $b/\lambda_3=0.3$, $\varepsilon_1/\varepsilon_3=5.5$, $\varepsilon_2/\varepsilon_3=2.54$ (E wave) (B) $a/b=0.5$, $d/b=0.3$, $b/\lambda_3=0.5$, $\varepsilon_1/\varepsilon_3=5.5$, $\varepsilon_2/\varepsilon_3=2.54$ (H wave).

References

1. J.A. Stratton, Electromagnetic Theory, New York, McGraw-Hill (1941).
2. P.M. Morse and H. Feshbach, Methods of Theoretical Physics, New York, McGraw-Hill (1953).
3. H.A. Ragheb, L. Shafai and M. Hamid, Plane wave scattering by a conducting elliptic cylinder coated by a nonconfocal dielectric, IEEE Trans. Antennas Propagat. **39**, 218 (1991).
4. A. Sebak, H.A. Ragheb and L. Shafai, Plane wave scattering by dielectric elliptic cylinder coated with nonconfocal dielectric, Radio Sci. **29**, 1393 (1994).
5. J. A. Roumeliotis and S.P. Savaidis, Scattering by an infinite circular dielectric cylinder coating eccentrically an elliptic metallic one, IEEE Trans. Antennas Propagat. **44**, 757 (1996).
6. J. A. Roumeliotis and S.P. Savaidis, Cutoff frequencies of eccentric circular-elliptic metallic waveguides, IEEE Trans. Microwave Theory Tech. **42**, 2128 (1994).
7. N. Kakogiannos and J.A. Roumeliotis, Electromagnetic scattering from an infinite elliptic metallic cylinder coated by a circular dielectric one, IEEE Trans. Microwave Theory Tech. **38**, 1660 (1990).

SOLUTION OF THE INVERSE SCATTERING PROBLEM USING REFLECTION TRAVEL TIME DATA

S. A. SOFIANOS, M. BRAUN, AND B. R. MABUZA *

Physics Department, University of South Africa, Pretoria 0003, South Africa

Aspects of the inverse scattering problem and the transformation of the wave equation for elastic displacement into a Marchenko integral equation are discussed. The latter equation provides the required connection with the scattering data and from its solution the potential ('profile') that caused the scattering can be reconstructed. As an application of our method, we use reflection travel time data for synthetic seismic experiments in layered substrate models to recover the characteristic impedances of the medium.

1. Introduction

The solution of the quantum mechanical inverse scattering problem on the line has been the subject of several investigations in the past [1]. The solution of the problem was based on the Marchenko integral equation requiring as input the reflection coefficient and therefore the phase problem [2] always was a major obstacle in using inverse scattering in practical applications.

In the present work we are concerned with the classical inverse scattering problem and the reconstruction of the characteristic impedance of the scatterer using reflection travel time data. Similarly to the quantum inversion, this can be achieved by transforming the classical wave equation into an integral equation (of Marchenko form), the solution of which can provide us with the information on the scatterer.

We apply the method to the seismic inverse scattering problem where seismic waves are reflected by interfaces of the medium and recorded by receivers. Assuming that the waveform remains unchanged during transmission through the medium (this is only valid for a limited range of frequencies) [3], then from the solution of the integral equation and the use

---

*Ppermanent address: Faculty of Applied and Computer Sciences, Vaal University of Technology, Vanderbijlpark, South Africa

of reflection travel time data, we can obtain the structural details of the (reflective) layers of the medium.

2. The inverse scattering method

2.1. *The Marchenko Integral Equation*

We shall outline here the derivation of the Marchenko integral equation for classical inversion. A more rigorous mathematical derivation of this equation can be found, for example, in Ref. [4] The time–dependent wave equation for the classical amplitude $u(x, y)$ is given by [5]

$$\frac{\partial^2 u(x, y)}{\partial x^2} - \frac{\partial^2 u(x, y)}{\partial y^2} = V(x)\, u(x, y) \tag{1}$$

where $x = v\tau$, $y = vt$, v being the velocity of the wave, τ and t represent times, and $V(x)$ is the scattering potential. We may rewrite this equation as a one-dimensional Schrödinger-like equation by assuming $u(x, y) = e^{-iky}\, \psi(k, x)$, where $k^2 = E$. Then we obtain

$$\frac{\partial^2 \psi(k, x)}{\partial x^2} + (k^2 - V(x))\psi(k, x) = 0\,, \tag{2}$$

the boundary conditions being

$$\psi(k, x) = \begin{cases} e^{ikx} + R(k)\, e^{-ikx} & \text{for } x \to -\infty \\ T(k)\, e^{ikx} & \text{for } x \to +\infty \end{cases} \tag{3}$$

Here, $R(k)$ and $T(k)$ are the reflection and transmission coefficients, respectively, for an incident signal from left.

For a δ–function pulse,

$$\delta(x - y) = \frac{1}{2\pi} \int_{-\infty}^{+\infty} e^{-ik(x-y)}\, dk\,, \tag{4}$$

impinging from the left and reaching the origin at $t = 0$ one may construct a solution of Eq. (1) by a linear superposition

$$u(x, y) = \frac{1}{2\pi} \int_{-\infty}^{+\infty} \psi(k, x) e^{-iky}\, dk\,. \tag{5}$$

We assume that $V(x) = 0$ for $x < 0$, and that reflection under such boundary conditions will only affect the solution of the wave equation for the region $y > |x|$. One then can show that for $x < 0$ the solution of Eq. (1) is given by

$$u_0(x, y) = \delta(x - y) + B(x + y) \tag{6}$$

where $B(x+y)$ is the *reflected wave* which can be expressed in terms of the reflection coefficient $R(k)$,

$$B(x+y) = \frac{1}{2\pi} \int_{-\infty}^{+\infty} e^{-ik(x+y)} R(k) \, dk. \tag{7}$$

Due to causality, $B(x+y)$ vanishes for $x+y < 0$ and we have thus obtained a solution for $x < 0$ consistent with our boundary conditions. We assume that the wave amplitude $u(x,y)$ in the region $x > 0$ can be expressed in terms of the amplitude $u_0(x,y)$ via a linear transformation of the form

$$u(x,y) = u_0(x,y) + \int_{-\infty}^{+x} K(x,y') u_0(y',y) dy'. \tag{8}$$

Substituting in the latter equation the expression for $u_0(x,y)$, defined by (6), and using the fact that $u(x,y) = 0$ for $y < x$, we obtain the integral equation

$$K(x,y) + B(x+y) + \int_{-x}^{+x} B(y+y') \, K(x,y') \, dy' = 0, \tag{9}$$

with $-x \le y \le x$ and $K(x,y) = 0$ for $x < |y|$. Eq. (9) is known as the Marchenko integral equation. and has to be solved for the function $K(x,y)$ on the triangular region, $-x \le y \le x$, in the xy-plane. The potential $V(x)$ is then given by

$$V(x) = 2\frac{dK(x,x)}{dx}. \tag{10}$$

Thus, given the reflection coefficient $R(k)$ one may evaluate $B(x + y)$ to obtain $K(x,y)$ and thereby reconstruct the details of the scatterer.

2.2. *The One-dimensional Seismic Wave Equation*

The one-dimensional seismic wave equation for the elastic displacement ϕ is given by [6]

$$\rho\frac{\partial^2 \phi}{\partial t^2} = \frac{\partial}{\partial z}\left(\rho c^2 \frac{\partial \phi}{\partial z}\right), \tag{11}$$

where t is the time, z the space coordinate along the direction of propagation, $\rho = \rho(z)$ is density of the medium, and $c = c(z)$ is the speed of the seismic wave. We are considering here a longitudinal displacement in the z-direction. We may define the travel time τ for a pulse to move from the origin to position z by

$$\tau(z) = \int_0^z \frac{1}{c(z')} \, dz', \tag{12}$$

from which $\partial\tau(z)/\partial z = 1/c(z)$. Using this relation (we shall, from now on, suppress the explicit arguments of the various functions when they are obvious, for convenience) we may rewrite the wave equation (11) as

$$\eta\frac{\partial^2\phi}{\partial t^2} = \frac{\partial}{\partial\tau}\left(\eta\frac{\partial\phi}{\partial\tau}\right). \tag{13}$$

where $\eta(\tau) \equiv \rho c$ is the characteristic impedance of the medium [7]. Letting $U = \eta^{1/2}\phi$ we obtain

$$\frac{\partial^2 U}{\partial\tau^2} - \frac{\partial^2 U}{\partial t^2} = V U \tag{14}$$

where V is given by

$$V(\tau) = \frac{1}{\eta^{\frac{1}{2}}}\frac{\partial^2\eta^{\frac{1}{2}}}{\partial\tau^2}. \tag{15}$$

Eq. (14) has the same form as (1) and thus it can be transformed into the Marchenko equation (9) (in t, τ) from the solution of which we can obtain the potential through Eq. (10).

The impedance η can be calculated from the potential $V(\tau)$ or directly from the relation [8]

$$\eta^{\frac{1}{2}}(\tau) = \eta^{\frac{1}{2}}(0)\left[1 + \int_{-\tau}^{+\tau} K(\tau,\tau')\,d\tau'\right]. \tag{16}$$

In other words, given the $\eta(0)$, the characteristic impedance $\eta(\tau)$ for $\tau > 0$ can be recovered from the knowledge of the kernel $K(\tau, t)$.

2.3. Reflections in a multi-layered Medium

The medium usually consists of N layers and therefore one may have multiple reflections. In Fig. 1 a three-layer model is shown in which the third layer is assumed to be extended all the way to $z = \infty$ i.e, no reflections are generated beyond the point $z = d_2$. It is further assumed that no reflections are created at $z = 0$, or alternatively, that they have been removed from the signal. The waves in the regions 1, 2 and 3 are given by [8]

$$\psi_1 = A_1\left(e^{ik_1 z} + \tilde{R}_{12}\,e^{ik_1(2d_1 - z)}\right), \tag{17}$$

$$\psi_2 = A_2(e^{ik_2 z} + R_{23}\,e^{ik_2(2d_2 - z)}), \tag{18}$$

$$\psi_3 = A_3\,e^{ik_3 z}. \tag{19}$$

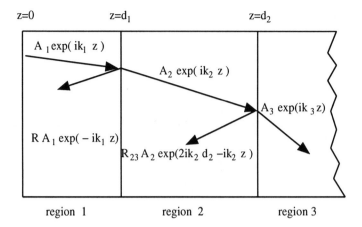

Figure 1. Multiple transmissions and reflections in a three-layered medium.

The $\tilde{R}_{12}$ is the generalised reflection coefficient due to reflections on interfaces 2 and 3. Matching the various waves as usual one obtains

$$\tilde{R}_{12} = R_{12} + \frac{T_{12}R_{23}T_{21}\,e^{2ik_2(d_2-d_1)}}{1 - R_{21}R_{23}\,e^{2ik_2(d_2-d_1)}} \tag{20}$$

which can be expanded in a geometric series,

$$\tilde{R}_{12} = R_{12} + T_{12}R_{23}T_{21}\,e^{2ik_2(d_2-d_1)} + T_{12}R_{23}^2 R_{21}T_{21}\,e^{4ik_2(d_2-d_1)} + \cdots ,$$

in which the n−th term represents the n−th reflection.

In the Marchenko formalism we need to work with the overall reflection coefficient $R(k)$ for the whole half-space which is defined via

$$\psi_1 = A_1\left(e^{ik_1 z} + R(k)\,e^{-ik_1 z}\right) . \tag{21}$$

Thus, by comparison with Eq. (17) we obtain

$$R(k) = e^{2ik_1 d_1}\left(R_{12} + T_{12}R_{23}T_{21}\,e^{2ik_2(d_2-d_1)} \right.$$
$$\left. + T_{12}R_{23}^2 R_{21}T_{21}\,e^{4ik_2(d_2-d_1)} + \cdots\right) . \tag{22}$$

In the travel time coordinate τ, the interfaces are located at

$$\tau_1 = \frac{d_1}{c_1}, \qquad \tau_2 = \frac{d_1}{c_1} + \frac{d_2-d_1}{c_2} = \frac{d_1}{c_1} + \frac{k_2}{k_1}\frac{d_2-d_1}{c_1} \tag{23}$$

where c_i is the speed of the wave in the i-th layer. Since $k_1 z = k_1 c_1 z/c_1$ the reflection coefficient is written as

$$R(k) = R_{12}e^{2ik\tau_1} + T_{12}R_{23}T_{21}\,e^{2ik\tau_2} + T_{12}R_{23}^2 R_{21}T_{21}\,e^{4ik\tau_2 - 2ik\tau_1}$$
$$+ T_{12}R_{23}^3 R_{21}^2 T_{21}\,e^{6ik\tau_2 - 4ik\tau_1} + T_{12}R_{23}^4 R_{21}^3 T_{21}\,e^{8ik\tau_2 - 6ik\tau_1} . \tag{24}$$

In seismic inversion it is customary to assume that the various reflection coefficients R_{ij} are constant. Thus by taking the Fourier transform of Eq. (24), we obtain, for a δ-function impulse,

$$B(\tau) = B_1 + B_2 + B_3 + \cdots \tag{25}$$

where $B_1(\tau) = R_{12}\delta(\tau - 2\tau_1)$, $B_2(\tau) = T_{12}R_{23}T_{21}\delta(\tau - 2\tau_2)$, $B_3(\tau) = T_{12}R_{23}^2 R_{21}T_{21}\delta(\tau - (4\tau_2 - 2\tau_1))$ etc.

2.4. A Simulated Reflected Signal

A synthetic reflected signal generated by a seismic source placed on the surface of a layer can be simulated by assuming that the reflective layer is situated at travel time τ_1. Then an ideal reflected pulse would take the form $R_{12}\delta(\tau - 2\tau_1)$. Since, however, the reflective interfaces are not sharply defined, $B(\tau)$ is stretched over a time interval $\Delta\tau$ and it can be modeled by the Gaussian form

$$B(\tau) = R_{12}\beta\, e^{-(\tau - 2\tau_1)^2/b^2}, \tag{26}$$

where b is the width of the Gaussian (of the order $c\Delta\tau$) and $\beta = b/\sqrt{\pi}$ is the normalization constant. As can be easily shown by inverting Eq. (7) and identifying the factor $\exp(2ik\tau_1)$ with the one already present in Eq. (24), this signal corresponds to k-dependent reflection coefficient given by

$$\bar{R}_{12}(k) = R_{12}\, e^{-b^2 k^2/4}, \tag{27}$$

where R_{12} is given in terms of the impedances of the medium,

$$R_{12} = \frac{\eta_1 - \eta_2}{\eta_1 + \eta_2}. \tag{28}$$

The corresponding transformation for the k-dependent classical transmission coefficient is given by $\bar{T}_{12}(k) = 1 + \bar{R}_{12}(k)$.

If more than two layers are present the subsequent multiple scattering contributions of the series Eq. (24) contain a product of reflection coefficients describing the multiple scattering each with a Gaussian dependence. For example, the third term of the series for $R(k)$ will be given by

$$R_3(k) = \left(1 - R_{12}^2 e^{-2b^2\frac{k^2}{4}}\right) R_{23}^2 R_{21} e^{-3b^2 k^2/4} e^{4ik\tau_2 - 2ik\tau_1}. \tag{29}$$

Thus after transformation to the travel time domain, two Gaussians of width $3b$ and $5b$ are obtained yielding

$$\begin{aligned}
B_{\text{Gauss 3}} = {} & \frac{1}{\sqrt{3\pi}b} R_{23}^2 R_{21}\, e^{-(\tau - (4\tau_2 - 2\tau_1))^2/(3b^2)} \\
& - \frac{1}{\sqrt{5\pi}b} R_{12}^2 R_{23}^2 R_{21}\, e^{-(\tau - (4\tau_2 - 2\tau_1))^2/(5b^2)}.
\end{aligned} \tag{30}$$

3. Implementation of the method

Since we have no experimental data at our disposal, we demonstrated the reliability of the proposed method by using model seismic experiments. As a first example we consider a single reflection in a two-layered medium with impedances $\eta_1 = 1$ and $\eta_2 = 1.5$. The reflected signal is assumed to be of Gaussian form with $\tau_1 = 3$ and is the result of one bounce at the (first) interface. The $B(\tau)$ is given by Eq. (26) and the R_{12} by Eq. (28). The b in this case was chosen as 0.5.

The signal used and the recovered impedance are shown in Fig. 2. It is seen that the reproduction of the impedance of the reflective medium is excellent. Well outside the transition region, the values of η_1 and η_2 are recovered to about 4 decimals.

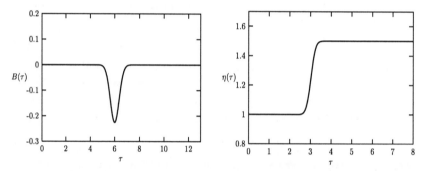

Figure 2. The reflected signal in a two-layered medium (left) and the recovered impedance (right). The input impedances are $\eta_1 = 1$ and $\eta_2 = 1.5$.

The second example considered is the one with a three-layered medium with interfaces at travel times $\tau_1 = 3$ and $\tau_2 = 5$, resulting in the reflected Gaussians being centered on $t_1 = 2\tau_1 = 6$, $t_2 = 2\tau_2 = 10$, $t_3 = 4\tau_2 - 2\tau_1 = 14$, $t_4 = 6\tau_2 - 4\tau_1 = 18$, and $t_5 = 8\tau_2 - 6\tau_1 = 22$. The impedances of the layers are assumed to be $\eta_1 = 1$, $\eta_2 = 1.5$, and $\eta_3 = 0.9$. The reflected signal together with the recovered impedance. are shown in Fig. 3. The recovery of the impedances is, once more, excellent.

4. Conclusions

Our conclusions can be short. We, firstly, presented a method to obtain the characteristic impedances of a multi-layered medium. The method is based on the one-dimensional Marchenko integral equation and the use of

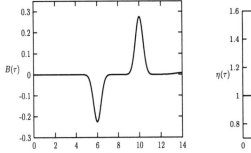
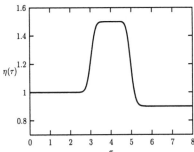

Figure 3. The reflected signal in a three-layered medium (left) and the recovered impedance (right). The input impedances are $\eta_1 = 1$, $\eta_2 = 1.5$, and $\eta_3 = 0.9$.

reflection travel-time data. We, secondly, successfully applied the proposed method to model seismic cases, namely, to a two-layer medium with a single reflection at the interface and to a three-layered medium in which multiple reflections can occur. The reconstruction of the impedances in both cases is excellent which implies that the proposed procedure is reliable and promising in field applications.

References

1. R. Lipperheide, G. Reiss, H. Leeb, H. Fiedeldey , and S.A. Sofianos, Phys. Rev. B, **51**, 11032 (1995) and references therein.
2. J.M. Cowley, *Diffraction Physics*, North Holland, Amsterdam (1975).
3. M.T. Silvia and E.A. Robinson, *Deconvolution of Geophysical Time Series in the Exploration for oil and natural gas*, Elsevier Scientific Publishing Company, Amsterdam (1979).
4. Z.S. Agranovich and V.A. Marchenko, *The Inverse Problem of Scattering Theory*, Gordon & Breach, New York (1963).
5. S. Nettel, *Wave Physics, Oscillations - Solitons - Chaos*, Springer-Verlag, Berlin, Heidelberg (1992).
6. R. Burridge, Wave Motion 2, 305 (1980).
7. K.P Bube and R. Burridge, *The one-dimensional inverse problem of reflection seismology*, Society for industrial and Applied Mathematics, 497 (1983).
8. W.C. Chew, *Waves and Fields in Inhomogeneous Media*, Van Nostrand Reinhold, New York (1990).

SCATTERING OF PLANE ACOUSTIC WAVES BY MOVING SOFT SPHERE IN FREE SPACE

GEORGE VENKOV

Technical University of Sofia, Faculty of Applied Mathematics and Informatics,

8, 'St. Kliment Ohridski' Str., 1000 Sofia,Sofia, Bulgaria

YANI ARNAOUDOV

Faculty of Applied Mathematics and Informatics,

8, 'St. Kliment Ohridski' Str., 1000 Sofia,Sofia, Bulgaria

We investigate the propagation of a plane acoustic wave in the exterior of a moving obstacle. Under the assumption that the obstacle moves with uniform velocity and more slowly than the speed of sound, we apply the Lorentz transformation. In the object's frame, where the scatterer is stationary, we introduce the low-frequency approximation technique. Moreover, we obtain a formula, which represent the effect of Lorentz transformation on the scattering amplitude. Finally, the results are applied to the recovering of the scattering cross-section, which plays an important role in solving numerically the inverse scattering problems.

Introduction

Scattering of acoustic waves from a moving obstacle has been investigated to some extend in the literature [1], although it is by far less studied than the scattering problem for electromagnetic waves. In this paper is considered the problem of scattering of a plane wave by an acoustically soft sphere in uniform motion, relative to the observation point. To treat this problem we present a rather general technique, using a relativistic notation. Since the scatterer is moving with uniform velocity we introduce an analogue of the space-time Lorentz transformation restricted to the case of acoustic propagation [1].

Immediately arises the question: Is the application of the Lorentz transformation correct when the speed of light is replaced by the speed of sound? From the mathematical point of view, the Lorentz transformation is an imaginary rotation in the pseudoeuclidean Minkowski space and transforms the space-time components from the observer's frame K to the object's frame K'. As a linear transformation its application is allowable, which follows from the theory of the continuous group of transformations (see Eisenhart [8]). In the modern electrodynamics and in the theory of relativity the application of this acoustical analogue of the Lorentz transformation is a matter of philosophical discussions. The main reason is that it contradicts with the Einstein principle of the constant light velocity. On the other hand, the acoustical Lorentz transformation appears in many works [1], [9] and the derived theoretical and numerical conclusions are taken for sure-footed.

Using the fact that d'Alembertian operator is a Lorentz invariant and that in the frame K' the scatterer remains stationary, suppressing the harmonic time-dependence we arrive to the scalar Helmholtz equation. Unfortunately, the Lorentz transformation perturbs the geometry of the obstacle, as well as the form of the incident field with factors depending on the velocity and the direction of motion. Under the assumption that in the object's frame K' the perturbed characteristic dimension of the obstacle is much smaller than the new wavelength of the incident field we apply the low-frequency approximation technique [3]. Then the first two approximations for the near, as well as for the far field are evaluated in the object's frame. The relativistic result stated in Section 3 is used for obtaining the above measurements in the initial space-time domain. Finally, as the velocity of motion tends to zero, the relative results recover all the known expressions for low-frequency scattering of acoustic wave by a stationary soft sphere [3].

1.   Lorentz transformation and the acoustic wave incidence

Let in the observer's frame K the scatterer $S_a(t)$ be an acoustically soft sphere of radius a moving in free space at a uniform velocity $\mathbf{v} = V\hat{\mathbf{n}}$ in direction $\hat{\mathbf{n}}$. Let the location of the center of the sphere $\mathbf{r}_0(t)$ be chosen at $x = y = z = 0$ when $t = 0$ and the surface of the scatterer is given by

$$S_a(t) = \left\{ \mathbf{r} \in \mathbf{R}^3 ; \left| \mathbf{r} - \mathbf{r}_0(t) \right| = a \right\} \quad \mathbf{r}_0(t) = V\hat{\mathbf{n}}t \tag{1.1}$$

Let assume that the sphere disturbs the incident plane acoustic field u^i that emanates in direction $\hat{\mathbf{d}}$, expressed by

$$u^i(\mathbf{r}, t) = e^{ik_0(\mathbf{r} \cdot \hat{\mathbf{d}}) - i\omega_0 t} \tag{1.2}$$

where ω_0 is the wave frequency, $k_0 = \omega_0/c$, $c = speed\ of\ sound$ in free space. Then the scattering problem is formulated as follows. Find the total field $u = u^i + u^s$, which solves the wave equation

$$\frac{\partial_{tt}}{c^2} u(\mathbf{r}, t) - \Delta u(\mathbf{r}, t) = 0 \tag{1.3}$$

and satisfies the Dirichlet condition

$$u = 0, \quad \mathbf{r} \in S_a(t)$$

(1.4)

on the boundary of the obstacle.

If we assume that $V < c$ we may define an analogue of the Lorentz transformation, related to the sound propagation in free space [1].

Let us represent the position vector $\mathbf{r}$ as a sum $\mathbf{r}_\perp + \mathbf{r}_\parallel$, where $\mathbf{r}_\perp, \mathbf{r}_\parallel$ denote the components of $\mathbf{r}$ that are respectively orthogonal to, and parallel to, the direction of motion $\hat{\mathbf{n}}$. Then we define the four-vectors transformation in the Minkowski space $\mathbf{R}^{(3+1)}$ as follows

$$\mathbf{r'}_\perp = \mathbf{r}_\perp, \quad \mathbf{r'}_\parallel = \frac{(\mathbf{r} \cdot \hat{\mathbf{n}}) - Vt}{\sqrt{1-\beta^2}} \hat{\mathbf{n}}, \quad t' = \frac{t}{\sqrt{1-\beta^2}} - \frac{V(\mathbf{r} \cdot \hat{\mathbf{n}})}{c^2\sqrt{1-\beta^2}},$$

(1.5)

where $\beta = V/c$. From (1.5) we may derive the inverse transformation between the two reference frames K and K' as

$$\mathbf{r} = \mathbf{r'}_\perp + \frac{|\mathbf{r'}_\parallel| + Vt'}{\sqrt{1-\beta^2}} \hat{\mathbf{n}}, \quad t = \frac{t'}{\sqrt{1-\beta^2}} + \frac{V|\mathbf{r'}_\parallel|}{c^2\sqrt{1-\beta^2}}.$$

(1.6)

Using the fact that d'Alembertian operator is a Lorentz invariant, the scattering problem $(1.3) - (1.4)$ in the frame K' becomes

$$\frac{\partial_{t't'}}{c^2} u(\mathbf{r'}, t') - \Delta u(\mathbf{r'}, t') = 0$$

(1.7)

$$u = 0, \quad \mathbf{r'} \in \widetilde{S}_a$$

(1.8)

Substituting (1.6) into (1.2), for the incident field we obtain

$$u^i(\mathbf{r'}, t') = \exp\left\{ ik_0 \left[(\mathbf{r'}_\perp \cdot \hat{\mathbf{d}}) + \mathbf{r'}_\parallel \frac{(\hat{\mathbf{n}} \cdot \hat{\mathbf{d}}) - \beta}{\sqrt{1-\beta^2}} \right] - i\omega t' \right\}$$

(1.9)

where the new wave frequency is $\omega = \omega_0 (1 - \beta(\hat{\mathbf{n}} \cdot \hat{\mathbf{d}}))/\sqrt{1-\beta^2}$. In the frame K' the moving sphere $S_a(t)$ becomes a stationary prolate spheroid $\widetilde{S}_a = \left\{ \mathbf{r'} \in \mathbf{R}^3; \left| \mathbf{r'}_\perp + \sqrt{1-\beta^2} \mathbf{r'}_\parallel \right| = a \right\}$ with its major axis along the direction of motion $\hat{\mathbf{n}}$.

Since we shall investigate the scattering problem in the object's frame K', to simplify the notation we shall denote the space-time variables as not primed $(\mathbf{r}, t)$ hereinafter. Suppressing the harmonic time-dependence $e^{-i\omega t}$ in (1.9) we obtain the stationary acoustic scattering problem for the space dependent part. The total acoustic field $u = u^i + u^s$ solves the Helmholtz equation

$$\Delta u(\mathbf{r}) + k^2 u(\mathbf{r}) = 0, \quad k = \omega/c$$

(1.10)

and satisfies the Dirichlet boundary condition

$$u(\mathbf{r}) = 0, \quad \mathbf{r} \in \widetilde{S}_a$$

(1.11)

while the scattered field u^s satisfies the Sommerfield radiation condition

$$\lim_{r \to \infty} r \left(\frac{\partial}{\partial r} u^s(\mathbf{r}) - iku^s(\mathbf{r}) \right) = 0$$

(1.12)

The solution u of the above scattering problem enjoys the integral representation

$$u(\mathbf{r}) = u^i(\mathbf{r}) - \frac{1}{4\pi} \int_{\widetilde{S}_a} \Phi(\mathbf{r}, \mathbf{r}') \frac{\partial}{\partial \mathbf{v}'} u(\mathbf{r}') ds(\mathbf{r}')$$

(1.13)

where Φ is the fundamental solution of the Helmholtz operator and $\mathbf{v}$ is the outward unit normal to the obstacle surface.

2. Low-frequency approximations and the prolate spheroidal coordinates

When the characteristic dimension of the scatterer is much smaller than the wavelength of the incident wave, all the fields involved are analytic function of the wavenumber [3]. In the frame K' the obstacle $\widetilde{S}_a$ is a prolate spheroid with dimensional characteristic equals to its major radius $a/\sqrt{1-\beta^2}$. In order to apply the low-frequency approximation technique we shall assume that $k\,a/\sqrt{1-\beta^2} \ll 1$. Thus, the total field agrees the low-frequency expansion with respect to the new wave number k

$$u(\mathbf{r}) = \sum_{n=0}^{\infty} \frac{(ik)^n}{n!} u_n(\mathbf{r})$$

(2.1)

where the approximation coefficients u_n solve the equation

$$\Delta u_n(\mathbf{r}) = n(n-1)u_{n-2}(\mathbf{r}),$$

(2.2)

and the Dirichlet boundary condition

$$u_n(\mathbf{r}) = 0, \quad \mathbf{r} \in \widetilde{S}_a$$

(2.3)

for every $n = 0, 1, 2, \ldots$. The approximation coefficients of the scattered field u_n^s enjoy the integral representation formula

$$u_n^s(\mathbf{r}) = -\frac{1}{4\pi} \sum_{l=1}^{n} \binom{n}{l} \int_{S_a} |\mathbf{r} - \mathbf{r'}|^{l-1} \frac{\partial u_{n-l}(\mathbf{r'})}{\partial v'} ds(\mathbf{r'}) + O\left(\frac{1}{r}\right)$$

(2.4)

Without loss of generality, we assume that the velocity direction $\hat{\mathbf{n}}$ coincides with $\hat{\mathbf{z}}$. Thus, in Cartesian coordinates the prolate spheroid $\widetilde{S}_a$ is given by the equation $\widetilde{S}_a = \{x^2 + y^2 + (1-\beta^2)z^2 = a^2\}$ with interfocal distance $c = 2a\beta/\sqrt{1-\beta^2}$ and foci at the points $z_{1/2} = \pm c/2$.

The prolate spheroidal system (ρ, θ, φ) is an orthogonal system [4], which is connected to the cartesian system (x, y, z) through the equations

$$x = \frac{c}{2}\sinh\rho\sin\theta\cos\varphi, \quad y = \frac{c}{2}\sinh\rho\sin\theta\sin\varphi, \quad z = \frac{c}{2}\cosh\rho\cos\theta,$$

(2.5)

for $\rho \in [0, +\infty), \theta \in [0, \pi], \varphi \in [0, 2\pi)$. If we demand that the spheroid $\widetilde{S}_a$ corresponds to the value $\rho = \rho_0$, than we need to specify a positive number ρ_0 in such a way as to satisfy the conditions

$$\frac{c}{2}\cosh\rho_0 = \frac{a}{\sqrt{1-\beta^2}}, \quad \frac{c}{2}\sinh\rho_0 = a$$

(2.6)

which yields $\cosh\rho_0 = 1/\beta$. The region where the scattered wave propagates corresponds to the domain $\widetilde{S}_a^{ext} = \{(\rho, \theta, \varphi); \rho \in (\rho_0, +\infty), \theta \in [0, \pi], \varphi \in [0, 2\pi)\}$.

The Rayleigh approximation
In prolate spheroidal coordinates the boundary value problem (2.2) –(2.3) leads to the axially symmetric solution u_0 of the Laplace equation, with asymptotic behavior given by

$$u_0(\rho,\theta) = 1 + O\left(\frac{2}{c\sqrt{\cosh^2\rho - \sin^2\theta}}\right), \quad \rho \to +\infty \tag{2.7}$$

The Laplace equation accepts separation of variables and we seek the Rayleigh approximation $u_0(\rho,\theta)$, independent of the azimuthal angle φ, in the form

$$u_0(\rho,\theta) = \sum_{n=0}^{\infty} A_n Q_n^0(\cosh\rho) P_n(\cos\theta) \tag{2.8}$$

where Q_n^m are the Legendre polynomials of second kind [4]. In view of the boundary condition (2.3) we arrive at the expression

$$u_0(\rho,\theta) = 1 - \frac{Q_0^0(\cosh\rho)}{Q_0^0(\cosh\rho_0)} \tag{2.9}$$

The first-order approximation
Because of the asymptotic expansion (2.4), the first-order low-frequency coefficient u_1 is not azimuthal independent anymore. It is a solution to the Laplace equation, satisfies the boundary condition (2.3) and the asymptotic form

$$u_1(\rho,\theta,\varphi) = \eta\left[(\mathbf{r}_\perp \cdot \hat{\mathbf{d}}) + \mathbf{r}_\parallel \frac{(\hat{\mathbf{n}}\cdot\hat{\mathbf{d}}) - \beta}{\sqrt{1-\beta^2}}\right] - \frac{1}{4\pi}\int_{S_a} \frac{\partial u_0(\mathbf{r}')}{\partial\nu'} ds(\mathbf{r}') + O\left(\frac{1}{r}\right), \tag{2.10}$$

with $\eta = \omega_0/\omega$. We shall seek the first-order approximation in the general separable form

$$u_1(\rho,\theta,\varphi) = \sum_{n=0}^{\infty}\sum_{m=0}^{n}[A_{nm}\cos(m\varphi) + B_{nm}\sin(m\varphi)]Q_n^m(\cosh\rho)P_n^m(\cos\theta) \tag{2.11}$$

Translating this boundary value problem into the prolate spheroidal language, adding the asymptotic behavior (2.10) and calculating the surface integral, we find the coefficients A_{nm}, B_{nm} which are

$$A_{00} = \frac{c}{2}\frac{1}{[Q_0^0(\cosh\rho_0)^2]}, \qquad A_{10} = -\frac{c}{2}\left(\frac{d_3-\beta}{1-\beta d_3}\right)\frac{P_1^0(\cosh\rho_0)}{Q_1^0(\cosh\rho_0)},$$

$$A_{11} = -\frac{cd_1\eta}{2}\frac{P_1^1(\cosh\rho_0)}{Q_1^1(\cosh\rho_0)}, \qquad B_{11} = -\frac{cd_2\eta}{2}\frac{P_1^1(\cosh\rho_0)}{Q_1^1(\cosh\rho_0)} \tag{2.12}$$

Thus, the explicit form of the approximation u_1 becomes

$$
\begin{aligned}
u_1(\rho, \theta, \varphi) = & \left[A_{00} Q_0^0(\cosh \rho) - \frac{c}{2 Q_0^0(\cosh \rho_0)} \right] P_0(\cos \theta) \\
& + \left[A_{10} Q_1^0(\cosh \rho) + \frac{c}{2} \left(\frac{d_3 - \beta}{1 - \beta d_3} \right) P_1^0(\cosh \rho_0) \right] P_1(\cos \theta) \\
& + i \left\{ \left[A_{11} Q_1^1(\cosh \rho) + \frac{c d_1 \eta}{2} P_1^1(\cosh \rho) \right] \cos \varphi \right. \\
& \left. + \left[B_{11} Q_1^1(\cosh \rho) + \frac{c d_2 \eta}{2} P_1^1(\cosh \rho) \right] \sin \varphi \right\} P_1^1(\cos \theta).
\end{aligned}
$$

$$(2.13)$$

The far field

Once u_0 and u_1 are known, the low-frequency analysis of the scattering amplitude (see [3]) leads to the approximation

$$
\begin{aligned}
g(\hat{\mathbf{r}}) = & -\frac{ik}{4\pi} \int_{S_a} \frac{\partial u_0(\mathbf{r}')}{\partial \nu'} ds(\mathbf{r}') \\
& + \frac{k^2}{4\pi} \int_{S_a} \left[\frac{\partial u_1(\mathbf{r}')}{\partial \nu'} - (\hat{\mathbf{r}} \cdot \mathbf{r}') \frac{\partial u_0(\mathbf{r}')}{\partial \nu'} \right] ds(\mathbf{r}') + O(k^3)
\end{aligned}
$$

$$(2.14)$$

In view of the expression for the normal derivative

$$
\frac{\partial}{\partial \nu} = \frac{2}{c \sqrt{\cosh^2 \rho - \cos^2 \theta}} \frac{\partial}{\partial \rho}
$$

$$(2.15)$$

and for the surface element on the prolate spheroidal surface

$$
ds(\mathbf{r}) = \frac{c^2}{4} \sqrt{\cosh^2 \rho - \cos^2 \theta} \, \sinh \rho \sin \theta \, d\theta d\varphi
$$

$$(2.16)$$

we calculate the three integrals in (2.14) and obtain the following low-frequency approximation of the scattering amplitude in the object's frame K'

$$
\begin{aligned}
g'(\hat{\mathbf{r}}) = & -i \left(\frac{kc}{2 Q_0^0(\cosh \rho_0)} \right) - \left(\frac{kc}{2 Q_0^0(\cosh \rho_0)} \right)^2 \\
& + O\left(\left(\frac{kc}{2} \right)^3 \right), \qquad \left(\frac{kc}{2} \right) \to 0.
\end{aligned}
$$

$$(2.17)$$

If we substitute the above approximation into the formula defining the scattering cross-section σ_s as the L^2-norm of g on the unit sphere [3], we obtain

$$\sigma'_s = 4\pi \left(\frac{c}{2Q_0^0(\cosh \rho_0)} \right)^2 \left[1 + \left(\frac{kc}{2Q_0^0(\cosh \rho_0)} \right)^2 \right]$$

$$+ O\left(\left(\frac{kc}{2} \right)^4 \right), \qquad \left(\frac{kc}{2} \right) \to 0. \tag{2.18}$$

By using the relativistic result (3.6), obtained in Section 3 and turning back to the space-time variables of the frame K, we can derive the approximations of the scattering amplitude g

$$g(\hat{r}, k_0) = -i(1 - \beta d_3) \left(\frac{k_0 c}{2Q_0^0(\cosh \rho_0)} \right) - \frac{(1 - \beta d_3)^2}{\sqrt{1 - \beta^2}} \left(\frac{k_0 c}{2Q_0^0(\cosh \rho_0)} \right)^2$$

$$+ O\left(\left(\frac{k_0 c}{2} \right)^3 \right), \qquad \left(\frac{k_0 c}{2} \right) \to 0, \tag{2.19}$$

as well as for the scattering cross-section σ_s

$$\sigma_s = 4\pi \left(\frac{c(1 - \beta d_3)}{2Q_0^0(\cosh \rho_0)} \right)^2 \left[1 + \left(\frac{1 - \beta d_3}{\sqrt{1 - \beta^2}} \right)^2 \left(\frac{k_0 c}{2Q_0^0(\cosh \rho_0)} \right)^2 \right]$$

$$+ O\left(\left(\frac{k_0 c}{2} \right)^4 \right), \qquad \left(\frac{k_0 c}{2} \right) \to 0. \tag{2.20}$$

By using that $c = 2a\beta/\sqrt{1 - \beta^2}$ we analyze the behavior of the scattering cross-section and conclude that in the case of $(\hat{n} \cdot \hat{d}) = d_3 < 0$ it increases when the velocity of motion V increases, while in the case of $(\hat{n} \cdot \hat{d}) = d_3 > 0$ the scattering cross-section decreases with respect to V. Moreover, when $V = 0$, i.e. in the stationary case, the scattering cross-section becomes

$$\sigma_s = 4\pi a^2 \left[1 + (ka)^2 \right] + O\left((ka)^4 \right), \qquad (ka) \to 0 \tag{2.21}$$

which is a well-known result for the low-frequency acoustic scattering by a small soft sphere.

3. Scattering amplitude and Lorentz transformation

In this part of the paper, which is a continuation of the works of Cooper and Strauss [5], Majda [6] and Georgiev [7], we study the effect of Lorentz transformation on the scattering amplitude g. Let Q be an open subset of $\mathbf{R}^{(3+1)}$ with a smooth boundary Σ and $\Omega(t) = \{\mathbf{r} \in \mathbf{R}^3 : (\mathbf{r}, t) \in Q\}$ be the open cross-section of Q at time t. We assume that the obstacle $S_a(t) = \mathbf{R}^3 \setminus \Omega(t)$ remains in a fixed bounded set and its speed of motion is less than the speed of sound c, which means that the normal vector to Σ is spacelike.

Let L be a Lorentz transformation of $\mathbf{R}^{(3+1)}$, which preserves the sense of time. We write $X = (ct, x, y, z)$ and denote $X' = L(X)$. We introduce the dual variables $\Theta = (\xi_0, \xi_1, \xi_2, \xi_3)$ and the Lorentz inner product

$$\langle \Theta, X \rangle = \xi_0 ct - \xi_1 x - \xi_2 y - \xi_3 z \tag{3.1}$$

With $\Theta' = L(\Theta)$, we have $\langle \Theta', X' \rangle = \langle \Theta, X \rangle$. Next we consider a sphere S_a moving with constant velocity $\mathbf{v}$, where $\mathbf{v} \in \mathbf{R}^3$, $|\mathbf{v}| < c$. Thus $S_a(t) = S_a + \mathbf{v}t$. We make the Lorentz transformation (1.5) such that in the new coordinates the body is stationary.

The scattering amplitude $g(\hat{\mathbf{r}}, \hat{\mathbf{d}}, k_0)$ is defined as the Fourier transform of the scattering (echo) kernel $S_\#(\hat{\mathbf{r}}, \hat{\mathbf{d}}, s, \tilde{s})$ at $\tilde{s} = 0$. By using the results in [5], [7] the scattering kernel can be given in the following form

$$S_\#(\hat{\mathbf{r}}, \hat{\mathbf{d}}, s, \tilde{s}) = \sum_{j=1,2} \int_\Sigma f_j(s - \langle \Theta, X \rangle) u^s(X, \tilde{s}) d\Sigma \tag{3.2}$$

where the function u^s is the unique solution to the mixed problem

$$\left(\frac{\partial_{tt}}{c^2} - \Delta\right) u^s = -\left(\frac{\partial_{tt}}{c^2} - \Delta\right) u^i = 0,$$
$$\mathbf{B}_\Sigma\left[u^s + \delta(\tilde{s} - \langle \Theta, X \rangle)\right] = 0,$$
$$u^s = 0, \quad t \ll 0 \tag{3.3}$$

and $\mathbf{B}_\Sigma$ is an energy preserving boundary condition on the surface of the obstacle. The functions $f_j = \delta^{(j)}, j = 1,2$ are the derivatives of the Dirac measure on $\mathbf{R}$ and the integral (3.2) is taken in the sense of distribution.

Because of the invariance of the Lorentz inner product $\langle .,. \rangle$ the scattering kernels $S_\#$ and $S'_\#$ differ in the same way as do the two surface elements $d\Sigma$ and $d\Sigma' = d[L(\Sigma)]$. In the observer's frame K we have

$$d\Sigma = \sqrt{\frac{a^2}{a^2 - x^2 - y^2} - \beta^2}\, dxdydt$$

(3.4)

In the moving frame K' the surface element is

$$d\Sigma' = \frac{1}{\sqrt{1 - \beta^2}} \sqrt{\frac{a^2}{a^2 - x'^2 - y'^2} - \beta^2}\, dx'\, dy'\, dt'$$

(3.5)

Finally, in view of (3.4) and (3.5) we conclude that

$$g(\hat{\mathbf{r}}, \hat{\mathbf{d}}, k_0) = \sqrt{1 - \beta^2}\, g'(\hat{\mathbf{r}}', \hat{\mathbf{d}}', k)$$

(3.6)

It is clear that the scattering amplitude of the acoustic field is not an invariant. As many other quantities in relativity, its magnitude, measured in the object's frame K' is perturbed by the factor $\sqrt{1 - \beta^2}$.

References

1. Morse, P. and Ingard, U.: Theoretical Acoustics, Mc Graw-Hill, N.Y., (1968).
2. Colton, D., Kress, R.: Inverse Acoustic and Electromagnetic Scattering Theory. 2.ed., Springer-Verlag, Berlin Heidelberg NewYork, (1998).
3. Dassios, G., Kleinman, R.: Low-frequency Scattering, Oxford Mathematical Monographs, Oxford University Press, Clarenton, (2000).
4. Morse, P. and Feshbach, H.: Methods of Theoretical Physics. I, II, McGraw-Hill, New York, (1953).
5. Cooper, J. and Strauss, W.: *Representation of the Scattering Operator for Moving Obstacles*, Indiana Univ. Math. J., 28, (1979).
6. Majda, A.: A Representation Formula for the Scattering Operator and the Inverse Problem for Arbitrary Bodies, Comm. Pure Appl. Math., 30, (1977).

7. Georgiev, V.: Inverse Scattering Problem for the Maxwell Equations Outside Moving Body, Ann. Inst. Henri Poincaré, Vol. 50, 1, (1989).

8. Eisenhart, L. P.: Continuous Groups of Transformations, Princeton, (1943).

9. Marechal, F., Dubois, F., Duceau, E., Terrasse, I.: *Couche Limite Absorbante pour l'Acoustique Convective*, Congres National d'Analyse Numerique, (2000).

APPLIED MATHEMATICS

WAVE DISPERSION PHENOMENA IN CONCRETE

D. G. AGGELIS, D. POLYZOS

Department of Mechanical Engineering and Aeronautics,University of Patras, P.O. Box 1401, Patras 26500, Greece and Institute of Chemical Engineering and High Temperature Chemical Processes (FORTH/ICE-HT) Patras 26500, Greece

Experiments shown that longitudinal pulses propagating through concrete and mortar specimens exhibit a dispersion behavior at low frequencies. In the present work the dipolar gradient elastic theory of Mindlin is employed and the experimentally observed dispersion in concrete and mortar is successfully explained. Both concrete and mortar are considered as homogeneous isotropic solids with microstructural effects. These microstructural effects are taken into account with the aid of the simple dipolar gradient elastic theory, which is the simplest possible version of the general higher order gradient elastic theory proposed by Mindlin. The agreement between experimental observations and theoretical predictions is accomplished by controlling the two new microstructural material constants introduced by the dipolar gradient elastic theory.

1. Introduction

Concrete is a highly non-homogeneous material with a complex microstructure containing random inhomogeneities over a wide range of length scales. Its structure can be considered as a composite material where large aggregates (3 to 30mm) are embedded in a mortar matrix. Similarly, the mortar consists of small aggregates (0.1 to 3mm) dispersed in a cement paste medium. Understanding of how a stress wave propagates through such a medium is of paramount importance for many non-destructive testing techniques like ultrasonics and acoustic emission [1,2].

Experiments performed in [3] with the aid of an experimental setup explained in [4], show that traveling longitudinal waves in concrete as well as in mortar undergo dispersion only at low frequencies where the material microstructure is much smaller than the wavelength of the incident wave. On the contrary, longitudinal pulses propagating in cement paste do not exhibit any dispersion behavior. This fact reveals that microstructural effects on the propagating pulses appear to be more dominant in concrete and mortar than in cement paste specimens. On the other hand, it has been shown in [3] that the scattering of traveling waves by the embedded inhomogeneities [5,6] is not the mechanism of the observed dispersion.

Considering concrete as a linear elastic material with microstructure, its dynamic mechanical behavior cannot be described adequately by the classical theory of linear elasticity, which is associated with concepts of homogeneity and locality of stresses. When the material exhibits a non-homogeneous behavior, microstructural effects become important and the state of stress has to be defined in a non-local manner. These microstructural effects can be successfully modeled in a macroscopic framework by employing higher order gradient elastic theories like those proposed by Cosserat brothers, Mindlin, Eringen, Aifantis and Vardoulakis. For a literature review on the subject of these theories one can consult the review articles of Tiersten and Bleustein [7], Eringen [8] and Exadaktylos and Vardoulakis [9], the literature review in the recent paper by Tsepoura et al. [10] and the book of Vardoulakis and Sulem [11].

During the last fifteen years, a variety of elastodynamic boundary value problems were solved either analytically or numerically by employing mainly simplified forms of the above mentioned enhanced continuum theories. Here one can mention the simple gradient elastic theory of Altan and Aifantis [12] used by Altan et al. [13], Tsepoura et al. [10] and Polyzos et al. [14] to treat one and three dimensional vibration problems in gradient elastic bars, the gradient elastic theory with surface energy proposed by Vardoulakis and Sulem [11] employed by Georgiadis and Vardoulakis [15] to explain the dispersion behavior of Rayleigh waves propagating on the free surface of an elastic half space with microstructure, and the dipolar gradient elastic theory of Mindlin [16] used by Georgiadis [17] to solve Mode-III crack problems in dynamic gradient elasticity.

In the present work, the dipolar gradient elastic theory introduced by Mindlin [16] is employed in order to explain the dispersion of longitudinal ultrasonic pulses observed experimentally in [3]. The dipolar gradient elastic theory is the simplest possible dynamic form of Mindlin's higher order gradient theory. This theory is called dipolar since besides the classical Lame constants, two new material constants are introduced. Both correlate the microstructure with the macrostructure of the considered gradient elastic continuum and they are responsible for the dispersive behavior of longitudinal and transverse waves propagating through it. The key idea of the present study is that a proper determination of the two microstructural material constants enables one to explain remarkably well the low frequency dispersive nature of concrete and mortar.

2. Dipolar Gradient Elastic Theory and Wave Propagation

Taking into account the non-local nature of microstructural effects, Mindlin [16] considered that the density of strain density is not only a function of strains, as in the classical case, but also a function of the gradients of the strains. In the dipolar version of his theory, this is expressed as

$$W = \frac{1}{2}\lambda(tr\widetilde{\mathbf{e}})^2 + \mu\widetilde{\mathbf{e}} : \widetilde{\mathbf{e}} + \frac{1}{2}\lambda g^2 \nabla(tr\widetilde{\mathbf{e}}) \cdot \nabla(tr\widetilde{\mathbf{e}}) + \mu g^2 \nabla\widetilde{\mathbf{e}} : \nabla\widetilde{\mathbf{e}} \qquad (1)$$

where $\widetilde{\mathbf{e}}$ and $tr\widetilde{\mathbf{e}}$ are the classical strain tensor and its trace, respectively, ∇ represents the gradient operator, the dot, the double dots and the column of three dots indicate inner product between vectors and tensors of second and third order, respectively, (λ, μ) are the classical Lame constants and g^2 is a new material constant (units of m^2) called volumetric strain gradient energy coefficient, which correlates the microstructure with macrostructure.

Extending the idea of non-locality to the inertia of the continuum with microstructure, Mindlin proposed a new expression for the kinetic energy density function [16] where the gradients of the velocities are taken into account, i.e.

$$T = \frac{1}{2}\rho\dot{\mathbf{u}} \cdot \dot{\mathbf{u}} + \frac{1}{2}\rho h^2 \nabla\dot{\mathbf{u}} : \nabla\dot{\mathbf{u}} \qquad (2)$$

where ρ is the mass density, $\mathbf{u}$ is the displacement vector, $\dot{\mathbf{u}} = d\mathbf{u}/dt$ and h^2 is the second new material constant (units of m^2) called velocity gradient coefficient, which is always smaller than the volumetric strain gradient energy coefficient g^2.

Taking the variation of strain and kinetic energy, according to the Hamilton's principle, one concludes to the equation of motion of a continuum with microstructure, which in terms of displacements is written as follows:

$$\mu\nabla^2\mathbf{u} + (\lambda + \mu)\nabla\nabla \cdot \mathbf{u} + g^2\nabla^2\big(\mu\nabla^2\mathbf{u} + (\lambda + \mu)\nabla\nabla \cdot \mathbf{u}\big) = \rho\ddot{\mathbf{u}} - \rho h^2\nabla^2\ddot{\mathbf{u}} \qquad (3)$$

The Helmholtz vector decomposition implies that $\mathbf{u}$ can be written as a sum of irrotational and solenoidal fields according to the relation:

$$\mathbf{u} = \nabla\phi + \nabla \times \mathbf{A} \qquad (4)$$

with $\nabla\phi$, $\nabla \times \mathbf{A}$ deonoting volumetric and shape with no volume changes, respectively. In terms of wave propagation this means that $\nabla\phi$ corresponds to longitudinal waves, while $\nabla \times \mathbf{A}$ represents shear waves propagating through the medium, i.e.

$$\nabla \phi = \hat{\mathbf{k}} e^{i(k_p \hat{\mathbf{k}} \cdot \mathbf{r} - \omega t)}$$

$$\nabla \times \mathbf{A} = \hat{\mathbf{b}} e^{i(k_s \hat{\mathbf{k}} \cdot \mathbf{r} - \omega t)}$$

(5)

where $\hat{\mathbf{k}}$ represents the direction of incidence, $\hat{\mathbf{b}}$ is the polarization vector for the shear wave, $\mathbf{r}$ stands for position vector, k_p, k_s are the wave numbers of the longitudinal and shear disturbances, respectively, while ω is the frequency of the propagating waves.

Representing by C_p, C_s the classical phase velocities of longitudinal and shear waves, respectively, and inserting Eqs.(5) into Eq. (3) one obtains the following two dispersion relations:

$$\omega^2 = C_p^2 \frac{k_p^2 \left(1 + g^2 k_p^2\right)}{1 + h^2 k_p^2}, \quad C_p^2 = \frac{\lambda + 2\mu}{\rho}$$

(6)

for longitudinal waves and

$$\omega^2 = C_s^2 \frac{k_s^2 \left(1 + g^2 k_s^2\right)}{1 + h^2 k_s^2}, \quad C_s^2 = \frac{\mu}{\rho}$$

(7)

when shear waves are considered.

Eqs (6) and (7) reveal that both longitudinal and shear stress waves undergo dispersion when they travel in solids with microstructure. Their dispersion is entirely due to the presence of the two microstructural material constants g^2 and h^2. It is easy to see that by zeroing g^2 and h^2, Eqs (6) and (7) lead to the linear expressions:

$$\omega^2 = C_p^2 \cdot k_p^2$$

$$\omega^2 = C_s^2 \cdot k_s^2$$

(8)

which characterize propagation of non dispersive waves in a classical elastic medium.

Solving Eqs (6) and (7) for the wave numbers k_p and k_s, respectively, one yields the following relation:

$$k_{p,s}^2 = \frac{-\left(C_{p,s}^2 - \omega^2 h^2\right) + \sqrt{\left(C_{p,s}^2 - \omega^2 h^2\right)^2 + 4 \cdot C_{p,s}^2 g^2 \omega^2}}{2 C_{p,s}^2 g^2}$$

(9)

Then the phase velocity $V_{p,s}$ of longitudinal and shear plane waves propagating in a dipolar gradient elastic continuum has the following form:

(10)

$$V_{p,s} = \frac{\omega}{k_{p,s}} = \frac{\omega}{\sqrt{\dfrac{-\left(C_{p,s}^2 - \omega^2 h^2\right) + \sqrt{\left(C_{p,s}^2 - \omega^2 h^2\right)^2 + 4 \cdot C_{p,s}^2 g^2 \omega^2}}{2 C_{p,s}^2 g^2}}}$$

3. Results and Discussion

For the first case of concrete with w/c=0.375, taking into account the density of concrete (approximately 2.3gr/cm^3) and that the poisson ratio does not vary much around 0.2, the C_p parameter was set to C_p=4400m/s which is considered very reasonable. In Figure 1 the experimental curve of wave velocity vs frequency for this concrete is depicted, along with the theoretically obtained curve. It is seen that the phase velocity predicted by the dipolar gradient elastic theory describes very closely the experimental velocities. The volumetric and inertia microstructural material constants obtained the values g^2=0.0001 and h^2=g^2/1.2321, respectively. As to the h^2 parameter, which controls mainly the elevation of the curves, it was seen that being proportional to g^2 served the fitting of experimental curves. The constant of proportionality was obtained in order the theoretical values for high frequencies to coincide with experimental ones. The modeling of w/c=0.50 was conducted using the same g^2 and h^2 values, as also seen in Figure 1, while the C_p was set to 4050m/s, a value expected since a water content increase has a definite negative impact on stiffness. Therefore, it is seen that the dispersion behavior of materials sharing the same microstructure exhibits similarities and can be modeled using the same volumetric and acceleration parameters, adjusting only the stiffness of the matrix material.

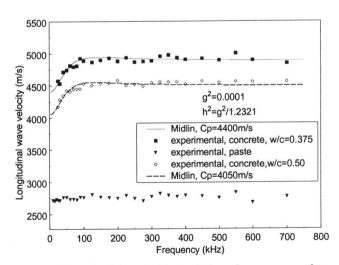

Figure 1. Experimental and theoretical wave velocity vs frequency curves for concrete.

Results of cement paste of the w/c =0.375 are also supplied for comparison purposes. The curve is positioned much lower due to the less stiff nature of the material, while it exhibits no serious dispersion probably due to the lack of appreciable microstructural effects.

In Figure 2 the experimental and theoretical curves concerning a mortar of w/c=0.375 and sand content of approximately 40% are presented. The close modeling requires a C_p of 4100m/s. This is the result of the decrease in aggregate content, compared to the concrete of w/c=0.375, which certainly decreases also the stiffness of the material. The volumetric parameter g^2 obtained the value 0.000049 while the h^2 parameter was set to $g^2/1.113025$

The agreement between theoretical and experimental curves is very good in this case also, as seen in Figure 2.

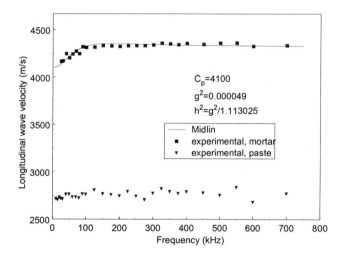

Figure 2. Experimental and theoretical wave velocity vs frequency curves for mortar.

The lower value of the volumetric parameter, used to fit the experimental data is likely to be related to the smaller inclusion size (sand) in the case of mortar in comparison to the bigger aggregate of concrete

4. Conclusions

Considering concrete and mortar as solids characterized by microstructural effects, theoretical predictions made by the dipolar gradient elastic theory of Mindlin are very close to the experimental observations.

Besides the classical Lame constants, Mindlin's theory introduces two new material constants, namely the volumetric strain gradient energy coefficient g^2 and the velocity gradient coefficient h^2, which correlate the microstructure with the dynamic macrostructural behavior of the considered gradient elastic continuum. Correlations between experimental observations and theoretical predictions have shown that h^2 is always smaller than g^2 while both constants depend on the aggregate size.

Acknowledgments

The financial support of the present work by PENED 01EΔ 420 is gratefully acknowledged.

References

1. V. M. Malhotra, N. J. Carino, (Eds.), CRC Handbook on Nondestructive Testing of Concrete. CRC Press, Florida (1991).
2. T. P. Philippidis, D. G. Aggelis, Cement Concrete Res **33**, 525 (2003)
3. D. G. Aggelis, D. Polyzos, submitted (2003).
4. T. P. Philippidis, D. G. Aggelis, submitted (2003)
5. S. V. Tsinopoulos, J. T. Verbis, D. Polyzos, Adv. Compos. Lett. **9**, 193 (2000)
6. J.T. Verbis, S.E. Kattis, S.V. Tsinopoulos, D. Polyzos, Comput. Mech. **27**, 244 (2001)
7. H.F. Tiersten, J.L. Bleustein, in: R.D. Midlin and Applied Mechanics, (Hermann, G., ed.). Pergamon Press, New York, 67 (1974)
8. A.C. Eringen,. Vistas of Nonlocal Continuum Physics. International Journal of Engineering Science **30**, 1551 (1992)
9. G.E. Exadaktylos, I. Vardoulakis, Tectonophysics **335**, 81 (2001)
10. K.G. Tsepoura, S. Papargyri-Beskou, D. Polyzos, A boundary element method for solving 3D static gradient elastic problems with surface energy. Comput Mech **29**, 361 (2002)
11. I. Vardoulakis, J. Sulem, Bifurcation Analysis in Geomechanics. Blackie/Chapman and Hall, London (1995)
12. B.S. Altan, E.C. Aifantis, Scripta Metallurgica et Materialia **26**, 319 (1992)
13. B.S. Altan, H.A. Evensen, E.C. Aifantis, Mech. Res. Commun. **23**, 35 (1996)
14. D.Polyzos, K.G. Tsepoura, S. Tsinopoulos, D.E. Beskos, Comput. Method. Appl. M. **192**, 2845 (2003)
15. H.G. Georgiadis, I. Vardoulakis, Wave Motion **28**, 353 (1998)
16. R.D. Midlin, Arch. Ration. Mech. An., **16**, 51 (1964)
17. H.G. Georgiadis, J. Appl. Mech. T. ASME **70**, 517 (2003)

SIMULATION-BASED ANALYSIS OF I/Q MISMATCH IN SDR RECEIVERS

C. T. ANGELIS, V. N. CHRISTOFILAKIS, P. KOSTARAKIS

Physics Department, University of Ioannina, Panepistimioupolis
Ioannina,45110,Greece

In conventional wireless receivers errors are introduced due to the Inphase-Quadrature (IQ) phase imbalances (mismatch) between the ADCs and gain and phase mismatch in the I/Q demodulation. Also the gain and phase errors in the local oscillators and in the filters account for the I/Q errors. Due to these imbalances the performance of the receivers and the quality of the received signal are degraded due to the gain and the phase imbalances in the I/Q phases. By performing the I/Q demodulation in the digital domain some of these errors are alleviated. In this paper many approaches for the correction of the IQ channel mismatches and the previous methods applied for correction of IQ mismatches using analog and digital domains have been reviewed. The trade-offs and the issues behind the design and implementation of the wireless receivers are described. The different modulation schemes of wireless receivers are also mentioned. Finally, a novel and a feasible adaptive digital signal processing method, using MUSIC algorithm, which gives a solution for the correcting IQ mismatches in a Direct Conversion Receiver, has been proposed and implemented. The algorithm that involves the MUSIC method has been found to improve the SNR by 15-40 dB.

1. Basic Concepts

A homodyne I/Q receiver is shown in Fig. 1.1. The I/Q demodulation is performed in the analog domain. The In-phase (I) and the Quadrature (Q) channels are necessary for any angle modulated signals because the two sidebands of the RF spectrum contain different information and may result in irreversible corruption if they overlap each other without being separated into two phases. The demodulator at the receiver has to be synchronous in nature. The mixers for the I and Q channel works at the same frequency, but with a phase difference of 90 degrees between them. The two channels should have matched gains between them as well. If these two requirements is not fulfilled for all frequencies within the bandwidth the dynamic range of the receiver is reduced. In additional, false targets may appear due to phase and gain mismatch. Another drawback with the conventional receiver is that two ADCs, one in each channel, is required. Mismatch between these ADCs will reduce the receiver performance further. The receiver must possess an oscillator, which is at the exactly same frequency, and phase as the carrier oscillator, at the transmitter end. The low pass filters at the two paths of the receiver must have identical characteristics, as any mismatch in their characteristics would lead to I/Q errors.

In this paper the I/Q mismatch that occurs in the Direct Conversion receivers is mentioned and a feasible DSP solution has been proposed. The direct conversion (Zero IF or Homodyne conversion) receivers converts a signal from RF to Base band. The band of interest is translated to zero and then the signal is low-pass filtered to suppress the interferences. The direct conversion receivers do not suffer from the problem of images as the intermediate frequencies are zero and hence they do not require image reject filters.

The main problem faced in the design of the direct conversion receivers is the I/Q mismatch. The errors that occur due to the phase shift between the oscillators and change in the coefficients/phase shift between the low pass filters in the I/Q paths, corrupt the signal to a large extent and severely distort the signal to noise ratio. The gain imbalance appears as a non-unity scale factor in the amplitude while the phase imbalance corrupts one channel with a fraction of data pulses in the other channel. Fig.1 shows the block diagram of a QPSK (Quadrature phase shift keying) receiver. Assuming no gain or phase imbalance between the I/Q paths then the signal after demodulation passes through the low pass filter and are received as undistorted signal. The amplitude and the phase mismatches are usually random and changes from time to time.

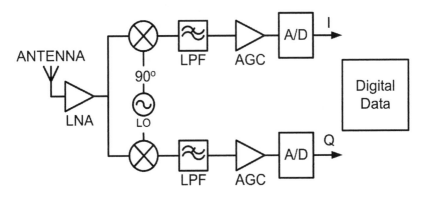

Fig. 1.1. Schematic diagram of a homodyne receiver.

Results

A software model of the SDR receiver structure has been developed with MATLAB. The modulation schemes that will be handled are used in the wireless systems standards E-GSM900, DCS1800, IS-136, and PDC. The purpose with the model is to evaluate the performance of the receiver as well as for identification of possible design trade-offs between the RF part, the ADC, and the digital filters. For the evaluation the receiver has been implemented in MATLAB.

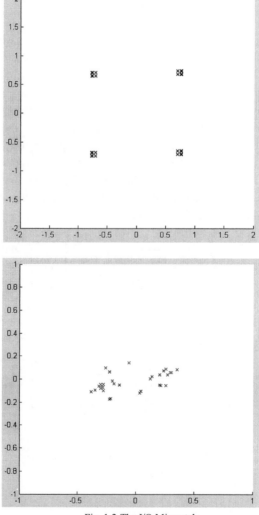

Fig. 1.2. The I/Q Mismatch.

Fig. 1.2 represents the I/Q plots with and without I/Q mismatch. In order to increase the SNR we have to minimize the errors that are caused due to the mismatch that exist between the filters and the oscillators in the two paths. In ideal case the oscillators in the two paths must very stringently oscillate at the same frequency and phase and the low-pass filters in both of the paths must have identical characteristics. Error occurs when there is mismatch (phase or gain imbalance) between the two oscillators and the low-pass filters. Compensation must be done in order to reduce the error.

A novel and a feasible adaptive digital signal processing method, which gives a solution for the IQ mismatches, has been proposed. This algorithm involves the MUSIC method and the SNR have been found to improve by 15-40 dB. MUSIC or Multiple SIgnal Classification is an algorithm that uses matrix manipulation to get the directions of arrival (DOAs). This is done by processing signals received at the antenna elements. This algorithm is useful in estimating the number of sources, the strength of the cross-correlation between the source signals, the DOAs and strength of noise. Since the two channels contain different signals, the mismatch could never be measured, especially if the mismatch is random. To avoid this difficulty, the two signals in the Inphase and the quadrature phase are made the same and then compared. They are now supposed to be identically equal and a filter using an MUSIC algorithm is used to compensate for the mismatch.

The basic principle for compensating for the filter mismatch is to make the input signal through the filters identical and then compensate for the error between the two outputs. The compensation filter is an adaptive filter using the MUSIC algorithm to reduce the errors. The calibration of the circuit is essentially done with a random input signal. Thus the circuit initially is trained so that the two signals at the filter output would become identical. After training, the weights are calculated and the transfer function of the new filter is obtained and added to the network.

The results after mismatch compensation are obtained and tabulated. The SNR, the signal to noise ratio has been found to increase and the errors have been decreased as the MUSIC filter showed improvement for all kinds of mismatches. The improvement in the SNR due to the both change in the odd coefficients and change in the phase of the filter has been investigated and calculated. The signal to noise ratio decreases (I/Q mismatch increases), as the percentage of change in the odd coefficients of the filter and as the phase change increases (Fig 1.3).

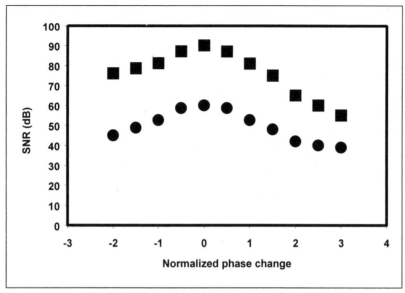

Fig. 1.3 The effect of the phase change in the LPF, ● before and ■ after the algorithm use.

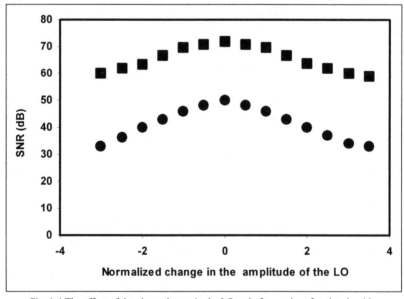

Fig. 1.4 The effect of the phase change in the LO, ● before and ■ after the algorithm use.

Another main factor that leads to the I/Q error is the mismatch due to the local oscillator. Another MUSIC algorithm similar to the previous case is applied to account for the mismatch. After calibration, the transfer function of the filter is added that compensates for the mismatch caused by the oscillators. The improvement in the SNR due to change in both the % amplitude and the phase of the local oscillator has been investigated. The signal to noise ratio decreases (I/Q mismatch increases), as the percentage change in the amplitude and the phase of the oscillator increases (fig 1.4, Fig. 1.5).

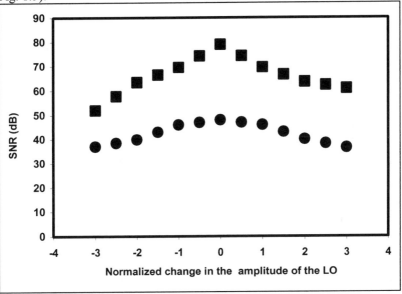

Fig. 1.5 The effect of the % of change in the amplitude of the LO, ● before and ■ after the algorithm use.

In this paper the I/Q mismatch that occurs in the Direct Conversion receivers due to the Inphase-Quadrature (IQ) phase imbalances (mismatch) between the ADCs and gain and phase mismatch in the I/Q demodulation mentioned and a feasible DSP solution has been proposed. The proposed and implemented novel and feasible adaptive digital signal processing method, using MUSIC algorithm, gives a solution for the correcting IQ mismatches in a Direct Conversion Receiver. The algorithm that involves the MUSIC method has been found to improve the SNR by 15-40 dB.

References

1. Behzad Razavi, " Design Considerations for Direct- Conversion Receivers" IEEE Transactions on Circuits and Systems-II: Analog and Digital Signal Processing, Vol.44, No.6.June 1997 pp.428-435.
2. A. Bateman and D.M. Haines, " Direct conversion Transreceiver design for compact low-cost portable mobile radio terminals" Proc.IEEE Veh.technol.Conf.May 1989 pp.57-62.
3. A. Wiesbauer and G.C.Temes, "Online Digital compensation of analog circuit imperfections for cascaded Sigma-Delta modulators, " in Proceedings. 1996 IEEE-Cas Region 8 workshop on analog and mixed IC Design, Pavia Italy, p.92.
4. S. Abdennadher et al, " Adaptive self-calibrating deltasigma modulators," Electronic Letters, vol.28, pp.1288-1289, July 1992.
5. A. Swami Nathan and M. Snelgrove et al, "A monolithic complex sigma-delta modulator for digital radio," in Proceedings.1996 IEEE-CAS Region 8 workshop on Analog and Mixed IC Design, Pavia, Italy, Sept 1996.p83.
6. S.A. Jantzi et.al, " The effects of mismatch in complex band pass sigma Delta modulators," in Proceedings. ISCAS 1996, Atlanta, GA, May 1996, p.227.
7. G. Cauwenberghs and G.C.Temes, " Adaptive calibration of multiple quantization over sampled A/D converters," in Proceedings. IEEE ISCAS'96, Atlanta, pp.512-516.
8. G. Schultes et. al, "Basic performance of a Directconversion DECT receiver". Electronic Letters, Vol.26, No.21, pp.1746-1748, Oct.1990.
9. D. Van Compernolle and S.Van Gervan, " Signal separation in a symmetric adaptive noise canceller by output décor relation, " IEEE Transactions in Signal Processing, Vol.43, p.1602, July 1995.

FINITE ELEMENT DISCRETIZATION OF THE 'PARABOLIC' EQUATION IN AN UNDERWATER VARIABLE BOTTOM ENVIRONMENT

D.C. ANTONOPOULOU AND V.A. DOUGALIS

Department of Mathematics,
University of Athens, 15784 Zographou, Greece,
and
Institute of Applied and Computational Mathematics, FO.R.T.H.
P.O. BOX 1527, 71110 Heraklion, Greece.

We use the standard 'parabolic' approximation to the 2D Helmholtz equation of underwater acoustics to model long-range propagation of sound in the sea in a domain with a rigid bottom of variable topography. The rigid bottom is modeled by an exact Neumann boundary condition and a paraxial approximation thereof due to Abrahamsson and Kreiss. The problem is solved by a change of variables that flattens the bottom and, subsequently, by a fully discrete finite element-Crank-Nicolson scheme. After outlining the main stability and convergence results for the numerical scheme, we present results of numerical experiments with various bottom topographies. These suggest that an apparent singularity develops in finite range in a case of interest where the existence theory of the p. d. e. fails.

1. Introduction

The standard Parabolic Equation (PE), [1], is the linear p.d.e. of Schrödinger type

$$\psi_r = (i/2k_0)\psi_{zz} + (ik_0/2)(\eta^2(z,r)-1)\psi, \qquad \text{(PE)}$$

where $\psi = \psi(z, r)$ is a complex-valued function of the two real variables z and r. It is widely used as a model for the simulation of one-way, long-range sound propagation near a horizontal plane, in inhomogeneous, weakly range-dependent marine environments. It may be derived as a narrow-angle paraxial approximation to the Helmholtz equation in the presence of cylindrical symmetry

$$\Delta p + k_0^2 \eta^2(z,r)p = 0 . \qquad \text{(HE)}$$

Here $r \geq 0$, the range, is the horizontal distance from a harmonic point source of frequency f_0 placed on the z axis. The depth variable $z \geq 0$ is increasing downward. For simplicity we shall assume that the medium consists of a single layer of water of constant density, occupying the region $0 \leq z \leq l(r)$, $r \geq 0$, between the free surface $z=0$ and the range-dependent bottom $z=l(r)$; $l(r)$ will be assumed to be smooth and positive. The function

$p=p(z,r)$ is the acoustic pressure field, $k_0 = 2\pi f_0 / c_0$ is a reference wave number, c_0 a reference sound speed, and $\eta(z,r)$, the index of refraction, is defined as $c_0 / c(z,r)$, where $c(z,r)$ is the speed of sound in the water. (HE) is supplemented by the surface pressure-release condition $p=0$ and the rigid bottom (Neumann) boundary condition (with $l' = dl / dr$)

$$p_z - l'(r)p_r = 0 \text{ at } z = l(r).\tag{B}$$

The (PE) may be derived from (HE) if we make the change of variable $p(z,r) = \psi(z,r)e^{ik_0 r} / \sqrt{k_0 r}$, assume that $|2ik_0\psi_r| \gg |\psi_{rr}|$ (paraxial approximation) and neglect $O(r^{-2})$ terms (far-field approximation). Analogously the condition (B) becomes

$$\psi_z - l'(r)\psi_r - g(r)l'(r)\psi = 0 \text{ at } z = l(r),\tag{PB}$$

where $g(r)$ is usually taken as $ik_0 - (1/2r)$ or simply as ik_0. The initial-boundary value problem (i-b.v.p.) to be solved then for the PE in the domain $0 \le z \le l(r), r \ge 0$, consists of the p.d.e. (PE), the surface boundary condition $\psi = 0$ for $z = 0, r \ge 0$, the bottom condition (PB), and an initial condition of the type $\psi(z,0) = \psi_0(z), 0 \le z \le l(0)$, representing the source at $r = 0$.

If we introduce the nondimensional variables $y = z/L, t = r/L, w = \psi/\psi_{ref}$, taking e.g. $L = 1/k_0$ and $\psi_{ref} = \max|\psi_0|$, and put $s(t) = k_0 l(k_0^{-1}t)$, etc., we obtain the following i-b.v.p.: Seek a complex-valued function $w=w(y,t)$ defined on $D = \{(y,t) \in \nabla^2, 0 \le y \le s(t), t \ge 0\}$ such that

$$\begin{aligned}
&w_t = iaw_{yy} + i(\beta_R(y,t) + i\beta_I(y,t))w \text{ in } D,\\
&w(0,t) = 0, \ t \ge 0,\\
&w_y(s(t),t) - \dot{s}(t)[w_t(s(t),t) + g_*(t)w(s(t),t)] = 0, \ t \ge 0,\\
&w(y,0) = w_0(y), \ 0 \le y \le s(0).
\end{aligned}\tag{1.1}$$

Here $\dot{s} = ds/dt$, a is a nonzero real constant, w_0 and g_* are complex-valued functions and $\beta_R, \beta_I : D \to \nabla$. (We take the index of refraction as complex in order to model attenuation in the water by its imaginary part β_I.)

The problem (1.1) may be transformed into an equivalent one, posed on a horizontal strip of unit depth, by the range-dependent change of depth variable $x = y/s(t)$. Specifically, (1.1) is transformed to the problem of seeking a complex-valued $u = u(x,t)$, $0 \leq x \leq 1$, $t \geq 0$ such that

$$u_t = i(\xi(t))^{-1} u_{xx} + x\mu(t)u_x + i(\gamma_R(x,t) + i\gamma_I(x,t))u , 0 \leq x \leq 1 , t \geq 0 ,\tag{1.2a}$$

$$u(0,t) = 0 , t \geq 0 , \tag{1.2b}$$

$$u_x(1,t) = S_1(t)u_t(1,t) + S_2(t)u(1,t) , t \geq 0 , \tag{1.2c}$$

$$u(x,0) = u_0(x) , 0 \leq x \leq 1 , \tag{1.2d}$$

where

$$u(x,t) = w(y,t) , \xi(t) = s^2(t)/a , \mu(t) = \dot{s}(t)/s(t) , S_1(t) =$$
$$\dot{s}(t)s(t)/(1+(\dot{s}(t))^2) ,$$
$$S_2(t) = g_*(t)S_1(t) \text{ and } \gamma_{R,I}(x,t) = \beta_{R,I}(y,t) , u_0(x) = w_0(xs(0)).$$

The i-b.v.p (1.2), for t in an arbitrary finite interval [0,T], has been studied by Abrahamsson and Kreiss, [2]; they establish an *a priori* H^2 estimate and, consequently, existence and uniqueness of solutions under the hypothesis that $|\dot{s}(t)| > 0$, i.e. when the bottom is strictly either increasing or decreasing. In [3] the authors establish a new H^1 estimate for $|\dot{s}(t)| > 0$ (or for $\dot{s}(t) \leq 0$ if a smooth solution exists) and use it to prove stability and convergence of a Crank-Nicolson type finite difference scheme for approximating solutions of (1.2). The difficulty of the analysis is due, of course, to the presence of the ψ_r term in (PB), equivalently of the $u_t(1,t)$ term in (1.2c).

In [4] Abrahamsson and Kreiss consider the PE over a rigid bottom and propose, in place of (PB), the 'paraxialized' boundary condition

$$\psi_z - ik_0 l'(r)\psi = 0 , \tag{PB*}$$

which appears to have several advantages over (PB) from a theoretical and a numerical point of view. After nondimensionalization and change of variables the resulting i-b.v.p. in the u, x, t variables assumes the form (1.2a), (1.2b), (1.2d) with (1.2c) replaced by

$$u_x(1,t) = i\xi(t)\mu(t)u(1,t)/2 . \tag{1.2c*}$$

The i–b.v.p. is now well-posed on [0,*T*] for any T > 0 with no restrictions on $s(t)$ other than $s(t) > 0$ on [0,T], [4]. The b.c. (PB*) has been used by Sturm, [5], in finite element discretizations of the 3D analog of the standard PE in problems with many layers. See also [3] for the analysis of a Crank-Nicolson type finite difference scheme in 2D, for this problem.

In the present note we discretize the i-b.v.p. (1.2a)-(1.2d) with the original b.c. (1.2c) using a standard finite element scheme, and state, in the case $\dot{s}(t) \leq 0$, $t \in [0,T]$, optimal-order L^2 error estimates for the associated semidiscrete problem and for a full discretization with the Crank-Nicolson scheme, assuming existence of a sufficiently smooth solution. We only comment briefly on the estimation techniques; full proofs will be found in [6] and [7]. We also briefly comment on the analogous results in the case of the paraxialized b.c. (1.2c*). We then present the results of some numerical experiments with the finite element scheme for general $s(t)$. We compare first the finite element numerical results with those of an analogous finite difference scheme. We then show the results of several computations in cases where $\dot{s}$ becomes zero at some point t^*. Our experiments suggest that the i-b.v.p. seems to remain well-posed if $\ddot{s}(t^*) \neq 0$. If $\dot{s}(t^*) = 0, \ddot{s}(t^*) = 0$, we observed the development of an apparent singularity at $t = t^*$ when the $\dddot{s}(t)$ is positive for $t > t^*$ and $\dot{s}(t) > 0$ for $t > t^*$.

2. The finite element method

We will solve numerically the problem

$$v_t = (i/\xi(t))v_{xx} + i(\widetilde{\beta}_R(x,t) + i\widetilde{\beta}_I(x,t))v, 0 \leq x \leq 1, 0 \leq t \leq T, \qquad (2.1a)$$

$$v(0,t) = 0, 0 \leq t \leq T, \qquad (2.1b)$$

$$v_x(1,t) = S_1(t)v_t(1,t) + m(t)v(t), 0 \leq t \leq T, \qquad (2.1c)$$

$$v(x,0) = v_0(x), 0 \leq x \leq 1, \qquad (2.1d)$$

where $m(t) = S_2(t) + i(S_1(t)\dot{\omega}(t) - 2\omega(t)) + i\lambda S_1(t)$, with $\omega(t) := \dot{s}(t)\xi(t)/(4s(t))$, and λ a real constant. The i-b.v.p. (2.1a)-(2.1d) is derived from the i-b.v.p. (1.2a)-(1.2d) by setting $v(x,t) = \exp(-i(\lambda t + x^2\omega(t)))u(x,t)$. The constant λ is chosen so that the quantity $\max_{x,t} \widetilde{\beta}_R$ is sufficiently small, as required by the technique of error estimation. We assume that $s(t)$ is C^2 and that (2.1a)-(2.1d) has sufficiently smooth solutions.

2.1. The semidiscretization of (2.1a)-(2.1d) in depth.

We assume that $\dot{s}(t) \leq 0$ for $t \in [0,T]$, let $(H^k(0,1), \|u\|_k)$ be the usual complex Sobolev spaces on $(0,1)$, define the space $H_0 := \{\phi \in H^1(0,1) : \phi(0) = 0\}$, and let $\|u\|^2 = (u,u)$, $(u,v) := \int_0^1 u\bar{v} \, dx$. Let, for $0 < h < 1$, S_h be a finite-dimensional subspace of H_0 with the property that for integer $r \geq 2$

$$\inf_{\chi \in S_h} \{\|u - \chi\| + h\|u - \chi\|_1\} \leq ch^s \|u\|_s, 1 \leq s \leq r, \qquad (2.1)$$

for any $u \in H^r(0,1) \cap H_0$. (Here c is a constant independent of u and h.)

We seek $v_h : [0,T] \to S_h$ such that

$$(v_{ht} - i\tilde{\beta}v_h, \phi) = i(\xi(t))^{-1}S_1(t)v_{ht}(1,t)\overline{\phi(1)} + im(t)(\xi(t))^{-1}v_h(1,t)\overline{\phi(1)}$$
$$- i(\xi(t))^{-1}(v_{hx}, \phi_x), \tag{2.2}$$

for any ϕ in S_h, $0 \le t \le T$

with $v_h(0)$ a suitable approximation of v_0 in S_h and $\tilde{\beta} = \tilde{\beta}_R + i\tilde{\beta}_I$. The system of o. d. e.'s represented by (2.2) has a unique solution in $[0,T]$, [6], and there holds:

Theorem 2.1. Let the solution v of i-b.v.p. (2.1a)-(2.1d) be sufficiently smooth and let v_h be the solution of the semidiscrete scheme (2.2). Then, for $0 \le t \le T$,

$$\|v_h - v\| \le c \|(R_h v_0 - v_h(0))_x\| + ch^r \{\|v\|_r + \|v_t\|_r$$
$$+ \int_0^t (\|v\|_r + \|v_t\|_r + \|v_{tt}\|_r)\, d\tau\} \tag{2.3}$$

where c is a positive constant independent of h and v, and R_h a suitable projection onto S_h, to be defined below.

Outline of proof : We set $v_h - v = \theta + \rho$, with $\theta := v_h - R_h(v) \in S_h$, $\rho := R_h(v) - v$, where $R_h(v)$ is the elliptic projection of v in S_h defined by $((R_h(v))_x, \phi_x) = (v_x, \phi_x) \; \forall \phi \in S_h$. In (2.2) we set $\phi := \theta_t - iL_h(\tilde{\beta}\theta) \in S_h$, where L_h is the L^2 projection operator onto S_h. The energy-type proof of the estimate (2.3) requires that $S_1(t) \le 0$ in $[0,T]$ or, equivalently, that $\dot{s}(t) \le 0$ in $[0,T]$. The proof depends on establishing suitable bounds for the quantities $|\rho(1,t)|, |L_h(\tilde{\beta}v_h(1))|, \|(L_h(\tilde{\beta}v_h))_x\|$. The conclusion is that $\|v_h - v\|$ is of optimal order if $v_h(0) = R_h v_0$.

2.2. The Crank-Nicolson scheme in range.

We discretize now the o. d. e. system (2.2) in r by a Crank-Nicolson type method. For this purpose, for N positive integer and $k = T/N$, we define $t^n := nk$ and denote the approximation of $v_h(t^n)$ by V^n. Let $t^{n+1/2} := t^n + k/2$. For V^n known we seek V^{n+1} in S_h such that

$$\xi(t^{n+1/2})(2(V^{n+1}-V^n)-ik\widetilde{\beta}(t^{n+1/2})(V^{n+1}+V^n),\phi)=$$
$$2iS_1(t^{n+1/2})(V^{n+1}(1)-V^n(1))\overline{\phi(1)} \tag{2.4}$$
$$+ikm(t^{n+1/2})(V^{n+1}(1)+V^n(1))\overline{\phi(1)}-ik(\nabla(V^{n+1}+V^n),\nabla\phi),$$

for any $\phi \in S_h$ and $0 \le n \le N-1$, $V^0 :=$ suitable projection of v_0 in S_h.

It may be shown that V^{n+1} exists uniquely and that the scheme is stable. Solving for V^{n+1} requires solving a J x J complex sparse linear system, where $J = \dim S_h$.

Theorem 2.2. If V^n is the solution of the scheme (2.4) and v the solution of the i-b.v.p. (2.1a)-(2.1d), we have, for k small enough

$$\max_{0 \le n \le N} \| V^n - v(.,t^n) \| \le c \| (V^0 - R_h(v_0))_x \| + c\{h^r + k^2\}, \tag{2.5}$$

where c is a positive constant independent of h, k.

Outline of proof: We suppose again that $\dot{s}(t) \le 0$ in $[0,T]$ and use estimates analogous to those of the semidiscrete case and the fact that the Crank-Nicolson discretization in t is stable and of $O(k^2)$ accuracy. (See [6] for details).

2.3. Abrahamsson-Kreiss boundary condition

If we make the change of variables $v = \exp(-ix^2\omega)u$ with $\omega = \omega(t)$ as before, the problem (1.2a,b,c*,d) becomes simply

$$v_t = i(\xi(t))^{-1}v_{xx} + i(\widetilde{\beta}_R(x,t) + i\widetilde{\beta}_I(x,t))v, \ 0 \le x \le 1, t \in [0,T],$$
$$v(0,t) = 0, \ 0 \le t \le T, \tag{2.6}$$
$$v_x(1,t) = 0, \ 0 \le t \le T,$$
$$v(x,0) = v_0(x), \ 0 \le x \le 1.$$

The finite element semidiscretization of (2.6) is now

$$(v_{ht} - i\widetilde{\beta}v_h, \phi) = -i(\xi(t))^{-1}(v_{hx}, \phi_x) \ \text{for} \ \phi \in S_h, \ 0 \le t \le T.$$

We may now prove that $\| v_h - v \| \le c \| v_h(0) - v_0 \| + O(h^r)$, without any assumption on the sign of $\dot{s}(t)$. The Crank-Nicolson scheme is again stable and of $O(h^r + k^2)$ accuracy in L^2. See [6]-[7] for details.

3. Numerical experiments

In this section, we show the results of some numerical experiments that we performed with our fully discrete Crank-Nicolson finite element (FEM) method (taking as S_h the subspace of H_0 consisting of piecewise linear continuous functions on a uniform depth discretization) in the case of the PE with the exact Neumann bottom b. c..

First we compare the results of the FEM with those of the Crank-Nicolson finite difference method FD of [3] (The FD code was kindly supplied to us by Dr. G. E. Zouraris). The test problem involves a slightly downsloping environment with linear bottom. The i-b. v. p. consists of the original physical problem (PE) with bottom b. c. (PB), where

$$f_0 = 80 \quad Hz \ , \quad c = c_0 = 1524.003 \quad m \, / \sec \ ,$$
$$l(r) = 874.89994 \quad \times 10^{-4} \ r + 30.48006 \quad m \ ,$$

$0 \le r \le 5 \ \times 10^4 m$. We took $g(r) = i \, k_0$, while the source was placed at

$z_s = 15.24003 \, m$ and modeled by

$$\psi_0(z) = \sqrt{k_0 / 2} \{ \exp(-(z - z_s)^2 k_0^2 / 4) - \exp(-(z + z_s)^2 k_0^2 / 4) \} .$$

Figure 1. shows the real and imaginary parts of the field variable ψ for both methods on the physical domain and also the Transmission Loss (TL) curve as a function of r at depth $z = 150 \, m$ (TL$= 10 \log_{10}(2 \, | \, \psi(z, r) \, |^2 \, / (\pi \, k_0 \, r))$. It is evident that we have very good agreement between the two codes.

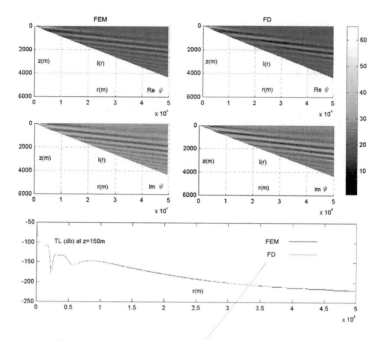

Figure 1. Comparison of FEM and FD for a model PE problem.

The next group of experiments (a - h, Figure 2.) shows results of FEM computations for the i-b. v. p. (2.1a-2.1d) with $a = 1$, $\widetilde{\beta} = 0$, $g_* = 0$, $\lambda = 0$, $v_0(x) = -x(x-1)^3$, in the case of eight bottom profiles $s(t)$, $0 \le t \le 1$, shown for each case on the left icon. The right icon shows the associated L^2 norm of the solution $\| v \|$ as a function of t. In all cases (except the last two) the theoretical question of existence-uniqueness of solutions is open, [2], since $\dot{s}(t)$ is zero or fails to exist at some t.

Our numerical experiments indicate that whenever $\dot{s}(t) \le 0$ (case (a)) the problem is apparently well posed. It also appears that the problem is well posed for seamounts (Fig (c)) or for trenches (Cases (d) and (b).) (In Figure (b) a t-node was placed at $t^* = 0.5$ where $\dot{s}$ does not exist.). We observe that $\| v \|$ grows if $s(t)$ is downsloping (see (c),(e)-(g)), as predicted by [4]. In cases (e), (f) we have $\dot{s}(t^*) = 0$, $\ddot{s}(t^*) = 0$ for $t^* = 0.5$, $t^* = 0$, respectively and $\dot{s}(t) > 0$, $\ddot{s}(t) > 0$ for $t > t^*$. In case (e) an apparent singularity develops at $t = t^*$, causing the L^2 norm of the solution to grow rapidly for $t > t^*$. Case (f) shows a weaker growth. This interesting phenomenon of blow-up (?) in cases like (e) is currently under further investigation. More results can be found in [6].

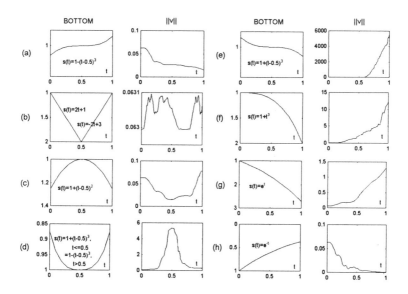

Figure 2. L^2 norm of v for various bottom profiles. Problem (2.1a-2.1d).

Acknowledgments

This work was supported by IACM, FORTH and a grant by the Research Committee of the University of Athens. The authors thank Dr. G. E. Zouraris for his comments and advice.

References

1. F. D. Tappert, *The parabolic approximation method*, in 'Wave Propagation and Underwater Acoustics', J. B. Keller and J. S. Papadakis, eds., Lecture Notes in Phys. 70, Springer-Verlag, Berlin, 1977, pp. 224-287.
2. L. Abrahamsson and H.-O. Kreiss, *The initial boundary value problem for the Schrödinger equation*, Math. Methods Appl. Sci., 13 (1990), pp. 385-390.
3. G. D. Akrivis, V. A. Dougalis, and G. E. Zouraris, *Finite difference schemes for the 'Parabolic' Equation in a variable depth environment*

with a rigid bottom boundary condition, SIAM J. Numer. Anal., 39 (2001), pp. 539-565.

4. L. Abrahamsson and H.-O. Kreiss, *Boundary conditions for the parabolic equation in a range-dependent duct*, J. Acoust. Soc. Amer., 87 (1990), pp. 2438-2441.

5. F. B. Sturm, *Modélisation mathématique et numérique d'un problème de propagation en acoustique sous-matine : prise en compte d'un environnement variable tridimensionnel*, Thèse de Docteur en Sciences, Université de Toulon et du Var, France, 1997.

6. D. C. Antonopoulou and V. A. Dougalis, *Finite element methods for the 'Parabolic' Equation in a variable depth environment with a rigid bottom boundary condition.* (to appear.)

7. D. C. Antonopoulou, Ph.D. Thesis (in preparation).

MLPG METHODS FOR DISCRETIZING
WEAKLY SINGULAR BIES

S. N. ATLURI AND Z. D. HAN

Center for Aerospace Research & Education, University of California, Irvine
5251 California Avenue, Suite 140Irvine, CA, 92612, USA

The general Meshless Local Petrov-Galerkin (MLPG) type weak-forms of the displacement & traction boundary integral equations are presented, for solids undergoing small deformations. Using the directly derived non-hyper singular integral equations for displacement gradients, simple and straight-forward derivations of weakly singular traction BIE's for solids undergoing small deformations are also presented. As a framework for meshless approaches, the MLPG weak forms provide the most general basis for the numerical solution of the non-hyper-singular displacement and traction BIEs. By employing the various types of test functions, several types of MLPG/BIEs are formulated. Numerical examples show that the present methods are very promising, especially for solving the elastic problems in which the singularities in displacemens, strains, and stresses, are of primary concern.

1. Introduction

The boundary integral equations (BIEs) have been developed for solving PDEs during the passed several decades. The BIE methods become even more powerful, when several fast algorithms are combined, such as the panel-clustering method, multi-pole expansions, fast Fourier-transforms, wavelet methods, and so on. However, it is well known that hyper-singularities are encountered, when the displacement BIEs are directly differentiated to obtain the traction BIEs, usually for solving crack problems. The hyper-singularities also make it very difficult for the fast algorithms to be applied to traction BIEs, even after the some de-singularization. In contrast, as far back as 1989, Okada, Rajiyah, and Atluri [9] have proposed a simple way to *directly derive* the integral equations for gradients of displacements. It simplified the derivation processes by taking the gradients of the displacements of the foundation solutions as the test functions, while writing the balance laws in their weak- form. It resulted in "non-hyper-singular" boundary integrals of the gradients of displacements. Recently, this concept has been followed and extended for a *directly-derived* traction BIE[7], which is also "non-hyper-singular" $[1/r^2]$, as opposed to being "hyper-singular" $[1/r^3]$, as in the most common literature for 3D tBIEs.

The meshless methods have been investigated in recent years, besides the traditional element-based methods for solving BIEs. As a systematic framework for developing various meshless methods, the Meshless Local

Petrov-Galerkin (MLPG) approach has been proposed as a fundamentally new concept[1-4]. The many research demonstrates that the MLPG method is one of the most promising alternative methods for computational mechanics. In this paper, we write the local weak-forms of non-hyper-singular displacement & traction BIEs, in the local sub-boundary surfaces, throught the MLPG approach. From this general MLPG approach, various boundary solution methods can be easily derived by choosing, a) a variety of the meshless interpolation schemes for the trial functions, b) a variety of test functions over the local sub-boundary surfaces, and c) a variety of numerical integration schemes. Such variants have been excellently summarized by Atluri and Shen[2,3] in the complementary formulation of the MLPG approach for domain- solutions. In this paper, we focus on developing the general MLPG/BIE approach, and demonstrate its variants as the collocation method (MLPG/BIE1); MLS interpolation with its weight function as the test function (MLPG/BIE2); and general interpolation with the nodal trial function as the nodal test function (MLPG/BIE6), similar to those studied in Atluri and Shen[2,3]. It should be pointed out that MLPG/BIE methods here are not limited to those variants that are presented here; and many other special suitable combinations are feasible, according to the problems to be solved. We implement the MLPG/BIE6 and solve some elastic problems, including fracture mechanics problems of non-planar crack-growth.

2. Non-Hyper-singular MLPG Displacement and Traction BIEs

The non-hypersingular displacement and traction BIEs for a linear elastic, homogeneous, isotropic solid, are summarized in this section. Consider a linear elastic, homogeneous, isotropic body in a domain Ω, with a boundary $\partial\Omega$. The Lame' constants of the linear elastic isotropic body are λ and μ; and the corresponding Young's modulus and Poisson's ratio are E and υ, respectively. We use Cartesian coordinates ξ_i, and the attendant base vectors $\mathbf{e}_i$, to describe the geometry in Ω. The solid is assumed to undergo infinitesimal deformations. The equations of balance of linear and angular momentum can be written as:

$$\nabla \cdot \boldsymbol{\sigma} + \mathbf{f} = \mathbf{0}; \quad \boldsymbol{\sigma} = \boldsymbol{\sigma}^{t} \qquad (1)$$

The constitutive relations of an isotropic elastic homogeneous solid are:

$$\boldsymbol{\sigma} = \lambda \mathbf{I}(\nabla \cdot \mathbf{u}) + 2\mu\boldsymbol{\varepsilon} \qquad (2)$$

It is well known that the displacement vector, which is a continuous function of ξ, can be derived, in general, from the Galerkin-vector-potential φ such that:

$$\mathbf{u} = \nabla^2 \varphi - \frac{1}{2(1-\upsilon)} \nabla(\nabla \cdot \varphi) \tag{3}$$

Consider a point unit load applied in an arbitrary direction $\mathbf{e}^p$ at a generic location $\mathbf{x}$ in a linear elastic isotropic homogeneous infinite medium. It is well-known that the displacement solution is given by the Galerkin-vector-displacement-potential:

$$\varphi^{*p} = (1-\upsilon)F^* \mathbf{e}^p \tag{4}$$

in which F^* is a scalar function[7]. The corresponding displacements are derived, using (3), as:

$$u_i^{*p}(\mathbf{x},\xi) = (1-\upsilon)\delta_{pi}F^*_{,kk} - 0.5 F^*_{,pi} \tag{5}$$

and the gradients of the displacements in (5) are:

$$u_{i,j}^{*p}(\mathbf{x},\xi) = (1-\upsilon)\delta_{pi}F^*_{,kkj} - 0.5 F^*_{,pij} \tag{6}$$

By taking the fundamental solution $u_i^{*p}(\mathbf{x},\xi)$ in Eq. (5) as the test functions, one may write the weak-form of the equilibrium Eq. (1). The traditional displacement BIE can be written as,

$$u_p(\mathbf{x}) = \int_{\partial\Omega} t_j(\xi) u_j^{*p}(\mathbf{x},\xi)\, dS - \int_{\partial\Omega} n_i(\xi) u_j(\xi) \sigma_{ij}^{*p}(\mathbf{x},\xi)\, dS \tag{7}$$

Instead of the **scalar** weak form of Eq. (1), as used for the displacement BIE, we may also write a **vector** weak form of Eq. (1), by using the tensor test functions $u_{i,j}^{*p}(\mathbf{x},\xi)$ in Eq. (6) [as originally proposed in Okada, Rajiyah, and Atluri[9], and derive a non-hypersingular integral equation for tractions in a linear elastic solid,

$$-t_b(\mathbf{x}) = \int_{\partial\Omega} t_q(\xi) n_a(\mathbf{x}) \sigma_{ab}^{*q}(\mathbf{x},\xi)\, dS + \int_{\partial\Omega} D_p u_q(\xi) n_a(\mathbf{x}) \Sigma_{abpq}^*(\mathbf{x},\xi)\, dS \tag{8}$$

where u_j^{*p}, σ_{ij}^{*p} and Σ_{abpq}^* are kernel functions, which were first given in Han and Atluri[7]; the surface tangential operator D_t is defined as,

$$D_t = n_r e_{rst} \frac{\partial}{\partial \xi_s} \tag{9}$$

It should be noted that the integral equations for $u_p(\mathbf{x})$ and $u_{p,k}(\mathbf{x})$ as in Eqs. (7) and (8) are derived independently of each other. On the other hand, if we derive the integral equation for the displacement-gradients, by directly differentiating $u_p(\mathbf{x})$ in Eq. (7), a hyper-singularity is clearly introduced.

3. MLPG Approaches

Following the general idea as presented in [4], one may consider a local sub-boundary surface $\partial \Omega_L$, with its boundary contour Γ_L, as a part of the whole boundary-surface, as shown in Figure 1, for a 3-D solid. Eqs. (7) and (8) may be satisfied in weak-forms over the sub-boundary surface $\partial \Omega_L$, by using a Local Petrov-Galerkin scheme, as:

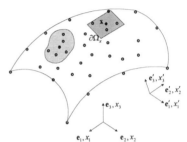

Figure 1   a local sub-part of the boundary around point x

$$\int_{\partial \Omega_L} w_p(\mathbf{x}) u_p(\mathbf{x}) dS_x = \int_{\partial \Omega_L} w_p(\mathbf{x}) dS_x \int_{\partial \Omega} t_j(\xi) u_j^{*p}(\mathbf{x},\xi) \, dS$$
$$- \int_{\partial \Omega_L} w_p(\mathbf{x}) dS_x \int_{\partial \Omega} n_i(\xi) u_j(\xi) \sigma_{ij}^{*p}(\mathbf{x},\xi) \, dS \tag{10a}$$

$$- \int_{\partial \Omega_L} w_b(\mathbf{x}) t_b(\mathbf{x}) dS_x = \int_{\partial \Omega_L} w_b(\mathbf{x}) dS_x \int_{\partial \Omega} t_q(\xi) n_a(\mathbf{x}) \sigma_{ab}^{*q}(\mathbf{x},\xi) \, dS_\xi$$
$$+ \int_{\partial \Omega_L} w_b(\mathbf{x}) dS_x \int_{\partial \Omega} D_p u_q(\xi) n_a(\mathbf{x}) \Sigma_{abpq}^{*}(\mathbf{x},\xi) \, dS_\xi \tag{10b}$$

where $w_b(\mathbf{x})$ is a test function. If $w_b(\mathbf{x})$ is chosen as a Dirac delta function, i.e. $w_b(\mathbf{x}) = \delta(\mathbf{x},\mathbf{x}_m)$ at $\partial \Omega_L$, we obtain the standard "collocation" method for displacement and traction BIEs, at the collocation point $\mathbf{x}_m$. With some basic identities of the fundamental solution, one may obtain the fully desingularized dBIE and tBIE for the standard "collocation" methods as,

$$0 = \int_{\partial\Omega} t_j(\xi) u_j^{*p}(\mathbf{x},\xi)\, dS - \int_{\partial\Omega} n_i(\xi)[u_j(\xi) - u_j(\mathbf{x})]\sigma_{ij}^{*p}(\mathbf{x},\xi)\, dS$$

(11)a

$$0 = \int_{\partial\Omega} [t_q(\xi) - n_p(\xi)\sigma_{pq}(\mathbf{x})] n_a(\mathbf{x})\sigma_{ab}^{*q}(\mathbf{x},\xi)\, dS$$
$$+ \int_{\partial\Omega} [D_p u_q(\xi) - (D_p u_q)(\mathbf{x})] n_a(\mathbf{x})\Sigma_{abpq}^{*}(\mathbf{x},\xi)\, dS$$

(11)b

If $w_b(\mathbf{x})$ is chosen such that it is continuous over the local sub boundary-surface $\partial\Omega_L$ and zero at the contour Γ_L, one may apply Stokes' theorem to Eq. (10), and re-write it as:

$$\frac{1}{2}\int_{\partial\Omega_L} w_p(\mathbf{x}) u_p(\mathbf{x}) dS_x = \int_{\partial\Omega_L} w_p(\mathbf{x}) dS_x \int_{\partial\Omega} t_j(\xi) u_j^{*p}(\mathbf{x},\xi)\, dS_\xi$$
$$+ \int_{\partial\Omega_L} w_p(\mathbf{x}) dS_x \int_{\partial\Omega} D_i(\xi) u_j(\xi) G_{ij}^{*p}(\mathbf{x},\xi)\, dS_\xi$$

(12)a

$$+ \int_{\partial\Omega_L} w_p(\mathbf{x}) dS_x \int_{\partial\Omega}^{CPV} n_i(\xi) u_j(\xi)\phi_{ij}^{*p}(\mathbf{x},\xi)\, dS_\xi$$

$$-\frac{1}{2}\int_{\partial\Omega_L} t_b(\mathbf{x}) w_b(\mathbf{x}) dS_x = \int_{\partial\Omega_L} D_a w_b(\mathbf{x}) dS_x \int_{\partial\Omega} t_q(\xi) G_{ab}^{*q}(\mathbf{x},\xi)\, dS_\xi$$
$$- \int_{\partial\Omega} t_q(\xi)\, dS_\xi \int_{\partial\Omega_L}^{CPV} n_a(\mathbf{x}) w_b(\mathbf{x})\phi_{ab}^{*q}(\mathbf{x},\xi) dS_x$$
$$+ \int_{\partial\Omega_L} D_a w_b(\mathbf{x}) dS_x \int_{\partial\Omega} D_p u_q(\xi) H_{abpq}^{*}(\mathbf{x},\xi)\, dS_\xi$$

(12)b

where G_{ab}^{*q}, ϕ_{ab}^{*q} and H_{abpq}^{*} are kernel functions[7].

If the test function $w_b(\mathbf{x})$ is chosen to be identical to a function that is energy-conjugate to u_p (for dBIE) and t_b (for tBIE), namely, the nodal trial function $\hat{t}_p(\mathbf{x})$ and $\hat{u}_b(\mathbf{x})$, respectively, we obtain the local weak-forms of the symmetric Galerkin dBIE and tBIE.

4. Numerical Implementation

4.1. *Moving Least Squares for 3D Surface*

It is very common to adopt the moving least squares (MLS) interpolation scheme for interpolating the trial functions over a 3-D surface, as it has been done successfully in meshless domain methods[4]. Unfortunately, the moment matrix in the MLS interpolation sometimes becomes singular, when it is applied to three-dimensional surface approximation, if Cartesian coordinates are used. An alternative choice is to use the curvilinear coordinates[5], or choose a varying polynomial basis instead of the complete basis[8]. However, these algorithms require the local geometry information, which hinder the truly meshless implementation. In the present study, the enhanced MLS approximation has been implemented, after reconditioning the singular moment matrix, while still using the global Cartesian coordinates to approximate the trial function over a surface.

Consider a local sub-part of the boundary $\partial \Omega$, of a 3-D solid, denoted as $\partial \Omega_x$, the neighborhood of a point $\mathbf{x}$, which is a local region in the global boundary $\partial \Omega$. To approximate the function u in $\partial \Omega_x$, over a number of scattered points $\{\mathbf{x}_I\}, (I = 1,2,...,n)$ (where $\mathbf{x}$ is given, in the global Cartesian coordinates by x_1, x_2 and x_3), the moving least squares approximation $u(\mathbf{x})$ of u, $\forall \mathbf{x} \in \partial \Omega_x$, can be defined by

$$u(\mathbf{x}) = \mathbf{p}^T(\mathbf{x})\mathbf{a}(\mathbf{x}) \quad \forall \mathbf{x} \in \partial \Omega_x \tag{13}$$

where $\mathbf{p}^T(\mathbf{x}) = [p_1(\mathbf{x}), p_2(\mathbf{x}),..., p_m(\mathbf{x})]$ is a monomial basis of order m; and $\mathbf{a}(\mathbf{x})$ is a vector containing coefficients, which are functions of the global Cartesian coordinates $[x_1, x_2, x_3]$, depending on the monomial basis. They are determined by minimizing a weighted discrete L_2 norm, defined, as:

$$J(\mathbf{x}) = \sum_{i=1}^{m} w_i(\mathbf{x})[\mathbf{p}^T(\mathbf{x}_i)\mathbf{a}(\mathbf{x}) - \hat{u}_i]^2 \equiv [\mathbf{P}\mathbf{a}(\mathbf{x}) - \hat{\mathbf{u}}]^T \mathbf{W}[\mathbf{P}\mathbf{a}(\mathbf{x}) - \hat{\mathbf{u}}]$$

$$\tag{14}$$

where $w_i(\mathbf{x})$ are the weight functions and $\hat{u}_i$ are the fictitious nodal values. The stationarity of J in Eq. (14), with respect to $\mathbf{a}(\mathbf{x})$ leads to following linear relation between $\mathbf{a}(\mathbf{x})$ and $\hat{\mathbf{u}}$,

$$\mathbf{A}(\mathbf{x})\mathbf{a}(\mathbf{x}) = \mathbf{B}(\mathbf{x})\hat{\mathbf{u}} , \quad \mathbf{A}(\mathbf{x}) = \mathbf{P}^T\mathbf{W}\mathbf{P} \quad \mathbf{B}(\mathbf{x}) = \mathbf{P}^T\mathbf{W} \quad \forall \mathbf{x} \in \partial \Omega_x$$

$$\tag{15}$$

The MLS approximation is well defined only when the matrix $\mathbf{A}(\mathbf{x})$ in Eq. (15) is non-singular. It needs to be reconditioned, if the monomial basis defined in the global Cartesian coordinate system for an approximation of u as in Eq. (13), becomes nearly linearly dependent on a 3-D surface. One may define a local set of orthogonal coordinates, x_i' as in Figure 1, on $\partial \Omega_x$. One may rewrite Eq. (13) as:

$$u = [1; x_1; x_2; x_3; x_1^2; x_2^2; x_3^2; x_1 x_2; x_2 x_3; x_3 x_1; ...]$$
$$[a_1(\mathbf{x}); a_2(\mathbf{x}); a_3(\mathbf{x}); a_4(\mathbf{x}); ...]^T$$
$$\equiv [1; x_1'; x_2'; x_3'; x_1'^2; x_2'^2; x_3'^2; x_1' x_2'; x_2' x_3'; x_3' x_1'; ...]$$
$$[a_1'(\mathbf{x}); a_2'(\mathbf{x}); a_3'(\mathbf{x}); a_4'(\mathbf{x}); ...]^T$$

for $\quad \forall \mathbf{x} \in \partial\Omega_x$

$$(16)$$

Suppose $\partial\Omega_x$ becomes nearly planar, which may be defined in the local-set of orthogonal coordinates x_i', for instance, as $x_3' = \text{constant}$. It is then clear that the monomial basis in Eq. (16), in terms of x_i' becomes linearly dependent. In fact, one may make the basis to be *linearly indepent* again in Eq. (16), for instance, for $x_3' = \text{constant}$, by setting the corresponding coefficients $a'(\mathbf{x})$ to be zero. When this is done, the order of the vector $\mathbf{p}'(\mathbf{x})$ is correspondingly reduced; and thus, correspondingly, the order of $\mathbf{A}(\mathbf{x})$ in Eq. (15) is reduced. Thus, it can be seen that if one proceeds with a full monomial basis, with m basis functions in x_i coordinates in Eq. (16), and if the points on $\partial\Omega_x$ are not all in the same plane, the matrix $\mathbf{A}(\mathbf{x})$ in Eq. (15) will have the full rank of m. On the other hand, if $\partial\Omega_x$ becomes almost planar, say normal to x_3', then the rank of $\mathbf{A}(\mathbf{x})$ is clearly only $(m - n)$, where n is the reduction in the number of basis due to the fact that $x_3' = \text{constant}$. Thus, by simply monitoring the eigen-values of $\mathbf{A}(\mathbf{x})$, and if a set of eigen-values becomes nearly or precisely zero, we automatically detect that $\partial\Omega_x$ is becoming nearly planar. In addition, it implies that the normal to the surface can be determined from the lowest eigenvalue of matrix $\mathbf{A}(\mathbf{x})$ when it is singular or nearly-singular, without the local geometry information. It makes the present method to be truly meshless, which does need any background cells to define the geometry as well as the normal direction, if the boundary integrals are handled based on the nodal influence domain.

Once coefficients $\mathbf{a}(\mathbf{x})$ in Eq. (15) are determined, one may obtain the approximation from the nodal values at the local scattered points, by substituting them into Eq. (13), as

$$u(\mathbf{x}) = \mathbf{\Phi}^T(\mathbf{x})\hat{\mathbf{u}} \quad \forall \mathbf{x} \in \partial\Omega_x \qquad (17)$$

where $\mathbf{\Phi}(\mathbf{x})$ is the so-called shape function of the MLS approximation, as,

$$\mathbf{\Phi}(\mathbf{x}) = \mathbf{p}^T(\mathbf{x})\mathbf{A}^{-1}(\mathbf{x})\mathbf{B}(\mathbf{x}) \qquad (18)$$

The weight function in Eq. (14) defines the range of influence of node I. Normally it has a compact support. The possible choices are the Gaussian and spline weight functions with compact supports.

It should be pointed out that the shape functions given in Eq. (18) are based on the fictitious nodal values. This introduces an additional complication, since all the nodal values in BIEs are the direct boundary values, a situation which is totally different from the domain meshless methods. As a practical way, a conversion matrix is used to map the fictitious values to true values and applied to the system equations.

4.2. Variants of the MLPG/BIE: Several Types of Interpolation (Trial) and Test functions, and Integration schemes

The general MLPG/BIE approaches are given through Eqs. (10) and (12). For a so-called meshless implementation, we need to introduce a meshless interpolation scheme to approximate the trial functions over the surface of a three-dimensional body. Considering a meshless approximation, one may define the non-overlapping cells for the whole boundary as,

$$\partial\Omega = \bigcup_{i=1}^{NC} Cell_i \quad \text{and} \quad Cell_i \cap Cell_j = 0 \text{ for } \forall i \neq j \tag{19}$$

where NC is the number of cells. Thus, one may obtain the boundary integrals, in the meshless form, as,

$$\int_{\partial\Omega} f(\mathbf{x}, \xi)\, dS = \sum_{i=1}^{NC} \int_{Cell_i} f(\mathbf{x}, \mathbf{s})\, ds \tag{20}$$

in which the curvilinear coordinates $\mathbf{s}$ are used.

In addition, if the variables are interpolated from the nodal ones as in Eq. (17) and the range of influence of each node is compact, one set of overlapped nodal domains may be defined as,

$$\partial\Omega = \bigcup_{i=1}^{NN} DOI_i \quad \text{and} \quad DOI_i \cap DOI_j \neq 0 \text{ for } \exists i \neq j \tag{21}$$

where NN is the number of nodes and DOI_i is the influence domain of node i. *The boundary integrals can be re-written in the truly meshless form, as,*

$$\int_{\partial\Omega} f(\mathbf{x},\xi) \, dS = \sum_{i=1}^{NN} \int_{DOI_i} f(\mathbf{x},\mathbf{s}) \, d\mathbf{s} \qquad (22)$$

This integration scheme has been applied to BNM for solving potential problems[5].

Besides the traditional collocation methods, the weakly-singular BIEs in Eq. (11) can also be easily used for developing the meshless BIEs in their numerically tractable weak-forms. All piece-wise continuous functions can be used here as the test functions. For example, all the weight functions in MLS approximation discussed above, are suitable for such purposes. In addition, one may also directly use the nodal shape function as the nodal test functions as in Eq. (18), which leads to the symmetric BIEs in Eq. (12). In summary, the several variants, of the general MLPG/BIE approach, are defined through a suitable combination of the trial-function interpolation scheme, the choice for the test functions, and the choices for the integration scheme.

5. Numerical Experiments

Several problems in three-dimensional linear elasticity are solved to illustrate the effectiveness of the present method. The numerical results of the MLPG/BIE6 method as applied to problems in 3D elasto-statics, specifically (i) a cube, (ii) a hollow sphere, (iii) a concentrated load on a semi-infinite space, and (iv) non-planar fatigue growth of an elliptical crack, are discussed.

5.1. Cube under Uniform Tension

The first example is the standard patch test, shown in Figure 2. A cube under the uniform tension is considered. The material parameters are taken as $E = 1.0$, and $\nu = 0.25$. All six faces are modeled with the same

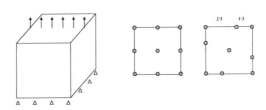

Figure 2 A cube under uniform tension, and its nodal configurations

configurations with 9 nodes. Two nodal configurations are used for the testing purpose: one is regular and another is irregular, as shown in Figure 2.

In the patch tests, the uniform tension stress is applied on the upper face and the proper displacement constraints are applied to the lower face.

The satisfaction of the patch test requires that the displacements are linear on the lateral faces, and are constant on the upper face; and the stresses are constant on all faces. It is found that the present method passes the patch tests. The maximum numerical errors are 1.7×10^{-7} and 3.5×10^{-7} for two nodal configurations, respectively, which may be limited by the computer.

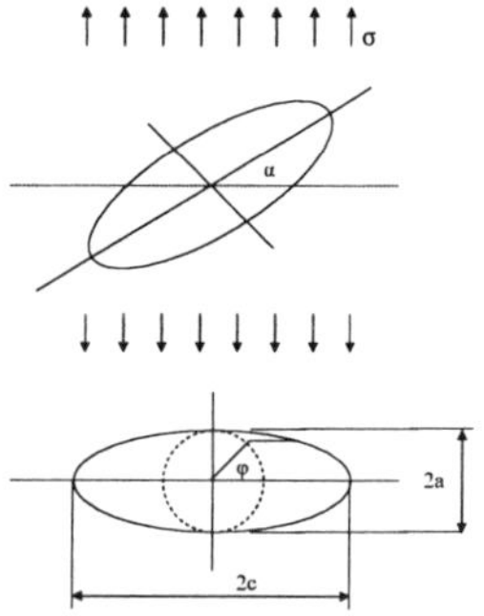

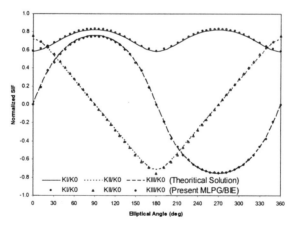

Figure 3   Inclined elliptical crack under tension

5.2. Non-planar Crack Growth

An inclined elliptical crack with semi-axes c and a, subjected to fatigue loading, is shown in Figure 3. Its orientation is characterized by an angle, α. The present meshless method is applied to solve this problem. The nodal configuration is used to model the crack inclined at 45 degrees with 249 nodes. The exact solution for a tensile loading σ is given in [10].

As a mixed-mode crack, the distribution of all three stress intensity factors, K_I, K_{II} and K_{III}, along the crack front are shown in Figure 4, after being normalized by $K_0 = \sigma \sqrt{\pi a} / 2$. It can be seen that a good agreement of the present numerical results with the theoretical solution is obtained.

Figure 4   Normalized stress intensity factors along the crack front of an inclined elliptical crack under tensile load

The fatigue growth is also performed for this inclined crack. The Paris model is used to simulate fatigue crack growth, as:

$$da / dN = C (\Delta K_{eff})^n \qquad (23)$$

in which the material parameters C and n are taken for 7075 Aluminum as $C = 1.49 \times 10^{-8}$ and $n = 3.21$ [6]. The crack growth is simulated by adding nodes along the crack front. The newly added points are determined through the K solutions. 7 increments are performed to grow the crack from the initial size $a = 1$ to the final size $a = 2.65$. The normalized stress intensity factors during the crack growing are given in Figure 5, which are also normalized by K_0. The results show that K_I keeps increasing while K_{II} and K_{III} are decreasing during the crack growth. It confirms that this mixed-mode crack becomes a mode-I dominated one, while growing. The shape of the final crack is shown in Figure 6. It is clear that while the crack, in its initial configuration, starts out as a mixed-mode crack; and after a substantial growth, the crack configuration is such that it is in a pure mode-I state.

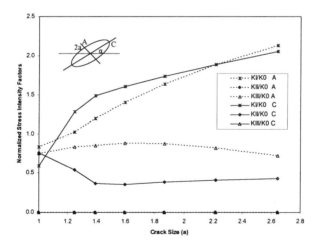

Figure 5   Normalized stress intensity factors for the mixed-mode fatigue growth of an inclined elliptical crack

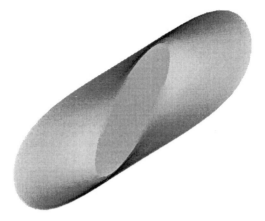

Figure 6   Final shape of an inclined elliptical crack after mixed-model growth

6. Closure

In this paper, we presented the "Meshless Local Petrov-Galerkin BIE Methods" (MLPG/BIE), by using the concept of the general meshless local Petrov-Galerkin (MLPG) approach developed in Atluri et al[1-4]. The several variants of the MLPG/BIE solution methods are also formulated, in terms of the varieties of the interpolation schemes for trial functions, the test functions, and the integration schemes. With the use of a nodal influence domain, truly meshless BIEs are also presented. The MLS surface-interpolation, with the use of Cartesian coordinates, is enhanced for the three dimensional surface without the requirement of a mesh or cells, to define the local geometry. It leads to the truly meshless BIE methods with the use of the nodal influence domain for the boundary integrations. The accuracy and efficiency of the present MLPG approach are demonstrated with numerical results.

References

1. Atluri, S. N.; Han, Z. D.; Shen, S.: Meshless Local Patrov-Galerkin (MLPG) approaches for weakly-singular traction & displacement boundary integral equations, CMES: *Computer Modeling in Engineering & Sciences*, vol. 4, no. 5, pp. 507-517, 2003.
2. Atluri, S. N.; Shen, S.: The meshless local Petrov-Galerkin (MLPG) method. Tech. Science Press, 440 pages, 2002.
3. Atluri, S. N.; Shen, S.: The meshless local Petrov-Galerkin (MLPG) method: A simple & less-costly alternative to the finite element and

boundary element methods. CMES: *Computer Modeling in Engineering & Sciences*, vol. 3, no. 1, pp. 11-52, 2002.

4. Atluri, S. N.; Zhu, T.: A new meshless local Petrov-Galerkin (MLPG) approach in computational mechanics. *Computational Mechanics.*, Vol. 22, pp. 117-127, 1998.

5. Gowrishankar, R.; Mukherjee S.: A 'pure' boundary node method for potential theory, *Comminucations in Numerical Methods,* vol. 18, pp. 411-427, 2002.

6. Han. Z. D.; Atluri, S. N.: SGBEM (for Cracked Local Subdomain) – FEM (for uncracked global Structure) Alternating Method for Analyzing 3D Surface Cracks and Their Fatigue-Growth, CMES: *Computer Modeling in Engineering & Sciences*, vol. 3, no. 6, pp. 699-716, 2002.

7. Han. Z. D.; Atluri, S. N.: On Simple Formulations of Weakly-Singular Traction & Displacement BIE, and Their Solutions through Petrov-Galerkin Approaches, CMES: *Computer Modeling in Engineering & Sciences*, vol. 4 no. 1, pp. 5-20, 2003.

8. Li, G.; Aluru, N. R.: A boundary cloud method with a cloud-by-cloud polynomial basis, *Engineering Analysis with Boundary Elements*, vol. 27, pp. 57-71, 2003.

9. Okada, H.; Rajiyah, H., Atluri, S.N.: Non-hyper-singular integral representations for velocity (displacement) gradients in elastic/plastic solids (small or finite deformations), *Computational Mechanics.*, vol. 4, pp. 165-175, 1989.

10. Tada, H.; Paris, P.C.; Irwin, G.R. *The Stress Analysis of Cracks Handbook*, ASME Press, 2000.

HOMOGENIZATION OF MAXWELL'S EQUATIONS IN LINEAR DISPERSIVE BIANISOTROPIC MEDIA

G. BARBATIS

Department of Mathematics, University of Ioannina
45110 Ioannina, Greece
E-mail: gbarbati@cc.uoi.gr

I.G. STRATIS

Department of Mathematics, University of Athens
Panepistimiopolis, 15784 Athens, Greece
E-mail: istratis@math.uoa.gr

We study the periodic homogenization of Maxwell's equations for linear dissipative bianisotropic media in the time domain. We consider general constitutive laws where dispersive effects are taken into account.

1. Introduction - Formulation

In many problems in Applied Mathematics and other sciences one is led to the study of boundary value problems in media with periodic structure. It is often the case that one is interested to the behaviour of the solutions as the period becomes very small. The aim of homogenization theory is precisely to describe the limit behaviour of such solutions[2,3]. Our aim here is to study the homogenization problem for Maxwell's equations in a general setting. We work in the time domain, and consider general linear bianisotropic constitutive laws, taking dispersive effects into account. We note that the proof of the main theorem has been sketched in[1], where we refer for more references on the homogenization of Maxwell's equations.

Let Ω be a domain in $\mathbf{R}^3$ with C^1 boundary $\partial\Omega$. We consider Maxwell's equations

$$\partial_t \mathbf{D} = \operatorname{curl} \mathbf{H} + \mathbf{F}(\mathbf{x}, t) \tag{1}$$

$$\partial_t \mathbf{B} = -\operatorname{curl} \mathbf{E} + \mathbf{G}(\mathbf{x}, t), \qquad \mathbf{x} \in \Omega, \ t > 0, \tag{2}$$

with initial conditions

$$\mathbf{E}(\mathbf{x}, 0) = \mathbf{0}, \quad \mathbf{H}(\mathbf{x}, 0) = \mathbf{0}, \qquad \mathbf{x} \in \Omega, \tag{3}$$

and the perfect conductor boundary condition

$$\mathbf{n} \times \mathbf{E} = \mathbf{0}, \qquad\qquad \mathbf{x} \in \partial\Omega,\ t > 0, \tag{4}$$

where $\mathbf{n}$ is the outward unit normal on $\partial\Omega$.

The constitutive relations for a linear bianisotropic medium have the following general form[7]:

$$\mathbf{D} = \eta\mathbf{E} + \xi\mathbf{H} + \eta_d * \mathbf{E} + \xi_d * \mathbf{H}$$
$$\mathbf{B} = \zeta\mathbf{E} + \mu\mathbf{H} + \zeta_d * \mathbf{E} + \mu_d * \mathbf{H}; \tag{5}$$

here $*$ stands for temporal convolution, i.e.

$$(u * v)(t) = \int_{-\infty}^{t} u(t - s)v(s)ds.$$

The functions η, ξ, ζ and μ belong in the space $M_3(L^\infty(\Omega))$ of 3×3 real matrices with entries in $L^\infty(\Omega)$. They describe the optical (instantaneous) response of the material. The susceptibility functions η_d, ξ_d, ζ_d and μ_d have an additional dependence on time, and we assume that for each fixed t they belong in $M_3(L^\infty(\Omega))$. We point out that they vanish for $t < 0$ due to causality. We do not include electric and magnetic current densities in Maxwell's equations, since such terms can be incorporated in the dispersion terms by a suitable gauge transformation[5].

Physical considerations related to dissipativity imply that the matrices

$$\mathbf{A}(\mathbf{x}) := \begin{pmatrix} \eta & \xi \\ \zeta & \mu \end{pmatrix}, \quad \mathbf{K}(\mathbf{x}, t) := \begin{pmatrix} \eta_d & \xi_d \\ \zeta_d & \mu_d \end{pmatrix},$$

are positive definite with respect to $x \in \Omega$, for each fixed $t > 0$ [5].

The symbols ϵ and ϵ_d are usually used instead of η and η_d, but, as is typical in homogenization problems, we reserve the letter ϵ to stand for a typical length at the microscopic scale.

Using the electromagnetic six-vector field $\mathcal{E}$ and the six-vector flux density $\mathcal{D}$, given, respectively, by

$$\mathcal{E} = \begin{pmatrix} \mathbf{E} \\ \mathbf{H} \end{pmatrix}, \quad \mathcal{D} = \begin{pmatrix} \mathbf{D} \\ \mathbf{B} \end{pmatrix},$$

the constitutive relations (5) are written as a single six-vector equation

$$\mathcal{D} = \mathbf{A}\mathcal{E} + \mathbf{K} * \mathcal{E}. \tag{6}$$

Let us now define the operator

$$\mathbf{Q} := -i \begin{pmatrix} 0 & \mathbf{curl} \\ -\mathbf{curl} & 0 \end{pmatrix}$$

with domain

$$D(\mathbf{Q}) = \left\{ (\mathbf{u}, \mathbf{v}) \; : \; \mathbf{u}, \, \mathbf{v} \in L^2(\Omega) \,, \; \mathbf{curl}\,\mathbf{u}, \; \mathbf{curl}\,\mathbf{v} \in L^2(\Omega) \,, \; \mathbf{n} \times \mathbf{u} = \mathbf{0} \right\}.$$

Here the boundary condition $\mathbf{n} \times \mathbf{curl}\,\mathbf{u} = \mathbf{0}$ is understood in the sense of the trace operator $H(\Omega, \mathbf{curl}) \to H^{-1/2}(\partial\Omega)$. It can be shown that $\mathbf{Q}$ is then self-adjoint on $L^2(\Omega)^4$. Hence the Maxwell system can be written as a non-local in time Cauchy problem in $L^2(\Omega)$,

$$\mathbf{A}\mathcal{E}' + (K * \mathcal{E})' = i\mathbf{Q}\mathcal{E} + \mathcal{F}, \qquad \mathcal{E}(0) = \mathbf{0}. \tag{7}$$

For sufficient conditions for the existence and uniqueness of solution to (7) we refer to[6].

2. Homogenization

Our aim is to study the periodic homogenization problem related to the Maxwell system (7). More precisely, for $\epsilon > 0$ we consider the system

$$\partial_t \mathbf{D}^\epsilon = \mathbf{curl}\,\mathbf{H}^\epsilon + \mathbf{F}(\mathbf{x}, t)$$
$$\partial_t \mathbf{B}^\epsilon = -\mathbf{curl}\,\mathbf{E}^\epsilon + \mathbf{G}(\mathbf{x}, t), \qquad \mathbf{x} \in \Omega \,, \; t > 0, \tag{8}$$
$$\mathbf{E}^\epsilon(\mathbf{x}, 0) = \mathbf{0}, \quad \mathbf{H}^\epsilon(\mathbf{x}, 0) = \mathbf{0},$$
$$\mathbf{n} \times \mathbf{E}^\epsilon = \mathbf{0}, \qquad \mathbf{x} \in \partial\Omega \,, \; t > 0,$$

subject to the periodic constitutive laws

$$\mathbf{D}^\epsilon = \eta^\epsilon \mathbf{E}^\epsilon + \xi^\epsilon \mathbf{H}^\epsilon + \eta_d^\epsilon * \mathbf{E}^\epsilon + \xi_d^\epsilon * \mathbf{H}^\epsilon$$
$$\mathbf{B}^\epsilon = \zeta^\epsilon \mathbf{E}^\epsilon + \mu^\epsilon \mathbf{H}^\epsilon + \zeta_d^\epsilon * \mathbf{E}^\epsilon + \mu_d^\epsilon * \mathbf{H}^\epsilon. \tag{9}$$

The functions $\eta^\epsilon(\mathbf{x}), \zeta^\epsilon(\mathbf{x}), \xi^\epsilon(\mathbf{x}), \mu^\epsilon(\mathbf{x})$ as well as the functions $\eta_d^\epsilon(\mathbf{x}, t)$, $\zeta_d^\epsilon(\mathbf{x}, t), \xi_d^\epsilon(\mathbf{x}, t), \mu_d^\epsilon(\mathbf{x}, t)$ are periodic in $\mathbf{x}$ of period ϵY, where $Y \subset \mathbf{R}^3$ is a parallelepiped which is fixed throughout this paper. Equivalently, the susceptibility (block) matrices $A^\epsilon(\mathbf{x})$ and $\mathbf{K}^\epsilon(\mathbf{x}, t)$ have the form

$$\mathbf{A}^\epsilon(\mathbf{x}) = \mathbf{A}(\mathbf{x}/\epsilon), \quad \mathbf{K}^\epsilon(\mathbf{x}, t) = \mathbf{K}(\mathbf{x}/\epsilon, t)$$

where the matrices $\mathbf{A}(\cdot)$ and $\mathbf{K}(\cdot, t)$ are periodic in $\mathbf{R}^3$ of period Y. Our aim is to describe the asymptotic behaviour of the solution $(\mathbf{E}^\epsilon, \mathbf{H}^\epsilon)$ of the above system in the limit $\epsilon \to 0$.

As it turns out, the functions $\mathbf{E}^\epsilon$ and $\mathbf{H}^\epsilon$ do indeed have limits $\mathbf{E}^*$ and $\mathbf{H}^*$ as $\epsilon \to 0$. These limits will be defined implicitly as solutions of a 'limit' Maxwell system of the same type, that is, using the six-vector notation (7),

$$(\mathbf{A}^h \mathcal{E}^* + \mathbf{K}^h * \mathcal{E}^*)' = i\mathbf{Q}\mathcal{E}^* + \mathcal{F}, \qquad \mathcal{E}^*(0) = \mathbf{0}. \tag{10}$$

As it is intuitively expected, the corresponding constitutive laws turn out to be homogeneous, that is the matrices $\mathbf{A}^h(\mathbf{x})$ and $\mathbf{K}^h(\cdot, \mathbf{x})$ are actually independent of $\mathbf{x} \in \Omega$. These matrices describe the electromagnetic properties of a fictitious homogeneous ('limit') material.

At this point let us recall that if $\phi : \mathbf{R}^3 \to \mathbf{R}$ is periodic of period ϵ and $\phi^\epsilon(\mathbf{x}) = \phi(\mathbf{x}/\epsilon)$, then

$$\phi_\epsilon \to < \phi > \qquad \text{weakly in } L^p(V), \quad 1 < p < +\infty,$$

for all bounded $V \subset \mathbf{R}^3$ (here and below $< \phi >$ denotes the average of ϕ over Y). In this respect it is remarkable (and against one's intuition) that the homogenous susceptibility matrices $\mathbf{A}^h$ and $\mathbf{K}^h(t)$ are in fact not obtained by simply averaging the functions $\mathbf{A}(\mathbf{y})$ and $\mathbf{K}(\mathbf{y}, t)$ over Y.

We recall the defininition of the Hilbert spaces

$$H(\Omega, \mathrm{div}) = \{\mathbf{u} \in L^2(\Omega) \,:\, \mathrm{div}\,\mathbf{u} \in L^2(\Omega)\}$$
$$H(\Omega, \mathbf{curl}) = \{\mathbf{u} \in L^2(\Omega) \,:\, \mathbf{curl}\,\mathbf{u} \in L^2(\Omega)\}$$

and we denote by $\mathcal{D}'(\Omega)$ the space of all distributions on Ω equipped with its standard topology.

To define the homogeneous susceptibility matrices we need some auxiliary definitions. We denote by $\hat{\alpha}(p)$, Re $p > 0$, the Laplace transform of a function $\alpha(t)$, $t > 0$, that is $\hat{\alpha}(p) = \int_0^\infty e^{-pt}\alpha(t)dt$. We make the following dissipativity assumption: there exists $c > 0$ such that the block matrix

$$\hat{\mathbf{A}}(\mathbf{y},p) := \begin{pmatrix} \eta + \hat{\eta}_d & \xi + \hat{\xi}_d \\ \zeta + \hat{\zeta}_d & \mu + \hat{\mu}_d \end{pmatrix} =: \begin{pmatrix} \bar{\eta} & \bar{\xi} \\ \bar{\zeta} & \bar{\mu} \end{pmatrix}$$

satisfies

$$\mathrm{Re}\, \langle \hat{\mathbf{A}}(\mathbf{y},p)\mathcal{U}, \mathcal{U} \rangle \geq c\|\mathcal{U}\|^2, \quad \mathbf{y} \in Y, \; p \in \mathbf{C}_+, \; \mathcal{U} \in \mathbf{R}^6. \tag{11}$$

This is for example satisfied if $\mathbf{K}(\mathbf{y},t)$ is small compared to $\mathbf{A}(\mathbf{y})$, $\mathbf{y} \in Y$. Let $H^1_{per}(Y)$ denote the closed subspace of $H^1(Y)$ that consists of periodic functions and for fixed $p \in \mathbf{C}_+$ we define the operator $L_{p,per} : H^1_{per}(Y) \to (H^1_{per}(Y))^*$ by

$$L_{p,per} = \begin{pmatrix} -\mathrm{div}(\bar{\eta}\,\mathbf{grad}\,) & -\mathrm{div}(\bar{\xi}\,\mathbf{grad}\,) \\ -\mathrm{div}(\bar{\xi}^T\,\mathbf{grad}\,) & -\mathrm{div}(\bar{\mu}\,\mathbf{grad}\,) \end{pmatrix}.$$

The coercivity assumption (11) implies that $L_{p,per}$ is invertible modulo constant functions; in particular we can define the functions u_1^j, u_2^j, v_1^j and v_2^j, $j = 1, 2, 3$, by the relations

$$L_{per}\begin{pmatrix} u_1^j \\ u_2^j \end{pmatrix} = \begin{pmatrix} \partial \bar{\eta}_{ij}/\partial y_i \\ \partial \bar{\zeta}_{ij}/\partial y_i \end{pmatrix}, \quad L_{per}\begin{pmatrix} v_1^j \\ v_2^j \end{pmatrix} = \begin{pmatrix} \partial \bar{\xi}_{ij}/\partial y_i \\ \partial \bar{\mu}_{ij}/\partial y_i \end{pmatrix}.$$

We define the matrices $\tilde{\eta}^h$, $\tilde{\xi}^h$, ζ^h and $\tilde{\mu}^h$ by

$$\tilde{\eta}_{ij}^h = \; < \bar{\eta}_{ij} + \bar{\eta}_{ik}\, \partial_{y_k} u_1^j + \bar{\xi}_{ik}\, \partial_{y_k} u_2^j >,$$

$$\tilde{\xi}_{ij}^h = \; < \bar{\xi}_{ij} + \bar{\xi}_{ik}\, \partial_{y_k} v_2^j + \bar{\eta}_{ik}\, \partial_{y_k} v_1^j >,$$

$$\tilde{\zeta}_{ij}^h = \; < \bar{\zeta}_{ij} + \bar{\zeta}_{ik}\, \partial_{y_k} u_1^j + \bar{\mu}_{ik}\, \partial_{y_k} u_2^j >,$$

$$\tilde{\mu}_{ij}^h = \; < \bar{\mu}_{ij} + \bar{\mu}_{ik}\, \partial_{y_k} v_2^j + \bar{\zeta}_{ik}\, \partial_{y_k} v_1^j > . \tag{12}$$

(Note that while the coefficients of L^h are spatially constant, they do depend on $p \in \mathbf{C}_+$.) We assume that the functions $\tilde{\eta}^h, \tilde{\xi}^h, \tilde{\zeta}^h$ and $\tilde{\mu}^h$, $p \in \mathbf{C}_+$, are Laplace transforms of functions $\eta^h, \xi^h, \zeta^h, \mu^h \in M_3(L^\infty(\Omega))$; see[6].

We assume that the matrix

$$\mathbf{K}^h(t) = \begin{pmatrix} \eta^h(t) & \xi^h(t) \\ \zeta^h(t) & \mu^h(t) \end{pmatrix}$$

has a positive definite real part. This is known to be true if the dispersive matrix $\mathbf{K}(\mathbf{y}, t)$ is zero[2] and will also hold if we view $\mathbf{K}(\mathbf{y}, t)$ as a small enough perturbation of the optical response part $\mathbf{A}(\mathbf{y})$.

Our main result is the following theorem:

Theorem 2.1. *Assume that the Maxwell system (8) - (9) is uniquely solvable for all $\epsilon > 0$ and that $\|\mathbf{E}^\epsilon\|_2, \|\mathbf{H}^\epsilon\|_2 \leq c$ for all $\epsilon, t > 0$. Then the solution $(\mathbf{E}^\epsilon, \mathbf{H}^\epsilon)$ of the above system satisfies*

$$\mathbf{E}^\epsilon \to \mathbf{E}^*, \quad \mathbf{H}^\epsilon \to \mathbf{H}^* \quad \text{*-weakly in } L^\infty((0, \infty), L^2(\Omega)),$$

where $(\mathbf{E}^*, \mathbf{H}^*)$ is the unique solution of the homogeneous Maxwell system

$$\begin{aligned} \partial_t \mathbf{D}^* &= \operatorname{curl} \mathbf{H}^* + \mathbf{F} \\ \partial_t \mathbf{B}^* &= -\operatorname{curl} \mathbf{E}^* + \mathbf{G}, \quad \mathbf{x} \in \Omega, \; t > 0, \\ \mathbf{E}^*(\mathbf{x}, 0) &= \mathbf{0}, \quad \mathbf{H}^*(\mathbf{x}, 0) = \mathbf{0}, \\ \mathbf{n} \times \mathbf{E}^* &= \mathbf{0}, \quad \mathbf{x} \in \partial\Omega, \; t > 0, \end{aligned} \tag{13}$$

subject to the constitutive laws

$$\begin{aligned} \mathbf{D}^* &= \eta^h * \mathbf{E}^* + \xi^h * \mathbf{H}^* \\ \mathbf{B}^* &= \zeta^h * \mathbf{E}^* + \mu^h * \mathbf{H}^*, \quad \mathbf{x} \in \Omega, \; t > 0. \end{aligned} \tag{14}$$

Note. As was indicated in (10), the homogenized coefficient matrix $\mathbf{K}^h(t)$ will have a singular δ-type component at $t = 0$ which corresponds to the optical response part of the homogenized laws; for the sake of simplicity of notation we do not make this explicit.

Proof. The assumption $\|\mathbf{E}^\epsilon\|_2, \|\mathbf{H}^\epsilon\|_2 \le c$ implies that $\mathbf{D}^\epsilon$ and $\mathbf{B}^\epsilon$ are bounded in $L^2(\Omega)$ uniformly in $\epsilon, t > 0$. Hence there exist $\mathbf{E}^*$, $\mathbf{H}^*$, $\mathbf{D}^*$ and $\mathbf{B}^*$ in the space $L^\infty((0,\infty), L^2(\Omega))$ such that, up to taking a subsequence $\epsilon \to 0$, there holds

$$\left. \begin{array}{l} \mathbf{E}^\epsilon \to \mathbf{E}^*, \ \mathbf{H}^\epsilon \to \mathbf{H}^* \\ \mathbf{D}^\epsilon \to \mathbf{D}^*, \ \mathbf{B}^\epsilon \to \mathbf{B}^* \end{array} \right\} \ *\text{- weakly in } L^\infty((0,\infty), L^2(\Omega)) \text{ as } \epsilon \to 0. \quad (15)$$

The ensuing arguments will identify $(\mathbf{E}^*, \mathbf{H}^*)$ and hence will show that *any* $*$-weakly convergent subsequence of $(\mathbf{E}^\epsilon, \mathbf{H}^\epsilon)$ has $(\mathbf{E}^*, \mathbf{H}^*)$ as its limit. This easily implies the convergence of the full sequence $(\mathbf{E}^\epsilon, \mathbf{H}^\epsilon)$.

Let us take the Laplace transform $g(t) \mapsto \hat{g}(p)$, $p \in \mathbf{C}_+ := \{\text{Re } p > 0\}$, of Maxwell's equations (8); we obtain

$$\begin{array}{l} p\hat{\mathbf{D}}^\epsilon = \mathbf{curl}\,\hat{\mathbf{H}}^\epsilon + \hat{\mathbf{F}} \\ p\hat{\mathbf{B}}^\epsilon = -\mathbf{curl}\,\hat{\mathbf{E}}^\epsilon + \hat{\mathbf{G}}, \qquad p \in \mathbf{C}_+, \ \mathbf{x} \in \Omega. \end{array} \quad (16)$$

Moreover (15) implies that

$$\left. \begin{array}{l} \hat{\mathbf{E}}^\epsilon \to \hat{\mathbf{E}}^*, \ \hat{\mathbf{H}}^\epsilon \to \hat{\mathbf{H}}^* \\ \hat{\mathbf{D}}^\epsilon \to \hat{\mathbf{D}}^*, \ \hat{\mathbf{B}}^\epsilon \to \hat{\mathbf{B}}^* \end{array} \right\} \text{ weakly in } L^2(\Omega) \quad (\text{fixed } p \in \mathbf{C}_+.) \quad (17)$$

Combining (16) and (17) implies that for fixed $p \in \mathbf{C}_+$ the vector fields $\mathbf{curl}\,\hat{\mathbf{E}}^\epsilon$ and $\mathbf{curl}\,\hat{\mathbf{H}}^\epsilon$ have L^2-norms that remain bounded as $\epsilon \to 0$. Hence, up to taking a subsequence, they have weak limits in $L^2(\Omega)$. It then follows from (17) that $\hat{\mathbf{E}}^*$ and $\hat{\mathbf{H}}^*$ belong to $H(\Omega, \mathbf{curl})$ and moreover

$$\hat{\mathbf{E}}^\epsilon \to \hat{\mathbf{E}}^*, \qquad \hat{\mathbf{H}}^\epsilon \to \hat{\mathbf{H}}^* \quad \text{weakly in } H(\Omega, \mathbf{curl}). \quad (18)$$

Letting $\epsilon \to 0$ in (16) thus yields

$$\begin{array}{l} p\hat{\mathbf{D}}^* = \mathbf{curl}\,\hat{\mathbf{H}}^* + \hat{\mathbf{F}} \\ p\hat{\mathbf{B}}^* = -\mathbf{curl}\,\hat{\mathbf{E}}^* + \hat{\mathbf{G}}, \qquad p \in \mathbf{C}_+, \ \mathbf{x} \in \Omega, \end{array} \quad (19)$$

which implies that $\mathbf{E}^*, \mathbf{H}^*, \mathbf{D}^*$ and $\mathbf{B}^*$ satisfy the Maxwell system

$$\partial_t \mathbf{D}^* = \mathbf{curl}\,\mathbf{H}^* + \mathbf{F}$$

$$\partial_t \mathbf{B}^* = -\mathbf{curl}\,\mathbf{E}^* + \mathbf{G}, \qquad \mathbf{x} \in \Omega, \ t > 0, \quad (20)$$

$$\mathbf{E}^*(\mathbf{x},0) = 0, \quad \mathbf{H}^*(\mathbf{x},0) = 0, \qquad \mathbf{x} \in \Omega. \quad (21)$$

Hence it remains to establish that the boundary condition $\mathbf{n} \times \mathbf{E}^* = \mathbf{0}$ is also satisfied and that the vector fields $\mathbf{E}^*, \mathbf{H}^*, \mathbf{D}^*$ and $\mathbf{B}^*$ are related by the constitutive laws (14).

Validity of the boundary condition: We first note that the boundary condition is understood in the sense of the trace operator $H(\mathbf{curl}, \Omega) \to H^{-\frac{1}{2}}(\partial\Omega)$, $u \mapsto u \times \mathbf{n}|_{\partial\Omega}$. Let us fix a function $\phi \in H^{\frac{1}{2}}(\partial\Omega)$. There exists[4] $\Phi \in H^1(\Omega)$ such that $\Phi|_{\partial\Omega} = \phi$. Now, for $\epsilon > 0$ there holds

$$\int_\Omega \mathbf{curl}\,\Phi \cdot \mathcal{E}^\epsilon = \int_\Omega \mathbf{curl}\,\mathbf{E}^\epsilon \cdot \Phi + \int_{\partial\Omega} \Phi(\mathbf{E}^\epsilon \times \mathbf{n}),$$

$$\int_\Omega \mathbf{curl}\,\Phi \cdot \mathcal{E}^* = \int_\Omega \mathbf{curl}\,\mathbf{E}^* \cdot \Phi + \int_{\partial\Omega} \Phi(\mathbf{E}^* \times \mathbf{n}).$$

Combining these with the fact that $\mathbf{E}^\epsilon \times \mathbf{n}|_{\partial\Omega} = \mathbf{0}$ and the relations

$$\int_\Omega \mathbf{curl}\,\Phi \cdot \mathbf{E}^\epsilon \to \int_\Omega \mathbf{curl}\,\Phi \cdot \mathbf{E}^*$$

$$\int_\Omega \mathbf{curl}\,\mathbf{E}^\epsilon \cdot \Phi \to \int_\Omega \mathbf{curl}\,\mathbf{E}^* \cdot \Phi, \qquad (\epsilon \to 0)$$

we conclude that

$$\int_{\partial\Omega} \phi(\mathbf{E}^* \times \mathbf{n}) = \int_{\partial\Omega} \Phi(\mathbf{E}^* \times \mathbf{n}) = 0.$$

Since $\phi \in H^{\frac{1}{2}}(\partial\Omega)$ was arbitrary, we conclude that $\mathbf{E}^* \times \mathbf{n} = \mathbf{0}$ on $\partial\Omega$.

Validity of the constitutive laws: Let us fix a subdomain $V \subset\subset \Omega$. Since div $\mathbf{curl} = 0$, (16) and (19) imply that $\mathrm{div}\hat{\mathbf{D}}^\epsilon = \mathrm{div}\hat{\mathbf{D}}^*$ and $\mathrm{div}\hat{\mathbf{B}}^\epsilon = \mathrm{div}\hat{\mathbf{B}}^*$, and (17) then yields

$$\hat{\mathbf{D}}^\epsilon \to \hat{\mathbf{D}}^*, \quad \hat{\mathbf{B}}^\epsilon \to \hat{\mathbf{B}}^* \quad \text{weakly in } H(V, \mathrm{div}). \tag{22}$$

For fixed $p \in \mathbf{C}_+$, let L^ϵ denote the elliptic operator $H_0^1(V) \to H^{-1}(V) := (H_0^1(V))^*$ given in block form by

$$L^\epsilon = \begin{pmatrix} -\mathrm{div}(\eta^\epsilon + \hat{\eta}_d^\epsilon)^T\mathbf{grad} & -\mathrm{div}(\eta^\epsilon + \hat{\eta}_d^\epsilon)^T\mathbf{grad} \\ -\mathrm{div}(\eta^\epsilon + \hat{\eta}_d^\epsilon)^T\mathbf{grad} & -\mathrm{div}(\eta^\epsilon + \hat{\eta}_d^\epsilon)^T\mathbf{grad} \end{pmatrix};$$

here M^T denotes the complex transpose of the matrix M. Then L^ϵ is invertible for all $\epsilon > 0$ by (11). Now, let $g_1, g_2 \in H^{-1}(V)$ be fixed and let $u^\epsilon, v^\epsilon \in H_0^1(V)$ solve the system

$$L^\epsilon \begin{pmatrix} u^\epsilon \\ v^\epsilon \end{pmatrix} = \begin{pmatrix} g_1 \\ g_2 \end{pmatrix}.$$

Moreover, for fixed $p \in \mathbf{C}_+$, let $L^h : H_0^1(V) \to H^{-1}(V)$ be the constant coefficient operator

$$L^h = \begin{pmatrix} -\mathrm{div}(\eta^{h\,T}\mathbf{grad}) & -\mathrm{div}(\zeta^{h\,T}\mathbf{grad}) \\ -\mathrm{div}(\xi^{h\,T}\mathbf{grad}) & -\mathrm{div}(\mu^{h\,T}\mathbf{grad}) \end{pmatrix}.$$

By standard homogenization theory L^h is the limit as $\epsilon \to 0$ of L^ϵ in the following sense[2]: if (u, v) is the unique solution of

$$L^h \begin{pmatrix} u \\ v \end{pmatrix} = \begin{pmatrix} g_1 \\ g_2 \end{pmatrix},$$

then

$$\left.\begin{aligned} \mathbf{grad}\, u^\epsilon &\to \mathbf{grad}\, u, \\ \mathbf{grad}\, v^\epsilon &\to \mathbf{grad}\, v, \end{aligned}\right\} \quad \text{weakly in } L^2(V) \tag{23}$$

and moreover

$$\left.\begin{aligned} (\eta^\epsilon + \hat{\eta}_d^\epsilon)^T\mathbf{grad}\, u^\epsilon + (\zeta^\epsilon + \hat{\zeta}_d^\epsilon)^T\mathbf{grad}\, v^\epsilon &\to \eta^{h\,T}\mathbf{grad}\, u + \zeta^{h\,T}\mathbf{grad}\, v, \\ (\xi^\epsilon + \hat{\xi}_d^\epsilon)^T\mathbf{grad}\, u^\epsilon + (\mu^\epsilon + \hat{\mu}_d^\epsilon)^T\mathbf{grad}\, v^\epsilon &\to \xi^{h\,T}\mathbf{grad}\, u + \mu^{h\,T}\mathbf{grad}\, v \end{aligned}\right\} \tag{24}$$

weakly in $L^2(V)$. Relations (23) together with the fact that $\mathbf{curl\,grad} = 0$ imply that

$$\mathbf{grad}\, u^\epsilon \to \mathbf{grad}\, u, \quad \mathbf{grad}\, v^\epsilon \to \mathbf{grad}\, v \quad \text{weakly in } H(V, \mathbf{curl}). \tag{25}$$

Combining (22) with (25) and applying a well known result of Tartar[2] we obtain

$$\hat{\mathbf{D}}^\epsilon \cdot \mathbf{grad}\, u^\epsilon \to \hat{\mathbf{D}}^* \cdot \mathbf{grad}\, u, \tag{26}$$

$$\hat{\mathbf{B}}^\epsilon \cdot \mathbf{grad}\, v^\epsilon \to \hat{\mathbf{B}}^* \cdot \mathbf{grad}\, v \tag{27}$$

in $\mathcal{D}'(V)$, for each fixed $p \in \mathbf{C}_+$. Moreover we have

$$-\mathrm{div}\{(\eta^\epsilon + \hat{\eta}_d^\epsilon)^T\mathbf{grad}\, u^\epsilon + (\zeta^\epsilon + \hat{\zeta}_d^\epsilon)^T\mathbf{grad}\, v^\epsilon\}$$
$$= g_1 = -\mathrm{div}(\eta^{h\,T}\mathbf{grad}\, u + \zeta^{h\,T}\mathbf{grad}\, v)$$
$$-\mathrm{div}\{(\xi^\epsilon + \hat{\xi}_d^\epsilon)^T\mathbf{grad}\, u^\epsilon + (\mu^\epsilon + \hat{\mu}_d^\epsilon)^T\mathbf{grad}\, v^\epsilon\}$$
$$= g_2 = -\mathrm{div}(\xi^{h\,T}\mathbf{grad}\, u + \mu^{h\,T}\mathbf{grad}\, v)$$

and these together with (24) imply

$$\left.\begin{aligned} (\eta^\epsilon + \hat{\eta}_d^\epsilon)^T\mathbf{grad}\, u^\epsilon + (\zeta^\epsilon + \hat{\zeta}_d^\epsilon)^T\mathbf{grad}\, v^\epsilon &\to \eta^{h\,T}\mathbf{grad}\, u + \zeta^{h\,T}\mathbf{grad}\, v, \\ (\xi^\epsilon + \hat{\xi}_d^\epsilon)^T\mathbf{grad}\, u^\epsilon + (\mu^\epsilon + \hat{\mu}_d^\epsilon)^T\mathbf{grad}\, v^\epsilon &\to \xi^{h\,T}\mathbf{grad}\, u + \mu^{h\,T}\mathbf{grad}\, v \end{aligned}\right\}$$

weakly in $H(V, \text{div})$. Combining these with (18) we obtain[2]

$$\{(\eta^\epsilon + \hat{\eta}_d^\epsilon)^T \mathbf{grad}\, u^\epsilon + (\zeta^\epsilon + \hat{\zeta}_d^\epsilon)^T \mathbf{grad}\, v^\epsilon\} \cdot \hat{\mathbf{E}}^\epsilon$$
$$\longrightarrow (\eta^{h\,T} \mathbf{grad}\, u + \zeta^{h\,T} \mathbf{grad}\, v) \cdot \hat{\mathbf{E}}^* \qquad (28)$$
$$\{(\xi^\epsilon + \hat{\xi}_d^\epsilon)^T \mathbf{grad}\, u^\epsilon + (\mu^\epsilon + \hat{\mu}_d^\epsilon)^T \mathbf{grad}\, v^\epsilon\} \cdot \hat{\mathbf{H}}^\epsilon$$
$$\longrightarrow (\xi^{h\,T} \mathbf{grad}\, u + \mu^{h\,T} \mathbf{grad}\, v) \cdot \hat{\mathbf{H}}^* \qquad (29)$$

in $\mathcal{D}'(V)$, for each fixed $p \in \mathbf{C}_+$.

Now we observe that the left hand side of the sum of (26) and (27) coincides with the left hand side of the sum of (28) and (29). Hence the corresponding right hand sides are equal, that is

$$\hat{\mathbf{D}}^* \cdot \mathbf{grad}\, u + \hat{\mathbf{B}}^* \cdot \mathbf{grad}\, v =$$
$$= (\eta^{h\,T} \mathbf{grad}\, u + \zeta^{h\,T} \mathbf{grad}\, v) \cdot \hat{\mathbf{E}}^* + (\xi^{h\,T} \mathbf{grad}\, u + \mu^{h\,T} \mathbf{grad}\, v) \cdot \hat{\mathbf{H}}^*.$$

The fact that g_1 and g_2 were arbitrary implies that

$$\hat{\mathbf{D}}^* = \eta^h \hat{\mathbf{E}}^* + \xi^h \hat{\mathbf{H}}^*$$
$$\hat{\mathbf{B}}^* = \zeta^h \hat{\mathbf{E}}^* + \mu^h \hat{\mathbf{H}}^*, \quad \mathbf{x} \in V, \, p \in \mathbf{C}_+.$$

Since $V \subset\subset \Omega$ is arbitrary we obtain the Laplace transforms of the stated constitutive laws; this completes the proof.

References

1. G. Barbatis and I.G. Stratis, *Math. Methods Appl. Sci.* **26**, 1241-1253 (2003).
2. A. Bensoussan, J.L. Lions and G. Papanicolaou, Asymptotic Analysis for Periodic Structures, North Holland, New York, 1978.
3. D. Cioranescu and P. Donato, An Introduction to Homogenization. Oxford University Press, Oxford 1999.
4. G. Duvaut and J.L. Lions, Inequalities in Mechanics and Physics, Springer, Berlin, 1976.
5. J. Fridén, G. Kristensson and A. Sihvola, *Electromagnetics* **17**, 251-267 (1997).
6. A. Ioannidis, I.G. Stratis and A.N. Yannacopoulos, Electromagnetic wave propagation in dispersive bianisotropic media, in the present volume.
7. A. Karlsson and G. Kristensson, *J. Electromagn. Waves Appl.* **6**, 537-551 (1992).

CAVITY IDENTIFICATION USING 3-D ELASTODYNAMIC BEM, SHAPE SENSITIVITY AND TOPOLOGICAL DERIVATIVE

M. BONNET

Laboratoire de Mécanique des Solides (UMR CNRS 7649),
Ecole Polytechnique, F-91128 Palaiseau Cedex, France
E-mail: bonnet@lms.polytechnique.fr

B. B. GUZINA AND S. NINTCHEU FATA

Dept. of Civil Engineering, University of Minnesota,
Minneapolis, MN 55455, USA
E-mail: guzina@wave.ce.umn.edu

The problem of mapping underground cavities from surface, i.e. using non-intrusive seismic measurements, is investigated via a regularized boundary integral equation method. With the ground modeled as a three-dimensional uniform, isotropic elastic half-space, the inverse analysis of seismic waves scattered by a three-dimensional void is formulated as a task of minimizing a cost function involving the misfit between experimental observations and theoretical (i.e forward) predictions. This conventional choice of setting is dictated by the very high computational cost of solving the forward elastodynamic scattering problem, which makes e.g. global search strategies infeasible. For an accurate treatment of the gradient search technique employed to solve the inverse problem, derivatives of the predictive boundary element model with respect to the cavity parameters are evaluated using an adjoint problem approach. Here as in most situations where conventional descent methods (here the quasi-Newton algorithm with BFGS formula) are used, results depend on the choice of initial guess and occasional lack of convergence occurs. This has prompted the authors to investigate the use of topological derivative as a tool for preliminary probing. The topological derivative field is computed via a relatively inexpensive procedure, and appears to yield useful indications as to the topology and approximate location of the cavity system. Numerical examples are included to illustrate the effectiveness of the various steps developed so far.

Introduction. Three-dimensional imaging of cavities embedded in a semi-infinite solid using elastic waves is a topic of intrinsic interest in a number of applications ranging from nondestructive material testing to oil prospecting and underground object detection. In situations when detailed mapping of buried objects (defense facilities, buried waste) is required and only a few measurements can be made, the use of surface discretization-based boundary integral equation (BIE) techniques provides the most direct link between the surface measurements and the buried geometrical objects. While such an approach is well established for acoustic problems [2], limited attention has so far been paid to the use of BIE methods in wave-based sensing of elastic solids. This communication reports the development of an analytical and computational framework for the identification of cavities in a semi-infinite solid from surface seismic measurements via an elastodynamic BIE method, as well as preliminary results on the investigation of the usefulness of the topological derivative (e.g. for choosing the initial guess).

Formulation and solution technique. The focus of this study is the inverse scattering problem, in the framework of linear elastodynamics in the frequency domain (with the implicit time factor $e^{i\omega t}$ omitted thoroughout) for an isotropic, homogeneous elastic half-space housing an internal void. With reference to a Cartesian frame $\{O; \xi_1, \xi_2, \xi_3\}$, the half-space $\Omega = \{(\xi_1, \xi_2, \xi_3) | \xi_3 > 0\}$ is characterized by the Lamé's constants λ and μ, mass density ρ, and is bounded on top by the free surface $S = \{(\xi_1, \xi_2, \xi_3) | \xi_3 = 0\}$. The cavity inside the half-space occupies a simply connected finite region $\Omega_c \subset \Omega$ bounded by a piecewise smooth closed surface Γ; the normal to Γ directed towards the interior of Ω_c will be denoted by n. The cavity is 'illuminated' by a time-harmonic seismic source, with the resulting surface motion monitored over a finite set of slightly embedded control points $\xi = x^m$ $(m = 1, 2, \dots M)$.

The total elastic displacement field u is governed by the boundary integral equation

$$\frac{1}{2} u_k(x) + \text{PV} \int_\Gamma \hat{\sigma}_{ij}^k(\xi, x; \omega) n_j(\xi) u_i(x) \, dS_\xi = u_k^F(x) \qquad (x \in \Gamma) \qquad (1)$$

(its regularized form [5] being used in the implementation) with the free-field defined by

$$u_k^F(x) = \int_\Omega f_i(\xi) \hat{u}_i^k(\xi, x, \omega) \, dS_\xi$$

The fundamental solution (displacement $\hat{u}^k$, stress $\hat{\sigma}^k$) satisfies the traction-free condition $\hat{\sigma}^k . n = 0$ on S. Then, the displacement at sensor locations is given by the representation formula:

$$u_k(x^m) = u_k^F(x^m) - \int_\Gamma \hat{\sigma}_{ij}^k(\xi, x^m; \omega) n_j(\xi) u_i(\xi) \, dS_\xi \qquad (x^m \in S) \qquad (2)$$

Inverse problem. The inverse problem of cavity identification is set here as the minimization of the least squares misfit function

$$\mathcal{J}(\Gamma) = \sum_{1 \leq m \leq M} \frac{1}{2} \overline{(u(x^m) - u^{\text{obs}}(x^m))} \cdot W \cdot (u(x^m) - u^{\text{obs}}(x^m)), \quad (3)$$

where u is the solution of the direct problem (and thus of course depends on Γ) and the over-bar symbol denotes the complex conjugation. In view of the significant computational effort required to evaluate u for elastodynamic problems, the minimization of $\mathcal{J}$ is here performed by means of a gradient-based quasi-Newton method with the BFGS updating formula. The gradients are evaluated from the analytical formula

$$\overset{\star}{\mathcal{J}}(\Gamma) = \text{Re}\left[\int_\Gamma \left\{ \rho\omega^2 \tilde{u} \cdot u - \frac{2\lambda\mu}{\lambda + 2\mu} (\text{div}_S \tilde{u})(\text{div}_S u) \right.\right.$$
$$\left.\left. - \mu (\nabla_S \tilde{u} + \nabla_S^T \tilde{u}) : \nabla_S u + \mu (n \cdot \nabla_S \tilde{u}) \cdot (n \cdot \nabla_S u) \right\} \theta_n \, dS \right] \quad (4)$$

where θ_n denotes the normal transformation velocity of Γ associated with a given parameter perturbation, and $\tilde{u}$ is the adjoint solution governed by the integral equation [1,5]:

$$\frac{1}{2} \tilde{u}_k(x) + \text{PV} \int_\Gamma \hat{\sigma}_{ij}^k(\xi, x; \omega) n_j(\xi) \tilde{u}_i(x) \, dS_\xi$$
$$= \sum_{1 \leq m \leq M} W_{ij}^m \overline{(u_j(x^m) - u_j^{\text{obs}}(x^m))} \hat{u}_i^k(\xi, x^m, \omega) \, dS_\xi \quad (x \in \Gamma) \quad (5)$$

Computational treatment and results. The boundary element solution of (1) is implemented in a standard fashion. In this investigation, eight-node quadratic boundary elements are used. The location and shape of Γ is taken to depend on a finite set of design parameters: $\xi = \xi(p)$, with $p = (p_1, p_2, \ldots, p_D)$. With such assumption, the sensitivities $\partial \mathcal{J}/\partial p_d$ required for the minimization of $\mathcal{J}$ can be obtained by setting $\theta_n = \partial \xi/\partial p_d \cdot n$ in (4). As long as the topological characteristics of Γ are independent of p, the evolving boundary element mesh representing $\Gamma(p)$ can be created by interpolating the parameter-dependent nodes $x^q(p)$ with fixed, i.e. predefined, mesh connectivity.

Cavity mapping. Figure 1 illustrates the iterative process of finding an ellipsoidal cavity defined by $p^{\text{true}} = (-4a, -2a, 4a, 1.8a, 0.9a, 0.6a)$, starting from the initial guess $p^0 = (-1.5a, -0.5a, 5a, a, a, a)$ (the cavity is parameterized in terms of its centroid coordinates c_i and semi-axes lengths α_i,

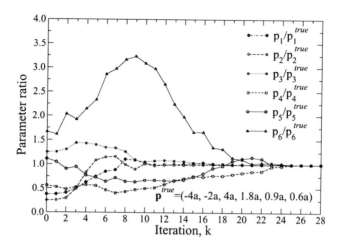

Figure 1. Evolution of design parameters in the minimization process.

$i = 1, 2, 3$ so that $\boldsymbol{p} = (c_1, c_2, c_3, \alpha_1, \alpha_2, \alpha_3)$. The cavity is illuminated in succession via nine point sources with respective magnitude $P = 0.2\mu a^2$. The testing configuration is a grid with 64 receivers. The shear wave length λ_s is approximately twice the diameter d of the cavity. As can be seen from Fig. 1, the iterative procedure converges after roughly 25 major iterations.

It should be noted, however, that the success of the foregoing method is strongly dependent on the choice of the starting point, a pitfall that is common to all gradient-based algorithms. This consideration led the authors to investigate the usefulness of the concept of *topological derivative* [3,9,4,6] in connection with the elastodynamic inverse problem.

Topological derivative. To search the semi-infinite domain Ω for cavities in the context of (3), let $B_a(\boldsymbol{x}^o) = \boldsymbol{x}^o + a\mathcal{B}$ define the cavity of size $a > 0$ and volume $a^3 |\mathcal{B}|$, where $\mathcal{B} \subset \mathbb{R}^3$ is a *fixed* and bounded open set of volume $|\mathcal{B}|$ containing the origin. Without loss of generality, $\mathcal{B}$ is chosen so that $B_a(\boldsymbol{x}^o)$ is contained inside the sphere of radius a centered at $\boldsymbol{x}^o$. With such definitions, the topological derivative of (3) can be defined as

$$\mathcal{T}(\boldsymbol{x}^o) = \lim_{a \to 0}\ (a^3 |\mathcal{B}|)^{-1} \left[\mathcal{J}(\Omega \setminus \overline{B}_a) - \mathcal{J}(\Omega) \right], \qquad \boldsymbol{x}^o \in B_a, \qquad (6)$$

which furnishes the information about the variation of $\mathcal{J}(\Omega)$ if a hole of prescribed shape $\mathcal{B}$ and infinitesimal characteristic size is created at $\boldsymbol{x}^o \in \Omega$. Within the framework of shape optimization, it was shown [9,4] that the elastostatic equivalent of (6) can be used as a powerful tool for the grid-based exploration of a solid for plausible void regions for a given functional $\mathcal{J}$. Here, this concept will be extended to elastic-wave imaging of semi-infinite solids on the basis of the elastodynamic fundamental solution for a homogeneous isotropic half-space.

In the present context, the topological derivative $\mathcal{T}(x^o)$ defined by (6) is found to be given by

$$\mathcal{T}(x^o) = \sum_{m=1}^{M} \mathrm{Re}\Big[\{\overline{u^{\mathrm{F}}(x^m)} - u^{\mathrm{obs}}(x^m)\}\cdot W^m\cdot e_k\left(\hat{\sigma}^k(x^o, x; \omega) : \mathcal{A} : \sigma^{\mathrm{F}}(x^o)\right.$$
$$\left. - \rho^2 \hat{u}^k(x^o, x; \omega)\cdot u^{\mathrm{F}}(x^o)\right)\Big] \quad (7)$$

where the constant tensor $\mathcal{A}$ depends in a known way on the shape of the infinitesimal cavity. When this shape is spherical, one has

$$\mathcal{A} = \frac{3(1-\nu)}{2\mu(7-5\nu)}\left[5\,\mathbf{I}_4^{sym} - \frac{1+5\nu}{2(1+\nu)}\,\mathbf{I}_2 \otimes \mathbf{I}_2\right]$$

and (7) is consistent with that given in Ref. 4 for elastostatic problems.

Numerical example. The configuration is as depicted in Fig. 2. The 'true' spherical cavity, of diameter $D = 0.4d$, is centered at $(d, 0, 3d)$ inside the half-space. In succession, the cavity is illuminated by sixteen axial point sources acting on the surface of a semi-infinite solid; for each source location, Cartesian components of the ground motion, u^{obs}, are monitored via twenty five sensors distributed over the square testing grid; here, this data is simulated using the BEM formulation of the forward problem. BEM. Four excitation frequencies $\bar{\omega} \equiv \omega d\sqrt{\rho/\mu} = 1, 2, 4, 8$ have been considered.

For this testing configuration, the values of $\mathcal{T}(x^o)$ are computed over the horizontal surface $S_h = \{\xi \in \Omega| - 5d < \xi_1 < 5d, -3d < \xi_2 < 3d, \xi_3 = 3d\}$

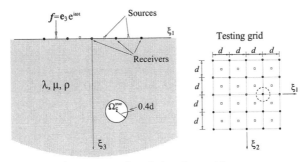

Figure 2.   Sample imaging problem.

210

passing through the centroid of the 'true' cavity and plotted in Fig. 3 for the above-defined set of frequencies. The computational grid is chosen so that the sampling points x^o are spaced by $0.25d$ in both ξ_1 and ξ_2 directions. In the display, the red tones indicate negative values of $\mathcal{T}$ and thus possible cavity location; for comparison, the true cavity is outlined in white in each of the diagrams. The results clearly demonstrate the usefulness of the topological derivative as a computationally efficient tool for exposing the approximate cavity location, with "higher" frequencies ($\bar{\omega} = 2, 4$) providing in general better resolution. From the diagram for $\bar{\omega} = 8$ where $\lambda_s/D \approx 1$, however, it is also evident that the infinitesimal-cavity assumption embedded in (7) performs best when used in conjunction with wave lengths exceeding the cavity diameter.

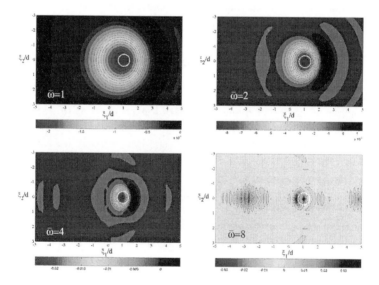

Figure 3.   Distribution of $(\mu d)^{-1}\mathcal{T}(x^o)$ in the $\xi_3 = 3d$ (horizontal) plane.

For completeness, the variation of $\mathcal{T}(x^o)$ across the vertical planar region $S_v = \{\boldsymbol{\xi} \in \Omega | -5d < \xi_1 < 5d,\ \xi_2 = 0,\ 0.25d < \xi_3 < 6d\}$ is given in Fig. 4. Similar to the earlier diagram, the sampling points x^o are spaced by $0.25d$ in the ξ_1 and ξ_3 directions. A diminished resolution relative to the previous result reflects the major limitation of the 'experimental' data set, that is, the fact that both source and receiver points are limited to a single planar surface. The contour plots for $\bar{\omega} = 2$ and 4 exhibit greater accuracy than that for $\bar{\omega} = 1$, but are also plagued with local minima that are absent in the former diagram. The non-informative distribution of $\mathcal{T}$ for $\bar{\omega} = 8$ indicates that the use of topological derivative in elastic-wave imaging is most effective at 'low' excitation frequencies, i.e. those inside the resonance region.

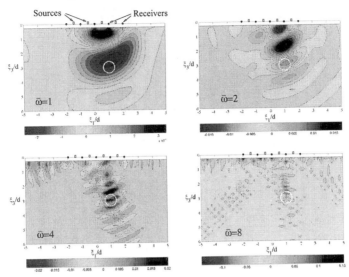

Figure 4.   Distribution of $(\mu d)^{-1}\mathcal{T}(\boldsymbol{x}^o)$ in the $\xi_2 = 0$ (vertical) plane.

With diagrams such as those in Figs. 3 and 4, an algorithm for identifying plausible cavity locations could be devised on the basis of the non-zero distribution of an auxiliary function

$$\overset{\star}{\mathcal{T}}(\boldsymbol{x}^o) = \begin{cases} \mathcal{T}(\boldsymbol{x}^o), \mathcal{T} < C, \\ 0, \quad \mathcal{T} \geq C, \end{cases} \tag{8}$$

with a suitable threshold value $C < 0$. With such definition, it is also possible to combine the individual advantages of different probing wavelengths by employing the product of (8) at several frequencies. As an illustration of the latter approach, Fig. 5 plots the distribution of the product of $\overset{\star}{\mathcal{T}}|_{\bar{\omega}=1}$ and $\overset{\star}{\mathcal{T}}|_{\bar{\omega}=2}$ in the vertical plane, with C set to approximately 40% of the global minima of the respective distributions in Fig. 4. Despite the limited accuracy and multiple minima characterizing respectively the individual solutions for $\bar{\omega} = 1$ and $\bar{\omega} = 2$, the combined result stemming from (8) points

to a single cavity with its centre and size closely approximating the true
void configuration.

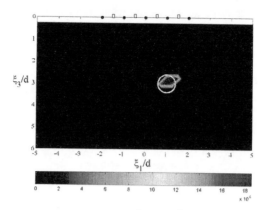

Figure 5.   Distribution of $(\mu d)^{-2}\, \overset{*}{\mathcal{T}}|_{\bar{\omega}=1} \times \overset{*}{\mathcal{T}}|_{\bar{\omega}=2}$ in the $\xi_2 = 0$ plane.

References

1. BONNET, M. BIE and material differentiation applied to the formulation of
   obstacle inverse problems. *Eng. Anal. with Bound. Elem.*, **15**, 121–136 (1995).
2. COLTON, D., KRESS, R. *Integral Equation Method in Scattering Theory*. Wiley, New York (1983).
3. ESCHENAUER, H. A., KOBELEV, V. V., SCHUMACHER, A. Bubble method
   for topology and shape optimization of structures. *Structural Optimization*, **8**,
   42–51 (1994).
4. GARREAU, S., GUILLAUME, P., MASMOUDI, M. The Topological Asymptotic
   for PDE Systems: The Elasticity Case. *SIAM J. Control Optim.*, **39**, 1756–
   1778 (2001).
5. GUZINA, B. B., NINTCHEU FATA, S., BONNET, M. On the stress-wave imaging
   of cavities in a semi-infinite solid. *Int. J. Solids Struct.*, **40**, 1505–1523 (2003).
6. GUZINA, B. B., BONNET, M. Topological derivative for the inverse scattering
   of elastic waves. *Quart. J. Mech. Appl. Math.*, to appear (2003).
7. NINTCHEU FATA, S., GUZINA, B. B., BONNET, M. A computational basis for
   elastodynamic cavity identification in a semi-infinite solid. *Comp. Mech.*, to
   appear (2003).
8. PAK, R. Y. S., GUZINA, B. B. Seismic soil-tructure interaction analysis by
   direct boundary element methods. *Int. J. Solids Struct.*, **36**, 4743–4766 (1999).
9. SOKOLOWSKI, J., ZOCHOWSKI, A. On the topological derivative in shape optimization. *SIAM J. Control Optim.*, **37**, 1251–1272 (1999).

RELATIVISTIC AND NON-RELATIVISTIC BOUNDARY CONDITIONS AND SCATTERING IN THE PRESENCE OF MOVING MEDIA AND OBJECTS

DAN CENSOR

Ben-Gurion University of the Negev,Department of Electrical and Computer Engineering,Beer Sheva, Israel 84105

Recently non-relativistic boundary conditions based on the Lorentz force formulas have been introduced. A general introduction and examples of scatterers moving in free space have been discussed [1]. In two follow-up articles the question of scattering in the presence of moving material media have been investigated: A statement of the model and scattering by a moving material cylinder, and scattering by a cylindrical region containing a material medium in motion [2]. Subsequently the problem of scattering by a sphere, the so-called Mie problem [3], has been considered for a sphere in motion [4]. Presently the method of solving such scattering problems is summarized, in order to provide a gateway for other configurations. It would seem to be interesting to solve such problems for other geometries and motional modalities, for example, to analyze the problem of a rotating elliptical object, as a way of simulating scattering from rotating machinery and aircraft elements. The formalism allows, in principle, to deal with arbitrary geometries and velocities.

1. Introduction

Let us start with the problem from the point of view of Einstein's Special Relativity theory (see e.g., [3], [5]). Accordingly, the theory stipulates the covariance of Maxwell's equations, i.e., they are functionally identical in all inertial system, written in terms of fields measured in each system and its appropriate, or native, spatiotemporal coordinates. Viewed from another system, we have to use the transformation formulas for the fields and represent boundary conditions in terms of the fields of the system in question. We limit our interest to first-order velocity effects. Using only terms of first order in the relative velocity v/c, the fields are related by

$$\mathbf{E}^{(2)} = \mathbf{E}^{(1)} + \mathbf{v}^{(2\mapsto1)} \times \mathbf{B}^{(1)}, \ \mathbf{H}^{(2)} = \mathbf{H}^{(1)} - \mathbf{v}^{(2\mapsto1)} \times \mathbf{D}^{(1)}$$

$$\mathbf{B}^{(2)} = \mathbf{B}^{(1)} - \mathbf{v}^{(2\mapsto1)} \times \mathbf{E}^{(1)} / c^2, \ \mathbf{D}^{(2)} = \mathbf{D}^{(1)} + \mathbf{v}^{(2\mapsto1)} \times \mathbf{H}^{(1)} / c^2 \tag{1}$$

where $\mathbf{v}^{(2\mapsto1)}$ denotes the velocity of system "2" as observed from "1". In (1) each of the fields is expressed in terms of its native coordinates, and the coordinate systems are related by the Lorentz transformation. Multiplying the first equation (1) by a test charge q_e, it can be rewritten as

$$\mathbf{f}_e^{(1)} = q_e^{(b)} \mathbf{E}_{eff}^{(b)}, \ \mathbf{E}_{eff}^{(b)} = \mathbf{E}^{(1)} + \mathbf{v}^{(b \mapsto 1)} \times \mathbf{B}^{(1)} \tag{2}$$

which is recognized as the Lorentz force formula for charges in motion, which can also be construed as the Coulomb force formula in the presence of the indicated effective field. In (2) $"2"$ is replaced by a point on the boundary $"b"$, with its appropriate relative velocity. This is therefore the way that boundary conditions can be introduced via forces acting on charges on the boundary, without resorting to relativistic theory. We know that the result is relativistically correct to the first order in v/c, but on the other hand we are not subjected to the restrictions imposed by the Lorentz transformation which applies to constant velocities only.

It follows that the force on a magnetic charge q_m will have a corresponding Lorentz force formula. Of course, magnetic charges have not been found to date, but forces can act on magnetic dipoles as if the magnetic dipole is composed of two very large and very close magnetic charges. Hence we conclude that the analog of (2) is

$$\mathbf{f}_m^{(1)} = q_m^{(b)} \mathbf{H}_{eff}^{(b)}, \ \mathbf{H}_{eff}^{(b)} = \mathbf{H}^{(1)} - \mathbf{v}^{(b \mapsto 1)} \times \mathbf{D}^{(1)} \tag{3}$$

The tangential forces cannot dissipate energy at the boundary, hence the forces are continuous across the boundary

$$\hat{\mathbf{n}}^{(b)} \times (\mathbf{f}_e^{(1)} - \mathbf{f}_e^{(2)}) = 0 \tag{4}$$

$$\hat{\mathbf{n}}^{(b)} \times (\mathbf{f}_m^{(1)} - \mathbf{f}_m^{(2)}) = 0 \tag{5}$$

The boundary conditions for the continuity of the effective tangential electric and magnetic fields at the boundary follow

$$\hat{\mathbf{n}}^{(b)} \times (\mathbf{E}_1^{(1)} + \mathbf{v}^{(b \mapsto 1)} \times \mathbf{B}_1^{(1)} - \mathbf{E}_2^{(2)} - \mathbf{v}^{(b \mapsto 2)} \times \mathbf{B}_2^{(2)}) = 0 \tag{6}$$

$$\hat{\mathbf{n}}^{(b)} \times (\mathbf{H}_1^{(1)} - \mathbf{v}^{(b \mapsto 1)} \times \mathbf{D}_1^{(1)} - \mathbf{H}_2^{(2)} + \mathbf{v}^{(b \mapsto 2)} \times \mathbf{D}_2^{(2)}) = 0 \tag{7}$$

In (6, 7) the formula applies to two moving media separated by the boundary.

At this point the problem of the mechanical flow of the two media must be considered. Most problems where the mechanical flow is taken into consideration are at the present time too complicated to deal with, and the compromise is to assume that the velocity is not affected by the presence of the boundary. This is the approach when the Fresnel drag is measured in the Fizeau experiment: In its classical form, a light beam is split and the two resulting beams propagate through two tubes of water, say, with transparent windows at their edges, then the beams are interfered, and depending on the beam shape an interference pattern is created, say on a screen. When the water in one arm of the experiment is set in motion, by injecting water at one end and draining it at the other, the interference pattern shifts, showing that the effective index of refraction is changed by the motion. All this is well known and appears in many textbooks. Our point here is that in the analysis of the experiment the flow is considered as uniform, although obviously this assumption is invalid in the regions where the fluid is injected and drained. This assumption of material motion which is not affected by the boundary is used here.

The implementation of (6, 7) is much more complicated. The approach used here is to analyze plane waves and represent the scattered field in the exterior domain as a superposition (integral) of such waves. Because of the motion, the waves posses different frequencies, and the representation of the integrals in terms of series of special functions, e.g., cylindrical waves, spherical waves, is a major difficulty. At moderate to high distances, the inverse distance differential operator representations are used. This technique was introduced by Twersky [9]-[12], and requires only minor changes to be adapted in to present class of problems, as shown below.

2. Cylinder and Sphere Moving in Free Space

These examples demonstrate the key elements of the method used here. Unlike the relativistic solutions, where one transforms waves from one inertial system to another (a technique dubbed by [6] as "train hopping"), all our calculations are in the original "laboratory" system of reference. Thus in the present context "non-relativistic" does not imply Galilean transformations between frames of reference.

We consider a circular cylinder of radius R whose axis is in the direction x, moving according to

$$z_T = z - vt \tag{8}$$

The excitation wave is taken as

$$\mathbf{E}_{ex} = \hat{\mathbf{x}} E_{ex} e^{i(k_{ex}z - \omega_{ex}t)}, \quad \mathbf{H}_{ex} = \hat{\mathbf{y}} H_{ex} e^{i(k_{ex}z - \omega_{ex}t)}$$
$$E_{ex} / H_{ex} = (\mu_0 / \varepsilon_0)^{1/2} = \zeta, \quad k_{ex} / \omega_{ex} = (\mu_0 \varepsilon_0)^{1/2} = 1 / v_{ph} = 1/c \tag{9}$$

At points parametrized by $\mathbf{r}_T = 0$, i.e., on the axis $z_T = 0$, the phase in (9) is $e^{-i\omega_T t}$, with

$$\omega_T / \omega_{ex} = k_T / k_{ex} = 1 - \beta, \ \omega_{ex} / k_{ex} = \omega_T / k_T = c, \ \beta = v / c \qquad (10)$$

which is recognized as the correct first order Doppler effect both for the Galilean and Lorentzian transformations. It follows that our technique can only be valid as a first order velocity approximation, but most problems allow this limitation. For any other point we have to include a phase shift factor relative to the center $\mathbf{r}_T = 0$. In particular on the cylinder rim we consider $z = R\cos\psi_T$, where the azimuthal angles are measured in the y, z plane, from the z axis in the right handed sense with respect to the cylindrical axis x. From (6, 7), the field *signals* (as opposed to *waves*) at the boundary are

$$\mathbf{E}_{exT} = \hat{\mathbf{x}} E_{0T} e^{i\varphi_{exT}}, \ \mathbf{H}_{exT} = \hat{\mathbf{y}} H_{0T} e^{i\varphi_{exT}}, \ E_{0T} / E_{ex} = H_{0T} / H_{ex} = 1 - \beta$$
$$\varphi_{exT} = k_T R C_{\psi_T} - i\omega_T t, \ C_\gamma = \cos\gamma, \ S_\gamma = \sin\gamma$$
$$\mathbf{E}_{exT} = \hat{\mathbf{x}} \bar{e} \Sigma_m i^m J_m(k_T R) e^{im\psi_T}, \ \bar{e} = E_{0T} e^{-i\omega_T t}, \ \Sigma_m = \Sigma_{m=-\infty}^{\infty}$$

$$(11)$$

which can be recast in a Fourier-Bessel series with J_m denoting the non-singular Bessel functions, as shown in (11).

Consider now a plane wave propagating in an arbitrary direction

$$\mathbf{E}_\alpha = \hat{\mathbf{x}} E_\alpha e^{i\varphi_\alpha}, \ \mathbf{H}_\alpha = \hat{\mathbf{k}}_\alpha \times \hat{\mathbf{x}} H_\alpha e^{i\varphi_\alpha}, \ \varphi_\alpha = \mathbf{k}_\alpha \cdot \mathbf{r} - \omega_\alpha t = k_\alpha r C_{\psi - \alpha} - \omega_\alpha t$$

$$(12)$$

Similarly to (10, 11), the phase at the boundary $r_T = R$ is

$$\varphi_{\alpha T} = k_{\alpha T} R C_{\psi_T - \alpha} - \omega_{\alpha T} t, \ \omega_{\alpha T} / \omega_\alpha = k_{\alpha T} / k_\alpha = 1 - \beta C_\alpha \qquad (13)$$

At every point on the boundary $r_T = R$, all the signals, corresponding to the excitation and scattering fields, must have the same frequency ω_T, in order to satisfy the boundary conditions, hence we have to impose

$$\omega_{\alpha T} = \omega_T, \ k_{\alpha T} = k_T \qquad (14)$$

Similarly to (11)

$$\mathbf{E}_{\alpha T} = \hat{\mathbf{x}} E_{\alpha T} e^{i\varphi_{\alpha T} t}, \ \mathbf{H}_{\alpha T} = \hat{\mathbf{k}}_\alpha \times \hat{\mathbf{x}} H_{\alpha T} e^{i\varphi_{\alpha T} t}$$
$$E_{exT} / E_{ex} = H_{exT} / H_{ex} = 1 - \beta C_\alpha$$

$$(15)$$

The scattered wave is represented as a superposition of waves (12), and recast in terms of the definitions in (13) and the constraint (14)

$$\mathbf{E}_{sc} = \hat{\mathbf{x}} E_{0T} \frac{1}{\pi} \int_C e^{ik_\alpha r \cos(\psi - \alpha) - i\omega_\alpha t} G(\alpha) d\alpha$$

$$= \hat{\mathbf{x}} \bar{e} \frac{1}{\pi} \int_{\psi_T - (\pi/2) + i\infty}^{\psi_T + (\pi/2) - i\infty} e^{ik_T R \cos(\psi_T - \alpha)} G(\alpha) d\alpha = \hat{\mathbf{x}} \bar{e} \Sigma_m i^m A_m H_m^{(1)}(k_T R) e^{im\psi_T}$$

$$G(\alpha) = \Sigma_m A_m e^{im\alpha}$$

$$(16)$$

The integration contour C (16) is chosen such that we have a Sommerfeld-type integral for $H_m^{(1)}$, the Hankel functions of the first kind, and together with the time factor $e^{-i\omega_T t}$ this corresponds to outgoing waves.

According to (13-15) the field at the boundary follows from (16)

$$\mathbf{E}_{scT} = \hat{\mathbf{x}} \bar{e} \frac{1}{\pi} \int_{\psi_T - (\pi/2) + i\infty}^{\psi_T + (\pi/2) - i\infty} e^{ik_T R \cos(\psi_T - \alpha)} G'(\alpha) d\alpha$$

$$(17)$$

$$= \hat{\mathbf{x}} \bar{e} \Sigma_m i^m A_m' H_m^{(1)}(k_T R) e^{im\psi_T}, \quad G'(\alpha) = \Sigma_m A'_m e^{im\alpha}$$

Upon comparing (11) and (17) it is realized that in terms of the new parameters, we are dealing with the boundary value problem for objects at rest, hence in (17)

$$G'(\alpha) = g(\alpha) = \Sigma_m a_m e^{im\alpha}$$

$$(18)$$

the velocity-independent scattering amplitude. It follows that

$$G'(\alpha) = g(\alpha) = \Sigma_m a_m e^{im\alpha}$$

$$G(\alpha) = g(\alpha)(1 + \beta C_\alpha) = \Sigma_m e^{im\alpha}[a_m + \tfrac{\beta}{2}(a_{m-1} + a_{m+1})]$$

$$\mathbf{E}_{sc} = \hat{\mathbf{x}} \bar{e} \Sigma_m i^m [a_m + \tfrac{\beta}{2}(a_{m-1} + a_{m+1})] H_m^{(1)}(k_T R) e^{im\psi_T}$$

$$(19)$$

$$= \hat{\mathbf{x}} \bar{e} \Sigma_m i^m a_m e^{im\psi_T} [H_m^{(1)} + \tfrac{i\beta}{2}(H_{m+1}^{(1)} e^{i\psi_T} - H_{m-1}^{(1)} e^{-i\psi_T})]$$

now applicable to any arbitrary point $\mathbf{r}_T$. Thus (19) demonstrates how scattering coefficients interact, and the way velocity-induced monopoles are created. One can substitute from (8) to finally derive (19) in terms of the initial coordinates $\mathbf{r}$.

The problem can be retraced for a three-dimensional object moving in free space. The mathematics is more complicated, involving vector spherical harmonics, vector spherical waves, and their representations in terms of Sommerfeld-type integrals.

The excitation wave (9) propagates now along the polar axis. Instead of (11) we have now $\varphi_{exT} = k_T R C_{\theta_T} - i\omega_T t$, where θ_T denotes the polar angle. The boundary conditions involve the effective fields (11). The excitation field signal (11) can be recast in terms of vector spherical waves as done by [3] (see p. 564), and the corresponding scattered field signal can be represented as a superposition (integral) of plane waves

$$\left\{ {\mathbf{E}_{scT} \atop \mathbf{H}_{scT}} \right\} = \left\{ {+\bar{e} \atop -\bar{h}} \right\} \frac{1}{2\pi} \int e^{ik_T R\hat{\mathbf{p}}\cdot\hat{\mathbf{r}}_T} \mathbf{g}_{\{{\mathbf{E} \atop \mathbf{H}}\}}(\hat{\mathbf{p}})d\Omega_{\hat{\mathbf{p}}}, \quad \int d\Omega_{\hat{\mathbf{p}}} = \int_{-\pi}^{\pi} d\beta \int_{0}^{(\pi/2)-i\infty} S_{\alpha} d\alpha$$

$$\mathbf{g}_{\mathbf{E}} = \Sigma_{nm}\delta_{m1}(d_{1nm}^{sc}\mathbf{C}_{1n}^{m} + d_{2nm}^{sc}\mathbf{B}_{Rn}^{m}), \quad \mathbf{g}_{\mathbf{H}} = \Sigma_{nm}\delta_{m1}(d_{2nm}^{sc}\mathbf{C}_{Rn}^{m} - d_{1nm}^{sc}\mathbf{B}_{1n}^{m})$$

$$\Sigma_{nm} = \Sigma_{n=0}^{\infty}\Sigma_{m}$$

$$(20)$$

In (20) δ_{m1} isolates the term $m = 1$. The notation and definition of the vector spherical harmonics appearing in (20) are detailed in [3], [4]. The coefficients d_{1nm}^{sc}, d_{2nm}^{sc} are developed in [4]. The problem of finding the scattered wave has already been discussed before [13]. Similar to the cylindrical case (16, 19), we now have

$$\left\{ {\mathbf{E}_{sc} \atop \mathbf{H}_{sc}} \right\} = \left\{ {+\bar{e} \atop -\bar{h}} \right\} \frac{1}{2\pi} \int e^{ik_T R\hat{\mathbf{p}}\cdot\hat{\mathbf{r}}_T} \mathbf{G}_{\{{\mathbf{E} \atop \mathbf{H}}\}}(\hat{\mathbf{p}})d\Omega_{\hat{\mathbf{p}}}$$

$$\mathbf{G}_{\{{\mathbf{E} \atop \mathbf{H}}\}} = \mathbf{g}_{\{{\mathbf{E} \atop \mathbf{H}}\}}(\hat{\mathbf{p}})(1 + \beta C_{\alpha}), \quad C_{\alpha} = \hat{\mathbf{p}}\cdot\hat{\mathbf{z}}$$

$$(21)$$

Using formulas from [4], where the details are given, the new scattering amplitudes $\mathbf{G}_{\{{\mathbf{E} \atop \mathbf{H}}\}}$ can be recast in spherical vector harmonics

$$\mathbf{G}_{\mathbf{E}} = \Sigma_{nm}\delta_{m1}(e_{1nm}\mathbf{C}_{1n}^{m} + e_{2nm}\mathbf{B}_{Rn}^{m}), \quad \mathbf{G}_{\mathbf{H}} = \Sigma_{nm}\delta_{m1}(e_{2nm}\mathbf{C}_{Rn}^{m} - e_{1nm}\mathbf{B}_{1n}^{m})$$

$$e_{1nm} = d_{1nm}^{sc} + \beta d_{5nm}^{sc}, \quad e_{2nm} = d_{2nm}^{sc} + \beta d_{6nm}^{sc}$$

$$(22)$$

Similarly to (19), the integral (21), for arbitrary coordinates $\mathbf{r}_T$, is recast in vector spherical waves. We quote the formula from [4], [12], relating the vector spherical harmonics and vector spherical waves, respectively

$$\left\{ i^{n}\mathbf{M}_{nm}(k\mathbf{r}), \ i^{n-1}\mathbf{N}_{nm}(k\mathbf{r}), \ i^{n-1}\mathbf{L}_{nm}(k\mathbf{r}) \right\}$$

$$= \frac{1}{2\pi} \int e^{ik r\hat{\mathbf{p}}\cdot\hat{\mathbf{r}}} \left\{ \mathbf{C}_{n}^{m}(\hat{\mathbf{p}}), \ \mathbf{B}_{n}^{m}(\hat{\mathbf{p}}), \ \mathbf{P}_{n}^{m}(\hat{\mathbf{p}}) \right\} d\Omega_{\hat{\mathbf{p}}}$$

$$(23)$$

Thus we have found the wave field produced by a three-dimensional object, e.g., a sphere, moving throughout free space.

3. Cylinder and Sphere Moving in a Material Medium

The scattering problems involving material media turn out to be much more complicated to manipulate. The simple technique that allowed us to use the representations (16, 17) with fails here, because (10, 13) are not applicable now. The reason stems from the fact that in (10, 13) we have assumed free space to display its attributes both for the initial excitation wave, and for the phase shifts in (11, 13). This is not applicable when dealing with material media in motion, where the Fresnel drag effect [5]-[7] (which vanishes in free space) must be taken into consideration.

We start with the excitation wave (9), but the phase velocity and wave impedance are now

$$v_{ph}^{(1)} = 1/(\mu_1 \varepsilon_1)^{1/2}, \ E_{ex}/H_{ex} = (\mu_1/\varepsilon_1)^{1/2} = \zeta^{(1)} \tag{24}$$

Instead of (10, 14), the frequency at a reference point $\mathbf{r}_T = 0$ and the frequency constraint now prescribe

$$\omega_{\alpha T} = \omega_\alpha (1 - \beta^{(1)} C_\alpha) = \omega_T = \omega_{ex}(1 - \beta^{(1)}), \ \beta^{(1)} = v/v_{ph}^{(1)}$$
$$\omega_\alpha = \omega_{ex}(1 - \beta^{(1)})/(1 - \beta^{(1)} C_\alpha) \approx \omega_{ex}\left(1 + \beta^{(1)}(C_\alpha - 1)\right) \tag{25}$$

where the last term (25) is the first order approximation.

To compute the phase shift and derive the analog of (11) we need to include the Fresnel drag effect [5]-[7]. It boils down to

$$\mathbf{k}^{(b)} = \mathbf{k}^{(1)} - \mathbf{v}^{(b \mapsto 1)} \omega^{(1)}/c^2 \tag{26}$$

which is also recognized as the first order relativistic formula relating propagation vectors in two reference systems. For the present case the phase in (11) becomes

$$\varphi_{exT} = k_T RC_{\psi_T} - \omega_T t, \ k_T = k_{ex}(1 - \beta^{(1)} A^{(1)}), \ A^{(1)} = (v_{ph}^{(1)}/c)^2 \tag{27}$$

Starting once again with plane waves (12) in the ambient material medium at rest, and using (25), we now have instead of (13)

$$\varphi_{\alpha T} = k_{\alpha T} RC_{\psi_T - \alpha} - \omega_{\alpha T} t = k_\alpha \left(C_\alpha - \beta^{(1)} A^{(1)}\right) RC_{\psi_T} + k_\alpha RS_\alpha S_{\psi_T} - \omega_T t$$
$$= k_\alpha RC_{\psi_T - \alpha} - \omega_T t - \beta^{(1)} k_{ex} A^{(1)} RC_{\psi_T}$$
$$\tag{28}$$

where the zero order approximation $\omega_{\alpha T} = \omega_{ex}$ is used in the Fresnel drag effect term. Substituting for $k_\alpha = \omega_\alpha / v_{ph}^{(1)}$ from (25), we recast (28) in the form

$$\varphi_{\alpha T} = k_\alpha R C_{\psi_T - \alpha} - \omega_T t - \beta^{(1)} K_{ex} A^{(1)} C_{\psi_T} = K_{ex} C_{\psi_T - \alpha} - \omega_T t + \beta^{(1)} K_{ex} B$$

$$K_{ex} = k_{ex} R, \; B = (C_\alpha - 1)C_{\psi_T - \alpha} - A^{(1)} C_{\psi_T}$$

$$(29)$$

The scattered wave is once again constructed as in (16). The analog of (17) is now

$$\mathbf{E}_{scT} = \hat{\mathbf{x}} \bar{e} \frac{1}{\pi} \int_{\psi_T - (\pi/2) + i\infty}^{\psi_T + (\pi/2) - i\infty} e^{iK_{ex} \cos(\psi_T - \alpha) + i\beta^{(1)} K_{ex} B} G'(\alpha) d\alpha \qquad (30)$$

where (30) includes the velocity-dependent terms prescribed by (1). Note that (18) is not valid for the present case, because of the extra term $\beta^{(1)} B$ in the exponent. Assuming that for the pertinent factors $\beta^{(1)}$, K_{ex}, this term is small, we approximate using $e^\gamma \approx 1 + \gamma$, and derive

$$\mathbf{E}_{scT} = \hat{\mathbf{x}} \bar{e} \frac{1}{\pi} \int_{\psi_T - (\pi/2) + i\infty}^{\psi_T + (\pi/2) - i\infty} e^{iK_{ex} \cos(\psi_T - \alpha)} (G + \beta^{(1)} gh) d\alpha$$

$$h = iK_{ex} B - C_\alpha = D_1 + D_2 C_\alpha + D_3 C_\alpha S_\alpha + D_4 S_\alpha + D_5 C_\alpha^2 \qquad (31)$$

$$D_1 = -iK_{ex} C_{\psi_T} A^{(1)}, \; D_2 = -(1 + iK_{ex} C_{\psi_T}), \; D_3 = iK_{ex} S_{\psi_T}$$

$$D_4 = -iK_{ex} S_{\psi_T}, \; D_5 = iK_{ex} C_{\psi_T}$$

In (31) there are factors depending on C_α, S_α. These trigonometric functions are recast in exponentials, and in the series g the sum indices are judiciously shifted up and down as we did in (19), in order to recast the integrals in series of cylindrical functions. Thus we have

$$gh = D_1 g_0 + D_2(g_{+1} + g_{-1}) + D_3(g_{+2} - g_{-2})$$

$$+ \tfrac{i}{2} D_4(g_{+1} - g_{-1}) + \tfrac{1}{4} D_5(g_{+2} + 2g_0 + g_{-2})$$

$$e^{\mp in\alpha} \Sigma_m a_m e^{ima} = \Sigma_m a_{m \pm n} e^{ima} = g_{\pm n}$$

$$\mathbf{E}_{scT} = \hat{\mathbf{x}} \bar{e} \Sigma_m i^m H_m(K) e^{im\psi_T} \{A_m + \beta^{(1)} [\tfrac{1}{2} D_1 a_m + D_2(a_{m+1} + a_{m-1})$$

$$+ \tfrac{i}{4} D_3(a_{m+2} - a_{m-2}) + \tfrac{i}{2} D_4(a_{m+1} - a_{m-1}) + \tfrac{1}{4} D_5(a_{m+2} + 2a_m + a_{m-2})]\}$$

$$(32)$$

However this does not suffice for exploiting the orthogonality of the series in the calculation of the scattering coefficients. The factors C_{ψ_T}, S_{ψ_T}

must also be recast in exponentials and the series indices judiciously shifted up and down, until we get a series of cylindrical functions involving $e^{im\psi_T}$, facilitating the implementation of orthogonality properties. This is carried out, yielding

$$\mathbf{E}_{scT} = \hat{\mathbf{x}} \bar{e} \Sigma_m i^m e^{im\psi_T} [A_m H_m(K) + \beta^{(1)} F_m(K)]$$

$$F_m(K) = D_6 + D_7 + D_8 + D_9 + D_{10}$$

$$D_6 = KA^{(1)} \tfrac{1}{2}(a_{m+1}H_{m+1} - a_{m-1}H_{m-1})$$

$$D_7 = -\tfrac{1}{2}H_m(a_{m+1} + a_{m-1}) + K\tfrac{1}{4}[H_{m+1}(a_{m+2} + a_m) - H_{m-1}(a_m + a_{m-2})]$$

$$D_8 = K\tfrac{1}{8}[H_{m-1}(a_{m+1} - a_{m-3}) + H_{m+1}(a_{m+3} - a_{m-1})]$$

$$D_9 = -K\tfrac{1}{4}[H_{m+1}(a_{m+2} - a_m) + H_{m-1}(a_m - a_{m-2})]$$

$$D_{10} = K\tfrac{1}{8}[H_{m-1}(a_{m+1} + 2a_{m-1} + a_{m-3}) - H_{m+1}(a_{m+3} + 2a_{m+1} + a_{m-1})]$$

$$(33)$$

The orthogonality of (33) with respect to $e^{im\psi_T}$ facilitates the computation of the scattering coefficients A_m, therefore the problem is considered as solved. For more details see [2].

The corresponding Mie problem for the moving sphere is developed in detail in [4]. Here we bring the main results. We have the same excitation wave (9, 24). In (25) α is now a polar angle, and similarly in (27) we replace ψ_T by θ_T. Instead of (28, 29) we have

$$\varphi_{pT} = K_{ex}\hat{\mathbf{k}}_p \cdot \hat{\mathbf{r}}_T - \omega_T t + \beta^{(1)} K_{ex}(C_\alpha - 1)\hat{\mathbf{k}}_p \cdot \hat{\mathbf{r}}_T - \beta^{(1)} K_{ex}A^{(1)}C_{\theta_T}$$

$$(34)$$

with $\mathbf{k}_p$ denoting the propagation vector. We have (20) and the first line of (21), and similarly to (31) we now have

$$\begin{Bmatrix} \mathbf{E}_{scT} \\ \mathbf{H}_{scT} \end{Bmatrix} = \begin{Bmatrix} +\bar{e} \\ -\bar{h} \end{Bmatrix} \frac{1}{2\pi} \int_c e^{iK_{ex}\hat{\mathbf{p}}\cdot\hat{\mathbf{r}}_T} (\mathbf{G}_{\{\substack{E\\H}\}}(\hat{\mathbf{p}}) + \beta^{(1)}\mathbf{f}_{\{\substack{E\\H}\}}(\hat{\mathbf{p}}))d\Omega_{\hat{\mathbf{p}}}$$

$$\mathbf{f}_{\{\substack{E\\H}\}}(\hat{\mathbf{p}}) = iK_{ex}(C_\alpha - 1)(\hat{\mathbf{p}}\cdot\hat{\mathbf{r}}_T)\mathbf{g}_{\{\substack{E\\H}\}} + (\hat{\mathbf{z}}\cdot\mathbf{g}_{\{\substack{E\\H}\}})\hat{\mathbf{p}} - (\hat{\mathbf{z}}\cdot\hat{\mathbf{p}})\mathbf{g}_{\{\substack{E\\H}\}} + A\mathbf{g}_{\{\substack{E\\H}\}}$$

$$= iK_{ex}(\hat{\mathbf{p}}\cdot\hat{\mathbf{r}}_T)\mathbf{h}_{\{\substack{E\\H}\}} + (\hat{\mathbf{z}}\cdot\mathbf{g}_{\{\substack{E\\H}\}})\hat{\mathbf{p}} - \mathbf{g}^c_{\{\substack{E\\H}\}} + A\mathbf{g}_{\{\substack{E\\H}\}}, \quad A = -K_{ex}A^{(1)}C_{\theta_T}$$

$$\mathbf{h}_{\{\substack{E\\H}\}} = \mathbf{g}^c_{\{\substack{E\\H}\}} - \mathbf{g}_{\{\substack{E\\H}\}}, \quad \mathbf{g}^c_{\{\substack{E\\H}\}}(\hat{\mathbf{p}}) = C_\alpha\mathbf{g}_{\{\substack{E\\H}\}} = (\hat{\mathbf{z}}\cdot\hat{\mathbf{p}})\mathbf{g}_{\{\substack{E\\H}\}}, \quad \mathbf{h}_H = -\hat{\mathbf{p}}\times\mathbf{h}_E$$

$$\mathbf{f}_E = \hat{\mathbf{p}}\times\mathbf{h}_H + (\hat{\mathbf{z}}\cdot\mathbf{g}_E)\hat{\mathbf{p}}, \quad \mathbf{f}_H = -\hat{\mathbf{p}}\times\mathbf{h}_E + (\hat{\mathbf{z}}\cdot\mathbf{g}_H)\hat{\mathbf{p}}$$

$$(35)$$

The tedious details of expressing $\mathbf{f}$ in terms of vector spherical harmonics leads to

$$\mathbf{f_E} = \Sigma(d_{15v\mu}^{sc}\mathbf{C}_{1\nu}^{\mu} + d_{16v\mu}^{sc}\mathbf{B}_{R\nu}^{\mu} + d_{17v\mu}^{sc}\mathbf{C}_{R\nu}^{\mu} + d_{18v\mu}^{sc}\mathbf{B}_{1\nu}^{\mu} + d_{19v\mu}^{sc}\mathbf{P}_{R\nu}^{\mu})$$

$$\mathbf{f_H} = \Sigma(-d_{15v\mu}^{sc}\mathbf{B}_{1\nu}^{\mu} + d_{16v\mu}^{sc}\mathbf{C}_{R\nu}^{\mu} - d_{17v\mu}^{sc}\mathbf{B}_{R\nu}^{\mu} + d_{18v\mu}^{sc}\mathbf{C}_{1\nu}^{\mu} + d_{20v\mu}^{sc}\mathbf{P}_{1\nu}^{\mu})$$

$$(36)$$

where the various coefficients are computed and defined in [4].

As was the case in (33), the coefficients in (36) are still functions of the coordinates $\hat{\mathbf{r}}_T$. This does not affect the integral (35), which is integrated according to coordinates $\hat{\mathbf{p}}$. Corresponding to (36), the velocity-dependent part of (35) is obtained in terms of vector spherical waves

$$\begin{Bmatrix} \mathbf{E}_{\mathbf{f}}(K_{ex}\hat{\mathbf{r}}_T) \\ \mathbf{H}_{\mathbf{f}}(K_{ex}\hat{\mathbf{r}}_T) \end{Bmatrix} = \begin{Bmatrix} +\bar{e} \\ -\bar{h} \end{Bmatrix} \frac{1}{2\pi} \int e^{iK_{ex}\hat{\mathbf{p}}\cdot\hat{\mathbf{r}}_T} \beta^{(1)} \mathbf{f}_{\begin{Bmatrix} \mathbf{E} \\ \mathbf{H} \end{Bmatrix}}(\hat{\mathbf{p}}) d\Omega_{\hat{\mathbf{p}}}$$

$$\mathbf{E}_{\mathbf{f}} = \bar{e}\beta^{(1)}\Sigma i^{\nu}(d_{15v\mu}^{sc}\mathbf{M}_{1\nu\mu}^{scT} - id_{16v\mu}^{sc}\mathbf{N}_{R\nu\mu}^{scT} + d_{17v\mu}^{sc}\mathbf{M}_{R\nu\mu}^{scT} - id_{18v\mu}^{sc}\mathbf{N}_{1\nu\mu}^{scT}$$

$$-id_{19v\mu}^{sc}\mathbf{L}_{R\nu\mu}^{scT}) = \bar{e}\beta^{(1)}\Sigma i^{\nu}(d_{21v\mu}^{sc}\mathbf{C}_{R\nu}^{\mu} + d_{22v\mu}^{sc}\mathbf{C}_{1\nu}^{\mu} + d_{23v\mu}^{sc}\mathbf{B}_{R\nu}^{\mu}$$

$$+d_{24v\mu}^{sc}\mathbf{B}_{1\nu}^{\mu} + d_{25v\mu}^{sc}\mathbf{P}_{R\nu}^{\mu} + d_{26v\mu}^{sc}\mathbf{P}_{1\nu}^{\mu})$$

$$\mathbf{H}_{\mathbf{f}} = -\bar{h}\beta^{(1)}\Sigma i^{\nu}(id_{15v\mu}^{sc}\mathbf{N}_{1\nu\mu}^{scT} + d_{16v\mu}^{sc}\mathbf{M}_{R\nu\mu}^{scT} + id_{17v\mu}^{sc}\mathbf{N}_{R\nu\mu}^{scT} + d_{18v\mu}^{sc}\mathbf{M}_{1\nu\mu}^{scT}$$

$$-id_{20v\mu}^{sc}\mathbf{L}_{1\nu\mu}^{scT}) = -\bar{h}\beta^{(1)}\Sigma i^{\nu}(d_{27v\mu}^{sc}\mathbf{C}_{R\nu}^{\mu} + d_{28v\mu}^{sc}\mathbf{C}_{1\nu}^{\mu} + d_{29v\mu}^{sc}\mathbf{B}_{R\nu}^{\mu}$$

$$+d_{30v\mu}^{sc}\mathbf{B}_{1\nu}^{\mu} + d_{31v\mu}^{sc}\mathbf{P}_{R\nu}^{\mu} + d_{32v\mu}^{sc}\mathbf{P}_{1\nu}^{\mu})$$

$$(37)$$

and recast in terms of their appropriate vector spherical harmonics. The coefficients are still dependent on coordinates $\hat{\mathbf{r}}_T$. From (37) the components parallel to the spherical boundary surface are extracted

$$\mathbf{E}_{\mathbf{f}\|} = \bar{e}\beta^{(1)}\Sigma i^{\nu}(d_{21v\mu}^{sc}\mathbf{B}_{R\nu}^{\mu} + d_{22v\mu}^{sc}\mathbf{B}_{1\nu}^{\mu} - d_{23v\mu}^{sc}\mathbf{C}_{R\nu}^{\mu} - d_{24v\mu}^{sc}\mathbf{C}_{1\nu}^{\mu})$$

$$\mathbf{H}_{\mathbf{f}\|} = -\bar{h}\beta^{(1)}\Sigma i^{\nu}(d_{27v\mu}^{sc}\mathbf{B}_{R\nu}^{\mu} + d_{28v\mu}^{sc}\mathbf{B}_{1\nu}^{\mu} - d_{29v\mu}^{sc}\mathbf{C}_{R\nu}^{\mu} - d_{30v\mu}^{sc}\mathbf{C}_{1\nu}^{\mu}) \quad (38)$$

$$\mathbf{E}_{\mathbf{f}\|} = \hat{\mathbf{r}}_T \times \mathbf{E}_{\mathbf{f}}, \quad \mathbf{H}_{\mathbf{f}\|} = \hat{\mathbf{r}}_T \times \mathbf{H}_{\mathbf{f}}$$

Finally the coefficients times the vector spherical harmonics in (38) are recast in spherical harmonics

$$\mathbf{E}_{\mathbf{f}\|} = \bar{e}\beta^{(1)}\Sigma'(d_{33v'\mu}^{sc}\mathbf{B}_{1\nu'}^{\mu'} - d_{34v'\mu}^{sc}\mathbf{C}_{R\nu'}^{\mu'})$$

$$\mathbf{H}_{\mathbf{f}\|} = -\bar{h}\beta^{(1)}\Sigma'(d_{39v'\mu}^{sc}\mathbf{B}_{R\nu'}^{\mu'} - d_{40v'\mu}^{sc}\mathbf{C}_{1\nu'}^{\mu'})$$

$$(39)$$

where the coefficients in (39) are independent of the angles involved in $\hat{\mathbf{r}}_T$. In this form the orthogonality of the series with respect to the vector spherical harmonics can be exploited, and scattering coefficients can be calculated. Thus the problem is considered as solved.

4. Calculation of the Scattered Wave

From (30, 31), by omitting the velocity-dependent term introduced by the boundary conditions (1), and for arbitrary points $\mathbf{r}_T$, we have

$$\mathbf{E}_{sc} = \hat{\mathbf{x}}\bar{e}\frac{1}{\pi} \int_{\psi_T-(\pi/2)+i\infty}^{\psi_T+(\pi/2)-i\infty} e^{iK'_{ex}\cos(\psi_T-\alpha)}\bar{G}(\alpha)d\alpha$$

$$\bar{G}(\alpha) = G(\alpha)e^{i\beta^{(1)}K'_{ex}[(C_\alpha-1)C_{\psi_T-\alpha}-A^{(1)}C_{\psi_T}]}, \quad K'_{ex} = k_{ex}r_T$$

(40)

where $\bar{G}(\alpha)$ is now a known function. The corresponding expression for three-dimensional objects is

$$\begin{Bmatrix}\mathbf{E}_{sc}\\\mathbf{H}_{sc}\end{Bmatrix} = \begin{Bmatrix}+\bar{e}\\-\bar{h}\end{Bmatrix}\frac{1}{2\pi} \int e^{iK'_{ex}\hat{\mathbf{p}}\cdot\hat{\mathbf{r}}_T}\bar{\mathbf{G}}_{\{\begin{smallmatrix}E\\H\end{smallmatrix}\}}(\hat{\mathbf{p}})d\Omega_{\hat{\mathbf{p}}}$$

$$\bar{\mathbf{G}}_{\{\begin{smallmatrix}E\\H\end{smallmatrix}\}}(\hat{\mathbf{p}}) = \mathbf{G}_{\{\begin{smallmatrix}E\\H\end{smallmatrix}\}}(\hat{\mathbf{p}})e^{i\beta^{(1)}K'_{ex}[(C_\alpha-1)(\hat{\mathbf{p}}\cdot\hat{\mathbf{r}}_T)-A^{(1)}C_{\theta_T}]}$$

(41)

We wish to express (40, 41) in an explicit way that can be used for calculations. A convenient representation in terms of an inverse distance power series has been proposed by Twersky [9]-[12], and with minor modifications applies to our problem. We present here an asymptotic series form [9], the exact series is available too [10]. Thus for the two-dimensional case we have

$$\mathbf{E}_{sc}(\mathbf{r}_T, t) = \hat{\mathbf{z}}E_{0T}e^{-i\omega_T t}D_\alpha\left\{\bar{G}(\alpha, \theta_T, r_T)\right\}$$

$$D_\alpha\{\bar{G}\} = H\left(1 + \frac{1+4\partial_\alpha^2}{i8\kappa} - \frac{9+40\partial_\alpha^2+16\partial_\alpha^4}{128\kappa^2}\cdots\right)\bar{G}(\alpha, \theta_T, r_T)\Big|_{\alpha=\theta_T}$$

$$= H\sum_{\mu=0}\frac{(1+4\partial_\alpha^2)(9+4\partial_\alpha^2)\cdots([2\mu-1]^2+4\partial_\alpha^2)}{(i8\kappa)^\mu\mu!}\bar{G}(\alpha, \theta_T, r_T)\Big|_{\alpha=\theta_T}$$

$$H = H(\kappa) = (2/(i\pi\kappa))^{1/2}e^{i\kappa}$$

(42)

The corresponding three-dimensional form is

$$\left\{ {\mathbf{E}_{sc} \atop \mathbf{H}_{sc}} \right\} = \left\{ {+\bar{e} \atop -\bar{h}} \right\} h_0(K'_{ex}) \tilde{\mathbf{O}}(K'_{ex}, \tilde{\mathbf{D}}) \cdot \bar{\mathbf{G}}(\hat{\mathbf{p}}) \big|_{\hat{\mathbf{p}} = \hat{\mathbf{r}}_T}$$

$$\tilde{\mathbf{O}} = \sum_{\nu=0}^{\infty} \rho_\nu \tilde{\mathbf{D}} \cdot (\tilde{\mathbf{D}} - 1 \cdot 2\tilde{\mathbf{I}}) \cdot (\tilde{\mathbf{D}} - 2 \cdot 3\tilde{\mathbf{I}}) \cdots (\tilde{\mathbf{D}} - (\nu - 1)\nu\tilde{\mathbf{I}})$$

$$= \tilde{\mathbf{I}} + \rho_1 \tilde{\mathbf{D}} + \rho_2 \tilde{\mathbf{D}} \cdot (\tilde{\mathbf{D}} - 2\tilde{\mathbf{I}}) + \rho_n \tilde{\mathbf{D}} \cdot (\tilde{\mathbf{D}} - 1 \cdot 2\tilde{\mathbf{I}}) \cdots (\tilde{\mathbf{D}} - (n-1)n\tilde{\mathbf{I}})$$

$$\tilde{\mathbf{D}} = \hat{\mathbf{r}}_T(D+2)\hat{\mathbf{r}}_T + \hat{\mathbf{r}}_T 2S_\alpha^{-1} \partial_\alpha (S_\alpha \hat{\boldsymbol{\alpha}}) + \hat{\mathbf{r}}_T \partial_\beta \hat{\boldsymbol{\beta}} + \hat{\boldsymbol{\alpha}}(D + S_\alpha^{-2})\hat{\boldsymbol{\alpha}}$$

$$+\hat{\boldsymbol{\alpha}} 2S_\alpha^{-2} C_\alpha \partial_\beta \hat{\boldsymbol{\beta}} - \hat{\boldsymbol{\alpha}} \partial_\alpha \hat{\mathbf{r}}_T + \hat{\boldsymbol{\beta}}(D + S_\alpha^{-2})\hat{\boldsymbol{\beta}} - \hat{\boldsymbol{\beta}} 2S_\alpha^{-2} C_\alpha \partial_\beta \hat{\boldsymbol{\alpha}} - \hat{\boldsymbol{\beta}} 2S_\alpha^{-1} \partial_\beta \hat{\mathbf{r}}_T$$

$$\rho_\nu = (i/(2K'_{ex}))^\nu / \nu!, \quad D = S_\alpha^{-2}[\partial_\beta^2 + S_\alpha \partial_\alpha (S_\alpha \partial_\alpha)]$$

$$\tag{43}$$

where in (43) the operator is a dyadic. After performing the operations, the initial coordinates $\mathbf{r}$ can be substituted from (8).

References

1. D. Censor, "Non-relativistic electromagnetic scattering: "Reverse engineering" using the Lorentz force formulas", PIER-Progress In Electromagnetic Research, Vol. *38*, pp. 199-221, 2002.
2. D. Censor, "Non-relativistic boundary conditions and scattering in the presence of arbitrarily moving media and objects: cylindrical problems", submitted.
3. J. A. Stratton, *Electromagnetic Theory*, McGraw-Hill, 1941.
4. D. Censor, "Non-relativistic scattering in the presence moving objects: the Mie problem for a moving sphere", submitted.
5. W. Pauli, *Theory of Relativity*, Pergamon, 1958, also Dover Publications.
6. J. Van Bladel, *Relativity and Engineering*, Springer, 1984.
7. J. A. Kong, *Electromagnetic Wave Theory*, Wiley, 1986.
8. D. Censor, "Non-relativistic scattering in the presence moving media: the Fresnel-Fizeau experiment for a sphere", in process.
9. V. Twersky, "Scattering of waves by two objects", pp. 361-389, in *Electromagnetic Waves*, Ed. R.E. Langer, Proc. Symp. in Univ. of Wisconsin, Madison, April 10-12, 1961, The University of Wisconsin Press, 1962.
10. J.E. Burke, D. Censor, and V. Twersky, "Exact inverse-separation series for multiple scattering in two dimensions", The Journal of Acoustical Society of America, Vol. 37, pp. 5-13, 1965.

11. V. Twersky, "Multiple scattering by arbitrary configurations in three dimensions", Journal of Mathematical Physics, Vol. 3, pp. 83-91, 1962.

12. V. Twersky, "Multiple scattering of electromagnetic waves by arbitrary configurations", Journal of Mathematical Physics, Vol. 8, pp. 589-610, 1967.

13. D. Censor, "Scattering in velocity dependent systems", Radio Science, Vol. 7, pp. 331 - 337, 1972.

THE LINEAR SAMPLING METHOD FOR N-BODIES IN 2-DIMENSIONAL LINEAR ELASTICITY

ANTONIOS CHARALAMBOPOULOS

Department of Material Science and Engineering,
University of Ioannina, 45110 Ioannina, Greece

DROSSOS GINTIDES

Department of Mathematics
Hellenic Naval Academy, Xatzikiriakio, Pireus, Greece

KIRIAKIE KIRIAKI

Department of Mathematics
National Technical University of Athens
Zografou Campus, 15780 Athens, Greece

In this paper the linear sampling method for the shape reconstruction of N - bodies with different boundary conditions in two dimensional linear elasticity is examined. We formulate the governing differential equations of the problem in dyadic form in order to acquire a symmetric and uniform representation for the underlying elastic fields. Assuming that K - bodies are penetrable non - dissipative inclusions, L - are rigid bodies and the rest M of them are cavities we establish the main theorem for the shape reconstruction. The inversion scheme which is proposed is based on the unboundedness of the solution of an equation of first kind having as known term the far-field of the free-space Green's dyadic, generated by sources inside the inclusions approaching the boundary. Numerical results are presented concerning two circular inclusions.

1. Introduction

One of the most significant methods for solving the inverse scattering problem is the so-called "linear sampling method" inspired by D. Colton, A. Kirsch [5]. It is characterized by its remarkable simplicity and mainly by the fact that it is not necessary to know a priori the type of the boundary conditions on scatterer's surface. An excellent presentation is given in [6]. Results for the 2-D case in linear elasticity are given in [1], and for rigid bodies or cavities in 3-D in [7]. The absorbing elastic transmission problem is presented in [2] while the non-dissipative case is investigated in [3].

In this work, the linear sampling method for the shape reconstruction of N - bodies with different boundary conditions in two dimensional linear elasticity is examined. In section 2, we formulate the scattering problem in dyadic form. The differential equations, the boundary conditions, the free-space Green's dyadic as well as its asymptotic form are presented. In section 3, the theorem for the inversion algorithm is formulated. It is remarkable that, there exists a Herglotz dyadic which is close enough to all approximating the solutions of specific problems Herglotz dyadics, in any simple inclusion. The proposed inversion scheme is based on the unboundedness of the solution of the approximate far-field equation with known term the far-field pattern of the free-space Green's dyadic, generated by sources inside the inclusions. Finally, in section 4, numerical results are presented for two circular inclusions, assuring the simple and efficient implementation of the algorithm, using synthetic far-field data derived from the boundary element method developed by D. Polyzos *et als* [9, 10].

2. Formulation of the scattering problem

We consider the scattering process in a homogeneous and isotropic elastic medium in $\mathbf{R}^2$ with real Lamé constants λ, μ where $\mu > 0$, $\lambda + 2\mu > 0$ and mass density ρ . Inside the medium there exist a finite number of disjoint bounded convex and closed subsets, having smooth boundaries ∂D_j, $\partial D = \partial D_1 \cup \partial D_2 \cup ... \cup \partial D_N$, $D_- = D_1 \cup D_2 \cup ... \cup D_N$ and D_+ be the complement of D_- in $\mathbf{R}^2$. We assume that the $K-$ subsets of them are penetrable bodies, that is they are filled up with homogeneous isotropic elastic media with Lamé constants λ_j, μ_j, where $\mu_j > 0$, $\lambda_j + 2\mu_j > 0$ and mass density ρ_j, $j = 1,...,K$, the next L bodies are rigid and the rest M are cavities. Obviously $K + L + M = N$. We denote the unit outward normal vector .on the boundary ∂D_j by $\tilde{\mathbf{v}}_j$, $j = 1,...N$.

Stating the problem in dyadic form, we have a unified representation, independent of the specific polarization. More precisely, the incident field $\tilde{\mathbf{u}}_{in}$ has the following dyadic plane wave form

$$\tilde{\mathbf{u}}_{in}(\mathbf{x}) = \hat{\mathbf{d}} \otimes \hat{\mathbf{d}}\, e^{ik_p \mathbf{x} \cdot \hat{\mathbf{d}}} + \hat{\boldsymbol{\phi}}_d \otimes \hat{\boldsymbol{\phi}}_d\, e^{ik_s \mathbf{x} \cdot \hat{\mathbf{d}}} \qquad (1)$$

$\hat{\boldsymbol{\phi}}_d$ is the orthogonal to $\hat{\mathbf{d}}$ vector of the system of polar coordinates, $k_\alpha, \alpha = p,s$ are the elastic wavenumbers for the exterior domain for longitudinal and transverse waves.

The differential equation satisfied by the displacement dyadic field $\tilde{\mathbf{u}}$ for the exterior region D_+ is the time-reduced Navier equation

$$\left(\Delta^* + \rho\omega^2\right)\tilde{\mathbf{u}}(\mathbf{x}) = \tilde{\mathbf{0}}, \ \mathbf{x} \in D_+ \tag{2}$$

where $\omega \in \mathbf{R}^+$ is the angular frequency. The operator Δ^* is defined as

$$\Delta^* = \mu\Delta + \left(\lambda + \mu\right)\nabla\nabla\cdot \tag{3}$$

The same Navier equation with the corresponding Lamé constants and densities is satisfied by the interior displacement fields in the penetrable inclusions.

The boundary conditions for the penetrable inclusions are the transmission conditions on ∂D_j and are written as

$$\tilde{\mathbf{u}}(\mathbf{x})\left(= \left(\tilde{\mathbf{u}}_{sc} + \tilde{\mathbf{u}}_{in}\right)(\mathbf{x})\right) = \tilde{\mathbf{u}}_j(\mathbf{x}), \ \mathbf{x} \in \partial D_j, \ j = 1, ..., K$$
$$T\tilde{\mathbf{u}}(\mathbf{x})\left(= T\left(\tilde{\mathbf{u}}_{sc} + \tilde{\mathbf{u}}_{in}\right)(\mathbf{x})\right) = T_j\tilde{\mathbf{u}}_j(\mathbf{x}), \ \mathbf{x} \in \partial D_j, \ j = 1, ..., K \tag{4}$$

$\tilde{\mathbf{u}}_{sc}$ is the scattered field, $\tilde{\mathbf{u}}_{sc}$, $\tilde{\mathbf{u}}_{in}$ satisfy Eq.(2), T is the surface stress operator on $\partial D_j, j = 1, ...K$ given by

$$T = 2\mu\hat{\mathbf{v}}_j \cdot \nabla + \lambda\hat{\mathbf{v}}_j \nabla \cdot + \mu\hat{\mathbf{v}}_j \times \nabla\times \tag{5}$$

and T_j are surface stress operators on $\partial D_j, j = 1, ...K$ given by (5) with the corresponding to the interior domains Lamé constants.

The Dirichlet conditions are satisfied on the boundaries of the next L rigid bodies, that is

$$\tilde{\mathbf{u}}(\mathbf{x})\left(= \left(\tilde{\mathbf{u}}_{sc} + \tilde{\mathbf{u}}_{in}\right)(\mathbf{x})\right) = \tilde{\mathbf{0}}, \ \mathbf{x} \in \partial D_j, \ j = K+1, ..., K+L \tag{6}$$

The Neumann conditions are satisfied on the boundaries of the next M cavities, that is

$$T\tilde{\mathbf{u}}(\mathbf{x})\left(= T\left(\tilde{\mathbf{u}}_{sc} + \tilde{\mathbf{u}}_{in}\right)(\mathbf{x})\right) = \tilde{\mathbf{0}}, \ \mathbf{x} \in \partial D_j, \ j = K+L+1, ..., N \tag{7}$$

For the well-posedness of the problem, radiation conditions due to Kupradze should also be satisfied by the scattered field.

The free-space Green's dyadic for the two-dimensional linear elasticity with Lamé constants those of the exterior domain is given by

$$\tilde{\Gamma}(\mathbf{x},\mathbf{y}) = -\frac{i}{2\mu}\{H_0^{(1)}(k_s|\mathbf{x}-\mathbf{y}|)\tilde{I}$$

$$+\frac{1}{(k_s)^2}\nabla_x\nabla_x\left(H_0^{(1)}(k_s|\mathbf{x}-\mathbf{y}|)-H_0^{(1)}(k_p|\mathbf{x}-\mathbf{y}|)\right)\}$$

(8)

The subscript $\mathbf{x}$ indicates differentiation with respect to the $\mathbf{x}$ variable. Using asymptotic analysis we take the following asymptotic relations for the P and S-parts in the Helmholtz decomposition of the Green's dyadic for the exterior domain

$$\tilde{\Gamma}^p(\mathbf{x},\mathbf{y}) = \frac{1}{\lambda+2\mu}\frac{e^{i\pi/4}}{\sqrt{2\pi k_p}}e^{-ik_p\hat{x}\cdot\mathbf{y}}\frac{e^{ik_p x}}{\sqrt{x}}\hat{x}\otimes\hat{x}+O(x^{-3/2}),\quad x=|\mathbf{x}|\to+\infty$$

(9)

$$\tilde{\Gamma}^s(\mathbf{x},\mathbf{y}) = \frac{1}{\mu}\frac{e^{i\pi/4}}{\sqrt{2\pi k_s}}e^{-ik_s\hat{x}\cdot\mathbf{y}}\frac{e^{ik_s x}}{\sqrt{x}}\hat{\phi}_x\otimes\hat{\phi}_x+O(x^{-3/2}),\quad x\to+\infty$$

(10)

3. The inversion scheme

The application of the sampling method is based on the treatment of the far-field equation

$$(F\tilde{g})(\hat{x}) = \tilde{\Gamma}^\infty(\hat{x},\mathbf{z}),\ \mathbf{z}\in D,\hat{x}\in\Omega$$

(11)

where F is the elastic far-field operator [2, 3, 4] and

$$\tilde{\Gamma}^\infty(\hat{x},\mathbf{z}) = \left(\tilde{\Gamma}_p^\infty(\hat{x},\mathbf{z}),\tilde{\Gamma}_s^\infty(\hat{x},\mathbf{z})\right) = \left(\frac{1}{\lambda_e+2\mu_e}\frac{e^{i\pi/4}}{\sqrt{2\pi k_p^e}}\hat{x}\otimes\hat{x}e^{-ik_p^e\hat{x}\cdot\mathbf{z}},\frac{1}{\mu_e}\frac{e^{i\pi/4}}{\sqrt{2\pi k_s^e}}\hat{\phi}_x\otimes\hat{\phi}_x e^{-ik_s^e\hat{x}\cdot\mathbf{z}}\right)$$

is the far-field pattern of the free-space Green's dyadic . As in previous works concerning the sampling method in elasticity the following theorem sets the theoretical framework for the method:

Theorem:*Let D_- satisfies the assumptions given in Section 2 and ω be neither an eigenvalue of the corresponding interior transmission problems, nor of the interior Dirichlet or Neumann problems. Then for every $\varepsilon>0$ and $\mathbf{z}\in D_-$ there exist*

$$\tilde{g}_p\in L_p^2(\Omega) = \left\{\tilde{g}_p:\Omega\to C^4:\hat{d}\times\tilde{g}_p(\hat{d})=\tilde{0},|\tilde{g}_p|\in L^2(\Omega)\right\}$$

and

$$\tilde{\mathbf{g}}_s \in L^2_s(\Omega) = \left\{ \tilde{\mathbf{g}}_s : \Omega \to C^4 : \hat{\mathbf{d}} \cdot \tilde{\mathbf{g}}_s(\hat{\mathbf{d}}) = \tilde{\mathbf{0}}, |\tilde{\mathbf{g}}_s| \in L^2(\Omega) \right\} \quad such\ that:$$

$$\left\| (F\tilde{\mathbf{g}})(\hat{\mathbf{x}}) - \tilde{\Gamma}^\infty(\hat{\mathbf{x}}, \mathbf{z}) \right\|_{L^2(\Omega)} < \varepsilon \tag{12}$$

where $\tilde{\mathbf{g}} = (\tilde{\mathbf{g}}_p, \tilde{\mathbf{g}}_s)$. It also holds

$$\lim_{\mathbf{z} \to \partial D} \left\| \tilde{\mathbf{g}} \right\|_{L^2(\Omega)} = +\infty. \tag{13}$$

Proof

By hypothesis avoiding the transmission eigenvalues, we consider the unique weak solutions of the following boundary value problems [3]

$$\Delta^*_j \tilde{\mathbf{w}}_j(\mathbf{x}) + \rho_j \omega^2 \tilde{\mathbf{w}}_j(\mathbf{x}) = \tilde{\mathbf{0}}, \mathbf{x} \in D_j, \quad j = 1, ..., K \tag{14}$$

$$\Delta^* \tilde{\mathbf{v}}_j(\mathbf{x}) + \rho \omega^2 \tilde{\mathbf{v}}_j(\mathbf{x}) = \tilde{\mathbf{0}}, \mathbf{x} \in D_j, \quad j = 1, ..., K \tag{15}$$

$$\tilde{\mathbf{w}}_j(\mathbf{x}) - \tilde{\mathbf{v}}_j(\mathbf{x}) = \tilde{\Gamma}(\mathbf{x}, \mathbf{z}), \mathbf{x} \in \partial D_j, \mathbf{z} \in D_j, \quad j = 1, ..., K \tag{16}$$

$$T_j \tilde{\mathbf{w}}_j(\mathbf{x}) - T \tilde{\mathbf{v}}_j(\mathbf{x}) = T \tilde{\Gamma}(\mathbf{x}, \mathbf{z}), \mathbf{x} \in \partial D_j, \mathbf{z} \in D_j, \quad j = 1, ..., K \tag{17}$$

Similarly, excluding the Dirichlet and Neumann eigenvalues, we also consider the unique solutions of the boundary value problems

$$\Delta^* \tilde{\mathbf{v}}_j(\mathbf{x}) + \rho \omega^2 \tilde{\mathbf{v}}_j(\mathbf{x}) = \tilde{\mathbf{0}}, \mathbf{x} \in D_j, \quad j = K+1, ..., L \tag{18}$$

$$\tilde{\mathbf{v}}_j(\mathbf{x}) + \tilde{\Gamma}(\mathbf{x}, \mathbf{z}) = \tilde{\mathbf{0}}, \mathbf{x} \in \partial D_j, \mathbf{z} \in D_j, \quad j = K+1, ..., L \tag{19}$$

and

$$\Delta^* \tilde{\mathbf{v}}_j(\mathbf{x}) + \rho \omega^2 \tilde{\mathbf{v}}_j(\mathbf{x}) = \tilde{\mathbf{0}}, \mathbf{x} \in D_j, \quad j = K+L+1, ..., N \tag{20}$$

$$T \tilde{\mathbf{v}}_j(\mathbf{x}) + T \tilde{\Gamma}(\mathbf{x}, \mathbf{z}) = \tilde{\mathbf{0}}, \mathbf{x} \in \partial D_j, \mathbf{z} \in D_j, \quad j = K+L+1, ..., N \tag{21}$$

The functions $\tilde{\mathbf{v}}_j$, $j = 1, ..., N$ can be approximated by elastic Herglotz functions $\tilde{\mathbf{v}}_{\mathbf{g}_j}$ with kernel $\tilde{\mathbf{g}}_j = (\tilde{\mathbf{g}}_{p,j}, \tilde{\mathbf{g}}_{s,j})$

Following the same steps as in [2, 3], we infer that

$$\left\| \tilde{\Gamma}(\cdot, \mathbf{z}) \right\|_{C(\partial D_j)} \leq c \left(\left\| \tilde{\mathbf{v}}_{\mathbf{g}_j} \right\|_{H^1(D_j)} + \varepsilon_j \right), \quad \varepsilon_j > 0, \quad j = 1, ..., N \tag{22}$$

We would like to mention here that for the transmission problems as in [3], a weak formulation of the problems is employed and consequently in the estimates we use $H^1(D_j)$ norms for $j = 1,...,K$. For the Dirichlet and Neumann problems we use the results in [7] where a classical description setting is applied. A similar estimation as (22) holds, however the estimation is given via the $C^1(D_j)$ -norm. Since classical solutions are also weak solutions the estimate holds in the $H^1(D_j)$ norm.

In the sequel we will show that there exists a Herglotz dyadic $\tilde{\mathbf{v}}_{\tilde{g}}$ which approximates $\tilde{\mathbf{v}}_{\tilde{g}_j}$ in $H^1(D_j)$. We consider a collection of domains $D_1^*, D_2^*, ..., D_N^*$ such that $D_j^* \supset D_j$, ∂D_j^*, $j = 1,...N$ is smooth and ω^2 is not a Dirichlet eigenvalue for the Lamé equation. This is possible because the eigenvalues of the Lamé operator are monotonically dependent on the domain [8].

Using Green's functions we can have a representation of $\tilde{v}_{\tilde{g}_j}$ in D_j^* with the aid of Betti's formulas. From the regularity properties of solutions of elliptic partial differential equations, it suffices to show that each $\tilde{v}_{\tilde{g}_j}$ can be approximated in $L^2(\partial D^*)$, where $D^* = D_1^* \cup D_2^* \cup ... \cup D_N^*$, by a Herglotz wave function $\tilde{\mathbf{v}}_{\tilde{g}}$. Since $\tilde{\Gamma}_e(\cdot, \mathbf{z})$ has a singularity of logarithmic type, we obtain that $\|\tilde{v}_{\tilde{g}}\|_{H^1(D)} \to +\infty$ (as $\mathbf{z}$ tends to $\mathbf{x}$), that is $\lim_{\mathbf{z} \to \partial D} \|\tilde{g}\|_{L^2(\Omega)} = +\infty$, since if the kernel remains bounded then the corresponding Herglotz wave function would be bounded in $H^1(D)$ as well. So, the points of the boundary of the scatterers constitute limits of point sources, rendering the Herglotz density L^2 -norm unbounded.

4. Numerical results

In the sequel we present numerical results for the case of two circular penetrable inclusions, with non-absorbing character. We follow the same method as in [3]. The synthetic far-field data are produced by the boundary element method presented in [9, 10]. In all cases we used 72 angles of incidence. The code developed, covers exactly the range of our elastic scattering problem for isotropic inclusions. All plots in Figures 1, 2, and 3 are produced by Matlab and are contour plots of the function $\frac{1}{|G_\gamma|^2}$ where

$$|\mathbf{G}_\gamma|^2 = \frac{4\pi^2 k_p}{N^2 \omega} \sum_{j=1}^N \left|\tilde{\mathbf{g}}_p(\hat{\mathbf{d}}_j)\right|^2 + \frac{4\pi^2 k_s}{N^2 \omega} \sum_{j=1}^N \left|\tilde{\mathbf{g}}_s(\hat{\mathbf{d}}_j)\right|^2 .$$

We consider the following cases:

Different distance between the two inclusions
In Figure 1 we present the reconstruction of two circular inclusions of the same radius. The right scatterer has radius a=1.2, k_p 2a=2.19 and k_s 2a=4. The left scatterer has radius b=1.2, k_p 2b=2.19 and k_s 2b=4. In subplot (a) the distance between the scatterers d=12.67. In (b) the distance between the scatterers is d=5.14 In (c) d=4.19 and in (d) d=0.01. In cases (a) and (b) a very clear image of the scatterers is produced. In (c) the information is enough to distinguish the two inclusions. Case (d) of course corresponds to a high interference between the inclusions and is not possible to derive different shapes.

Different radius of the one inclusion
In Figure 2 we present reconstruction images from two circular inclusions changing the radius of the one inclusion. In all cases the parameters for the isotropic medium are the same and the right scatterer has radius a=1.2, k_p 2a=2.19 and k_s 2a=4. In subplot (a) the left scatterer has radius b=1.2, k_p 2b=2.19 and k_s 2b=4. The distance between the scatterers d=10.79. In (b) the left scatterer has radius b=2.2, k_p 2b=4.01 and k_s 2b=7.33 and the distance between the scatterers d=8.79. In (c), similarly, b=3, k_p 2b=5.44, k_s 2b=10 and d=5.79. In (d) b=4, k_p 2b=7.3, k_s 2b=13.33 and d=3.79. In cases (a), (b) and (c) we have clear resolution and in (d) although we cannot have clear shapes we have an estimate of the position and the different area occupied by each inclusion.

Different elastic parameters of the one inclusion
We have for the left scatterer radius b=1.2, k_p 2b=2.19, k_s 2b=4 and for the right scatterer b=1.2. The distance between the scatterers is fixed in all experiments d=10.79. In (a) the right scatterer has k_p 2a=2.19, k_s 2a=4. In (b) k_p 2a=1.73, k_s 2a=3.1. In (c) k_p 2a=0.59, k_s 2a=1.44 and in (d) k_p 2a=0.45, k_s 2a=1.55. In this experiment we have satisfying inversions in all cases. It seems that the power of the Linear Sampling Method is rather independent of the different physical properties of the inclusions.

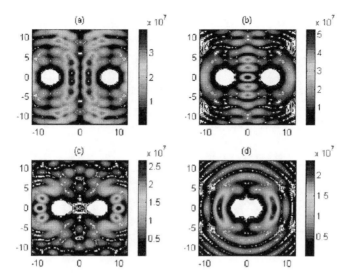

Figure 1. Different distance between the two inclusions

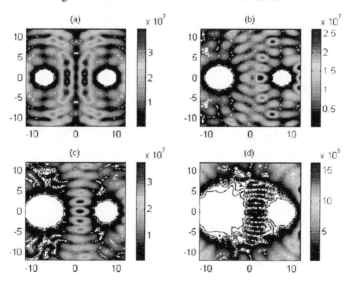

Figure 2. Different radius of the one inclusion

234

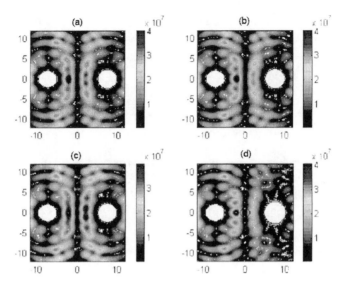

Figure 3. Different elastic parameters of the one inclusion

Acknowledgements

The authors would like to thank Prof. D. Polyzos and Dr. S. Tsinopoulos for providing their codes for the solution via the boundary element method of the direct elastic scattering problem. They also want to thank Mr. K. Anagnostopoulos for providing his code which solves the direct scattering problem for (generally anisotropic) elastic circular inclusions.

References

1. Arens, "Linear sampling methods for 2D inverse elastic wave scattering", *Inverse Problems* 17 pp.1445-1464 (2001).
2. Charalambopoulos, D. Gintides and K. Kiriaki, "The linear sampling method for the transmission problem in three-dimensional linear elasticity", *Inverse Problems* **19**, pp. 549-561 (2003).

3. Charalambopoulos, D. Gintides and K. Kiriaki, "The linear sampling method for non-absorbing penetrable elastic bodies elasticity", *Inverse Problems* **18**, pp.1-12 (2002).
4. D. Colton and R. Kress, "Inverse Acoustic and Electromagnetic Scattering Theory", Springer Verlag, (1992).
5. D. Colton and A. Kirsch, "A simple method for solving the inverse scattering problems in the resonance region", *Inverse Problems* **12** pp.383-393, (1996).
6. D. Colton, J. Coyle and P. Monk, "Recent developments in inverse acoustic scattering theory", *SIAM Review* **42**, No 3, pp. 369-414, (2000).
7. D. Gintides and K. Kiriaki, "The far-field equations in linear elasticity -An inversion scheme", *ZAMM*, Vol.: **81**, 5, pp. 305-316 (2001).
8. D. Gilbarg and N. S. Trudinger, "Elliptic Partial Differential Equations of Second Order", Springer Verlag, (1977).
9. D. Polyzos, S. V. Tsinopoulos and D. E. Beskos, "Static and dynamic boundary element analysis in incompressible linear elasticity", *Eur. J. Mech. A/Solids*, Vol. **17**, No 3, pp. 515-536 (1998).
10. J. T. Verbis, S. E. Kattis, S. V. Tsinopoulos and D. Polyzos, "Wave dispersion and attenuation in fiber composites", *Comp. Mech.*, Vol. **27**, pp. 244-252 (2001).

RELAXED DISCRETIZATION METHODS FOR NONCONVEX SEMILINEAR PARABOLIC OPTIMAL CONTROL PROBLEMS

I. CHRYSSOVERGHI

Department of Mathematics

National Technical University of Athens (NTUA)

Zografou Campus,15780 Athens, Greece

We consider an optimal control problem involving semilinear parabolic partial differential equations, with control constraints. Since no convexity assumptions are made, the problem is reformulated in relaxed form. The state equation is discretized using a finite element method in space and a Crank-Nicolson method in time, while the controls are approximated by blockwise constant relaxed controls. The first result is that accumulation points of discrete optimal (resp. extremal) controls are optimal (resp. extremal) for the continuous relaxed problem. We also propose a conditional gradient method for solving each discrete problem, and a progressively refining discrete conditional gradient method, both generating discrete relaxed controls, for solving the continuous relaxed problem. The second method has the advantage of reducing computations and memory. Relaxed controls computed by these methods can then be simulated by piecewise constant classical controls using a simple approximation procedure. Finally, a numerical example is given.

1. Introduction

Optimal control problems, without strong and often unrealistic convexity assumptions, have no classical solutions in general. For this reason they are reformulated in the so-called relaxed form, which in turn has a solution under mild assumptions. Relaxation theory has been extensively used to develop existence theory, necessary conditions for optimality, optimization methods and approximation methods (see [1]-[10], and references in [8]). Here we consider an optimal distributed control problem involving semilinear parabolic partial differential equations, with control constraints. Since no convexity assumptions are made, the problem is reformulated in relaxed form. The state equation is then discretized using a finite element method in space and a Crank-Nicolson method in time, while the controls are approximated by blockwise constant relaxed controls. The first result is that, under appropriate assumptions, the properties of optimality and extremality carry over in the limit to the corresponding properties for the relaxed continuous problem. In addition, we propose a conditional gradient method for solving each discrete problem, and a progressively refining discrete conditional gradient method, both generating relaxed controls, for solving the continuous relaxed problem. The second method has the advantage of reducing computations and memory. The use of relaxed controls exploits the nonconvex structure of the problem. The computed relaxed controls can then be simulated by piecewise constant classical controls using a simple approximation procedure. Finally, a numerical example is given.

2. The continuous optimal control problem

Let Ω be a bounded domain in $\mathbf{R}^d$ with a Lipschitz boundary Γ, and let $I = (0,T)$. Consider the following semilinear parabolic state equation

$$\partial y / \partial t + A(t)y = f(x,t,y(x,t),w(x,t)) \quad \text{in } Q = \Omega \times I,$$

$$y(x,t) = 0 \text{ in } \Sigma = \Gamma \times I \quad \text{and} \quad y(x,0) = y^0(x) \text{ in } \Omega,$$

where $A(t)$ is the second order elliptic differential operator

$$A(t)y = -\sum_{j=1}^{d}\sum_{i=1}^{d}(\partial / \partial x_i)[a_{ij}(x,t)(\partial y / \partial x_j)].$$

The constraints on the control variable w are

$$w(x,t) \in U \text{ in } Q,$$

where U is a compact, not necessarily convex, subset of $\mathbf{R}^d$, and the cost functional to be minimized is

$$J(w) = \int_Q g(x,t,y,w)dxdt.$$

Define the set of *classical controls*

$$W = \{w : (x,t) \mapsto w(x,t) | w \text{ measurable from } Q \text{ to } U\},$$

and the set of *relaxed controls* (Young measures, for relevant theory, see [8], [9]) by

$$R := \{r : Q \to M_1(U) | r \text{ weakly measurable}\}$$

$$\subset L_w^\infty(Q, M(U)) \equiv L^1(Q, C(U))^*,$$

where $M_1(U)$ is the set of probability measures on U. The set W (resp. R) is endowed with the relative strong (resp. weak star) topology, and R is convex, metrizable and compact. If we identify each classical control $w(\cdot)$

with its associated Dirac relaxed control $r(\cdot) := \delta_{w(\cdot)}$, then W may be considered as a subset of R, and W is thus dense in R. For given $\phi \in L^1(Q, C(U))$ and $r \in R$, we write for simplicity

$$\phi(x, t, r(x, t)) = \int_U \phi(x, t, u) r(x, t)(du).$$

The relaxed formulation of the above control problem is the following

$$< y_t, v > + a(t, y, v) = \int_\Omega f(t, y(x, t), r(x, t)) v(x) dx,$$

for every $v \in V$, a.e. in I,

$$y(x, t) = 0 \text{ in } \Sigma = \Gamma \times I \quad \text{and} \quad y(x, 0) = y^0(x) \text{ in } \Omega,$$

where $V = H_0^1(\Omega)$, $a(t, \cdot, \cdot)$ the usual bilinear form associated with A, $< \cdot, \cdot >$ the duality between V and $V*$, with control constraint $r \in R$, and cost to be minimized

$$J(r) = \int_Q g(x, t, y(x, t), r(x, t)) dxdt.$$

We suppose in the sequel that f is Lipschitz and sublinear w.r.t. y, the functions g_m are subquadratic w.r.t. y, and g_{my} sublinear w.r.t. y.

Theorem 2.1. The mappings $r \mapsto y \in L^2(Q)$ and $r \mapsto J(r)$ are continuous.

Theorem 2.2. There exists an optimal relaxed control.

Since there are no state constraints here, and W is dense in R, we have

$$\min_{r \in R} J(r) = \inf_{w \in W} J(w).$$

It can be shown (see [2]) that, for given controls $r, r' \in R$, the directional derivative of a functional J (we drop the index m) is given by

$$DJ(r, r'-r) = \int_Q H(x, t, y, z, r'-r) dxdt,$$

where the Hamiltonian is

$$H(x, t, y, z, u) = zf(x, t, y, u) + g(x, t, y, u),$$

and the *general adjoint* z satisfies the equation

$$-z_t + A*(t)z = f_y(y, r)z + g_y(y, r) \text{ in } Q,$$

$$z(x,t) = 0 \text{ in } \Sigma = \Gamma \times I \quad \text{and} \quad z(x,T) = 0 \text{ in } \Omega.$$

We have the following relaxed necessary conditions for optimality.

Theorem 2.3. If r is optimal, then it is *extremal*, i.e. it satisfies the inequalities

$$DJ(r, r'-r) \geq 0, \text{ for every } r' \in R,$$

or equivalently, the pointwise minimum principle

$$H(x,t,y,z,r) = \min_{u \in U} H(x,t,y,z,u) \text{ a.e. in } Q.$$

Theorem 2.4. The mappings $r \mapsto z \in L^2(Q)$ and $(r, r') \mapsto DJ(r, r'-r)$ are continuous.

3. Discretization

We suppose here that Ω is a polyhedron and that the bilinear form a is independent of t. For each $n \geq 0$, consider a finite element discretization, with piecewise affine basis functions of a subspace V^n of V, w.r.t. an admissible and regular triangulation T^n of Ω into simplices S_i^n, $i = 1, ..., M := M^n$, of maximum diameter h^n, and a time partition P^n of I into equal subintervals I_j^n, $j = 0, ..., N-1$, of length Δt^n. The set R^n of discrete relaxed controls is the set of relaxed controls that are equal to a constant measure in $M_1(U)$ on each block $S_i^n \times I_j^n$. Consider the following discrete problems, with state equation defined by the implicit Crank-Nicolson scheme

$$(1/\Delta t^n)(y_{j+1}^n - y_j^n, v) + a(\overline{y}_j^n, v) = (f(\overline{t}_j^n, \overline{y}_j^n, r^n), v),$$

for every $v \in V^n$, $j = 0, ..., N-1$, y_0^n given,

$$\overline{t}_j^n = (t_j^n + t_{j+1}^n)/2, \quad \overline{y}_j^n = (y_j^n + y_{j+1}^n)/2,$$

control constraint $r^n \in R^n$, and discrete cost

$$J^n(r^n) = \sum_{j=0}^{N-1} \int_\Omega g(\overline{t}^n, \overline{y}^n, r^n) dx.$$

Theorem 3.1. The mappings $r^n \mapsto z^n$ and $r^n \mapsto J^n(r^n)$ are continuous.

Theorem 3.2. There exists an optimal discrete control.

Dropping m, the general discrete adjoint equation is here

$$-(1/\Delta t)(z_{j+1}^n - z_j^n, v) + a(v, \overline{z}_j^n) = (f_y(\overline{t}_j^n, \overline{y}_j^n, r_j^n)\overline{z}_j^n, v) + (g_y(\overline{t}_j^n, \overline{y}_j^n, r_j^n), v)$$

for every $v \in V^n$, $j = N-1, \ldots, 0$,

$$\overline{z}_j^n = (z_j^n + z_{j+1}^n)/2, \; z_N^n = 0,$$

and the directional derivative of J^n

$$DJ^n(r^n, r^m - r^n) = \Delta t^n \sum_{j=0}^{N-1} H(\overline{t}_j^n, \overline{y}_j^n, \overline{z}_j^n, r^m_j - r_n^j).$$

Theorem 3.3. If r^n is optimal, then it is *extremal*, i.e. it satisfies the inequalities

$$DJ_m^n(r^n, r^m - r^n) \geq 0, \text{ for every } r^m \in R,$$

or equivalently

$$\int_\Omega H(x, \overline{t}_j^n, \overline{y}_j^n, \overline{z}_j^n, r_j^n)dx = \min_{u \in U} \int_\Omega H(x, \overline{t}_j^n, \overline{y}_j^n, \overline{z}_j^n, u)dx, \; j = 0, \ldots, N-1.$$

Theorem 3.4. The mappings $r^n \mapsto z^n$ and $(r^n, r^m) \mapsto DJ^n(r^n, r^m - r^n)$ are continuous.

4. Behavior in the limit

Theorem 4.1. (Control approximation) For each $r \in R$, there exists a sequence $(r^n \in R^n)$ that converges to r in R (the controls r^n can even be chosen to be blockwise constant classical ones).

We suppose that $y_0^n \to y^0$ in V strongly and that $\Delta t^n \leq c(h^n)^2$, for an appropriate c.

Theorem 4.2. (Consistency) If $r^n \to r$, $r^m \to r'$ in R, then

$$y^n \to y, \; z^n \to z, \; J^n(r^n) \to J(r), \; DJ^n(r^n, r^m - r^n) \to DJ(r, r' - r).$$

The two following results concern the behavior in the limit of the properties of optimality, and of extremality and admissibility.

Theorem 4.3. Let $(r^n \in R^n)$ be a sequence of optimal controls for the discrete problems. Every accumulation point of (r^n) is optimal for the continuous relaxed problem.

Theorem 4.4. Let $(r^n \in R^n)$ be a sequence of extremal controls for the discrete problems. Every accumulation point of (r^n) is extremal and admissible for the continuous relaxed problem.

5. Discrete optimization methods

In our first algorithm, we apply a relaxed conditional gradient method directly to the discrete problem, for some sufficiently large n. The use of relaxed controls exploits the nonconvex structure of the problem.

Algorithm 1: Discrete Relaxed Conditional Gradient Method (DRCGM)

Step 1. Set $k = 0$ and choose $r_0^n \in R^n$.
Step 2. Find r_k^n such that $d_k = DJ_n^n(r_k^n, \bar{r}_k^n - r_k^n) = \min_{r'''\in R^n} DJ_n^n(r_k^n, r''' - r_k^n)$.
Step 4. Either
(a) (Optimal step option) Find $\alpha_k \in [0,1]$ such that
$J_n^n(r_k^n + \alpha_k(\bar{r}_k^n - r_k^n)) = \min_{\alpha\in[0,1]} J_n^n(r_k^n + \alpha(\bar{r}_k^n - r_k^n))$, or
(b) (Armijo step option) Find the smallest positive integer s such that $\alpha_k = c^s$ satisfies the inequality $J_n^n(r_k^n + \alpha(\bar{r}_k^n - r_k^n)) - J_n^n(r_k^n) \le \alpha_k b d_k$.
Step 5. Choose any $r''_k \in R^n$ such that $J_n^n(r''_k) \le J_n^n(r_k^n + \alpha_k(\bar{r}_k^n - r_k^n))$.
Set $r_{k+1}^n = r''_k$, $k = k+1$ and go to Step 2.

Theorem 5.1. The accumulation points of sequences generated by Algorithm 1 are extremal for the discrete relaxed problem.

We next propose a relaxed discrete, progressively refining, conditional gradient method, where the discretization is refined according to some convergence criterion. The refining procedure has the advantage of reducing computations and memory. We suppose now that, for every n, either $T^{n+1} \equiv T^n$ and $P^{n+1} \equiv P^n$, or each simplex S_i^{n+1} is a subset of some S_i^n and each interval $I_{j\cdot}^{n+1}$ a subset of some I_j^n. Let (γ^n) be a positive decreasing sequence converging to zero, let $b, c \in (0,1)$, and consider the following algorithm.

Algorithm 2: Progressively Refining DRCGM

Step 1. Set $k = 0$ and choose $r_0^n \in R^n$.
Step 2. Find r_k^n such that $d_k = DJ_n^n(r_k^n, \bar{r}_k^n - r_k^n) = \min_{r'''\in R^n} DJ_n^n(r_k^n, r''' - r_k^n)$.

Step 3. If $|d_k| \le \gamma^n$, set $r^n = r_k^n$, $d^n = d_k$, $n = n+1$ and go to Step 2.

Step 4. Either

(a) (Optimal step option) Find $\alpha_k \in [0,1]$ such that
$$J^n(r_k^n + \alpha_k(\bar{r}_k^n - r_k^n)) = \min_{\alpha \in [0,1]} J^n(r_k^n + \alpha(\bar{r}_k^n - r_k^n)), \text{ or}$$

(b) (Armijo step option) Find the smallest positive integer s such that $\alpha_k = c^s$ satisfies the inequality $J^n(r_k^n + \alpha_k(\bar{r}_k^n - r_k^n)) - J^n(r_k^n) \le \alpha_k b d_k$.

Step 5. Choose any $r''_k \in R^n$ such that $J^n(r''_k) \le J^n(r_k^n + \alpha_k(\bar{r}_k^n - r_k^n))$.

Set $r_{k+1}^n = r''_k$, $k = k+1$ and go to Step 2.

Theorem 5.2. The accumulation points of the sequence (r^n) generated by Algorithm 2 are extremal for the continuous relaxed problem.

In the above algorithms, the Armijo step option is a finite procedure and is practically faster than the optimal step option (use of golden section search). For the implementation of algorithms generating relaxed controls, see [5]. The approximate discrete relaxed controls computed by the algorithms can then be simulated by piecewise constant classical controls using a simple approximation procedure (see [2]).

6. Numerical example

Let $\Omega = (0,\pi)$, $I = (0,1)$, $Q = \Omega \times I$, and define the functions

$$\bar{y}(x,t) = -e^{-t} \sin x + x(\pi - x)/2, \quad \bar{w}(x,t) = \begin{cases} 1/2 + t, & t \in [0,1/2) \\ 1, & t \in [1/2,1] \end{cases}.$$

Consider the following optimal control problem, with state equation

$$y_t - y_{xx} = 1 + w - \bar{w} + y - \bar{y}, \quad (x,t) \in Q,$$

$$y(0,t) = y(\pi,t) = 0, \quad t \in I, \quad y(x,0) = -\sin x + x(\pi - x)/2, \quad x \in \Omega,$$

nonconvex control constraints

$$w(x,t) \in U := [-1,-1/2] \cup \{1\}, \quad (x,t) \in Q,$$

and nonconvex cost functional

$$J_0(w) = \int_Q [\frac{1}{2}(y - \bar{y})^2 - w^2] dx dt.$$

It can be verified that the optimal relaxed control is given by

$$r^*(x,t)(\{1\}) = 1/2 + t, \quad r^*(x,t)(\{-1\}) = 1 - r^*(x,t)(\{1\}),$$

$$t \in [0,1/2) \quad \text{(non-classical part)},$$

$$r*(x,t)(\{1\}) = 1, \quad t \in [1/2,1] \quad \text{(classical part)},$$

with optimal state $y^* = \bar{y}$, and optimal cost $J(r^*) = -\pi$. Since the optimal relaxed cost value $-\pi$ cannot be attained by classical controls due to the constraints, but can be approximated as close as possible since W is dense in R, the classical problem cannot have a solution.

After 100 iterations of the relaxed conditional gradient method, with optimal step option and with three successive pairs of step sizes $h = \pi/30$, $\pi/60$, $\pi/120$, $\Delta t = 1/10$, $1/20$, $1/40$, we obtained

$$J_0(r^n) = -3.141574, \quad |d^n| = 0.60 \cdot 10^{-4}.$$

References

1. I. Chryssoverghi and A. Bacopoulos, *J. Optim. Theory and Appl.*, **65**, 395 (1990).
2. I. Chryssoverghi and A. Bacopoulos, *JOTA*, **77**, 31 (1993).
3. I. Chryssoverghi, A. Bacopoulos, B. Kokkinis, and J. Coletsos, *JOTA*, **94**, 311 (1997).
4. I. Chryssoverghi, A. Bacopoulos, J. Coletsos, and B. Kokkinis, *Control & Cybernetics*, **27**, 29 (1998).
5. I. Chryssoverghi, J. Coletsos, and B. Kokkinis, *Control & Cybern.*, **28**, 157 (1999).
6. I. Chryssoverghi, J. Coletsos, and B. Kokkinis, *Control & Cybern.*, **30**, 385 (2001).
7. T. Roubíček, *JOTA*, **69**, 589 (1991).
8. T. Roubíček, Relaxation in Optimization Theory and Variational Calculus, Walter de Gruyter, Berlin (1997).
9. J. Warga, Optimal Control of Differential and Functional Equations, Academic Press, New York (1972).
10. J. Warga, *SIAM J. on Control*, **15**, 674 (1977).

A GREEDY METHOD FOR MIXTURE-BASED CLASSIFICATION

CONSTANTINOS CONSTANTINOPOULOS AND ARISTIDIS LIKAS

Dept. of Computer Science
Univ. of Ioannina,
GR 45110, Ioannina, Greece
E-mail: {ccostas, arly}@cs.uoi.gr

The typical approach of mixture-based learning for classification employs the Expectation – Maximization (EM) algorithm. This widely used method suffers from two important drawbacks, namely it cannot estimate the true number of mixing components, and depends strongly on the initial parameter values. The Greedy EM algorithm is a recently proposed algorithm that tries to overcome these drawbacks for the density estimation task. The direct application of this method for the classification task is obvious, as long as we are interested in typical mixture models. In this work we propose a greedy learning method adapted for a generalized mixture model (called the Probabilistic RBF network) with the particular characteristic that it allows the sharing of the mixture components among all classes, in contrast to the convetional model that suggests mixture components describing only one class. The proposed algorithm starts with a single component and adds components sequentially, using a combination of global and local search. The addition of the new component is based on criteria for detecting a region in the data space that is crucial for the classification task. Experimental results using several well-known classification datasets indicate that the greedy method provides solutions of superior classification performance.

1. Introduction

In pattern recognition it is well-known that a convenient way to construct a classifier is on the basis of inferring the posterior probability of each class. From the statistical point of view this inference can be achieved by first evaluating the class conditional densities $p(x|k)$ and the corresponding prior probabilities $P(k)$ and then making optimal decisions for new data points by combining these quantities through Bayes theorem

$$P(k|x) = \frac{p(x|k)P(k)}{\sum_\ell p(x|\ell)P(\ell)},$$

(1)

and then selecting the class with maximum $P(k|x)$. In the traditional statistical approach each class density $p(x|k)$ is estimated using a separate mixture model and considering only the data points of the specific class, therefore the density of each class is estimated independently from the other classes. We will refer to this approach as the *separate mixtures* model.

The probabilistic RBF network[6,7] constitutes an alternative approach for class conditional density estimation. It is an RBF-like neural network[4] adapted to provide output values corresponding to the class conditional densities $p(x|k)$. Since the network is RBF,[4] the kernels (hidden units) are shared among classes and each class conditional density is evaluated using not only the corresponding class data points (as in the traditional statistical approach[5]), but using all the available data points. In order to train the PRBF network, an Expectation - Maximization (EM) algorithm can be applied.[1,2,7] In addition, it has been found[8] that the generalization performance is improved if after training the kernels are split. Considering the problem of EM initialization and its effect to the performance of the algorithm, we propose a greedy training method for the network. Adopting the greedy and split methods offers significant improvement in generalization. The effectiveness of the proposed method is demonstrated using several data sets and the experimental results indicate that the method leads to performance improvement over the classical PRBF training method.

2. The Probabilistic RBF Network (PRBF)

Consider a classification problem with K classes. We are given a training set $X = \{(x^{(n)}, y^{(n)}), n = 1, \ldots, N\}$ where $x^{(n)}$ is a d-dimensional pattern, and $y^{(n)}$ is a label $k = 1, \ldots, K$ indicating the class of pattern $x^{(n)}$. The original set X can be easily partitioned into K independent subsets X_k, so that each subset contains only the data of the corresponding class. Let N_k denote the number of patterns of class k, ie. $N_k = |X_k|$.

Assume that we have a number of M kernel functions (hidden units), which are probability densities, and we would like to utilize them for estimating the conditional densities of all classes by considering the kernels as a common pool.[6,7] Thus, each class conditional density function $p(x|k)$ is modeled as

$$p(x|k) = \sum_{j=1}^{M} \pi_{jk} p(x|j), \quad k = 1, \ldots, K \tag{2}$$

where $p(x|j)$ denotes the kernel function j, while the mixing coefficient π_{jk} represents the prior probability that a pattern has been generated from kernel j, given that it belongs to class k. The priors take positive values and satisfy the following constraint:

$$\sum_{j=1}^{M} \pi_{jk} = 1, \quad k = 1, \ldots, K. \tag{3}$$

It is also useful to introduce the posterior probabilities expressing our posterior belief that kernel j generated a pattern x given its class C_k. This probability is obtained using the Bayes' theorem

$$P(j|x, k) = \frac{\pi_{jk} p(x|j)}{\sum_{i=1}^{M} \pi_{ik} p(x|i)}. \tag{4}$$

In the following, we assume that the kernel densities are Gaussians of the general form

$$p(x|j) = \frac{1}{(2\pi)^{d/2} |\Sigma_j|^{1/2}} \exp \left\{ -\frac{1}{2}(x - \mu_j)^T \Sigma_j^{-1} (x - \mu_j) \right\} \tag{5}$$

where $\mu_j \in \Re^d$ represents the center of kernel j, while Σ_j represents the corresponding $d \times d$ covariance matrix. The whole adjustable parameter vector of the model consists of the priors and the component parameters (means and covariances), and we denote it by Θ, while we use θ_j to denote the parameter vector of each component.

It is apparent that the PRBF model is a special case of the RBF network,[4] where the outputs correspond to probability density functions and the second layer weights are constrained to represent prior probabilities. Furthermore, it can be shown that the separate mixtures model[5] can be derived as a special case of PRBF.

The typical training method of the network computes the Maximum Likelihood estimates of the parameters using the EM algorithm.[1,2,6,7] At the same time it is widely known that the effectiveness of the EM depends highly on the initial conditions. To avoid this we propose the *greedy* training method, which is a determimnistic two–stage algorithm that overcomes the initialization problem of EM. During the first stage we incrementaly add componenets to the network, and during the second stage we split all its components.

3. The Greedy Training Method

The procedure of component addition involves global and local search in the parameter space, for each new component. During the global search the algorithm searches over all data, to locate candidate components. The new component is selected among the candidates, according to some criteria. During the local search the algorithm adds the new component to the current model, and estimates the parameters of the resulting larger model using EM. This procedure is repeated until a maximum number of components is reached.

3.1. Global Search for Candidate Components

In order to incrementally add components, we need at each stage to search for candidate positions of the new component. To locate such positions we partition the data of each component to six subsets, using the kd-tree structure. The statistics of each subset define the parameters (mean, cavariance and mixing weights) of the candidate components. If the class conditional density of class k using M components is $p(x|k;\Theta_M)$ and the density of the component we add is $p(x|j = M + 1)$, then the new class conditional density is

$$p(x|k;\Theta_{M+1}) = (1 - \alpha_k)p(x|k;\Theta_M) + \alpha_k P(x|j = M + 1) \qquad (6)$$

where α_k is the mixing weight of the new component, and the log-likelihood of the data is

$$L = \sum_{k=1}^{K} \sum_{x \in X_k} \log\left\{(1 - \alpha_k)p(x|k;\Theta_M) + \alpha_k P(x|j = M + 1)\right\} \qquad (7)$$

We optimize the parameters α_k, μ_{M+1} and Σ_{M+1} by applying partial EM. During the *Expectation* step we compute the posterior probabilities $P^{(t)}(j = M + 1|x, k)$ using the current estimates of $\alpha_k^{(t)}$, $\mu_{M+1}^{(t)}$ and $\Sigma_{M+1}^{(t)}$, according to:

$$P^{(t)}(j = M + 1|x, k) = \frac{\alpha_k^{(t)}p(x|j = M + 1)}{(1 - \alpha_k^{(t)})p(x|k;\Theta_M) + \alpha_k^{(t)}p(x|j = M + 1)} \qquad (8)$$

During the *Maximization* step we compute the new estimates of the parameters, according to:

$$\mu_{M+1}^{(t+1)} = \frac{\sum_{k=1}^{k} \sum_{x \in X_k} P^{(t)}(j = M + 1|x, k)x}{\sum_{k=1}^{k} \sum_{x \in X_k} P^{(t)}(j = M + 1|x, k)} \qquad (9)$$

$$\Sigma_{M+1}^{(t+1)} = \frac{\sum_{k=1}^{k} \sum_{x \in X_k} P^{(t)}(j = M + 1|x, k)(x - \mu_{M+1}^{(t+1)})(x - \mu_{M+1}^{(t+1)})^T}{\sum_{k=1}^{k} \sum_{x \in X_k} P^{(t)}(j = M + 1|x, k)} \qquad (10)$$

$$\alpha_k^{(t+1)} = \frac{1}{|X_k|} \sum_{x \in X_k} P^{(t)}(j = M + 1|x, k), \quad k = 1. \ldots, K \qquad (11)$$

From the candidates we select the most appropriate, according to some criteria that we present in the next section. The above procedure is depicted in Fig 1.

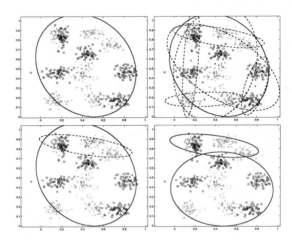

Figure 1. Addition of the first two components. The components of the network are drawn with solid lines, and the candidate components with dotted lines.

3.2. *Selection Criteria for Component M+1*

An obvious criterion for the determination of an appropriate candidate is the increase in the log-likelihood of the data. However in a classification problem we are interested in the generative model of each class individually. For this reason we introduce two more criteria related to the log-likelihood of each class. We are interested in the change of log-likelihood for each class, as we add a new component according to (6). So we define the change ΔL_k for class k as:

$$
\begin{aligned}
\Delta L_k &= L_k(\Theta_{M+1}) - L_k(\Theta_M) \\
&= \sum_{x \in X_k} \log \left\{ 1 + \alpha_k \left[\frac{p(x|j = M + 1)}{p(x|k; \Theta)} - 1 \right] \right\}
\end{aligned}
\tag{12}
$$

and suggest to search between candidate components to find the one that mostly increases the log-likelihood of two classes. Such a candidate lies in a region with high overlapping among classes, consequently on a decision boundary.

Experimental results revealed that, after the incremental addition of few initial components, there are no candidates that increase the log-likelihood of two classes at the same time. We can explain this attitude observing that:

$$\text{if } \sum_{x \in X_k} \frac{p(x|j = M + 1)}{p(x|k; \Theta_M)} < |X_k|, \text{ then } \Delta L_k < 0. \tag{13}$$

According to (13), the log-likelihood of one class decreases if for a sufficient number of its patterns the likelihood $p(x|j = M + 1)$ according to the new component is less than the likelihood $p(x|k; \Theta_M)$ according to the current model. Consequently there is a limit in the number of components we can add, while at the same time the log-ilkelihood of two classes increases. Beyond this limit we need a less strict criterion.

The next criterion we introduce, in order to continue the addition of components, searches for the candidate that mostly increases the log-likelihood of all data. This is a natural choice, in accordance with the Maximum Likelihood estimates we are looking for.

If there are no candidates that increase the log-likelihood of all data, we search for the candidate that mostly increases the log-likelihood of only one class. Such candidates lie away from the decision boundary, but are important for the generative model of the class. We cannot assure that this choice results in a model that increases the log-likelihood of the data compared to the previous model. If this is the case, we stop the addition of components.

3.3. Component Splitting

After the completion of the first stage of the training algorithm, there may be components of the network located to regions with significant overlapping among classes. This happens if we underestimate the maximum allowed number of components. In order to increase the generalization performance of the network we plit all the components.[8] Evaluating the posterior probability $P(j|x, k)$ for a component we can define if it is responsible for patterns of more than one class. If this is true, then we split it in K new components. The result is that each component descibes only one class. Each new component $p(x|j, k)$ is a Gaussian p.d.f., with mean μ_{kj}, covariance matrix Σ_{kj} and mixing weights π_{jk}. These parameters are estimated according to:

$$\pi_{jk} = \frac{\sum_{x \in X_k} P(j|x,k)}{|X_k|} \tag{14}$$

$$\mu_{kj} = \frac{\sum_{x \in X_k} P(j|x,k)x}{\sum_{x \in X_k} P(j|x,k)} \tag{15}$$

$$\Sigma_{kj} = \frac{\sum_{x \in X_k} P(j|x,k)(x - \mu_{kj})(x - \mu_{kj})^T}{\sum_{x \in X_k} P(j|x,k)}. \tag{16}$$

4. Experimental Results

The proposed greedy training method for the PRBF network is compared with the standard EM training followed by a split stage.[8] We considered four benchmark data sets from the UCI repository, namely Bupa Liver Disorder (bld), Clouds (cld), Ionosphere (ion) and Cleveland Heart Disease (clev) data sets. For each data set, in order to obtain an estimation of the generalization error, we employed 10-fold cross – validation. We evaluated the number of components of the network using a validation set for each of the ten trials. Table 1 provides the obtained results for both methods. For two of the data sets the two methods gave similar generalization error, but for these the typical approach exhibits already near the optimum performance and the greedy method preserved this behaviour. As regards the other data sets the greedy training clearly exhibits better generalization performance.

Table 1. The generalization error of greedy training compared to the typical training approach.

	bld	cld	ion	clev
PRBF & Split	0.31	0.10	0.07	0.23
Greedy PRBF	0.28	0.10	0.06	0.18

5. Future Work

We presented a greedy training of the PRBF network, that overcomes the initialization problem of the standard EM algorithm, and provides networks with superior generalization performance. In our future work we are going to extend the experiments, and compare the greedy PRBF network with Support Vector Machines. Also we are especially interested in Bayesian methods for estimating the number of components the network utilizes. Finaly we are going top examine the altrernative use of Probabilistic Principal Component Analyzers instead of Gaussian components.

References

1. A. P. Dempster, N. M. Laird and D. B. Rubin, "Maximum Likelihood Estimation from Incomplete Data via the EM Algorithm", *Journal of the Royal Statistical Society B*, vol. 39, pp. 1-38, 1977.
2. G. McLachlan, T. Krishnan, *The Em Algorithm and Extensions*, Wiley, 1997.
3. T. Mitchell, *Machine Learning*, McGraw-Hill, 1997.
4. C. M. Bishop, *Neural Networks for Pattern Recognition*, Oxford University Press, 1995.
5. G. McLachlan, D. Peel, *Finite Mixture Models*, Wiley, 2000.
6. M. Titsias, A. Likas, "A Probabilistic RBF network for Classification", *Proc. of International Joint Conference on Neural Networks*, Como, Italy, July 2000.
7. M. Titsias, A. Likas, "Shared Kernel Models for Class Conditional Density Estimation", *IEEE Trans. on Neural Networks*, vol. 12, no. 5, pp. 987-997, Sept. 2001.
8. M. Titsias, A. Likas,"Mixture of Experts Classification Using a Hierarchical Mixture Model", Neural Computation, vol. 14, no. 9, Sept. 2002.
9. N. A. Vlassis and A. Likas, "A Greedy-EM Algorithm for Gaussian Mixture Learning", *Neural Processing Letters*, vol. 15, pp. 77-87, 2002.

FROM D' ALEMBERT AND FOURIER TO GELFAND AND FOKAS VIA LAX. LINEARITY REVISITED

GEORGE DASSIOS

University of Patras, ICEHT-FORTH and Hellenic Open University

Two and a half centuries ago d'Alembert introduced the method of separation of variables, which provided the foundations on which Fourier constructed perhaps the most useful mathematical theory of approximation. This theory gave birth to almost all linear theories in mathematics, including Linear Algebra and Linear Functional Analysis. These days mathematicians use linear methods very effectively in every application of mathematics, but no major conceptual breakthroughs have been proposed for many-many decades.It is no more than thirty-five years ago when Peter Lax showed how one can combine linear methods, in a very elegant way, to solve via decomposition some nonlinear problems as well. This was a really important moment in the recent theory of Differential Equations, since it introduces a completely new way to look at the nonlinear world. Since then the method is known as the Lax-Pairs technique. Gelfand and Fokas, two eminent methamaticians in the theory of nonlinear equation, looked at the effects that the theory of Lax-Pairs could have on linear problems. The results of their observation were amazing and offered to the mathematical community a solution that was considered for many-many years impossible. The Gelfand–Fokas theory tells us how to obtain separable solutions even when separation of variables can not be performed. Not in an easy way, but it can be done! Unfortunately, this theory, which is shaped day-by-day by Fokas and his collaborators, is not more than five years old and because of its demanding analytical manipulations (a qualification not very common among young mathematicians today) we can not find as many researchers working on this fascinating topic as its future prejudges. I consider myself a "linear victim" of this fascination and I would like to make a few introductory comments and remarks on the way Gelfand and Fokas showed to us, after two and half centuries, of the right way to look at linearity. Professor Fokas will explain to us in the sequel the more sophisticated aspects of his theory.

I have to admit that I am feeling a little uncomfortable today, since I am going to talk about a new method which I appreciate very much, but it is not fully developed yet. This is the Gelfand – Fokas (G-F) transform theory which offers a method for solving differential equations, a method that comes from nonlinear mathematics, but which seems much more appropriate for linear problems. Let me start with some historical thoughts.

In trying to solve differential equations, the mathematicians of the late 17th and early 18th Century invented many tricks (today we call them transformations) with which they were able to bring their equations to a solvable, by their techniques, form. But this approach restricted their activities to those differential equations that accept solutions in terms of elementary functions. A real breakthrough occurred when Euler suggested to look for solutions in terms of power series. This idea enlarged the set of candidate solutions from the elementary to the analytic functions. In fact, it was this method that led Euler to assume that every linear homogeneous equation with constant coefficients has solutions expressible as powers of exponentials. Today we know that this is true since the exponential function is the unique eigenfuction of differentiation.

Since then, the theory of differential equations had a few other opportunities for breakthroughs. D' Alembert's method of separation of variables for partial differential equations in the 1750's was the wick for the blasting of Fourier analysis a few years later. Referring to these two instances as "breakthroughs" is the least we can say about these two fundamental branches of mathematics.

Two Centuries later, in the 1950's, the theory of distributions led to the idea of weak solutions, bringing the set of functions, where someone looks for solutions of differential equations, to the largest set known today.

Since I am one of those old fashion mathematicians who still look for solutions of linear equations, I would like to share with you today a breakthrough which, in my opinion, we experience these years in our search for solutions of linear partial differential equations.

But what is this idea and where it comes from? It comes from the theory of non-linear integrable systems as it was formulated by Peter Lax in 1968 [5]. Here is the basic idea of this theory. Lax's method is based on the determination of a pair of equations, known as Lax-Pair, which actually splits an initial – boundary value problem for a non-linear partial differential equation into a linear eigenvalue problem and a second, also linear, evolution equation. Where the non-linearity goes? It is hidden in the coefficients of the two linear operators which are polynomial functions of the solution of the non-linear equation we started with. The motive that led Lax to develop his theory was the famous paper by Gardner, Greene, Kruskal and Muira [4] which solves the KdV equation and proves mathematically the existence of solitary wave. Of course, this is not always possible, but when it is possible, the results are amazingly simple.

What Gelfand and Fokas did, less than ten years ago [2], is to look at linear problems via Lax's method. They proved that, for linear problems, this is always possible, they demonstrated what the corresponding Lax – Pair for every equation is, and they provided us with a fresh view of linear problems. This theory was considerably improved three years later by Fokas [3]. The G-F method is actually a generalization of the method of separation of variables and the method of Fourier transform, with two important differences.

a) The Lax – Pair splits the equation but at the same time unifies rather than separates the variables so that at each step, each one of the variables is aware of how you proceed with the other.

b) The generalized transform that is introduced is a tailor-made transform appropriate for the particular partial differential equations at hand.

Are there any important new results obtained with this new method? Well, I do not want to take the fun out of Professor Fokas talk, and I will let him answer this question in his subsequent talk. What I want to do next is to try to explain this method, with the help of a very simple example and to make some comments on this procedure. I hope that this will serve as an introduction to the talk of professor Fokas that follows mine.

Consider the 1-D linear Schrödinger equation

$$iq_t + q_{xx} = 0, \quad x \in I\!R, \quad t > 0 \tag{1}$$

and the Dirichlet problem

$$q(x,0) = q_1(x), \quad x \in I\!R \tag{2}$$

for the half plane t>0. Fourier transform theory implies the solution

$$q(x,t) = \frac{1}{2\pi} \int_{-\infty}^{+\infty} e^{-ikx - ik^2 t} \hat{q}_1(k) dk \tag{3}$$

where

$$\hat{q}_1(k) = \int_{-\infty}^{+\infty} e^{ikx} q_1(x) dx . \tag{4}$$

Similarly, the same problem for the quarter plane with conditions

$$q(x,0) = q_1(x), \quad x > 0 \tag{5}$$
$$q(0,t) = q_2(t), \quad t > 0 \tag{6}$$

leads to the solution

$$q(x,t) = \frac{2}{\pi} \int_0^{+\infty} \sin kx \cdot e^{-ik^2 t} \cdot \left[\hat{q}_1(k) + ik \int_0^t e^{-ik^2(t-\tau)} q_2(\tau) d\tau \right] dk \tag{7}$$

where

$$\hat{q}_1(k) = \int_0^{+\infty} q_1(x) \sin kx dx . \tag{8}$$

Comparing (3) and (7) we observe that mere change of the fundamental domain from the half to the quarter plane not only complicates the representation of the solution but destroys the separability between the time variable t and the spectral variable k. Why is that so? Because the Fourier sine transform is not the appropriate transform for the solution of this

Dirichlet problem. We will show in the sequel that the Gelfand-Fokas theory offers a more natural method to deal with this problem which leads to an expression as simple as (3).

The Lax-pair for equation (1) is given by

$$\mu_x + ik\mu = q \tag{9a}$$

$$\mu_t + ik^2\mu = iq_x + kq \tag{9b}$$

where the first equation (9a) is the same for every linear problem and introduces the Fourier transform and the second equation (9b) is dependent on (1) and actually modifies the Fourier transform to fit best the particular problem at hand. A simultaneous spectral analysis of (9a, 9b) asks for the solution $\mu(x,t;k)$ which is defined on the entire complex plane $\mathcal{C}$ and it is analytic and bounded within certain sectors. A careful but straightforward analysis provides the solution

$$\mu(x,t;k) = \begin{cases} \mu^{I+II}(x,t;k), & k \in \mathcal{C}(I) \cup \mathcal{C}(II) \\ \mu^{III}(x,t;k), & k \in \mathcal{C}(III) \\ \mu^{IV}(x,t;k), & k \in \mathcal{C}(IV) \end{cases} \tag{10}$$

where $\mathcal{C}(I), \mathcal{C}(II), \mathcal{C}(III), \mathcal{C}(IV)$ denote the four quadrants of the complex plane and

$$\mu^{I+II}(x,t;k) = -\int_x^{+\infty} e^{-ik(x-x')}q(x',t)dx', \quad k \in \mathcal{C}(I) \cup \mathcal{C}(II) \tag{11}$$

$$\mu^{III}(x,t;k) = \int_0^x e^{-ik(x-x')}q(x',t)dx'$$

$$-e^{-ikx}\int_t^{+\infty} e^{-ik^2(t-t')}[iq_x(0,t') + kq(0,t')]dt', \quad k \in \mathcal{C}(III) \tag{12}$$

$$\mu^{IV}(x,t;k) = \int_0^x e^{-ik(x-x')}q(x',t)dx'$$

$$+e^{-ikx}\int_0^t e^{-ik^2(t-t')}[iq_x(0,t') + kq(0,t')]dt', \quad k \in \mathcal{C}(IV) \tag{13}$$

The function μ^{I+II} is holomorphic in the open upper half plane and continuous in the closed upper half plane. Similarly μ^{III} is holomorphic in the open and continuous in the closed third quadrant and μ^{IV} is holomorphic in the open and continuous in the closed fourth quadrant. Let L_1 be the positive real axis directed from 0 to $+\infty$, L_2 be the negative real axis directed from 0 to $-\infty$ and L_3 the negative imaginary axis directed from $-i\infty$ to 0.

Then, the contour $L = L_1 \cup L_2 \cup L_3$ specifies the set of non analyticity of μ and the following jump conditions occur

$$\mu^{I+II} - \mu^{IV} = -e^{-ikx-ik^2t}s(k) \qquad\qquad k \in L_1 \qquad\qquad (14a)$$

$$\mu^{III} - \mu^{I+II} = -e^{-ikx-ik^2t}[\upsilon(k) - s(k)] \quad k \in L_2 \qquad (14b)$$

$$\mu^{III} - \mu^{IV} = -e^{-ikx-ik^2t}\upsilon(k) \qquad\qquad k \in L_3 \qquad (14c)$$

where

$$s(k) = \int_0^{+\infty} e^{ikx}q(x,0)dx \qquad\qquad\qquad\qquad (15)$$

$$\upsilon(x) = \int_0^{+\infty} e^{ik^2t}[iq_x(0,t) + kq(0,t)]dt . \qquad\qquad (16)$$

Integration by parts justifies the asymptotic form

$$\mu(x,t;k) = \frac{q(x,t)}{ik} + O\left(\frac{1}{k^2}\right), \ k \to \infty \qquad\qquad (17)$$

which combined with the above properties of μ leads to a Riemann-Hilbert [1] problem for μ having the solution

$$\mu(x,t;k) = -\frac{1}{2\pi i} \int_L \frac{e^{ik'x-ik'^2t}}{k'-k} \rho(k')dk' \qquad\qquad (18)$$

where

$$\rho(k) = \begin{cases} s(k), & k \in L_1 \\ \upsilon(k) - s(k), & k \in L_2 \\ \upsilon(k), & k \in L_3 \end{cases} \qquad\qquad (19)$$

Applying the differential operator $\partial_x + ik$ on μ, as it is dictated by (9a), we arrive at the solution

$$q(x,t) = \frac{1}{2\pi} \int_L e^{-ikx-ik^2t} \rho(k)dk \qquad\qquad (20)$$

which is exactly the same with (3) but with a richer contour. In other words the complicated formula (7) is due to the fact that $\mathbb{R}$ is not enough, as a spectrum, to represent the quarter-plane solution. We need the dissipative oscillations for corresponding to $k \in L_3$ as well.

Fokas proposed in [3] the global relation

$$\int_0^{+\infty} e^{ik^2t}[iq_x(0,t) + kq(0,t)]dt = \int_0^{+\infty} e^{ikx}q(x,0)dx \qquad (21)$$

which connects the initial with the boundary conditions in a global way. The amazing thing is that the global relation (21) can be used to solve any boundary value problem by means of algebraic manipulations. Indeed, for the Dirichlet problem with

$$q(x,0) = q_1(x), \qquad\qquad x > 0 \qquad\qquad (22a)$$

$$q(0,t) = q_2(t) \qquad\qquad t > 0 \qquad\qquad (22b)$$

the transformed data provide

$$\hat{q}_1(k) = \int_0^{+\infty} e^{ikx} q_1(x) dx \qquad\qquad (23a)$$

$$\hat{q}_2(k) = \int_0^{+\infty} e^{ik^2 t} q_2(t) dx \qquad\qquad (23b)$$

Now, the definition of $\upsilon(k)$ as

$$\upsilon(k) = \int_0^{+\infty} e^{ik^2 t} [ip_x(0,t) + kq(0,t)] dt \qquad\qquad (24)$$

and the global relation (21), with k replaced by $-k$, can be used to eliminate the integral involving the Neumann data $q_x(0,t)$, $t > 0$. That leads to

$$\upsilon(k) = 2k\hat{q}_2(k) + \hat{q}_1(-k) \qquad\qquad (25)$$

and finally to the solution

$$q(x,t) = \frac{1}{2\pi} \int_{-\infty}^{+\infty} e^{-ikx - ik^2 t} \hat{q}_1(k) dk$$

$$+ \frac{1}{2\pi} \int_{L_2 \cup L_3} e^{-ikx - ik^2 t} [2k\hat{q}_2(k) + \hat{q}_1(-k)] dk . \qquad\qquad (26)$$

We end this short communication with the following comments:

1. The G-F theory provides a tailored-made transform theory for individual problems. In fact the solution S is expressed as

$$S = \int_G e^E D dk$$

   where the exponent E is specified by the equation, the integrand D carries the given data in global form and the contour G is determined by the geometry of the fundamental domain.

2. Separation of variables depends on the geometry of the fundamental domain. The G-F theory is more general, since it is *not* based on the *geometry* of the domain, but on the *linearity* of the equation.

3. Fourier transform allows only for real frequencies. The G-F transform allows for *any* complex frequency. In other words, a much larger availability of "base" functions is at our disposal.

4. The fundamental domain dictates the contour in $\mathcal{C}$. It picks up the plane waves with the appropriate decaying, growing or constant

amplitudes, as well as the frequencies that are needed to construct (represent) the particular solution.

5. If the differential equation accepts separable solutions, the Lax-Pair construction will provide them, independently of whether there exists a coordinate system that effects separation or not.

6. Separation of variables is based on independent spectral analysis in each variable. The G-F method performs simultaneous spectral analysis i.e. it actually "unifies", it does not "separate" the variables.

7. The G-F theory is a transform theory for the whole partial differential equation not a transform theory for each separate ordinary differential equation.

8. The theory of Ehrenpreis uses Fourier analysis with complex frequencies but it does not use the importance of the Global relation that connects the spectral form of the given data.

9. The first equation of the Lax-Pair introduces the contour $\mathbb{R}$. The second equation splits $\mathbb{C}$ into more sectors and introduces more branches into the total contour.

10. The function $\mu(x,t;k)$ depends on k while $q(x,t)$ does not. The operator $(\partial_x + ik)$ annihilates k from μ.

11. The method is still very young to be considered "simple" but when the calculations are over the result is amazing!

12. This method offers another instance of the fact that no matter how "good" separation of variables is, a partial differential equation is something more than a set of other differential equations.

Solution of Laplace's equation in the interior of an equilateral triangle for Dirichlet data, as well as the same problem for the Modified Helmholtz's equation, is under current investigation. In both cases, we arrive at separable solutions without having to separate the variables. These solutions are obtained by purely algebraic means.

References

1. Ablowitz, M.J. and Fokas, A.S., *Complex Variables. Introduction and Applications,* Cambridge University Press, Cambridge, 1997.

2. Fokas, A.S. and Gelfand, I.M., Integrability of Linear and Nonlinear Evolution Equations and the Associated Nonlinear Fourier Transforms, *Letters in Mathematical Physics,* **32**, pp. 189-210, 1994.

3. Fokas, A.S., A Unified Transform Method for Solving Linear and Certain Nonlinear PDE's, *Proceedings of the Royal Society of London A,* **453**, pp. 1411-1443, 1997.

4. Gardner, C.S., Green, J.M., Kruskal, M.D. and Miura, R.M., Method for Solving the Korteweg-de Vries Equation, *Physical Review Letters*, **19**, pp. 1095-1097, 1967.
5. Lax, P.D., Integrals of Nonlinear Equations of Evolution and Solitary Waves, *Communication on Pure and Applied Mathematics*, **21**, pp. 467-490, 1968.

A GENERALIZATION OF BOTH THE METHOD OF IMAGES AND OF THE CLASSICAL INTEGRAL TRANSFORMS

ATHANASSIOS FOKAS[1]

lInstitute for Nonlinear Studies, Clarkson University,
Potsdam, NY 13699-5805.

DANIEL BEN - AVRAHAM

Physics Department, Clarkson University, Potsdam,
NY 13699-5820.

The relation of a new method, recently introduced by Fokas, with the method of images and with the classical integral transforms is discussed. The new method is easy to implement, yet it is applicable to problems for which the classical approaches apparently fail. As illustrative examples, initial-boundary value problems for the diffusion-convection equation and for the linearized Korteweg-de Vries equation with the space variables on the half-line are solved. The suitability of the new method for the analysis of the long-time asymptotics is ellucidated.PACS numbers: 02.30.Jr, 02.30.Uu, 05.60.Cd

I. Introduction

The goal of this paper is to discuss the relation of a new method, recently introduced by Fokas [1, 2], with the classical methods of images and of integral transforms. It will be shown that the new method provides an appropriate generalization of the classical approaches.

For simplicity, we will limit our discussion to evolution PDEs on the half line. Evolution PDEs on a finite domain are discussed in [3, 4]. The new method can also be applied to elliptic PDEs [5], such as the Laplace [6], Helmholtz [7], and biharmonic equations [8]. The extension of this method to nonlinear integrable evolution PDEs is discussed in [1, 9].

In order to help the reader become familiar with the new method, rather than discussing general initial-boundary value (IBV) problems, we will concentrate on the following two concrete, physically signi_cant problems:

1. The diffusion-convection equation,

---

* Permanent address: Department of Applied Mathematics and Theoretical Physics, Cambridge University, Cambridge CB3 0WA, UK.

$$q_t = q_{xxx} + \alpha q_x, \qquad 0 < x < \infty, \quad t > 0, \qquad (1.1a)$$

$$q(x,0) = q_0(x), \qquad 0 < x < \infty, \qquad (1.1b)$$

$$q(0,t) = g_0(t), \qquad t > 0, \qquad (1.1c)$$

where α is a real constant.

2. The linearized Korteweg - de Vries equation (with dominant surface tension),

$$q_t + q_x - q_{xxx} = 0, \qquad 0 < x < \infty, \qquad t > 0, \qquad (1.2a)$$

$$q(x,0) = q_0(x), \quad 0 < x < \infty, \qquad (1.2b)$$

$$q(0,t) = g_0(t), \qquad q_x(0,t) = q_1(t), \qquad t > 0 \qquad (1.2c)$$

The discussion of the physical significance of the above IBV problems, as well as the derivation of their solution is presented in sections II and III.

The proper transform in (x, t)

The proper transform of a given IBV problem is specified by the PDE, the domain, and the boundary conditions. For simple IBV problems there exists an algorithmic procedure for deriving the associated transform (see, for example, [10, 11]). This procedure is based on separating variables and on analyzing *one* of the resulting eigenvalue equations. Thus, for simple IBV problems in (x, t) there exists a proper x-transform and a proper t-transform. Sometimes these transforms can be found by inspection. For example, for the IBV (1.1) with $\alpha = 0$ the proper x-transform is the sine transform, and if $0 < t < \infty$ the proper t-transform is the Laplace transform.

For a general evolution equation in $\{0 < x < \infty, t > 0\}$, the x-transform is more convenient than the t-transform. For example, looking for a solution of the form $\exp[ikx - \omega(k)t]$ in Eqs. (1.1a) and (1.2a), we find that $\omega(k)$ is given *explicitly* by

$$\omega(\kappa) = \kappa^2 - i\alpha\kappa, \qquad \omega(\kappa) = i(\kappa + \kappa^3). \qquad (1.3)$$

On the other hand, looking for solutions of the form $\exp[-st + \lambda(s)x]$ we find that $\lambda(s)$ is given only implicitly, by

$$\lambda^2 + \alpha\lambda + s = 0, \qquad \lambda^3 - \lambda + s = 0. \qquad (1.4)$$

The advantage of the x-transform for an evolution equation becomes clear when the domain is the infinite line $-\infty < x < \infty$, in which case the x-Fourier transform yields an elegant representation for the initial value problem of an arbitrary evolution equation. Furthermore, if an *x-transform* exists, it provides a convenient method for the solution of problems define on the half line $0 < x < \infty$. However, in general an x-transform does not exist; this is, for example, the situation for the IBV problem (1.2). In this case, until recently one had no choice but to attempt to use the *t-Laplace* transform. For example, Eqs. (1.2) yield

$$-\tilde{q}_{xxx} + \tilde{q}_x + s\tilde{q} = q_0(x), \tag{1.5a}$$

$$\tilde{q}(0,s) = \tilde{g}_0(s), \qquad \tilde{q}_x(0,s) = \tilde{g}_1(s), \tag{1.5b}$$

where $\tilde{q}(x,s), \tilde{g}_0(s), \tilde{g}_1(s)$, denote the Laplace transforms of $q(x,t)$, $g_0(t), g_1(t)$, respectively. However, this approach is rather problematic: (a) The problem is posed for infite t, say $0 < t < T$, thus the given boundary condition can have arbitrary growth as $t \to \infty$. Hence, instead of the Laplace transform, the correct transform is

$$\tilde{q}(x,s) = \int_0^T dt \ e^{-st} g(x,t).$$

This yields the additional term $-q(x,T) \exp(-st)$ in the r.h.s. of Eq. (1.5a). (b) If $T = \infty$ and if $g_0(t), g_1(t)$ decay for large t, then the application of the Laplace transform can be justified. However, since the homogeneous version of (1.5a) involves $\exp(-\lambda(s)x)$, where λ solves the cubic equation (1.4b), the investigation of Eqs. (1.5) is cumbersome.

The method of images

A broad class of IBV problems is often approached by the method of images. Suppose that [12]

$$q_t + \omega(i\partial_x)q = 0, \qquad 0 < x < \infty, \qquad t > 0, \tag{1.6a}$$

$$q(x,0) = q_0(x), \qquad 0 < x < \infty, \tag{1.6b}$$

$$q(0,t) = g_0(t), \qquad t > 0 \tag{1.6c}$$

where $\omega(\cdot)$ is an *even* polynomial of its argument. The method of images then yields a solution as follows. Let $G(x,x',t)$ be the Greens function that satisfies

$$G_t + \omega(i\partial_x)G = 0, \qquad -\infty < x < \infty, \qquad t > 0,$$

$$G(x,0) = \delta(x-x'), \qquad -\infty < x < \infty, \qquad 0 < x',$$

$$G(0,t) = 0, \qquad t > 0,$$

and let $Q(x,t)$ satisfy

$$Q_t + \omega(i\partial_x)Q = 0, \qquad -\infty < x < \infty, \qquad t > 0,$$
$$Q(x,0) = 0, \qquad -\infty < x < \infty, \qquad 0 < x',$$
$$Q(0,t) = g_0(t), \qquad t > 0.$$

Note that both $G(x,x',t)$ and $Q(x,t)$ can be readily obtained by means of the Fourier x-transform. Then, the solution to Eqs. (1.6) is given by

$$q(x,t) = \int_0^\infty dx'\{G(x,x',t) - G(x,-x',t)\}q_0(x') + Q(x,t), x > 0. \quad (1.7)$$

Indeed, $G(x,x',t)$, $G(x,-x',t)$, $Q(x,t)$ satisfy Eq. (1.6a) and so does their linear superposition, Eq. (1.7). The initial condition (1.6b) is satisfied by the term involving $G(x,x',t)$, since $G(x,-x',t)$, and $Q(x,t)$ do not contribute to $q(x,t)$ at $t = 0$. Finally, the boundary condition *(1.6c)* is satisfied by $Q(x,t)$ alone, since the terms involving G cancel out at $x = 0$. For the latter to be true, it is crucial that $\omega(\cdot)$ be an even polynomial of its argument. For example, Eqs. (1.2) *cannot* be treated by the method of images, because $\omega(l) = i(l + l^3)$ contains only odd powers of l. Neither does the method of images apply for Eqs. (1.1), unless $\alpha = 0$, since $\omega(l) = -i\alpha l + l^2$.

Sometimes it is possible to apply the method of images *after* using a suitable transformation. For example, such a transformation for Eq. *(1.1a)* is

$$u(x,t) = q(x,t)e^{ax/2}. \tag{1.8}$$

Using this transformation, Eqs. (1.1) become

$$u_t = u_{xx} - \frac{a^2}{4}u, \qquad 0 < x < \infty, \qquad t > 0, \tag{1.9a}$$

$$u(x,0) = q_0(x)e^{ax/2}, \qquad 0 < x < \infty \tag{1.9b}$$

$$u(0,t) = g_0(t), \qquad t > 0. \tag{1.9c}$$

Eqs. (1.9) *can* be solved both by an x-sine transform and by the method of images, provided that $u(x,0)$ decays as $x \to \infty$; since the r.h.s. of Eq. *(1.9b)* involves $\exp(ax/2)$ it follows that, for a general initial condition $q_0(x)$, this is the case if $\alpha < 0$.

The method of images may also work for von Neumann boundary conditions. However, for mixed boundary conditions, such as $q_x(x,0) + \beta q(x,0) = g_0(t)$, the application of the method of images is far from straightforward. For the application of the method of images to a large class of physically important problems, see [14].

The new method

For equations in one spatial dimension the new method constructs $q(x,t)$ as an integral in the complex k-plane, involving an x-transform of the initial condition and a t-transform of the boundary conditions.

For example, for the IBV problem (1.1), it will be shown in section II that this integral representation is

$$q(x,t) = \frac{1}{2\pi} \int_{-\infty}^{\infty} dk e^{ikx-\omega(k)t} \hat{q}_0(k) + \frac{1}{2\pi} \int_{\partial D_+} dk e^{ikx-\omega(k)t} \hat{g}(k,t), \quad (1.10)$$

where $\omega(k) = k^2 - i\alpha k$, the oriented contour ∂D_+ is the curve in the complex k-plane defined by

$$k_I = \frac{\alpha}{2} + \sqrt{k_R^2 + (\frac{\alpha}{2})^2}, \qquad k = k_R + ik_I, \quad (1.11)$$

see Fig. 1, and the functions $\hat{q}_0(k)$, $\hat{g}(k,t)$ are defined in terms of the given initial and boundary conditions as follows:

$$\hat{q}_0(k) = \int_0^{\infty} dx e^{-ikx} q_0(x), \qquad \text{Im} k \le 0, \quad (1.12)$$

$$\hat{g}(k,t) = -\hat{q}_0(i\alpha - k) - (2ik + \alpha) \int_0^t d\tau e^{\omega(k)\tau} g_0(\tau). \quad (1.13)$$

We note that $\hat{g}(k,t)$ involves $\hat{q}_0$ evaluated at $i\alpha - k$.

The long-time asymptotics

The representation obtained by the new method is convenient for computing the long-time asymptotics of the solution: Suppose that a given evolution PDE is valid for $0 < t < T$, where T is a positive constant. It can be shown (see section II) that the representation for $q(x,t)$ is equivalent to the representation obtained by replacing $\hat{g}(k,t)$ with $\hat{g}(k,T)$. In particular, if $0 < t < \infty$, then $\hat{g}(k,t)$ can be replaced by $\hat{g}(k) \equiv \hat{g}(k,\infty)$. For example, the solution of the IBV (1.1) with $0 < t < \infty$ is given by (1.10) with $\hat{g}(k,t)$ replaced by $\hat{g}(k)$,

$$\hat{g}(k) = -\hat{q}_0(i\alpha - k) - (2ik + \alpha)\hat{g}_0(k), \quad \hat{g}_0(k) = \int_0^{\infty} dt e^{\omega(k)t} g_0(t) \quad (1.14)$$

Thus, the only time-dependence of $q(x,t)$ appears in the form $\exp[ikx - \omega(k)t]$; hence it is straightforward to obtain the long-time asymptotics, using the steepest descent method.

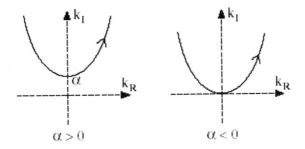

FIG. 1: The contour ∂D_+ associated with Eq. (1.1a).

From the complex plane to the real axis

In case that the given IBV problem can be solved by an x-transform, the relevant representation can be obtained by deforming the integral representation of the solution obtained by the new method, from the complex k-plane to the real axis. Consider for example the integral in the r.h.s. of Eq. (1.10) for the case that $\alpha < 0$: It can be verified that the functions $\exp(ikx)$, $\hat{q}_0(i\alpha - k)$, and $\exp[\omega(k)(T-t)]$ are bounded in the region of the complex k-plane above the real axis and below the curve ∂D_+. Thus, the integral along ∂D_+ can be deformed to an integral along the real axis,

$$q(x,t) = \frac{1}{2\pi} \int_{-\infty}^{\infty} dk e^{ikx - \omega(k)t} [\hat{q}_0(k) - \hat{q}_0(i\alpha - k) - (2ik + \alpha) \int_0^t d\tau e^{\omega(k)\tau} g_0(\tau)] \quad (1.15)$$

It can be shown that this formula is equivalent to the solution of Eqs. (1.8), (1.9) using the sine transform. However, if $\alpha > 0$ this deformation is not possible.

Indeed, $\hat{q}_0(i\alpha - k)$ involves

$$\exp[-ix(i\alpha - k)] = \exp[ik_R x] \exp[-x(k_I - \alpha)];$$

this term is bounded for $k_I \geq \alpha$ (and in particular is bounded on ∂D_+), but it is not bounded in the region of the complex k-plane below the curve ∂D_+.

We note that even in the cases when it is possible to deform the contour to the real axis, the representation in the complex k-plane has certain advantages. For example, the integral involving the sine transform is not uniformly convergent at $x = 0$ (unless $q(0,t) = 0$). Also, a convenient way to study the long-time behavior of the representation involving the sine transform is to *transform* it to the associated representation in the complex k-plane.

II. The diffusion Convection Equation

Physical significance

Eq. (1.1a) arises, for example, from the diffusion-convection equation

$$q_\tau = Dq_{\xi\xi} + \alpha u q_\xi, \qquad \alpha \pm 1, \quad 0 < \xi < \infty, \qquad \tau > 0, \quad (2.1)$$

where $q(\xi, \tau)$ is a probability density function in the spatial and time variables ξ and τ, D is a diffusion coefficient with dimensions of $(\text{length})^2 / (\text{time})$, and u is a convection field with dimensions of $(\text{length}) / (\text{time})$. Eq. (1.1a) is the normalized form of (2.1), expressed in terms of the dimensionless variables $t = (D/u^2)\tau$ and $x = (D/u)\xi$. The dependence upon the single spatial variable x is justified in systems that are homogeneous in the transverse directions to x. The choice $\alpha = +1$ corresponds to convection (a background drift velocity v) directed toward the origin, while $\alpha = -1$ corresponds to a background drift away from the origin. In the long-time asymptotic limit convection dominates diffusion and determines the fate of the system. Suppose for example that the boundary condition is $q(0,t) = 0$, corresponding to an ideal sink at the origin, for the case, of say, of particle flow. Then, if $\alpha = +1$ the particles will flow to the sink at typical speed u, and the probability density is expected to decay to zero exponentially with time. If $\alpha = -1$, the particles are drifting away from the sink; in this case one expects a depleted zone near the origin, which grows linearly with time, and exponential convergence to a finite level of survival. The case of generic boundary conditions is harder to predict by such heuristic arguments.

The method of images

We have seen that the method of images is applicable in the singular case of $\alpha = 0$, and even then the solution is straightforward only with pure Dirichlet or von Neumann boundary conditions. For $\alpha < 0$ the method of images can be used after a suitable transform, see Eqs. (1.9), and also provided that the boundary conditions are simple enough. For $\alpha > 0$, the method of images fails to provide a solution for generic initial conditions. We now show how the new method serves as a natural extension of the methods of images and of classical integral transforms, even as these methods fail. For pedagogical reasons, we expose the new method simultaneously for problems (1.1) and (1.2). The physical interpretation of (1.2) is discussed in Section III.

The new method

Suppose that a linear evolution PDE in one space variable admits the solution $\exp[ikx - \omega(k)t]$. For well posedness we assume that

$$\text{Re}\,\omega(k) > 0, \quad \text{for } k \text{ real.} \tag{2.2}$$

The starting point of the new method is to rewrite this PDE in the form

$$(e^{-ikx+\omega(k)t}q)_t + (e^{-ikx+\omega(k)t}x)_x = 0, \tag{2.3}$$

where

$$X = -i(\frac{\omega(k) - \omega(l)}{k - l})q, \qquad l = -i\partial_x. \tag{2.4}$$

For Eq. (1.1a), $\omega(k) = k^2 - i\alpha k$, thus

$$X = -i(\frac{k^2 - l^2 - i\alpha(k - l)}{k - l})q = -[i(k + l) + \alpha]q.$$

For Eq. (1.2a), thus

$$X = (\frac{(k - l) + (k^3 - l^3)}{k - l})q = (1 + k^2 + l^2 + kl)q.$$

Hence Eqs. (1.1a) and (1.2a) can be written in the form (2.3), where X is given, respectively, by

$$X = -q_x - (ik + \alpha)q, \quad X = -q_{xx} - ikq_z + (1 + k^2)q. \tag{2.5}$$

(a) The Fourier transform representation

The solution of Eq. (2.3) can be expressed in the form

$$q(x,t) = \frac{1}{2\pi} \int_{-\infty}^{\infty} dke^{ikz - \omega(k)\tau} \hat{q}(k) + \frac{1}{2\pi} \int_{-\infty}^{\infty} dke^{ikz - \omega(k)\tau} \hat{g}(k,t), \tag{2.6}$$

where

$$\hat{q}_0(k) = \int_0^{\infty} dxe^{-ikx}q_0(x), \qquad \text{Im}\,k \le 0, \tag{2.7}$$

$$\hat{g}(k,t) = \int_0^t d\tau e^{\omega(k)\tau} X(0,\tau,k), \quad k \in C \tag{2.8}$$

Furthermore, the functions $\hat{q}_0(k)$ and $\hat{g}(k,t)$ satisfy the *global relation*

$$\hat{g}(k,t) + \hat{q}_0(k) = e^{\omega(k)\tau} \hat{q}(k,t), \quad \text{Im}\,k \le 0, \tag{2.9}$$

where $\hat{q}(k,t)$ denotes the x-Fourier transform of $q(x,t)$.

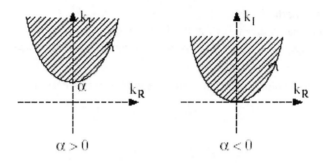

FIG. 2: The domain D_+ for Eq. (1.1a).

Indeed, using (2.3) it is straightforward to compute the time evolution of $\hat{q}(k,t)$,

$$(e^{\omega(k)t}\hat{q}(k,t))_t = \int_0^\infty dx(e^{-ikx+\omega(k)t}q(x,t))_t = -\int_0^\infty dx(e^{-ikx+\omega(k)t}X)_x = e^{\omega(k)t}X(0,t,k).$$

Integrating this equation we find Eq. (2.9). Solving Eq. (2.9) for $\hat{q}(k,t)$ and then using the inverse Fourier transform, we find Eq. (2.6).

(b)   An integral representation in the complex k-plane

The first crucial step of the new method is to replace the second integral on the r.h.s. of Eq. (2.6) by an integral along the oriented curve ∂D_+. This curve is the boundary of the domain D_+,

$$D_+ = \{k \in C, \operatorname{Im} k > 0, \operatorname{Re}\omega(k) < 0\}, \tag{2.10}$$

oriented so that D_+ is on the left of ∂D_+. For example, for Eqs. (1.1a) and (1.2a) the domains D_+ are the shaded regions in Figs. 2 and 3, respectively. Each of the curves in Fig. 2 is defined by Eq. (1.11), while the curve in Fig. 3 is defined by

$$k_I = \sqrt{1+3k_R^2}.$$

Indeed, for Eq. (1.1a),

$$\operatorname{Re}\omega(k) = \operatorname{Re}\{(k_R + ik_I)^2 - i\alpha(k_R + ik_I)\} = k_R^2 - k_I^2 + \alpha k_I.$$

Thus, D_+ is the domain of the upper half complex k-plane specified by

$$(k_I - \frac{\alpha}{2})^2 - (k_R^2 + (\frac{\alpha}{2})^2) > 0.$$

For Eq. (1.2a),

$$\operatorname{Re}\omega(k) = \operatorname{Re}\{i(k_R + ik_I)^3 + i(k_R + ik_I)\} = k_I(k_I^2 - 3k_R^2 - 1).$$

Thus, D+ is the domain of the upper half complex k-plane specified by

$$k_I^2 - 3k_R^2 - 1 < 0.$$

The deformation of the integral from the real axis to the integral along the curve ∂D_+ is a direct consequence of Cauchy's theorem. Indeed, the term

$$e^{ikx-\omega(k)t}\hat{g}(k,t) = e^{ikx}\int_0^t d\tau e^{-\omega(k)(t-\tau)}X(0,\tau,k)$$

is analytic and bounded in the domain E_+,

$$E_+ = \{k \in C, \operatorname{Im} k > 0, \operatorname{Re}\omega(k) > 0\}.$$

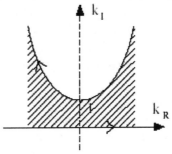

FIG. 3: The domain D_+ for Eq. (1.2a).

Thus, using Jordan's lemma in E_+, the integral along the boundary of E_+ vanishes.

For example, for Eq. (1.1a) E_+ is the domain above the real axis and below the curve ∂D_+, thus the integral along the real axis can be deformed to the integral along ∂D_+. For Eq. (1.2a), E_+ is the domain above the curve $\{k_I = \sqrt{1+3k_R^2}\}$, thus the integral along this curve vanishes; hence ∂D_+ is the union of the real axis and the curve $\{k_I = \sqrt{1+3k_R^2}\}$.

(c) Analysis of the global relation

The second crucial step in the new method is the analysis of the global relation (2.9): This yields $\hat{g}(k,t)$ in terms of $\hat{q}_0(k)$ and the *t*-transform of the given boundary conditions.

Before implementing this step, we note that we have already used Eq. (2.9) in the derivation of Eq. (2.6). However, while Eq. (2.9) was used earlier only for real *k*, in what follows it will be used in the complex *k*-plane.

Recall that $\hat{g}(k,t)$ is defined by Eq. (2.8), where for Eq. (1.1a) X is given by (2.5a). Thus

$$\hat{g}(k,t) = -\hat{g}_1(\omega(k),t) - (ik+\alpha)\hat{g}_0(\omega(k),t),$$
$$\omega(k) = k^2 - i\alpha k, \tag{2.11}$$

where

$$-\hat{g}_1(\omega(k),t) = \int_0^t d\tau e^{\omega(k)\tau} q_x(0,\tau),$$

$$\hat{g}_0(\omega(k),t) = \int_0^t d\tau e^{\omega(k)\tau} g_0(\tau);$$

we have used the notation $\hat{g}_1(\omega(k),t)$ and $\hat{g}_0(\omega(k),t)$ to emphasize that $\hat{g}_1$ *and* $\hat{g}_0$ *depend on k only through* $\omega(k)$.

Substituting the expression for $\hat{g}(k,t)$ from (2.11) into the global relation (2.9) we find

$$-\hat{g}_1(\omega(k),t) = (ik+\alpha)\hat{g}_0(\omega(k),t) - \hat{q}_0(k) + e^{\omega(k)t}\hat{q}(k,t), \quad \text{Im}\,k \le 0. \tag{2.12}$$

Our task is to compute $\hat{g}(k,t)$ on the curve ∂D_+; since $\hat{g}(k,t)$ is analytic in D_+ this is equivalent to computing $\hat{g}(k,t)$ for $k \in D_+$. Eq. (2.11) shows that $\hat{g}(k,t)$ involves $\hat{g}_0$, which is known in terms of the given boundary condition $g_0(t)$, as well as $\hat{g}_1$, which involves the *unknown* boundary value $q_x(0,t)$. In order to compute $\hat{g}_1$ using the global relation(2.12), we first transform Eq. (2.12) from the lower half complex k-plane to the domain D_+. In this respect, it is üimportant to observe that since $\hat{g}_1$ depends on k only through $\omega(k)$, $\hat{g}_1$ *remains invariant by those transformations* $k \to v(k)$ *which preserve* $\omega(k)$. For Eq. (1.1a), the equation $\omega(k) = \omega(v(k))$ has one non-trivial root (the trivial root is $v(k) = k$),

$$v^2 - i\alpha v = k^2 - i\alpha k, \quad v(k) = i\alpha - k.$$

Let D_- be defined by

$$D_- \{k \in C, \text{Im}\,k < 0, \text{Re}\,\omega(k) < 0\}. \tag{2.13}$$

For $\omega(k) = k^2 - i\alpha k$, the domains D_+ and D_- are the shaded regions depicted in Fig. 4. The transformations that leave $\omega(k)$ invariant map the domain $\{D_+ \oplus D_-\}$ onto itself. Thus, if $k \in D_+$, then $i\alpha - k \in D_-$. Hence, replacing k by $i\alpha - k$ in Eq. (2.12), we find an equation that is valid for k in D_+:

$$-\hat{g}_1 = -ik\hat{g}_0 - \hat{q}_0(i\alpha - k) + e^{\omega(k)t}\hat{q}(i\alpha - k,t), \quad k \in D_+ \tag{2.14}$$

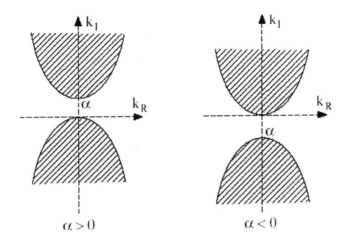

FIG. 4: The domains D_+ and associated D_- with Eq. (1.1a).

Replacing $-\hat{g}_1$ in (2.11) by the r.h.s. of the above equation, we find

$$\hat{g}(k,t) = -(2ik+\alpha)\hat{g}_0(\omega(k),t) - \hat{q}_0(i\alpha-k) + e^{\omega(k)t}\hat{q}(i\alpha-k,t).$$

The term $e^{\omega(k)t}\hat{q}(i\alpha-k,t)$ does not contribute to $q(x,t)$. Indeed, this term gives rise to the integral

$$\int_{\partial D_+} dk e^{ikx}\hat{q}(i\alpha-k,t),$$

for which we note: $\exp(ikx)$ is bounded and analytic for $\operatorname{Im} k > 0$; the term $\hat{q}(i\alpha-k,t)$ involves $\exp[-i(i\alpha-k)x] = \exp(ik_R x)\exp[-x(k_I-\alpha)]$, which is bounded and analytic for $k_I > \alpha$, i.e., in D_+. Thus using Jordan's lemma in D_+ it follows that the above integral vanishes, hence the effective part of $\hat{g}(k,t)$ is given by (1.13).

(d) The long-time asymptotics

Suppose that a given evolution PDE is valid for $0 < t < T$, where T is a positive constant. Then, as mentioned earlier, we can replace $\hat{g}(k,t)$ by $\hat{g}(k,T)$. Indeed, the two representations associated with $\hat{g}(k,t)$ and $\hat{g}(k,T)$ differ by

$$\int_{\partial D_+} dk e^{ikx-\omega(k)t} \int_t^T d\tau e^{\omega(k)\tau} X(0,\tau,k).$$

Since $T \geq t$, the coefficient $(\tau-t)$ of $\omega(k)$ is non-negative, thus $\exp[ikx+\omega(k)(\tau-t)]$ is bounded and analytic in D_+, and Jordan's lemma implies that the above integral vanishes.

In the case of Eq. (1.1a) the solution is given by Eqs. (1.10)-(1.12), Eq. (1.14). The long-time asymptotics of $q(x,t)$ can be analyzed by the method of steepest descent. For $x = vt$, the critical point occurs at $d\omega / dk = iv$, or

$$k = \frac{i}{2}(\alpha + v) \equiv i\gamma.$$

Let $v > \alpha > 0$, then both integrals on the r.h.s. of (1.10) contribute to the long-time asymptotics, since the path of integration can be modified to pass through the critical point in each case. The behaviour of the argument of the relevant exponentials near the critical point is given by

$$ikvt - \omega(k)t = -\gamma^2 t - (k - i\gamma)^2 t,$$

hence the leading-order asymptotics to $q(vt,t)$ is

$$\frac{1}{2\pi}\hat{q}_0(i\gamma)e^{-\gamma^2 t}\int_{-\infty}^{\infty} e^{-(k-i\gamma)^2 t}dk + \frac{1}{2\pi}\hat{g}(i\gamma)e^{-\gamma^2 t}\int_{\partial D_+} e^{-(k-i\gamma)^2 t}dk = \frac{1}{2\sqrt{\pi t}}[\hat{q}_0(i\gamma) + \hat{g}(i\gamma)]e^{-\gamma^2 t}.$$

The numerical value of $\hat{q}_0(i\gamma)$ and $\hat{g}(i\gamma)$ are given explicitly in terms of the initial and boundary conditions, see Eqs. (1.12) and (1.14).

For $\alpha > v > 0$ the second integral on the r.h.s. of (1.10) does not contribute, since the critical point lies outside of the domain D_+. In this case the leading-order asymptotics is

$$q(vt,t) \sim \frac{1}{2\sqrt{\pi t}}\hat{q}(i\gamma)e^{-\gamma^2 t}.$$

For $\alpha < 0$ and $v > -\alpha$ both integrals contribute and the answer is identical with that of the case $v > \alpha > 0$. Finally, for $\alpha < 0$ and $v < -\alpha$ neither integral contributes and the leading-order asymptotic behavior is zero. The latter is expected on physical grounds, since the (rightward) drift sweeps the probability density $q(vt,t)$ past $x = vt$.

III. The linearized Korteweg-de Vries Equation

Eq. (1.2a) is the linear limit of the celebrated Korteweg - de Vries equation:

$$q_t + q_x + \lambda q_{xxx} + \partial q q_x = 0, \qquad \lambda = \pm 1 \qquad (3.1)$$

for the case of $\lambda = -1$; this corresponds to dominant surface tension. Eq. (3.1) is the normalized form of

$$\frac{\vartheta \eta}{\vartheta \tau} = \frac{3}{2}\sqrt{\frac{g}{h}}\frac{\vartheta}{\vartheta \xi}(\frac{1}{2}\eta^2 + \eta + \frac{1}{3}\sigma\frac{\vartheta^2 \eta}{\vartheta \xi^2}), \quad \sigma = \frac{1}{8}h^3 - \frac{Th}{\rho g},$$

$$(3.2)$$

where η is the elevation of the water above the equilibrium level h, T is the surface tension, ρ is the density of the medium, and g is the free-fall acceleration constant. This equation is the small amplitude, long wave limit of the equations describing idealized (inviscid) water waves under the assumption of irrotationality. Eq. (3.1) is obtained after transforming to the dimensionless variables

$$t = \frac{1}{2}\sqrt{\frac{g}{h\sigma}}T, x = -\sigma^{1/2}\xi, q = \eta/2.$$

Eq. (3.1) usually appears without the qx term. This is because the Korteweg - de Vries equation is usually studied on the full line and then the term q_x can be eliminated by means of a Galilean transformation. However, for the half-line this transformation would change the domain from a quarter-plane to a wedge.

Laboratory experiments with water waves typically involve Eq. (3.1) with $\lambda = 1, q(x,0) = 0$, and $q(0,t)$ a periodic function of t. Thus the linearized version of Eq. (3.1) with $\lambda = 1$ is valid until small amplitude waves reach the opposite end of the water tank. Eq. (1.2a) is valid under similar circumstances, in the case of dominant surface tension.

It is interesting to note that while in the case of $\lambda = 1$ the problem is well posed with only one boundary condition at $x = 0$, in the case of $\lambda = -1$ the problem is well posed with two boundary conditions. The case of $\lambda = 1$ is solved in [2]. Here we solve the case of $\lambda = -1$.

The solution of the IBV problem (1.2) is given by Eq. (1.10), where $\omega(k) = i(k + k^3)$, $\hat{q}_0(k)$ is the Fourier transform of $q_0(x)$ (see Eq. (1.12)), $\hat{g}(k,t)$ is defined by

$$\hat{g}(k,t) = -\hat{q}_0(v) + i(v-k)\int_0 d\tau e^{\omega(k)\tau}g_1(\tau) + (k^2 - v^2)\int_0 d\tau e^{\omega(k)\tau}g_0(\tau), \quad (3.3)$$

$$v = \frac{-k - i\sqrt{3k^2 + 4}}{2} \text{ for } k \in D_+^{(1)}, \quad v = \frac{-k + i\sqrt{3k^2 + 4}}{2} \text{ for } k \in D_+^{(2)},$$

and ∂D_+ is the union of the boundaries of $D_+^{(1)}$ and $D_+^{(2)}$ depicted in Fig. 5 (the relevant curve is $\{k_I = \sqrt{3k_R^2 + 1}, k_I > 0\}$).

Indeed, the representation (1.10) was derived in section II. Using the definition of $\hat{g}(k,t)$, Eq. (2.8), and recalling that X is given by Eq. (2.5b), we find

$$\hat{g}(k,t) = -\hat{g}_2(\omega(k),t) - ik\hat{g}_1(\omega(k),t) + (1 + k^2)\hat{g}_0(\omega(k),t), \quad \omega(k) = i(k + k^3), (3.4)$$

where $\hat{g}_1$, $\hat{g}_0$ are the first, second integrals appearing in the r.h.s. of Eq. (3.3) and $\hat{g}_2$ involves the unknown boundary value $q_{xxx}(0,t)$,

$$\hat{g}_2(\omega(k),t) = \int_0^t d\tau e^{\omega(k)\tau} q_{xx}(0,\tau).$$

Substituting the expression for ^g(k; t) from Eq. (3.4) into the global relation (2.9), we find

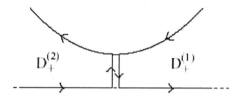

FIG. 5: The contours $\partial D_+^{(1)}$ and associated $\partial D_+^{(2)}$ with Eq. (1.2a).

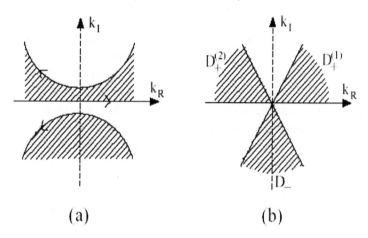

FIG. 6: The domains D_+ and D_- for Eq. (1.2a).

$$-\hat{g}_2(\omega(k),t) = ik\hat{g}_1(\omega(k),t) - (1+k^2)\hat{g}_0(\omega(k),t) - \hat{q}_0(k) + e^{\omega(k)t}\hat{q}(k,t), \quad \text{Im}\, k \le 0.$$
$$(3.5)$$

The equation $\omega(k) = \omega(v(k))$ has two nontrivial roots,

$$v + v^3 = k + k^3, \quad v^2 + kv + k^2 + 1 = 0, \quad v_{1,2} = \frac{-k \mp i\sqrt{3k^2 + 4}}{2}; \qquad (3.6)$$

$$v_1(k) \sim e^{4ix/3}k, \quad v_2(k) \sim e^{2i\pi/3}k, k \to \infty.$$

The domains D_+ and D_- (see Eqs. (2.10) and (2.13)) are the shaded regions depicted in Fig. 6a, where the two relevant curves are

$k_1 = \pm\sqrt{3k_R^2 + 1}$; the limit of the domains as $k \to \infty$ are depicted in Fig. 6b, where each of the relevant angles is $\pi / 3$.

If $\dfrac{2\pi}{3} < \arg k < \dfrac{4\pi}{3}$ then $\dfrac{4\pi}{3} < \arg k + \dfrac{2\pi}{3} < 2\pi$, and

$2\pi < \arg k + \dfrac{4\pi}{3} < \dfrac{8\pi}{3}$. Thus

$$k \in D_+^{(2)}, \qquad v_2(k) \in D_-, \qquad v_1(k) \in D_+^{(1)}.$$

Similarly,

$$k \in D_+^{(1)}, \qquad v_2(k) \in D_+^{(2)}, \qquad v_1(k) \in D_-.$$

Thus, replacing k by $v_1(k)$ and $v_2(k)$ in (3.5), we find

$$-\hat{g}_2 = iv_1\hat{g}_1 - (1+v_1^2)\hat{g}_0 - \hat{q}_0(v_1) + e^{\omega(k)t}\hat{q}(v_1(k),t), \qquad k \in D_+^{(1)};$$

$$-\hat{g}_2 = iv_2\hat{g}_1 - (1+v_2^2)\hat{g}_0 - \hat{q}_0(v_2) + e^{\omega(k)t}\hat{q}(v_2(k),t), \qquad k \in D_+^{(2)}.$$

Replacing $-\hat{g}_2$ in (3.4) by the r.h.s. of the above equations, we find

$$\hat{g}(k,t) = i(v-k)\hat{g}_1 + (k^2 - v^2)\hat{g}_0 - \hat{q}_0(v) + e^{\omega(k)t}\hat{q}(v,t), \qquad k \in D_+.$$

Due to analiticity considerations the term $\hat{q}(v,t)$ does not contribute to $q(x,t)$, see section II; thus the effective part of $\hat{g}(k,t)$ is given by (3.3).

The long-time asymptotics

The long-time asymptotics can again be analyzed by replacing $\hat{g}(k,t)$ with $\hat{g}(k,\infty) = \hat{g}(k)$ and following standard steepest descent or stationary phase expansions. One thus finds [13]: For $v = x/t > 1$, $q(vt,t)$ satisfies

$$q(vt,t) = \frac{1}{\sqrt{12\pi\gamma t}}[(P(\gamma) - P(v(\gamma))e^{i\varphi} + (P(-\gamma) - P(v(-\gamma))e^{i\varphi}] + O(t^{-3/2}),$$

as $t \to \infty$, where

$$\gamma = \sqrt{\frac{v-1}{3}},$$

$$\varphi(t) = 2\gamma^3 t - \pi/4,$$

$$v(k) = -\frac{1}{2}(k + i\sqrt{3k^2 + 4}),$$

$$P(k) = \hat{q}_0(k) + (k^2 + 1)\hat{g}_0(k) - ik\hat{g}_1(k),$$

$\hat{q}_0(k)$ is defined in terms of the initial condition (Eq. (1.14)), and $\hat{g}_1(k)$, $\hat{g}_0(k)$ are the first, second integrals appearing in the r.h.s. of Eq. (3.3) evaluated with $t \to \infty$. For $v < 1, q(vt, t)$ decays faster than any algebraic power of t, as $t \to \infty$.

Acknowledgments

We gratefully acknowledge partial support of this work from the NSF, under contract no. PHY-0140094 (D.b.-A.), and the EPRSC (A.S.F.).

References

[1] A. S. Fokas \A unified transform Method for linear and certain nonlinear PDEs," Proc. R. Soc. 53, 1411 (1997).

[2] A. S. Fokas, \A new transform method for evolution PDEs," IMA J. Appl. Math. 67, 559 (2002).

[3] A. S. Fokas and B. Pelloni, \A transform method for evolution PDEs on a finite interval," (preprint).

[4] B. Pelloni, \Well-posed boundary value problems for linear evolution equations on the half-line," Math. Proc. Camb. Phil. Soc. (in press).

[5] A. S. Fokas, \Two dimensional linear PDEs in a convex polygon," Proc. R. Soc. 457 371 (2001).

[6] A. S. Fokas and A. A. Kapaev, \On a transform method for the Laplace equation in a polygon," IMA J. Appl. Math. 68 (2003).

[7] D. ben-Avraham and A. S. Fokas, \The solution of the modified Helmholtz equation in a wedge and an application to diffusion-limited coalescence," Phys. Lett. A 263, 355{359 (1999); \The modified Helmholtz equation in a triangular domain and an application to diffusion-limited coalescence," Phys. Rev. E 64, 016114 (2001).

[8] D. Crowdy and A. S. Fokas, \Explicit integral solutions for the plane elastostatic semi-strip," Proc. R. Soc. (in press).

[9] A. S. Fokas, \Integrable nonlinear evolution equations on the half line," Comm. Math. Phys. 230, 1 (2002).

[10] G. Roach, Green's functions, CUP (1982).

[11] I. Stakgold, Green's functions and boundary value problems, Wiley-Interscience, NY (1979).

[12] For simplicity, we are assuming that $\omega(\cdot)$ is a polynomial of order less or equal 2. When $\omega(\cdot)$ is a polynomial of order 2n one needs n boundary conditions and the method of images is complicated further.

[13] A. S. Fokas and P. F. Schultz, \The long-time asymptotics of moving boundary problems using an Ehrenpreis-type representation and its Riemann-Hilbert nonlinearization," Comm. Pure Appl. Math. LVI 0517 (2003).

[14] I.V. Lindell, Methods for electromagnetic field analysis, Clarendon Press, Oxford (1992).

THE HAPPEL MODEL FOR AN ELLIPSOID VIA THE PAPKOVICH – NEUBER REPRESENTATION

GEORGE DASSIOS

University of Patras, Hellenic Open University and ICEHT-FORTH Patras, GR 26504

PANAYIOTIS VAFEAS

University of Patras and ICEHT-FORTH Patras, GR 26504

Particle–fluid systems are encountered in many important applications and the study of the flow through a swarm of particles is of a considerable practical and theoretical interest concerning heat and mass transport problems. The general consideration consists of rigid particles of three–dimensional ellipsoidal shapes embedded within viscous, incompressible fluids, which characterize the steady, creeping (low Reynolds number) and non–axisymmetric Stokes flow for interior and exterior domains. The ellipsoid–in–cell model with the proper boundary conditions, physically identical to those of the Happel (self–sufficient in mechanical energy) sphere–in–cell model, assuming two confocal ellipsoids, is employed in order to analyze the flow of a swarm of ellipsoidal particles in a quiescent Newtonian fluid. The particles move with constant uniform velocity in an arbitrary direction and rotate with an arbitrary constant angular velocity. The solid internal ellipsoid, where the non–slip conditions are applied on its surface, represents a particle of the swarm. The external ellipsoid contains the ellipsoidal particle and the amount of fluid required to match the fluid volume fraction of the swarm and the boundary conditions here assumes nil normal velocity component and shear stress. The boundary value problem is solved with the help of the Papkovich – Neuber differential representation for Stokes flow, which is valid in non–axisymmetric geometries and provides us with the velocity and the total pressure fields in terms of harmonic ellipsoidal eigenfunctions. The necessity of extra conditions during the solution is satisfied by the use of this representation, where its flexibility is demonstrated. The velocity to the first degree, which represents the leading term of the series, is sufficient for most engineering applications.

1. Introduction

The motion of aggregates of small particles relative to viscous fluids covers a wide range of heat and mass transfer phenomena. The steady and creeping flow of an incompressible, viscous fluid is described by the well–known Stokes equations (1851), which connect a biharmonic vector velocity with a harmonic scalar pressure field [1]. In order to retain our interest to realistic models, where the orientation of the particle must be taken into account, we introduce complicated geometries in the absence of symmetry. According to this aspect, ellipsoidal geometry [2] provides the most widely used framework for representing small particles of arbitrary shape.

In order to obtain an approximate solution in close form we are led to the idea of particle–in–cell models, whereas the technique of cell models is adopted. More specific, the mathematical treatment of each problem is based on the assumption that a three–dimensional assemblage may be considered to consist of a number of identical unit cells. Each of these cells contains a particle surrounded by a fluid envelope, containing a volume of fluid sufficient to make the fractional void volume in the cell identical to that in the entire assemblage. Uchida [3] proposed a cell model where a spherical particle is surrounded by a fluid envelope with cubic outer boundary. Happel [4] and Kuwabara [5] proposed cell models in which both the particle and the outer envelope are spherical, having the significant advantage that this formulation preserves the axial symmetry of the flow. On the other hand, these models, which are identical, are not space filling and their difference lie on the boundary conditions, caused by the relative movement of the particle–fluid system. Nevertheless, it is the Happel model which is slightly superior to the Kuwabara one, since it does not allow any energy exchange with the environment. Neale and Nader [6] improved the formulation of Happel and Kuwabara, whereas Epstein and Masliyah [7] proposed a useful generalization by considering a spheroid–in–cell, instead of a sphere–in–cell, model for swarms of spheroidal particles. However, they had to solve the creeping flow problem numerically since the well–known equation of motion in spheroidal coordinates is not separable. This difficulty was resolved by Dassios et al. [8] with the introduction of semiseparable solutions which are based on an appropriate finite dimensional spectral decomposition of the operator of motion and the solution for the spheroidal Kuwabara model was obtained as a demonstration of this method. Under this project, Dassios et al. [9] solved the Happel model in spheroidal coordinates analytically and the results were compared with those obtained by using the Kuwabara–type boundary conditions. Moreover, the problem of the space filling when we refer to the assemblage of particles was discussed in [9].

However, in three–dimensional cases as ours the introduction of three–dimensional representations [10] of the flow fields is necessary, since they use more than one potential to represent the physical fields allowing in that way more flexibility [11]. Papkovich and Neuber proposed a differential representation [10,12] of the flow fields in terms of harmonic functions [13].

In the present work, the solution to the Stokes flow problem in an ellipsoid–in–cell with Happel–type boundary conditions is obtained with the aim of the Papkovich – Neuber representation. Under the assumption of very small Reynolds number and pseudosteady state, a three–dimensional (3–D) Stokes flow in an ellipsoidal envelope of appropriate shape and dimensions is adopted as a fair approximation to the flow around a typical particle of the swarm, in accordance with the concept introduced by Happel [4]. The inner ellipsoid, which represents a particle in the assemblage, is solid, moves with a constant arbitrary velocity and rotates arbitrarily with a constant angular velocity, whereas the outer ellipsoid represents a fictitious fluid envelope identifying the surface of a unit cell (ellipsoid–in–cell). The volume of the fluid cell is chosen so that the solid volume fraction in the cell coincides with the volume fraction of the swarm. The appropriate boundary conditions, resulting from these assumptions, are imposed: non–slip flow on the inner ellipsoid, no normal flow and nil tangential stresses on the outer ellipsoidal envelope

2. Mathematical Formulation

Under the assumption of pseudosteady, non–axisymmetric, creeping flow (Reynolds number $Re \ll 1$) which characterizes the Stokes flow [1], we introduce the governing equations of motion of an incompressible, viscous fluid, around particles embedded within smooth, bounded domains $\Omega(I\!R^3)$, with dynamic viscosity μ_0 and mass density ρ_0. These connect the biharmonic velocity field $\mathbf{v}(\mathbf{r})$ with the harmonic total pressure field $\mathbf{P}(\mathbf{r})$,

$$\mu_0 \Delta \mathbf{v}(\mathbf{r}) = \nabla \mathbf{P}(\mathbf{r}), \qquad \mathbf{r} \in \Omega(I\!R^3), \qquad (1)$$

$$\nabla \cdot \mathbf{v}(\mathbf{r}) = 0, \qquad \mathbf{r} \in \Omega(I\!R^3). \qquad (2)$$

Once the velocity is obtained, the harmonic vorticity field $\omega(\mathbf{r})$ is defined as

$$\omega(\mathbf{r}) = \frac{1}{2}\nabla \times \mathbf{v}(\mathbf{r}), \qquad \mathbf{r} \in \Omega(I\!R^3). \qquad (3)$$

Papkovich – Neuber [10,12] proposed the following (3–D) differential representation of the solution for Stokes flow, in terms of the harmonic potentials Φ and Φ_0

$$\mathbf{v}(\mathbf{r}) = \Phi(\mathbf{r}) - \frac{1}{2}\nabla\left(\mathbf{r} \cdot \Phi(\mathbf{r}) + \Phi_0(\mathbf{r})\right), \mathbf{r} \in \Omega(I\!R^3), \qquad (4)$$

$$\mathbf{P}(\mathbf{r}) = \mathbf{P}_0 - \mu_0 \nabla \cdot \Phi(\mathbf{r}), \qquad \mathbf{r} \in \Omega(I\!R^3), \qquad (5)$$

whereas $\mathbf{P}_0$ is a constant pressure of reference, ∇ and Δ are the gradient and the Laplacian differential operators, respectively, while

$$\Delta\Phi(\mathbf{r}) = 0, \quad \Delta\Phi_0(\mathbf{r}) = 0, \quad \mathbf{r} \in \Omega(I\!R^3).$$ (6)

If we define the thermodynamic pressure p, then the following relation gives us the total pressure as a function of the thermodynamic pressure

$$P(\mathbf{r}) = p(\mathbf{r}) + \rho_0 g x_1, \qquad \mathbf{r} \in \Omega(I\!R^3),$$ (7)

where the contribution of the term $\rho_0 g x_1$ (g is the acceleration of the gravity) refers to the gravitational pressure force, once we choose a height of reference.

The stress tensor $\tilde{\boldsymbol{\Pi}}(\mathbf{r})$ is defined as follows

$$\tilde{\boldsymbol{\Pi}}(\mathbf{r}) = -p(\mathbf{r})\tilde{\mathbf{I}} + \mu_0 \left[\nabla \otimes \mathbf{v}(\mathbf{r}) + \left(\nabla \otimes \mathbf{v}(\mathbf{r}) \right)^{\mathrm{T}} \right], \quad \mathbf{r} \in \Omega(I\!R^3),$$ (8)

while $\tilde{\mathbf{I}}$ stands for the unit dyadic and the symbol "T" denotes transposition.

The Cartesian coordinates (x_1, x_2, x_3) are related to the ellipsoidal ones (ρ, μ, ν), $\rho \in (h_2, +\infty)$, $\mu \in (h_3, h_2)$, $\nu \in (-h_3, h_3)$ through the equations [2]

$$x_\kappa = \frac{h_\kappa}{h_1 h_2 h_3} \sqrt{|\rho^2 - \alpha_1^2 + \alpha_\kappa^2|} \sqrt{|\mu^2 - \alpha_1^2 + \alpha_\kappa^2|} \sqrt{|\nu^2 - \alpha_1^2 + \alpha_\kappa^2|},$$

$$\kappa = 1, 2, 3$$ (9)

and h_κ, $\kappa = 1, 2, 3$ are the semifocal distances of the ellipsoid, which are written in terms of its main axis via the relations

$$h_1^2 = \alpha_2^2 - \alpha_3^2, \quad h_2^2 = \alpha_1^2 - \alpha_3^2 \quad \text{and} \quad h_3^2 = \alpha_1^2 - \alpha_2^2 = h_2^2 - h_1^2$$ (10)

for every $0 < \alpha_3 < \alpha_2 < \alpha_1 < +\infty$.

In ellipsoidal coordinates

$$\tilde{\mathbf{I}} = \hat{\mathbf{x}}_1 \otimes \hat{\mathbf{x}}_1 + \hat{\mathbf{x}}_2 \otimes \hat{\mathbf{x}}_2 + \hat{\mathbf{x}}_3 \otimes \hat{\mathbf{x}}_3 = \hat{\boldsymbol{\rho}} \otimes \hat{\boldsymbol{\rho}} + \hat{\boldsymbol{\mu}} \otimes \hat{\boldsymbol{\mu}} + \hat{\boldsymbol{\nu}} \otimes \hat{\boldsymbol{\nu}},$$ (11)

where $\hat{\boldsymbol{\rho}}, \hat{\boldsymbol{\mu}}, \hat{\boldsymbol{\nu}}$ stand for the normal unit vectors.

The gradient operator assume the following ellipsoidal representation,

$$\nabla = \hat{\boldsymbol{\rho}} \frac{\sqrt{\rho^2 - h_3^2} \sqrt{\rho^2 - h_2^2}}{\sqrt{\rho^2 - \mu^2} \sqrt{\rho^2 - \nu^2}} \frac{\partial}{\partial \rho} + \hat{\boldsymbol{\mu}} \frac{\sqrt{\mu^2 - h_3^2} \sqrt{h_2^2 - \mu^2}}{\sqrt{\rho^2 - \mu^2} \sqrt{\mu^2 - \nu^2}} \frac{\partial}{\partial \mu} + \hat{\boldsymbol{\nu}} \frac{\sqrt{h_3^2 - \nu^2} \sqrt{h_2^2 - \nu^2}}{\sqrt{\rho^2 - \nu^2} \sqrt{\mu^2 - \nu^2}} \frac{\partial}{\partial \nu}$$ (12)

for the values of $\rho \in (h_2, +\infty)$, $\mu \in (h_3, h_2)$, $\nu \in (-h_3, h_3)$. The outward unit normal vector on the surface of an ellipsoid coincides with the orthonormal vector $\hat{\boldsymbol{\rho}}$, which is given by

$$\hat{\mathbf{n}}(\mathbf{r}) = \frac{\rho \sqrt{\rho^2 - h_3^2} \sqrt{\rho^2 - h_2^2}}{\sqrt{\rho^2 - \mu^2} \sqrt{\rho^2 - \nu^2}} \left[\frac{x_1 \hat{\mathbf{x}}_1}{\rho^2} + \frac{x_2 \hat{\mathbf{x}}_2}{\rho^2 - h_3^2} + \frac{x_3 \hat{\mathbf{x}}_3}{\rho^2 - h_2^2} \right] = \hat{\boldsymbol{\rho}}, \quad \rho \in (h_2, +\infty)$$ (13)

Two confocal ellipsoids are considered. The inner one, indicated by S_a, at $\rho = \rho_a$, is solid and it is moving with a constant translational velocity U in an arbitrary direction (the relation between the velocity U

and the mean interstitial velocity through a swarm of ellipsoidal particles was discussed in [9]) and is rotating, also arbitrarily, with a constant angular velocity Ω. It lives within an otherwise quiescent fluid layer, which is confined by the outer surface indicated by S_b, at $\rho = \rho_b$. Following the formulation of Happel [4], the velocity component normal to S_b and the tangential stresses are assumed to vanish on S_b. These boundary conditions are supplemented by the obvious non–slip flow conditions on the surface of the particle. Thus, the general B.C.s for a three–dimensional consideration of the Happel–type boundary value problem are

B.C.(1): $\quad \mathbf{v}(\mathbf{r}) = U + \Omega \times \mathbf{r}$ $\qquad$ for $\qquad$ $\mathbf{r} \in S_a$, $\quad$ (14)

B.C.(2): $\quad \hat{\rho} \cdot \mathbf{v}(\mathbf{r}) = 0$ $\qquad$ for $\qquad$ $\mathbf{r} \in S_b$, $\quad$ (15)

B.C.(3): $\quad \hat{\rho} \cdot \tilde{\Pi}(\mathbf{r}) \cdot \left(\tilde{I} - \hat{\rho} \otimes \hat{\rho} \right) = 0$ $\qquad$ for $\qquad$ $\mathbf{r} \in S_b$, $\quad$ (16)

establishing that way the statement of a well–posed Happel–type boundary value problem within *3–D* domains, bounded by two ellipsoidal surfaces S_a and S_b. Our goal is to solve the Happel model with the aim of the Papkovich – Neuber differential representation, using the ellipsoidal system that represents the most complex geometrical orthogonal system, which embodies the complete anisotropy of the three–dimensional space. From now, we shall refer to domains $\mathbf{r} \in \Omega(I\!R^3)$ with $\rho_a \leq \rho \leq \rho_b$.

3. Parkovich-Neuber potentials and flow fields

We introduce the internal $I\!E_n^m$ and the external $I\!F_n^m$ ellipsoidal harmonics of degree n ($n = 0, 1, 2, \ldots$) and of order m ($m = 1, 2, \ldots, 2n+1$) in terms of the Lamé functions [13],

$$I\!E_n^m(\mathbf{r}) = E_n^m(\rho) E_n^m(\mu) E_n^m(\nu) , \quad \mathbf{r} \in \Omega(I\!R^3), \tag{17}$$

$$I\!F_n^m(\mathbf{r}) = F_n^m(\rho) E_n^m(\mu) E_n^m(\nu) = (2n+1) I\!E_n^m(\mathbf{r}) I_n^m(\rho) , \mathbf{r} \in \Omega(I\!R^3) \tag{18}$$

where the elliptic integrals I_n^m are given by the elliptic integrals

$$I_n^m(\rho) = \int\limits_{\rho}^{+\infty} \frac{du}{\left[E_n^m(u) \right]^2 \sqrt{u^2 - h_2^2} \sqrt{u^2 - h_3^2}} , \quad \rho \in (h_2, +\infty). \tag{19}$$

The surface ellipsoidal harmonics $E_n^m(\mu) E_n^m(\nu)$ satisfy

$$\iint\limits_{\rho = \rho_0} E_n^m(\mu) E_n^m(\nu) E_{n'}^{m'}(\mu) E_{n'}^{m'}(\nu) l_{\rho_0}(\mu, \nu) dS = \delta_{nn'} \delta_{mm'} \gamma_n^m \tag{20}$$

on the surface of a constant ellipsoid for $\rho = \rho_0$, where $n = 0, 1, 2, \ldots$, $m = 1, 2, \ldots, 2n+1$ and γ_n^m are the constants of orthonormalization [13], where the weighting function l_{ρ_0} is defined by the relation

$$l_{\rho_0}(\mu, \nu) = (\rho_0^2 - \mu^2)^{-1/2} (\rho_0^2 - \nu^2)^{-1/2} , \quad \mu \in (h_3, h_2), \quad \nu \in (-h_3, h_3). \tag{21}$$

In view of the above, we arrive at the following complete representation of the Papkovich – Neuber potentials which belong to the kernel space of Δ,

$$\Phi(\mathbf{r}) = \sum_{n=0}^{\infty} \sum_{m=1}^{2n+1} \left[e_n^{(i)m} \, I\!E_n^m(\mathbf{r}) + e_n^{(e)m} \, I\!F_n^m(\mathbf{r}) \right], \quad \mathbf{r} \in \Omega(I\!R^3), \tag{22}$$

$$\Phi_0(\mathbf{r}) = \sum_{n=0}^{\infty} \sum_{m=1}^{2n+1} \left[d_n^{(i)m} \, I\!E_n^m(\mathbf{r}) + d_n^{(e)m} \, I\!F_n^m(\mathbf{r}) \right], \quad \mathbf{r} \in \Omega(I\!R^3). \tag{23}$$

Substituting the harmonic potentials (26) and (27) to the flow fields (3), (4), (5) and (8), we derive relations in 3–D domains, furnished by the formulae

$$\mathbf{v}(\mathbf{r}) = \frac{1}{2} \sum_{n=0}^{\infty} \sum_{m=1}^{2n+1} \Big\{ e_n^{(i)m} \, I\!E_n^m(\mathbf{r}) - \left[\left(e_n^{(i)m} \cdot \mathbf{r} \right) + d_n^{(i)m} \right] \nabla I\!E_n^m(\mathbf{r})$$
$$+ e_n^{(e)m} \, I\!F_n^m(\mathbf{r}) - \left[\left(e_n^{(e)m} \cdot \mathbf{r} \right) + d_n^{(e)m} \right] \nabla I\!F_n^m(\mathbf{r}) \Big\}, \quad \mathbf{r} \in \Omega(I\!R^3) \tag{24}$$

for the velocity field, while for the total pressure field we conclude

$$P(\mathbf{r}) = P_0 - \mu_0 \sum_{n=0}^{\infty} \sum_{m=1}^{2n+1} \Big\{ e_n^{(i)m} \cdot \nabla I\!E_n^m(\mathbf{r}) + e_n^{(e)m} \cdot \nabla I\!F_n^m(\mathbf{r}) \Big\} = p(\mathbf{r}) + \rho_0 g x_1, \quad \mathbf{r} \in \Omega(I\!R^3), \tag{25}$$

where all kinds of singularities have been excluded. When the velocity is calculated, the stress tensor is provided by the form

$$\tilde{\Pi}(\mathbf{r}) = -p(\mathbf{r})\tilde{I} - \mu_0 \sum_{n=0}^{\infty} \sum_{m=1}^{2n+1} \Big\{ \left[\left(e_n^{(i)m} \cdot \mathbf{r} \right) + d_n^{(i)m} \right] \nabla \otimes \nabla I\!E_n^m(\mathbf{r})$$
$$+ \left[\left(e_n^{(e)m} \cdot \mathbf{r} \right) + d_n^{(e)m} \right] \nabla \otimes \nabla I\!F_n^m(\mathbf{r}) \Big\} \tag{26}$$

for every $\mathbf{r} \in \Omega(I\!R^3)$, while the vorticity field is expressed as

$$\omega(\mathbf{r}) = \frac{1}{2} \sum_{n=0}^{\infty} \sum_{m=1}^{2n+1} \Big\{ \nabla I\!E_n^m(\mathbf{r}) \times e_n^{(i)m} + \nabla I\!F_n^m(\mathbf{r}) \times e_n^{(e)m} \Big\}, \quad \mathbf{r} \in \Omega(I\!R^3). \tag{27}$$

These general expressions (24)–(27) consist of the basic ellipsoidal flow fields in terms of the known solid ellipsoidal harmonic eigenfunctions and their gradient. The unknown constant coefficients for $n \geq 0$,

$$e_n^{(i/e)m} = c_n^{(i/e)m} \hat{\mathbf{x}}_1 + a_n^{(i/e)m} \hat{\mathbf{x}}_2 + b_n^{(i/e)m} \hat{\mathbf{x}}_3, \quad d_n^{(i/e)m}, \quad m = 1, 2, ..., 2n+1, \tag{28}$$

must be determined from the appropriate boundary conditions (14)–(16).

For particles of a particular size that conform to physical reality, the first term of the series seems to be enough for most engineering applications. In this project, our aim is to calculate all terms of the velocity field (main and most important field), which correspond to the use of ellipsoidal harmonics of degree less or equal than two ($n \leq 2$). In order to achieve this, the potentials Φ and Φ_0 must be of degree one and two, respectively. Hence, the constant coefficients that correspond to higher degrees must be zero. Hence,

$$e_n^{(e)m} = e_n^{(i)m} = \mathbf{0}, \quad n = 2,3,..., \quad m = 1,2,...,2n+1, \tag{29}$$

$$d_n^{(i)m} = d_n^{(e)m} = 0, \quad n = 3,4,..., \quad m = 1,2,...,2n+1. \tag{30}$$

Then, this restriction leads us to a velocity of the first degree, which forms our approximation to the problem,

$$\mathbf{v}^{(0)}(\mathbf{r}) = \frac{1}{2}\left\{\sum_{n=0}^{1}\sum_{m=1}^{2n+1}\left[e_n^{(i)m}\cdot\left(\tilde{\mathbf{I}}\,\mathbb{E}_n^m(\mathbf{r})-\mathbf{r}\otimes\nabla\mathbb{E}_n^m(\mathbf{r})\right)+e_n^{(e)m}\cdot\left(\tilde{\mathbf{I}}\,\mathbb{F}_n^m(\mathbf{r})-\mathbf{r}\otimes\nabla\mathbb{F}_n^m(\mathbf{r})\right)\right]\right.$$

$$\left.-\sum_{n=0}^{2}\sum_{m=1}^{2n+1}\left[d_n^{(i)m}\nabla\mathbb{E}_n^m(\mathbf{r})+d_n^{(e)m}\nabla\mathbb{F}_n^m(\mathbf{r})\right]\right\}, \quad \mathbf{r}\in\Omega(I\!\!R^3). \tag{31}$$

Of course, the total pressure (25), the vorticity (27) and the stress tensor (26) follow the same simplification resulting the first–degree approximation.

Before we turn to the application of the boundary conditions it is necessary to make two important observations. The first one concerns the internal ellipsoidal harmonics, where in most cases are used in their Cartesian form for simplicity reasons. As a consequence, we provide the expressions that relate ellipsoidal harmonics up to the second degree with the Cartesian coordinates, i.e.

$$\mathbb{E}_0^1(\mathbf{r}) = 1, \ \mathbb{E}_1^\kappa(\mathbf{r}) = \frac{h_1 h_2 h_3}{h_\kappa}x_\kappa, \ \mathbb{E}_2^{\kappa+l}(\mathbf{r}) = \frac{h_1^2 h_2^2 h_3^2}{h_\kappa h_l}x_\kappa x_l, \ \kappa,l = 1,2,3, \ \kappa \neq l \tag{32}$$

$$\mathbb{E}_2^1(\mathbf{r}) = \left(\Lambda-\alpha_1^2\right)\left(\Lambda-\alpha_2^2\right)\left(\Lambda-\alpha_3^2\right)\left[\sum_{i=1}^{3}\frac{x_i^2}{\Lambda-\alpha_i^2}+1\right] \equiv \mathbb{E}_2^1(\Lambda), \tag{33}$$

$$\mathbb{E}_2^2(\mathbf{r}) = \mathbb{E}_2^1(\Lambda'), \quad \left.\begin{matrix}\Lambda\\\Lambda'\end{matrix}\right\} = \frac{1}{3}\sum_{i=1}^{3}\alpha_i^2 \pm \frac{1}{3}\left[\sum_{i=1}^{3}\left(\alpha_i^4 - \frac{\alpha_1^2\alpha_2^2\alpha_3^2}{\alpha_i^2}\right)\right]^{1/2}, \tag{34}$$

$$x_\kappa^2 = \frac{\left(E_1^\kappa(\rho)\right)^2}{3}\left[1 - \frac{E_2^1(\mu)E_2^1(\nu)}{\left(\Lambda-\Lambda'\right)\left(\Lambda-\alpha_\kappa^2\right)} + \frac{E_2^2(\mu)E_2^2(\nu)}{\left(\Lambda-\Lambda'\right)\left(\Lambda'-\alpha_\kappa^2\right)}\right], \tag{35}$$

where $\mathbf{r}\in\Omega(I\!\!R^3)$ and $\delta_{\kappa l}$, $\kappa,l = 1,2,3$ is the delta of Kronecker.

The second remark concerns the appearance of certain indeterminacies during the process of the calculation of the coefficients, due to the ellipsoidal factor

$$\mathbf{R}^{el}(\mathbf{r}) = \frac{\hat{\rho}}{\sqrt{\rho^2-\mu^2}\sqrt{\rho^2-v^2}}, \quad \rho\in(h_2,+\infty), \tag{36}$$

which makes impossible the application of the orthogonality relations (20). This difficulty was overcome by the use of the flexibility that our representation possesses according to which the Papkovich – Neuber potentials are not independent. In detail, we used these degrees of freedom in order to calculate the constant coefficients that eliminate the terms involved with the factor (36) on the boundaries. Once we do that, the problem is solved and the constant coefficients are calculated. For example, the first–degree approximation for the velocity field assume the form

$$\mathbf{v}^{(0)}(\mathbf{r}) = U + \boldsymbol{\Omega} \times \mathbf{r} + Z(\rho) + \sum_{j=1}^{3} H_j(\rho)\, \mathbb{E}_1^j(\mathbf{r})$$

$$+ \frac{\hat{\rho}}{2\sqrt{\rho^2 - \mu^2}\sqrt{\rho^2 - \nu^2}}\left[\sum_{j=1}^{3} \Theta_j(\rho)\, \mathbb{E}_1^j(\mathbf{r}) + \sum_{\substack{i,j=1 \\ i \neq j}}^{3} M_i^j(\rho)\, \mathbb{E}_2^{i+j}(\mathbf{r}) \right] \tag{37}$$

for every $\mathbf{r} \in \Omega(\mathbb{R}^3)$ with $\rho_a \leq \rho \leq \rho_b$, where we have defined
$Z(\rho) =$

$$-E_3^7(\rho_b)\, U \cdot \sum_{i=1}^{3}\left[\left(I_0^1(\rho) - I_0^1(\rho_a)\right) + \left(E_1^i(\rho_a)\right)^2\left(I_1^i(\rho) - I_1^i(\rho_a)\right)\right]\frac{\hat{\mathbf{x}}_i \otimes \hat{\mathbf{x}}_i}{N_i}$$

$$\tag{38}$$

$$\Theta_\kappa(\rho) = -\frac{2h_\kappa E_3^7(\rho_b)}{h_1 h_2 h_3}\frac{(U \cdot \hat{\mathbf{x}}_\kappa)}{N_\kappa}\left[1 - \left(\frac{E_1^\kappa(\rho_a)}{E_1^\kappa(\rho)}\right)^2\right], \tag{39}$$

$$H_\kappa(\rho) = \frac{3}{2}\sum_{\substack{i=1 \\ i \neq \kappa}}^{3}\left\{h_i\left(e_1^{(e)\kappa}\cdot\hat{\mathbf{x}}_i\right)\left(I_1^\kappa(\rho) - I_1^\kappa(\rho_a)\right) - h_\kappa\left(e_1^{(e)i}\cdot\hat{\mathbf{x}}_\kappa\right)\left(I_1^i(\rho) - I_1^i(\rho_a)\right)\right.$$

$$+\left.\left[h_i\left(E_1^i(\rho_a)\right)^2\left(e_1^{(e)\kappa}\cdot\hat{\mathbf{x}}_i\right) + h_\kappa\left(E_1^\kappa(\rho_a)\right)^2\left(e_1^{(e)i}\cdot\hat{\mathbf{x}}_\kappa\right)\right]\left(I_2^{i+\kappa}(\rho) - I_2^{i+\kappa}(\rho_a)\right)\right\}\frac{\hat{\mathbf{x}}_i}{h_i},$$

$$\tag{40}$$

$$M_\kappa^l(\rho) = \frac{3h_\kappa}{h_1 h_2 h_3}\left(e_1^{(e)l}\cdot\hat{\mathbf{x}}_\kappa\right)\left[1 - \left(\frac{E_1^\kappa(\rho_a)}{E_1^\kappa(\rho)}\right)^2\right]\frac{1}{\left(E_1^l(\rho)\right)^2}, \tag{41}$$

$$N_\kappa = E_3^7(\rho_b)\left[\left(I_0^1(\rho_b) - I_0^1(\rho_a)\right) + \left(E_1^\kappa(\rho_a)\right)^2\left(I_1^\kappa(\rho_b) - I_1^\kappa(\rho_a)\right)\right]$$

$$+\left(E_1^\kappa(\rho_b)\right)^2 - \left(E_1^\kappa(\rho_a)\right)^2$$

$$\tag{42}$$

for $\rho_a \leq \rho \leq \rho_b$ and $\kappa, l = 1, 2, 3$, $\kappa \neq l$. The coefficients $\left(e_1^{(e)l}\cdot\hat{\mathbf{x}}_\kappa\right)$ have known but complicated ellipsoidal expressions. Similar are the formulae for the rest of the flow fields given by equations (25), (26) and (27).

References

[1]   Happel J. and Brenner H., *Low Reynolds Number Hydrodynamics*, Prentice Hall, Englewood Cliffs, NJ (1965) and Martinus Nijholl Publishers, Dordrecht (1986).

[2]   Moon P. and Spencer E., *Field Theory Handbook*, Springer–Verlag, Second Edition, Berlin (1971).

[3]   Uchida S., Inst. Sci. Technol. Univ. Tokyo (in Japanese) **3**, 97 (1949); Abstract, "Slow Viscous Flow through a Mass of Particles", *Ind. Engng. Chem.*, **46**, 1194-1195 (1954) (translation by T. Motai).

[4]   Happel J., "Viscous Flow in Multiparticle Systems: Slow Motion of Fluids Relative to Beds of Spherical Particles", *A. I. Ch. E. Jl*, **4**, 197-201 (1958).

[5]   Kuwabara S., "The Forces Experienced by Randomly Distributed Parallel Circular Cylinders or Spheres in a Viscous Flow at Small Reynolds Number", *J. Phys. Soc. Japan*, **14**, 527-532 (1959).

[6]   Neale G.H. and Nader W.K., "Prediction of Transport Processes within Porous Media: Creeping Flow Relative to a Fixed Swarm of Spherical Particles", *A. I. Ch. E. Jl*, **20**, 530-538 (1974).

[7]   Epstein N. and Masliyah J.H., "Creeping Flow through Clusters of Spheroids and Elliptical Cylinders", *Chem. Engng. J.*, **3**, 169-175 (1972).

[8]   Dassios G., Hadjinicolaou M. and Payatakes A.C., "Generalized Eigenfunctions and Complete Semiseparable Solutions for Stokes Flow in Spheroidal Coordinates", *Quart. of Appl. Math.*, **52**, 157-191 (1994).

[9]   Dassios G., Hadjinicolaou M., Coutelieris F.A. and Payatakes A.C., "Stokes Flow in Spheroidal Particle–in–cell Models with Happel and Kuwabara Boundary Conditions", *Intern. Jl. Eng. Sci.*, **33**, 1465-1490 (1995).

[10]   Xu X. and Wang M., "General Complete Solutions of the Equations of Spatial and Axisymmetric Stokes Flow", *Quart. J. Mech. Appl. Math.*, **44**, 537-548 (1991).

[11]   Eubanks R.A. and Sternberg E., "On the Completeness of the Boussinesq – Papkovich Stress Functions", *J. Rational Mech. Anal.*, **5**, 735-746 (1956).

[12]   Neuber H., "Ein neuer Ansatz zur Lösung räumblicher Probleme der Elastizitätstheorie", *Z. Angew. Math. Mech.*, **14**, 203-212 (1934).

[13]   Hobson E.W., *The Theory of Spherical and Ellipsoidal Harmonics*, Chelsea Publishing Company, New York (1965).

SOLITARY WAVES OF THE BONA - SMITH SYSTEM

V.A. DOUGALIS AND D.E. MITSOTAKIS

Department of Mathematics,
University of Athens, 15784 Zographou, Greece,
and
Institute of Applied and Computational Mathematics, FO.R.T.H.,
P.O. Box 1527, 71110 Heraklion, Greece

We consider the Bona - Smith family of Boussinesq systems and show, following Toland's theory, that it possesses solitary wave solutions for each $k > 1$, where k is the speed of propagation of the wave. In addition, the solitary waves are shown to be unique if k is small enough. We also make a brief computational study of the stability of these solitary waves.

1. Introduction

The Bona - Smith systems are Boussinesq type systems [2], of the form

$$\eta_t + u_x + (\eta u)_x + a u_{xxx} - b \eta_{xxt} = 0$$
$$u_t + \eta_x + u u_x + c \eta_{xxx} - d u_{xxt} = 0, \tag{1}$$

where $a = 0$, $b = d = (3\theta^2 - 1)/6$, $c = 2/3 - \theta^2$, and $\theta > 0$ a parameter such that $\theta^2 \in (2/3, 1]$. They approximate the Euler equations of water wave theory and model one-dimensional, two-way propagation of long waves of small amplitude when the Stokes number is $O(1)$. In their dimensionless, unscaled form represented by the system (1), $x \in \mathbb{R}$ is a spatial variable along the channel of propagation, $t \geqslant 0$ is the time, $\eta = \eta(x,t)$ is the wave height above an undisturbed level of zero elevation, and $u = u(x,t)$ is the horizontal velocity at height θ above the bottom. (The (horizontal) bottom corresponds to an elevation equal to -1 in these variables.)

The initial-value problem for the system (1) posed for $x \in \mathbb{R}$, $t \geqslant 0$, with given smooth initial data $\eta(x,0) = \eta_0(x)$, $u(x,0) = u_0(x)$, $x \in \mathbb{R}$, has been studied in [3], in the case $\theta^2 = 1$. A straightforward extension of the theory of [3], yields that if $\eta_0 \in H^1(\mathbb{R}) \cap C_b^3(\mathbb{R})$, $u_0 \in L^2(\mathbb{R}) \cap C_b^2(\mathbb{R})$ are

286

such that $\eta_0(x) > -1$ for $x \in \mathbb{R}$ and

$$E(0) := \int_{-\infty}^{+\infty} [\eta_0^2 + (\theta^2 - 2/3)(\eta_0')^2 + (1 + \eta_0)u_0^2]dx < 2\sqrt{\theta^2 - 2/3},$$

then, the initial-value problem for (1) has a unique classical solution $(\eta, u) \in C(0, T; H^1(\mathbb{R}) \cap C_b^3(\mathbb{R})) \times C(0, T; L^2(\mathbb{R}) \cap C_b^2(\mathbb{R}))$ for each $T > 0$. The crucial step in the proof is to establish an a priori $H^1 \times L^2$ estimate that follows from the hypotheses on the initial data and the invariance of the Hamiltonian $E(t) = \int_{-\infty}^{+\infty} [\eta^2 + (\theta^2 - 2/3)\eta_x^2 + (1 + \eta)u^2]dx$ for $t \geqslant 0$.

We shall be particularly interested in the *solitary wave* solutions of the system (1). These correspond to travelling wave solutions of the form

$$\eta_s(x, t) = \eta_s(x + x_0 - kt), \quad u_s(x, t) = u_s(x + x_0 - kt), \quad x, x_0 \in \mathbb{R}, \ t \geqslant 0, \quad (2)$$

which may travel to the left or right, if the (constant) speed k is positive or negative, respectively. In (2) the univariate functions $\eta_s(\xi)$, $u_s(\xi)$, $\xi \in \mathbb{R}$, will be supposed to be positive, to have a single maximum located at $\xi = 0$, to be even about $\xi = 0$, and to decay to zero, along with their first derivatives, as $\xi \to \pm\infty$. Substituting (2) into (1) and dropping the subscript s, we obtain that $(u(\xi), \eta(\xi))$ satisfy the system of nonlinear ordinary differential equations (the derivative $'$ is with respect to ξ)

$$S_1 \mathbf{u}'' + S_2 \mathbf{u} + \nabla g(u, \eta) = 0, \quad \xi \in \mathbb{R}, \quad (3)$$

where $\mathbf{u} = (u, \eta)^T$, and

$$S_1 = \begin{pmatrix} -\frac{6a}{k} & -6b \\ -6d & -\frac{6c}{k} \end{pmatrix}, \quad S_2 = \begin{pmatrix} -\frac{6}{k} & 6 \\ 6 & -\frac{6}{k} \end{pmatrix}, \quad \text{and} \quad g(u, \eta) = -\frac{3}{k}u^2\eta. \quad (4)$$

(Note that if (u, η) is a solution of (3) corresponding to some $k > 0$, the $(-u, \eta)$ is also a solution that propagates with speed $-k$, i.e. to the left. Henceforth we shall take $k > 0$).

2. Existence of Solitary Wave Solutions

The *existence* of solitary wave solutions of (1) was established, in the case $\theta^2 = 1$ for any $k > 1$, by Toland, [10]. Subsequently, in [11], by an insightful geometric proof, Toland showed that more general o.d.e. systems of the form (3) possess such symmetric 'homoclinic' orbits in the first quadrant of the u, η-plane.

Toland's theorem may be formulated as follows (using the notation of [4]):

Theorem 2.1. *Let S_1 and S_2 be symmetric, and let $g \in C^2(\mathbb{R}^2, \mathbb{R})$ be such that $g, \nabla g$ and the second partial derivatives of g are all zero at $(0,0)$. Moreover, if $Q(\mathbf{u}) = \mathbf{u}^T S_1 \mathbf{u}$ and $f(u, \eta) = \mathbf{u}^T S_2 \mathbf{u} + 2g$, assume that*

(I) *$\det(S_1) < 0$, and there exist two linearly independent vectors $\mathbf{u}_1 = (u_1, \eta_1)^T$ and $\mathbf{u}_2 = (u_2, \eta_2)^T$ such that $Q(\mathbf{u}_1) = Q(\mathbf{u}_2) = 0$.*

(II) *There exists a closed plane curve $\mathcal{F}$ which passes through $(0,0)$ such that:*

   (i) *$f = 0$ on $\mathcal{F}$, and $\mathcal{F} \setminus \{(0,0)\}$ lies in the set $\{(u, \eta) : Q(u, \eta) < 0\}$,*

   (ii) *$f(u, \eta) > 0$ in the (non-empty) interior of $\mathcal{F}$,*

   (iii) *$\mathcal{F} \setminus \{(0,0)\}$ is strictly convex, i.e.*

$$D = f_{uu} f_\eta^2 - 2f_{u\eta} f_u f_\eta + f_{\eta\eta} f_u^2 < 0 \quad on \;\; \mathcal{F} \setminus \{(0,0)\},$$

   (iv) *$\nabla f(u, \eta) = 0$ on $\mathcal{F}$ if and only if $(u, \eta) = (0,0)$.*

Then, there exists an orbit γ of (3) which is homoclinic to the origin. Moreover

(a) *$(u(0), \eta(0)) \in \Gamma$, where Γ is the segment of $\mathcal{F}$ not including the origin between P_1 and P_2, with P_i satisfying $\nabla f(P_i) \cdot \mathbf{u}_i = 0$ for $i = 1, 2$,*

(b) *u, η are even functions on $\mathbb{R}$,*

(c) *$(u(\xi), \eta(\xi))$ is in the interior of $\mathcal{F}$ for all $\xi \in \mathbb{R} \setminus \{0\}$,*

(d) *γ is monotone in the sense that $u(\xi) \leqslant u(s)$, $\eta(\xi) \leqslant \eta(s)$ if $\xi \geqslant s \geqslant 0$.*

Toland's general theory was applied by Chen, [4], to several specific examples of Boussinesq systems. It may also be applied to establish existence of solitary wave solutions of the whole family of Bona - Smith systems, i.e. for each $\theta^2 \in (2/3, 1]$, for any $k > 1$. Specifically, in the case of the general Bona - Smith system (1) we have

$$S_1 = \begin{pmatrix} 0 & -6b \\ -6b & -\frac{6c}{k} \end{pmatrix}, \quad S_2 = \begin{pmatrix} -\frac{6}{k} & 6 \\ 6 & -\frac{6}{k} \end{pmatrix}, \quad \nabla g(u, \eta) = \begin{pmatrix} -\frac{6}{k} u\eta \\ -\frac{3}{k} u^2 \end{pmatrix}.$$

Hence, S_1, S_2 are symmetric and $\det(S_1) = -36b^2 < 0$. Also $Q(u, \eta) = -\frac{6}{k}(2bku\eta + c\eta^2)$ and

$$f(u, \eta) = -\frac{6}{k}(u^2(1 + \eta) + \eta^2 - 2k\eta u). \tag{5}$$

It is easy to check that Q vanishes on the directions of the vectors $\mathbf{u}_1 = (1, 0)^T$ and $\mathbf{u}_2 = (1, -\frac{2bk}{c})^T$. Hence, hypothesis (I) of Theorem 2.1 holds.

To verify hypothesis (II) we consider $\mathcal{F}^* = \{(u,\eta) \; : \; f(u,\eta) = 0\}$. If we factor f when $\eta \neq -1$ then $f(u,\eta) = -\frac{6}{k}(1+\eta)(u-u_-)(u-u_+)$, where $u_\pm = (k\eta \pm \sqrt{\eta^2(k^2-1-\eta)})/(1+\eta)$. Therefore, if $k > 1$, the set $\{f = 0\}$ represents a closed curve $\mathcal{F}$ which passes through the origin and lies on the first quadrant when $0 \leqslant \eta \leqslant k^2 - 1$ (actually $\mathcal{F}$ is the part of the curve $\mathcal{F}^*$ in the first quadrant). Moreover $f > 0$ in the interior of $\mathcal{F}$ since u is between u_- and u_+. Furthermore $\mathcal{F}$ is strictly convex since $D = -2(6/k)^3((u^2-\eta^2)^2 + u^4\eta)\eta^{-1} < 0$. Also, $\mathcal{F}$ lies in the interior of the cone K_1 which is formed by the straight lines $\eta = s_1 u$ and $\eta = s_2 u$, where $s_1 = k - \sqrt{k^2-1}$ and $s_2 = k + \sqrt{k^2-1}$. These lines are tangent to $\mathcal{F}$ at $(0,0)$. Also, the cone K_1 is in the interior of the cone K, which is formed by the vectors $\mathbf{u}_1$ and $\mathbf{u}_2$, and so $\mathcal{F}\backslash\{(0,0)\}$ is in the set $\{(u,\eta) \; : \; Q(u,\eta) < 0\}$, i.e. for each $\theta^2 \in (2/3, 1]$ and $k > 1$ $K_1 \subset K$. (It is interesting to observe that the direction of $\mathbf{u}_2$ tends to the line $\eta = s_2 u$ as $\theta^2 \uparrow 1$ and $k \to \infty$). Finally it is easy to check that, $\nabla f = 0$ on $\mathcal{F}$ if and only if $(u,\eta) = (0,0)$. Note that $P_1 = (k - 1/k, k^2 - 1)$ and $P_2 = (2(k-1), 2(k-1))$.

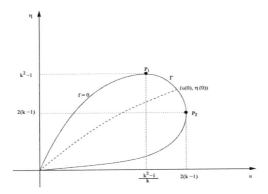

Figure 1.   Locus of possible $(u(0), \eta(0))$

In conclusion, all hypotheses of Theorem 2.1 are valid and so the general Bona - Smith system (1) possesses (single hump) solitary wave solutions. In particular, it follows from Toland's theory that the existence of a solitary wave implies that the pair of peaks $(u(0), \eta(0))$ lies in the open segment $\Gamma = P_1 P_2$ of $\mathcal{F}$ not including the origin, cf. Fig. 1. Furthermore, since $f(u(0), \eta(0)) = 0$, it follows that the speed k of a solitary wave corresponding to the peak $(u(0), \eta(0))$ satisfies the equation

$$k = \frac{\mu^2(1 + \eta(0)) + 1}{2\mu}, \tag{6}$$

where $\mu = u(0)/\eta(0)$. The curve $\mathcal{F}$ and, consequently, the relation (6) are independent of the constants a, b, c, d of the particular Boussinesq system, and, in particular, independent of the parameter θ^2 of the Bona - Smith systems. However, the shape of a solitary wave of an individual Bona - Smith system (i.e. the dashed line in Fig. 1 representing the solitary wave $(u(\xi), \eta(\xi))$, $0 \leqslant \xi$), depends of course on θ^2.

For each value θ^2 in the interval $(7/9, 1)$ one may find, following [5], *one* explicit solitary wave solution (u_s, η_s) of the form

$$\begin{aligned}
\eta_s(x, t) &= \eta_0 \operatorname{sech}^2\left(\lambda(x + x_0 - kt)\right), \\
u_s(x, t) &= B\eta_s(x, t),
\end{aligned} \tag{7}$$

where $x_0 \in \mathbb{R}$ is arbitrary, and the parameters η_0, k, B, λ are given in terms of θ^2 by the formulas

$$\begin{aligned}
\eta_0 &= \frac{9}{2} \cdot \frac{\theta^2 - 7/9}{1 - \theta^2}, & k &= \frac{4(\theta^2 - 2/3)}{\sqrt{2(1 - \theta^2)(\theta^2 - 1/3)}}, \\
\lambda &= \frac{1}{2}\sqrt{\frac{3(\theta^2 - 7/9)}{(\theta^2 - 1/3)(\theta^2 - 2/3)}}, & B &= \sqrt{\frac{2(1 - \theta^2)}{\theta^2 - 1/3}}.
\end{aligned} \tag{8}$$

3. Uniqueness of the Solitary Wave Solutions

The *uniqueness* of the solitary wave solutions of (1) may be studied again by the methods of Toland, [12]; he established their uniqueness in the case of the $\theta^2 = 1$ system, unconditionally if $u(0) \leqslant 1$, and in general if $1 < k \leqslant 3/2$ or $k \gg 1$. In the case of the general Bona - Smith system, following Toland's proof it is possible to prove uniqueness provided e.g. that $\frac{2 + \sqrt{0.2}}{3} < \theta^2 \leqslant 1$ and $1 < k \leqslant k_{\max}(\theta)$. We outline the proof below; full details may be found in [9].

The solitary waves are solutions of the boundary-value problem

$$\left.\begin{aligned}
& u(x),\ \eta(x),\ x \in \mathbb{R} \\
& k(\eta - b\eta'') = u + u\eta && \text{in } \mathbb{R} \\
& k(u - bu'') = c\eta'' + \eta + \tfrac{1}{2}u^2 && \text{in } \mathbb{R} \\
& u' < 0,\ \eta' < 0 && \text{in } (0, \infty) \\
& \lim_{|x| \to \infty} u(x) = \lim_{|x| \to \infty} \eta(x) = 0,\ u,\ \eta \text{ positive and even on } \mathbb{R}.
\end{aligned}\right\} \tag{9}$$

The proof of (conditional) uniqueness of solutions of (9) will be presented as a sequence of Lemmata below. The first result, concerns some needed *a priori* estimates and may be obtained by the fact that the peak $(u(0), \eta(0))$ must lie on the closed curve $\mathcal{F}$, so that $f(u(0), \eta(0)) = 0$, and by using a classical maximum principle for two-point boundary value problems.

Lemma 3.1. *If (u, η) satisfies (9), then*

$$0 < u(x) \leqslant u(0) \leqslant 2(k-1), \quad x \in \mathbb{R}, \tag{10}$$

$$0 < \eta(x) \leqslant \eta(0) \leqslant k^2 - 1, \quad x \in \mathbb{R}, \tag{11}$$

$$ku(x) > -\frac{c}{b}\eta(x) > 0, \quad x \in \mathbb{R}, \tag{12}$$

$$ku'(x) < -\frac{c}{b}\eta'(x) < 0, \quad x \in (0, \infty). \tag{13}$$

With the help of a nonlinear maximum principle, [7], we can prove as in [12], that $ku(x)$ is bounded from above by a multiple of $\eta(x)$ and vice versa for the derivatives. Specifically there holds:

Lemma 3.2. *If (u, η) satisfies (9) and $k^2 \leqslant \frac{c^2}{(b+c)b}$, $\frac{2+\sqrt{0.2}}{3} < \theta^2 \leqslant 1$, $\nu := -2\frac{c}{b} > 0$, then*

$$ku(x) \leqslant \left(\alpha_0 - \frac{c}{b}\right)\eta(x) \leqslant \nu\eta(x), \quad x \in \mathbb{R}, \tag{14}$$

and

$$0 > ku'(x) > \nu\eta'(x), \quad x \in (0, \infty), \tag{15}$$

where

$$\alpha_0 := \sup_{x \in \mathbb{R}} \left\{ \frac{2kb\eta + 2kc\eta - cu\eta}{2bu + bu\eta} \right\} < \infty. \tag{16}$$

The fact that $(u(0), \eta(0))$ lies on $\mathcal{F}$ is central to the analysis. Not only can we use it to estimate the speed k of a solitary wave with given $(u(0), \eta(0))$,(cf. (6)) but also to estimate the slope of $\mathcal{F}$ at that point:

Lemma 3.3. *If (u, η) satisfies (9), then $(u(0), \eta(0)) \in \mathcal{F}$ at a point, where the slope $du/d\eta$ of $\mathcal{F}$ satisfies the inequalities*

$$\frac{c}{kb} \leqslant \frac{du}{d\eta} \leqslant 0. \tag{17}$$

We finally state the uniqueness theorem which can be proved with the aid of the previous Lemmata in analogy with Theorem 3.2 of [12]:

Theorem 3.1. *If* $\frac{u(0)}{k} \leqslant \frac{3c+b}{3c}$ *and* $k^2 \leqslant \frac{c^2}{(b+c)b}$, $\frac{2+\sqrt{0.2}}{3} < \theta^2 \leqslant 1$, *then the solution of the problem (9) is unique.*

Remark: By (10), if $k \leqslant \frac{-6c}{b-3c}$, then $\frac{u(0)}{k} \leqslant \frac{3c+b}{3c}$. Hence, taking into account the assumption that $k^2 \leqslant \frac{c^2}{(b+c)b}$ it follows, from Theorem 3.1 that a sufficient condition for uniqueness is that $\theta^2 \in (\frac{2+\sqrt{0.2}}{3}, 1]$ and that

$$k \leqslant \min\left\{\frac{-6c}{b-3c}, \sqrt{\frac{c^2}{(b+c)b}}\right\} = \min\left\{\frac{12(3\theta^2-2)}{21\theta^2-13}, \frac{2(3\theta^2-2)}{\sqrt{(3-3\theta^2)(3\theta^2-1)}}\right\} =: k_{\max}(\theta).$$

If $\theta^2 = 1$, we recover the $k_{\max} = 3/2$ of [12].

4. Numerical Studies of the Stability of Solitary Waves

The question of stability of the solitary wave solutions of two-way propagation models, such as the Boussinesq systems is an important problem. No rigorous results exist yet, for example, about the nonlinear orbital stability of the solitary waves. In particular, the classical approach, cf. e.g. [1], [8], fails in the case of the Bona - Smith system (1) because the solitary wave, a stationary point of an appropriate constrained variational problem for a functional associated with the Hamiltonian structure of (1), turns out to be a saddle point of infinite index. It is of interest then to study numerically the effects of small and large perturbations on the evolution of solitary waves to provide clues and motivation for the analysis. Such a study has been carried out in [6], where initial solitary wave profiles have been perturbed in a wide collection of directions and magnitudes; the periodic initial-value problem for (1) has then been integrated numerically using a standard Galerkin finite element scheme with cubic splines in space and the classical explicit Runge-Kutta method for the temporal discretization.

The results indicate that the solitary waves are, apparently, *asymptotically stable*. For example, if a solitary wave of the form (7)–(8) for $\theta^2 = 9/11$, $x_0 = 100$ is perturbed so that $\eta_0(x) = r\eta_s(x, 0)$ with $r = 1.1$, there results one η-solitary wave of amplitude 1.06110 (note that the amplitude of the unperturbed initial profile $\eta_s(x, 0)$ was equal to one), that travels to the right with a speed $k = 1.4673$ plus right- and left-travelling oscillatory dispersive tails of small, decaying amplitude. A magnification of the solution at $t = 100$ appears in Fig. 2. The solitary wave is the big pulse located near $x = 50$ and followed by the part of the dispersive tail that is travelling to the right. The other part of the tail has wrapped itself around $x = 150$ by periodicity and is travelling to the left.

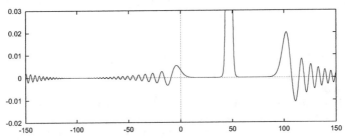

Figure 2.   Evolution of a perturbed η-solitary wave, $\theta^2 = 9/11$, $r = 1.1$, $t = 100$.

Larger initial perturbations may lead to more than one solitary waves plus dispersive tails. For example, if $r = 1.8$, in addition the main right-travelling solitary wave (of η-amplitude about 1.48043 and speed 1.6242), a second solitary wave (of η-amplitude about 0.251 and speed $k = -1.12$) emerges quite early at the head of the left-travelling dispersive tail. (By $t = 80$, cf. Fig. 3, the left-travelling wave train has wrapped itself around the right-hand boundary due to periodicity.)

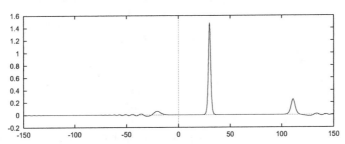

Figure 3.   Evolution of a perturbed η-solitary wave, $\theta^2 = 9/11$, $r = 1.8$, $t = 80$.

It is worthwhile to note that large perturbations, for example of the initial u-solitary wave profile, lead to apparent blow-up instabilities in finite time after η becomes less than -1 at some point (x^*, t^*). In an example where we took $u_0(x) = 3.8 u_s(x, 0)$, where u_s, η_s were given by (7)–(8) and $\theta^2 = 9/11$, $x_0 = 100$, η developed a negative excursion that became less than -1 at about $t = 3$. Subsequently, the solution apparently developed a singularity and blew up.

Acknowledgments

This work was supported in part by the Institute of Applied and Computational Mathematics of FO.R.T.H. and in part by a grant of the Research Committee of the University of Athens.

References

1. T. B. Benjamin, The stability of solitary waves, *Proc. Roy. Soc. Lond. A* **328**(1972), 153–183.
2. J. L. Bona, M. Chen, and J. -C. Saut, Boussinesq equations and other systems for small-amplitude long waves in nonlinear dispersive media. Pt I: Derivation and the linear theory, *J. Nonlin. Science* **12**(2002), 283–318.
3. J. L. Bona and R. Smith, A model for the two-way propagation of water waves in a channel, *Math. Proc. Camb. Phil. Soc.* **79**(1976), 167–182.
4. M. Chen, Solitary-wave and multi pulsed traveling-wave solutions of Boussinesq systems, *Applic. Analysis* **75**(2000), 213–240.
5. M. Chen, Exact traveling-wave solutions to bi-directional wave equations, *Int. J. Theor. Phys.* **37**(1998), 1547–1567.
6. V.A. Dougalis, A. Duran, M.A. Lopez-Marcos, and D.E. Mitsotakis, A numerical study of the stability of solitary waves of the Bona-Smith system, (to appear).
7. B. Gidas, W. -M. Ni, and L. Nirenberg, Symmetry and related properties via the maximum principle, *Commun. Math. Phys.* **68**(1979), 209–243.
8. M. Grillakis, J. Shatah, and W.A. Strauss, Stability of solitary waves in the presence of symmetry I, *J. Funct. Anal.* **74**(1987), 170–197.
9. D. E. Mitsotakis, Solitary waves for the Bona-Smith system: Existence, uniqueness and stability, (in Greek) *M.Sc. Thesis*, Mathematics Dept., Univ. of Athens, 2003.
10. J. F. Toland, Solitary wave solutions for a model of the two - way propagation of water waves in a channel, *Math. Proc. Camb. Phil. Soc.* **90**(1981), 343–360.
11. J. F. Toland, Existence of symmetric homoclinic orbits for systems of Euler-Lagrange equations, *A.M.S. Proceedings of Symposia in Pure Mathematics*, v.45, Pt. 2 (1986), 447–459.
12. J. F. Toland, Uniqueness and a priori bounds for certain homoclinic orbits of a Boussinesq system modelling solitary water waves, *Commun. Math. Phys.* **94**(1984), 239–254.

ELECTROMAGNETIC WAVE PROPAGATION IN DISPERSIVE BIANISOTROPIC MEDIA

A.D. IOANNIDIS, I.G. STRATIS

University of Athens
Department of Mathematics, Panepistimiopolis
157 84 Zografou, Athens, Greece

A.N. YANNACOPOULOS

University of the Aegean
Department of Statistics and Actuarial Science
83 200 Karlovassi, Samos, Greece

In this paper we treat the solvability of a general problem of linear electromagnetics. More precisely, we study the propagation (in the time domain) of electromagnetic fields in a bianisotropic medium, which presents dispersion. This is modeled by linear constitutive relations containing convolution terms. So the problem is represented by an integrodifferential equation in some appropriate Hilbert space and then is treated with the use of the vector-valued Laplace transform.

1. Introduction

A linear bianisotropic medium is a relatively simple example of an optically active medium. Such a medium displays both electric and magnetic polarization phenomena when excited by either electric or magnetic excitations. As it is well known[1], every electromagnetic phenomenon is specified by four vector quantities: the *electric field* $\mathbf{E}$, the *magnetic field* $\mathbf{H}$, the *electric displacement* $\mathbf{D}$ and the *magnetic induction* $\mathbf{B}$. These quantities are considered as time-dependent vector fields on the whole space $\mathbb{R}^3$, so they are functions of the spatial variable $x \in \mathbb{R}^3$ and the time variable $t \in \mathbb{R}$. The inter-dependence among these quantities is given by the celebrated *Maxwell system*

$$\frac{\partial \mathbf{D}}{\partial t} = \operatorname{curl} \mathbf{H}, \qquad \frac{\partial \mathbf{B}}{\partial t} = -\operatorname{curl} \mathbf{E} \tag{1}$$

with a possible inclusion of source terms in the right hand side. Assuming the absence of such sources we have, in addition

$$\operatorname{div} \mathbf{D} = \operatorname{div} \mathbf{B} = \mathbf{0} \qquad (2)$$

which are essentially equivalent with the equation of continuity. Equations (1) and (2) form a system of eight partial differential equations with twelve unknown functions. Indeed, equations (2) are partially redundant: it suffices to be valid for a certain time instance, say at $t = 0$, in order be valid for every t. So we focus our attention to the Maxwell system (1). Thus one needs six more equations to make the system well defined. These equations, known as *constitutive relations*, are given formally as follows

$$\begin{bmatrix} \mathbf{D} \\ \mathbf{B} \end{bmatrix} = C \begin{bmatrix} \mathbf{E} \\ \mathbf{H} \end{bmatrix} \qquad (3)$$

where C is an operator of known form. Roughly speaking, the constitutive relations (3) are the mathematical description of the considered medium where the phenomenon takes place. The operator C is specified by certain assumptions about the medium and by experimental data (measurements). In this paper we describe the exact form of (3) for linear, bianisotropic, time dispersive media and then we establish the solvability of system (1).

2. Notation and preliminaries. The Laplace transform

The object of this section is to formulate equations (1), (3) as an abstract problem. First of all, to simplify the notation, we introduce the *six-vector formulation*[2] $\mathbf{d} = (\mathbf{D}, \mathbf{B})^T$, $\mathbf{u} = (\mathbf{E}, \mathbf{H})^T$ and we agree to call $\mathbf{u}$ the *electromagnetic field* and $\mathbf{d}$ the *electromagnetic displacement*. Now equation (3) can be written as $\mathbf{d} = C\mathbf{u}$. Furthermore, we choose to work in the L^2-framework, that is we assume square integrable fields (as functions of the spatial variable). This assumption corresponds to finite energy solutions. After that, the fields $\mathbf{d}$, $\mathbf{u}$ can be regarded as functions $\mathbb{R} \to H$, where $H = L^2(\mathbb{R}^3; \mathbb{C}^3) \oplus L^2(\mathbb{R}^3; \mathbb{C}^3) \cong L^2(\mathbb{R}^3; \mathbb{C}^6)$ is the underlying state space for the problem. Equipped with the usual inner product $\langle g_1, g_2 \rangle = \int_{\mathbb{R}^3} g_1 \cdot \overline{g_2}\, dr$ and the corresponding norm $\| g \| = \int_{\mathbb{R}^3} |g|^2\, dr$ H becomes a Hilbert space. Here by $\cdot$ we denote the dot product in six-space and by overbar the complex conjugate. So far we have considered $\mathbf{d}$, $\mathbf{u}$ as elements of a function space X containing H-valued functions of the real variable t. Next, we consider the *Maxwell operator* $\mathcal{M}$ with domain

$$D(\mathcal{M}) = H(\operatorname{curl}, \mathbb{R}^3) \oplus H(\operatorname{curl}, \mathbb{R}^3)$$

where $H(\text{curl}, \mathbb{R}^3)$ denotes the corresponding Sobolev space for the curl operator. Operator $\mathcal{M}$ is defined formally by the "matrix"

$$\mathcal{M} = \begin{bmatrix} 0 & \text{curl} \\ -\text{curl} & 0 \end{bmatrix}$$

Now we are in position to formulate the abstract problem:

(**AP**) *Given initial data $u_0 \in H$, find $\mathbf{u} \in X$ that satisfies the differential equation*

$$\frac{d}{dt}(C\mathbf{u}) = \mathcal{M}\mathbf{u}$$

subject to the initial condition $\mathbf{u}(0) = u_0$.

For our approach, it suffices to take as X the linear space $L_\omega^\infty(H)$ of measurable, causal, ω-exponentially bounded functions where ω is a known constant. That is a measurable function (limit of H-valued simple functions) $\mathbf{f}$ belongs to $L_\omega^\infty(H)$ if and only if $\mathbf{f}(t) = \mathbf{0}$ for $t < 0$ and there is a positive constant K so as for almost all $t \geqslant 0$

$$\| \mathbf{f}(t) \| \leqslant K e^{\omega t} \tag{4}$$

The space $L_\omega^\infty(H)$, normed with the (essential) supremum norm

$$\| \mathbf{f}(t) \|_{\omega, \infty} = \inf \{ K \geqslant 0 : K \text{ is such that (4) is valid} \}$$

is a Banach space. Furthermore, it is evident that the correspondence

$$L_\omega^\infty(H) \ni \mathbf{f} \mapsto e^{-\omega(\cdot)}\mathbf{f}(\cdot) \in L^\infty([0, \infty); H)$$

establishes an isometric isomorphism. For a function $\mathbf{f} \in L_\omega^\infty(H)$ one can define the *vector-valued Laplace transform*[4,5] $\hat{\mathbf{f}}(\lambda) = \int_0^\infty e^{-\lambda t}\mathbf{f}(t)\, dt$ where the integral is understood in the Bochner sense. It is obvious that $\hat{\mathbf{f}}$ is an analytic function of the complex variable $\lambda \in (\omega, \infty) \times i\mathbb{R}$. In what follows we are interested for real $\lambda > \omega$. Therefore, $\hat{\mathbf{f}}$ is an infinitely differentiable function of λ, and a member of the *Widder class*[3]

$$W_\omega(H) = \left\{ \mathbf{r} \in C_\omega^\infty((\omega, \infty); H) : \sup_{\substack{\lambda > \omega \\ k \in \mathbb{N}_0}} \left\| \frac{(\lambda - \omega)^{k+1}\mathbf{r}^{(k)}(\lambda)}{k!} \right\| < \infty \right\}$$

The Widder class is obviously a vector space that, normed with the quantity in the right hand side, becomes a Banach space. So we may think of the Laplace transform as a mapping

$$L_\omega^\infty(H) \ni \mathbf{f} \mapsto \hat{\mathbf{f}} \in W_\omega(H)$$

The crucial properties of this mapping are collected in the following

Theorem 2.1. *a) (Post–Widder Theorem[4]) The Laplace transform is a linear isometric isomorphism. In particular, $\mathbf{r} \in W_\omega(H)$ if and only if $\mathbf{r} = \hat{\mathbf{f}}$ for a (unique) $\mathbf{f} \in L_\omega^\infty(H)$.*

b) (Peng–Chung[5]) $\mathbf{r} \in W_\omega(H)$ if and only if there is a positive constant M so that for all $\lambda > \omega$ and $k = 1, 2, \dots$

$$|(\lambda - \omega)\mathbf{r}(\lambda)| \leqslant M \quad and \quad \left| \sum_{j=1}^{\infty} \frac{(-1)^{j-1}}{(j-1)!} e^{jk} (\lambda - \omega) \mathbf{r}(j\lambda) \right| \leqslant M \quad (5)$$

Incidentally, the infimum of $M \geqslant 0$ in (5) equals to the norm of $\mathbf{r}$.

c) If $\mathbf{f} \in L_\omega^\infty(H)$ is continuously differentiable, then $\mathbf{f}' \in L_\omega^\infty(H)$ and $\hat{\mathbf{f}'}(\lambda) = \lambda \mathbf{f}(\lambda) - \mathbf{f}(0)$.

d) If $\mathcal{S} \in L^\infty(\mathcal{B}(H))$ and $\mathbf{f} \in L_\omega^\infty(H)$, then the convolution

$$(\mathcal{S} * \mathbf{f})(t) = \int_0^t \mathcal{S}(t - \tau)\mathbf{f}(\tau)\, dt$$

is well defined, $\mathcal{S} * \mathbf{f} \in L_\omega^\infty(H)$ and $\widehat{\mathcal{S} * \mathbf{f}}(\lambda) = \hat{\mathcal{S}}(\lambda)\hat{\mathbf{f}}(\lambda)$.

3. Linear bianisotropic dispersive media

Now we introduce the principles that govern the considered media following Refs. 2 and 6. These principles are based both on theoretical and experimental arguments. Our final goal is to derive the constitutive relations, that is to find the form of the operator C in (3). In particular, we suppose that

(**P1**) The medium reacts linearly to electromagnetic excitations.

(**P2**) The medium is causal with respect to electromagnetic excitations, that is if $\mathbf{u}(t) = \mathbf{0}$ for $t \in (-\infty, \tau]$ then $\mathbf{d}(t) = \mathbf{0}$ for $t \in (-\infty, \tau]$, too.

(**P3**) The electromagnetic quantities of the medium remain invariant with respect to time translations, that is if to the electromagnetic field $\mathbf{u}(t)$ corresponds an electromagnetic displacement $\mathbf{d}(t)$, then to $\mathbf{u}_\tau(t) = \mathbf{u}(t - \tau)$, for arbitrary $\tau > 0$, corresponds $\mathbf{d}_\tau(t) = \mathbf{d}(t - \tau)$.

(**P4**) The medium displays stability with respect to excitations, in the sense that to "small" perturbations on $\mathbf{u}$ correspond "small" perturbation on $\mathbf{d}$

Let us discuss these assumptions in some depth. First of all, in the light of equation (3), we consider $\mathbf{u}$ as the excitation (cause) and $\mathbf{d}$ as the the result. (**P1**) is stated under the condition that the excitation is relatively small in amplitude, so a linear dependence between $\mathbf{d}$ and $\mathbf{u}$ can be assumed. (**P2**), as already remarked, is a causality requirement: we cannot know $\mathbf{d}$ before we start the observation of $\mathbf{u}$. (**P3**) says that the medium displays some sort of memory in the sense that the result $\mathbf{d}$, at each time instant t_0, depends on the past of the excitation $\mathbf{u}$, that is on the values $\mathbf{u}(t)$ for $t \in (-\infty, t_0]$. In the sequel, we assume that $\mathbf{u} \in L_\omega^\infty(H)$. This means that the observation of $\mathbf{u}$ starts at $t = 0$ and we are not concerned about the past $t < 0$. Using (**P1**), (**P2**) and (**P4**) we have that C is a continuous linear operator on $L_\omega^\infty(H)$. Taking into account (**P4**), a form of (3), consistent with these assumptions is

$$\mathbf{d} = A\mathbf{u} + K * \mathbf{u}$$

Here A and K are 6×6 matrices of the general block-form

$$A = \begin{bmatrix} \varepsilon & \xi \\ \zeta & \mu \end{bmatrix} \qquad K = \begin{bmatrix} \varepsilon_d & \xi_d \\ \zeta_d & \mu_d \end{bmatrix}$$

A models the instantaneous response of the medium to the excitations and is known in the literature as the *optical response* part. The 3×3 matrices ε, ξ, ζ and μ have entries that are real, measurable, bounded functions of the spatial variable. There is a widely used classification of media with respect to these entries[7]. In particular, one calls the medium

(**M1**) *homogeneous* if ε, ξ, ζ and μ are constant matrices,

(**M2**) *anisotropic* if $\xi = \zeta = 0$, and *isotropic* if, furthermore, ε and μ are proportional to the identity matrix,

(**M3**) *bi-isotropic* if all ε, ξ, ζ and μ are all proportional to the identity matrix and

(**M4**) *bianisotropic* in the general case.

K models the memory (time-dispesive) phenomena and is known as the *susceptibility kernel*. Its entries are real, sufficiently smooth, bounded, causal functions of t. We also assume, for convenience, that $K(0) = \mathbf{0}$.

So, by inspection, we see that a form of operator C, consistent with the assumptions (**P1**)–(**P4**), is

$$C = A + K* \tag{6}$$

We call (6) a *linear dispersive law*[6]. Additional hypotheses can be made about the medium in order to model particular situations such as dissipation, reciprocity etc. These hypotheses lead to special forms of the matrices A and K. In this paper we suppose that the matrix A is coercive[2]; this means that there exists a positive constant a so that for all $r \in \mathbb{C}^6$

$$Ar \cdot \bar{r} \geqslant a \, |r|^2 \tag{7}$$

Eventually, (7) ensures that A is a symmetric matrix and therefore $\varepsilon = \varepsilon^T$, $\mu = \mu^T$ and $\xi = \zeta^T$. Furthermore, it is easy to see that A is invertible and A^{-1} is also symmetric.

4. Solvability of the problem

We are now in a position to proceed to solving the considered problem. For convenience, we will treat the following version of the problem

$$\frac{d}{dt}(A\mathbf{u} + K * \mathbf{u}) = \mathcal{M}\mathbf{u} + \mathbf{f}, \qquad \mathbf{u}(0) = \mathbf{0} \tag{8}$$

Thus we assume a homogeneous initial condition. The measurable, causal function $\mathbf{f}$ models the existent sources (currents) in the medium and presents the initial data of the problem. It is trivial to see that every problem with non-homogeneous initial condition $u_0 \in D(\mathcal{M})$ can be transformed in the form (8).

4.1. The optical response region

First we will assume the case where the dispersion phenomena are absent, that is the case $K = 0$. Then, using the invertibility of matrix A, (8) reduces to a usual Abstract Cauchy Problem

$$\frac{d\mathbf{u}}{dt} = \mathcal{M}_0\mathbf{u} + \mathbf{f}_0, \qquad \mathbf{u}(0) = \mathbf{0} \tag{9}$$

where $\mathcal{M}_0 = A^{-1}\mathcal{M}$ and $\mathbf{f}_0 = A^{-1}\mathbf{f}$. In the space H we consider the weighted inner product

$$\langle g_1, g_2 \rangle_A = \langle Ag_1, g_2 \rangle = \int_{\mathbb{R}^3} Ag_1 \cdot \overline{g_2} \, dr$$

Equipped with this, H becomes a Hilbert space. Thanks to (7), this inner product $\langle \cdot, \cdot \rangle_A$ is equivalent to the usual $\langle \cdot, \cdot \rangle$. It is well known[8] that the Maxwell operator $\mathcal{M}$ is skew-adjoint with respect to $\langle \cdot, \cdot \rangle$. Then, by using the fact that A and A^{-1} are Hermitian, one can easily deduce that $\mathcal{M}_0$ is also skew-adjoint with respect to $\langle \cdot, \cdot \rangle_A$. Using Stone's Theorem[9], $\mathcal{M}_0$ is thus the infinitesimal generator of a unitary C_0-semigroup, say $\mathcal{G}$.

Let us now examine the problem using the vector-valued Laplace transform. This approach is of formal nature and was proposed in the classic work of Hille and Yosida. Applying the Laplace transform to (9), we derive the *characteristic equation*

$$(\lambda \mathcal{I} - \mathcal{M}_0)\hat{\mathbf{u}}(\lambda) = \hat{\mathbf{f}}_0(\lambda) \tag{10}$$

The fact that $\mathcal{M}_0$ is skew-adjoint ensures that every $\lambda \neq 0$ is a regular value of $\mathcal{M}_0$, that is the resolvent $\mathcal{R}(\lambda) = (\lambda \mathcal{I} - \mathcal{M}_0)^{-1}$ is well defined on H and bounded. Indeed, the resolvent satisfies the inequality

$$\| \mathcal{R}(\lambda) \| \leqslant \frac{1}{\lambda} \tag{11}$$

Thus, for each $\omega \geqslant 0$, the interval (ω, ∞) contains regular values of $\mathcal{M}_0$ and thus equation (10) can be written for $\lambda > \omega$

$$\hat{\mathbf{u}}(\lambda) = \mathcal{R}(\lambda)\hat{\mathbf{f}}_0(\lambda) \tag{12}$$

The Hille-Yosida theorem states that a closed, densely defined operator on a Hilbert space is the infinitesimal generator of a C_0-semigroup if and only if the resolvent is a member of the Widder class $W_\omega (\mathcal{B}(H))$. In this case the resolvent is the Laplace transform of the semigroup[9]. Thus (12) takes the form

$$\hat{\mathbf{u}}(\lambda) = \hat{\mathcal{G}}(\lambda)\hat{\mathbf{f}}_0(\lambda)$$

and, after the inversion of the Laplace transform, we have

$$\mathbf{u}(t) = (\mathcal{G} * \mathbf{f}_0)(t) \tag{13}$$

So we have obtained an explicit formula for the solution. Indeed, the following result holds[8]

Theorem 4.1. *Let $\omega \geqslant 0$ and suppose $\mathbf{f} \in L_\omega^\infty(H)$ is continuous with $\mathbf{f}(t) \in A\,[D(\mathcal{M})]$. Then (9) admits a unique classical (i.e. continuously differentiable) solution $\mathbf{u} \in L_\omega^\infty(H)$, given by (13). In addition, the problem is well posed.*

4.2. The general case

We proceed now to the general problem (8) following Refs. 10 and 11. Applying the Laplace transform, we take

$$\lambda A\hat{\mathbf{u}}(\lambda) + \lambda \hat{K}(\lambda)\hat{\mathbf{u}}(\lambda) = \mathcal{M}\hat{\mathbf{u}}(\lambda) + \hat{\mathbf{f}}(\lambda)$$

Again, multiplying from the left by A^{-1} and setting $\hat{K}_0 = A^{-1}K$

$$(\lambda \mathcal{I} - \mathcal{M}_0)\hat{\mathbf{u}}(\lambda) = -\lambda \hat{K}_0(\lambda)\hat{\mathbf{u}}(\lambda) + \hat{\mathbf{f}}_0(\lambda)$$

Thus we obtain the equation

$$\hat{\mathbf{u}}(\lambda) = -\lambda \mathcal{R}(\lambda)\hat{K}_0(\lambda)\hat{\mathbf{u}}(\lambda) + \mathcal{R}(\lambda)\hat{\mathbf{f}}_0(\lambda) \qquad (14)$$

This is in the form of a fixed-point problem for an affine operator, dependent on the parameter λ. The linear part of this operator is

$$\hat{T}(\lambda) = -\lambda \mathcal{R}(\lambda)\hat{K}_0(\lambda) \in \mathcal{B}(H)$$

Taking into account that $\hat{K}_0$ belongs to the Widder class, and using (11), we see that

$$\lim_{\lambda \to \infty} \left\| \hat{T}(\lambda) \right\| = 0$$

This means that for sufficiently large λ, say for $\lambda > \omega_0 \geqslant 0$, we have

$$\left\| \hat{T}(\lambda) \right\| < 1/2$$

namely $\hat{T}(\lambda)$ is a contraction. By the Banach Fixed Point Theorem (14) has, for these λ's, a unique solution $\hat{\mathbf{u}}(\lambda)$. This way a function $\hat{\mathbf{u}}$ is defined on the interval (ω_0, ∞) which eventually is continuous[12]. We see that (14) can be written in terms of semigroups

$$\hat{\mathbf{u}}(\lambda) = \hat{\mathcal{G}}(\lambda) \left(\hat{\mathbf{f}}_0(\lambda) - \hat{L}(\lambda)\hat{\mathbf{u}} \right)$$

We have set, for convenience, $L = dK_0/dt$. Since $\hat{T}(\lambda) = -\hat{\mathcal{G}}(\lambda)\hat{L}(\lambda)$ is a contraction for $\lambda > \omega_0$, we know[12] that $\mathcal{I} - \hat{T}(\lambda)$ is invertible and the following estimate holds

$$\left\| \left(\mathcal{I} - \hat{T}(\lambda) \right)^{-1} \right\| \leqslant \frac{1}{1 - \left\| \hat{T}(\lambda) \right\|} < 2 \qquad (15)$$

Thus we take that

$$\hat{\mathbf{u}}(\lambda) = \left(\mathcal{I} - \hat{T}(\lambda) \right)^{-1} \hat{\mathcal{G}}(\lambda)\hat{\mathbf{f}}_0(\lambda) \qquad (16)$$

Furthermore, we know that the sequence

$$\hat{\mathbf{u}}_n(\lambda) = \hat{\mathcal{G}}(\lambda) \left(\hat{\mathbf{f}}_0(\lambda) - \hat{L}(\lambda)\hat{\mathbf{u}}_{n-1}(\lambda) \right), \qquad \hat{\mathbf{u}}_0 \equiv \mathbf{0}$$

converges to the value $\hat{\mathbf{u}}(\lambda)$. In other words, $(\hat{\mathbf{u}}_n)$ converges pointwise to $\hat{\mathbf{u}}$ on (ω_0, ∞). From the above recursive formula, one easily computes

$$\hat{\mathbf{u}}_n(\lambda) = \left(\mathcal{I} - \hat{T}(\lambda) \right)^{-1} \left(\mathcal{I} - \hat{T}(\lambda)^n \right) \hat{\mathcal{G}}(\lambda)\hat{\mathbf{f}}_0(\lambda) \qquad (17)$$

Theorem 4.2. *Let* $\omega \geqslant \omega_0$ *and suppose that* $\mathbf{f} \in L_\omega^\infty(H)$ *is continuous with* $\mathbf{f}(t) \in A[D(\mathcal{M})]$. *Then, in* $L_\omega^\infty(H)$, *the fixed-point problem*

$$\mathbf{u} = \mathcal{G} * (\mathbf{f}_0 - L * \mathbf{u}) \tag{18}$$

has a unique solution $\mathbf{u}$ *which is continuously differentiable. This* $\mathbf{u}$ *is also the (classical) solution of the problem (8) and this problem is well posed. Finally, the following* a priori *estimate holds*

$$\| \mathbf{u} \|_{\omega,\infty} \leqslant \frac{2}{a} \| \mathbf{f} \|_{\omega,\infty}$$

Proof. I. EXISTENCE: It is true that $\hat{\mathcal{G}}\hat{\mathbf{f}}_0 \in W_\omega(H)$. Using this fact, the estimate (15) and Theorem 2.1(b), we have from (16), on the one hand, that $\hat{\mathbf{u}} \in W_\omega(H)$ (this means that $\hat{\mathbf{u}}$ is *really* a Laplace transform) and from (17), on the other hand, that $(\hat{\mathbf{u}}_n)$ is a Cauchy sequence in $W_\omega(H)$. So it converges in norm and its limit is $\hat{\mathbf{u}}$. After that, we may consider (14) as an *identity* in $W_\omega(H)$. By inverting the Laplace transform, we obtain (18). Since the Laplace transform is an isometry (and thus a bi-continuous function), the sequence

$$\mathbf{u}_n = \mathcal{G} * (\mathbf{f}_0 - L * \mathbf{u}_{n-1}), \qquad \mathbf{u}_0 \equiv 0 \tag{19}$$

converges in $L_\omega^\infty(H)$ to a function $\mathbf{u}$, the inverse of $\hat{\mathbf{u}}$ with respect to the Laplace transform. This $\mathbf{u}$ is a solution of (18). By induction and standard semigroup theory[9] one can show that $\mathbf{u}_n$ is continuously differentiable and that $\mathbf{u}_n(t) \in D(\mathcal{M})$. Moreover, differentiating the relation (18) we obtain (8).

II. UNIQUENESS: Suppose $\mathbf{f} \equiv 0$. Then, by (16), $\hat{\mathbf{u}} \equiv 0$ is the unique solution of (14). The uniqueness follows from the injectivity of the Laplace transform.

III. STABILITY: Thanks to (16), the problem (14) is stable with respect to $\hat{\mathbf{f}}$. Using the continuity of of the inverse Laplace transform, it is clear that (18) is stable with respect to $\mathbf{f}$.

IV. ESTIMATE FOR THE SOLUTION: It follows immediately from (16), taking into account (7), (15) and the fact that the Laplace transform is an isometry.

Remark. In the proof of Theorem 4.2 it is evident that $\mathbf{u}_n$ is the solution of the Abstract Cauchy Problem

$$\frac{d\mathbf{u}_n}{dt} = \mathcal{M}_0\mathbf{u}_n + (\mathbf{f}_0 - L * \mathbf{u}_{n-1}), \qquad \mathbf{u}(0) = 0 \tag{20}$$

This means that the iterative scheme (20) converges to the solution of the original problem. This scheme can be further refined in order to be used for a numerical treatment of the problem, which avoids inversions of Laplace transforms.

Acknowledgements

This paper is part of one of the authors (A.D. Ioannidis) PhD thesis, that is financially supported by the IRAKLEITOS Research Scholarship.

References

1. J.D. Jackson, *Classical Electrodynamics*, 3rd ed. John Willey & Sons, New York (1999).
2. M. Gustafsson, *Time domain theory for the macroscopic Maxwell equations (technical report)*, Lund Institut of Technology, LUTEDX/(TEAT-7062)/1–24 (1997).
3. D.V. Widder, *The Laplace Transform*, Princeton University Press, Princeton (1946).
4. B. Bäumer and F. Neubrander, *Laplace transform methods for evolution equations*, Conferenze di Seminario di Matematica dell' Univerità di Bari **259**, 27–60 (1995).
5. J. Peng and S.-K. Chung, Proc. Amer. Math. Soc. **126**, 2407–2416 (1998)
6. A. Karlsson and G. Kristensson, J. Electro. Waves Applic. **6**, 537–551 (1992)
7. I.V. Lindell, A.H. Shivola, S.A. Tertyakov and A.J. Viitanen, *Electromagnetic Waves in Chiral and Bi-Isotropic Media*, Artech House, Boston (1994)
8. R. Dautray and J.-L. Lions, *Mathematical Analysis and Numerical Methods for Science and Technology, Vol. 5: Evolution Problems I*, Springer, Berlin (1992).
9. J.A. Goldstein, *Semigroups of Linear Operator and Applications*, Oxford University Press, Oxford (1985)
10. J. Sanchez-Hubert, Bolletino U.M.I. **16**, 857–875 (1979)
11. D.J. Frantzeskakis, A. Ioannidis, G.F. Roach, I.G. Stratis and A.N. Yannacopoulos, Appl. Anal. **82**, 839–856 (2003)
12. E. Zeidler, *Nonlinear Functional Analysis and its Application I: Fixed-Point Theorems*, Springer, New York (1986)

GLOBAL EXISTENCE AND BLOW-UP OF SOLUTIONS FOR A CLASS OF NONLOCAL PROBLEMS WITH NONLINEAR DIFFUSION *

N.I. KAVALLARIS

Department of Mathematics, Faculty of Applied Sciences, National Technical University of Athens, Zografou Campus, 157 80 Athens, Greece
E-mail: nkaval@math.ntua.gr

In this work, the behaviour of solutions for the Dirichlet problem of the nonlocal equation $u_t = \Delta(\kappa(u)) + \lambda f(u)/(\int_\Omega f(u)dx)^p$, $\Omega \subset \mathbb{R}^N$, $N = 1, 2$, is studied, mainly for the case where $f(s) = e^{\kappa(s)}$. More precisely, the interplay of exponent p of the nonlocal term and spatial dimension N, is investigated concerning the existence and nonexistence of solutions of the associated steady-state problem as well as the global existence and finite-time blow-up of the time-dependent solutions $u(x,t)$. The asymptotic stability of the steady-state solutions is also studied.

* An extended version of this work will appear in a Journal.

1. Introduction

In this paper, existence and nonexistence results are obtained for solutions of initial-boundary value problems to the nonlocal parabolic equation

$$u_t - \Delta(\kappa(u)) = \frac{\lambda f(u)}{\left(\displaystyle\int_\Omega f(u)dx\right)^p}, \quad x \in \Omega, \quad t > 0. \tag{1.1}$$

Here $f(u)$ is a Lipschitz continuous, positive, increasing function and $\kappa(u)$ is imposed to be a positive function with $\kappa \in C^2(\mathbb{R}_+)$. Ω is a smooth bounded subset of $\mathbb{R}^N$ and λ, p are positive parameters. Also for simplicity, it is assumed that $u(x,0) = \psi(x)$ is continuous with $\psi(x) = 0$, $x \in \partial\Omega$ and $\psi(x) \geq 0$, $x \in \Omega$.

Our original motivation for studying such problems comes from the plasma ohmic heating process. The plasma is an electrical conductor and so it could be heated by passing a current through it. This is called ohmic heating and it is the same kind of heating that occurs in thermistors. One naturally encounters, for one dimension, an equation of the form

$$u_t = (u^4)_{xx} + \frac{\lambda\, g(u)}{\left(\displaystyle\int_{-1}^{1} g(u)\, dx\right)^2}, \quad -1 < x < 1,\ t > 0, \tag{1.2}$$

where $u = u(x,t)$ stands for the dimensionless temperature of plasma, while $g(u)$ represents the temperature-dependent electrical resistance of plasma, see Refs. 6, 7. The nonlinear diffusion term $(u^4)_{xx}$ comes by assuming the Stefan-Boltzman law for emission of thermal radiation. However, it is possible that some more complicated expression, depending on the nature of heat transport, could replace $(u^4)_{xx}$. Moreover, the nonlocal term $\lambda\, g(u)/(\int_{-1}^{1} g(u)\, dx)^2$ arises due to the ohmic heating nature of the process. For more details concerning the derivation of mathematical model (1.2), see Ref. 6 and the references therein.

The key for the study of the asymptotic behaviour of the solutions to equation (1.1) is the study of the associated steady-state equation

$$-\Delta(\kappa(w)) = \frac{\lambda f(w)}{\left(\displaystyle\int_{\Omega} f(w)\, dx\right)^p}, \quad x \in \Omega. \tag{1.3}$$

The study of the asymptotic behaviour of solutions to equation (1.1), when f is a decreasing function can be done, since a maximum principle holds, by using comparison techniques, [6,7]. However, when f is an increasing function, comparison techniques fail, because there is not valid a maximum principle for equation (1.1). In this case, the asymptotic behaviour can be studied with a dynamical-system approach, [2]. We will place emphasis on cases where the spatial dimension N is 1 or 2, the nonlinearity $f(u) = e^{\kappa(u)}$ and the boundary conditions are of the Dirichlet type. Following a method due to M. Fila, Refs 3, 4 and some ideas of G. Lieberman, Ref. 5, we extend some of the results of Refs 6, 7, when $f(u)$ is now an increasing function and $(u^4)_{xx}$ is replaced by the more complicated nonlinear diffusion term $(\kappa(u))_{xx}$. Simultaneously, these results are extended to the case of spatial dimension $N = 2$.

2. Steady-State Problem

The corresponding steady-state problem with Dirichlet boundary conditions is

$$\Delta(\kappa(w)) + \frac{\lambda f(w)}{\left(\int_\Omega f(w)dx\right)^p} = 0, \quad x \in \Omega, \ \Omega \subset \mathbb{R}^N, \tag{2.1a}$$

$$w = 0, \quad x \in \partial\Omega. \tag{2.1b}$$

For the one-dimensional case the following existence result can be proved .

Prop 2.1. Let $N = 1$, $\Omega = (-1,1)$ and $p = 2$. Then problem (2.1), has a unique solution for every $\lambda > 0$.

Focusing now on the case where $f(s) = e^{\kappa(s)}$ we have.

Prop 2.2. Let $N = 1$, $f(s) = e^{\kappa(s)}$, $\kappa(0) = 0$ and $\Omega = (-1,1)$.

(a) If $p \geq 1$ then problem (2.1) has a unique solution for every $\lambda > 0$.
(b) If $0 < p < 1$ then there exists $\lambda^* > 0$ such that problem (2.1) has

    (i) at least two solutions for $\lambda < \lambda^*$,
    (ii) a unique solution for $\lambda = \lambda^*$,
    (iii) and no solution for $\lambda > \lambda^*$.

Prop 2.3. Let $N = 2$, $f(s) = e^{\kappa(s)}$, $\kappa(0) = 0$ and $\Omega = B(0,1) = \{x \in \mathbb{R}^2 : |x| \leq 1\}$.

(a) If $p > 1$, then problem (2.1) has a unique solution for every $\lambda > 0$.
(b) If $p = 1$, then problem (2.1) has a unique solution for $0 < \lambda < 8\pi$ and no solution for $\lambda \geq 8\pi$.
(c) If $0 < p < 1$, then there exists λ^* such that problem (2.1) has

    (i) at least two solutions for $0 < \lambda < \lambda^*$,
    (ii) a unique solution for $\lambda = \lambda^*$,
    (iii) no solution for $\lambda > \lambda^*$.

3. Global Existence and Asymptotic Stability

For the one-dimensional case and for functions $f(s)$ that are bounded away from 0, it is proved using maximum principle arguments, see Ref. 1, the following global-existence result.

Prop 3.1. Let $f(s) \geq c > 0$ and $p \geq 1$ then for every positive bounded function $\psi(x)$ and $\lambda > 0$ problem

$$u_t = (\kappa(u))_{xx} + \frac{\lambda f(u)}{\left(\int_\Omega f(u)dx\right)^p}, \quad x \in \Omega \subset \mathbb{R}, \quad t > 0, \tag{3.1a}$$

$$u(x,t) = 0, \quad x \in \partial\Omega, \quad t > 0, \tag{3.1b}$$

$$u(x,0) = \psi(x), \quad x \in \Omega, \tag{3.1c}$$

where Ω is a bounded subset of $\mathbb{R}$, has a (unique) uniformly bounded, classical solution in $Q_\infty = \bar{\Omega} \times [0, \infty)$.

For the two-dimensional case a result analogous to Proposition 3.1, is valid but only for $f(s) = e^{\kappa(s)}$. This restriction is due to the fact that a Lyapunov functional for this problem can be constructed only if $f(s) = e^{\kappa(s)}$.

Indeed, for spatial dimensions $N = 1, 2$ our problem with Dirichlet boundary conditions generates a local semiflow in $L^\infty(\Omega) \cap H_0^1(\Omega)$ defined by

$$S(t)\psi(\cdot) = u(\cdot, t; \psi(x)) \quad \text{for} \quad t > 0 \text{ and } \psi \in L^\infty(\Omega) \cap H_0^1(\Omega),$$

with the corresponding ω–limit set defined as

$$\omega(\psi) := \{ w \in L^\infty(\Omega) : \text{ there exists a sequence } (t_n)_{n \in \mathbb{N}}, \ t_n \to \infty \text{ such that }$$
$$S(t_n)\psi \to w \text{ in } C^1(\bar{\Omega}) \cap C^2(\Omega) \}$$

in $C^1(\bar{\Omega}) \cap C^2(\Omega)$. For $p \neq 1$, this local semiflow has a Lyapunov functional given by

$$J[v](t) = \frac{1}{2} \int_\Omega |\nabla\kappa(v)|^2 dx + \frac{\lambda}{p-1} \left(\int_\Omega e^{\kappa(v)}dx\right)^{1-p}$$

and the semiflow is gradient-like in the sence that for any $t \in [0, T_{max})$

$$\int_0^t \int_\Omega \kappa'(u)u_t{}^2 dx\, ds + J[u](t) = J[\psi]. \tag{3.2}$$

Using both the Gilbarg-Trudinger's and Young's inequality we can prove the following existence result.

Prop 3.2. Let $N = 2$ and $p > 1$ then problem

$$u_t = \Delta(\kappa(u)) + \frac{\lambda e^{\kappa(u)}}{\left(\int_\Omega e^{\kappa(u)}dx\right)^p}, \quad x \in \Omega, \quad t > 0, \tag{3.3a}$$

$$u(x,t) = 0, \quad x \in \partial\Omega, \quad t > 0, \tag{3.3b}$$

$$u(x,0) = \psi(x), \quad x \in \Omega, \tag{3.3c}$$

where Ω is a smooth bounded subset of $\mathbb{R}^2$ and $\kappa(0) = 0$ has a (unique) uniformly bounded solution in $Q_\infty = \bar{\Omega} \times [0, \infty)$, for every $\lambda > 0$, provided that $\kappa(\psi) \in H_0^1(\Omega)$.

Some stability results, when $f(s) = e^{\kappa(s)}$, can proved using the Lyapunov functional constructed above.

Prop 3.3. Let $f(s) = e^{\kappa(s)}$, $p \geq 1$ and $\Omega = (-1,1)$ then the solution $u(x,t)$ of problem (3.1) with $0 \leq u(x,0) = \psi(x) \in L^\infty(\Omega) \cap H_0^1(\Omega)$, converges in $C^1(\bar{\Omega}) \cap C^2(\Omega)$ to the unique solution $w(x)$ of problem (2.1), for every $\lambda > 0$.

Prop 3.4. Let $f(s) = e^{\kappa(s)}$ with $\kappa(0) = 0$.

1) If $N = 2$, $p > 1$ and $\Omega = B(0,1)$ then the solution $u(r,t)$ to (3.3), where $0 < u(r,0) = \psi(r) \in L^\infty(\Omega) \cap H_0^1(\Omega)$ is radial symmetric, converges in $C^1(\bar{\Omega}) \cap C^2(\Omega)$ to the unique solution $w(r)$ of (2.1) for every $\lambda > 0$.

2) If $N = 2$, $p = 1$ and $\Omega = B(0,1)$, then the solution $u(r,t)$ of (3.3), where $0 < u(r,0) = \psi(r) \in L^\infty(\Omega) \cap H_0^1(\Omega)$ is radial symmetric, converges in $C^1(\bar{\Omega}) \cap C^2(\Omega)$ to the unique solution $w(r)$ of (2.1) for every $0 < \lambda < 8\pi$.

4. Blow-up for the Exponential and $0 < p < 1$

In this section we investigate under which circumstances the solutions to problem

$$u_t = \Delta(\kappa(u)) + \frac{\lambda e^{\kappa(u)}}{\left(\displaystyle\int_\Omega e^{\kappa(u)} dx\right)^p}, \quad x \in \Omega, \quad t > 0, \tag{4.4a}$$

$$u(x,t) = 0, \quad x \in \partial\Omega, \quad t > 0, \tag{4.4b}$$

$$u(x,0) = \psi(x), \quad x \in \Omega, \tag{4.4c}$$

blow up in finite time, when $0 < p < 1$. For this investigation we use energy methods in contrast to comparison techniques used in the case of a decreasing function $f(s)$, [6,7].

By using a method due to M.Fila Refs. 3, 4, we first prove that if u is a global-in-time solution to (4.4), then $\kappa(u)$ should be uniformly $H_0^1(\Omega)$-bounded and afterwards we exclude the possibility to exist global-in-time solutions for $\lambda > \lambda^*$. In order to prove the first result we need a sequence of lemmas.

Lemma 1. *Let u be a global-in-time solution to problem (4.4), then there exists a constant $\nu = \nu(\psi, b)$ such that*

$$\int_\Omega K(u)dx \leq \nu \quad \text{for any} \quad t \geq 0,$$

where $K(s) = \int_0^s \kappa(s)ds$, provided that $\kappa(s) \geq C(K(s))^b$, $s \geq 0$, for some positive constants $C, b > 0$ and $\int_0^\infty \kappa(s)ds = \infty$.

Lemma 1 is essential in proving that the solution u to (4.4) blows up for "big enough" initial data.

Prop 4.1. *Let $\kappa(u) = u^m$, $m > 1$, then for every $\psi > 0$ with $\kappa(\psi) = \psi^m \in H_0^1(\Omega)$ there exists a μ^* such that for every $\mu > \mu^*$ the solution u to (4.4) with $u(x, 0) = \mu\psi(x)$ blows up in finite time.*

Lemma 2. *Let $\kappa(s)$ satisfy the hypotheses of Lemma 1. If $\int_\Omega |\nabla \kappa(u)|^2 dx \to \infty$ as $t \to t_{max}$, where t_{max} is the maximal existence time for the solution u to problem (4.4), then $t_{max} < \infty$.*

Lemma 3. *Let $u(x, t; \psi)$ be a global-in-time solution to problem (4.4) with $\omega(\psi) \neq \emptyset$. If $w \in \omega(\psi)$ is a steady–state solution to problem (4.4) then $\int_\Omega |\nabla \kappa(w)|^2 dx \leq M = M(\psi)$.*

Now, we introduce a functional $\Pi(t) \equiv \Pi[u](t) = \int_\Omega e^{\kappa(u(x,t))} dx$, which will be used to exclude the possibility

$$\lim_{t\to\infty} \sup \int_\Omega |\nabla \kappa(u(x,t))|^2 dx = \infty \quad \text{and} \quad \lim_{t\to\infty} \inf \int_\Omega |\nabla \kappa(u(x,t))|^2 dx < \infty. \tag{4.5}$$

The functional $\Pi(t)$ satisfies a result analogous to Th. 3.1 existing in [5].

Lemma 4. *Let $\Omega = (-1, 1)$ and $\kappa(s) \to \infty$ as $s \to \infty$. If $\tau > \sigma$ and β, B, γ are positive constants with $B > \gamma$, satisfying*

$$\Pi(\tau) = B^2, \quad \Pi(\sigma) = \gamma^2, \quad \Pi(t) \leq B^2 \quad \text{for} \quad t \in [\sigma, \tau], \tag{4.6}$$

$$H(\sigma) = \int_\Omega |\nabla \kappa(u(x, \sigma))|^2 dx \leq \beta^2, \tag{4.7}$$

then there exists a positive constant $\delta = \delta(B^2 - \gamma^2, \lambda, p)$ such that $\tau - \sigma \geq \delta$.

Lemma 5. *Let $\Omega = (-1, 1)$ and $\kappa(s) \to \infty$ as $s \to \infty$. If $u(x, t; \psi)$ is a global-in-time solution to problem (4.4) with*

$$\lim_{t \to \infty} \inf \|\kappa(u(\cdot, t))\|_{H_0^1(\Omega)} < +\infty \quad \text{and} \quad \lim_{t \to \infty} \sup \|\kappa(u(\cdot, t))\|_{H_0^1(\Omega)} = +\infty,$$
(4.8)

then for every big enough constant B there exists a steady–state solution $w \in \omega(\psi)$ such that $\Pi[w] = B^2$.

Now it can be proved that the global-in-time solutions of (4.4) are uniformly bounded with respect to $t > 0$. More precisely there holds.

Prop 4.2. Let $\Omega = (-1, 1)$, $\psi \in H_0^1(\Omega)$ and $\kappa(s)$ be a function such that Lemma 4 and Lemma 5 are valid. If $u(x, t; \psi)$ is a global-in-time solution to (4.4) then

 (i) $\sup_{t \geq 0} \|\kappa(u(\cdot, t))\|_{H_0^1(\Omega)} < \infty$,
 (ii) $\sup_{t \geq \tau} \|u(\cdot, t)\|_\infty < \infty$ for any $\tau > 0$.

For the two-dimensional case, a result analogous to Proposition 4.2 can be proved. In this case the proof of Lemma 4 should be modified and the Gilbarg–Trudinger's inequality should be used instead of Sobolev's one, [5].

Lemma 6. *Let $N = 2$, Ω be a bounded subset of $\mathbb{R}^2$ and $\kappa(s) \to \infty$ as $s \to \infty$. If $\tau > \sigma$ and β, B, γ are positive constants with $B > \gamma$, satisfying $\Pi(\tau) = B^2$, $\Pi(\sigma) = \gamma^2$, $\Pi(t) \leq B^2$ for $t \in [\sigma, \tau]$ and $H(\sigma) = \int_\Omega |\nabla \kappa(u(x, \sigma))|^2 dx \leq \beta^2$, then there exists a positive constant $\delta = \delta(B^2 - \gamma^2, \lambda, p)$ such that $\tau - \sigma \geq \delta$.*

Lemma 7. *Let $N = 2$, Ω be a bounded subset of $\mathbb{R}^2$ and $\kappa(s) \to \infty$ as $s \to \infty$. If $u(x, t; \psi)$ is a global-in-time solution of (4.4) with*

$$\lim_{t \to \infty} \inf \|\kappa(u(\cdot, t))\|_{H_0^1(\Omega)} < +\infty \quad \text{and} \quad \lim_{t \to \infty} \sup \|\kappa(u(\cdot, t))\|_{H_0^1(\Omega)} = +\infty,$$

then for every arbitrary big constant B there exists a steady–state solution $w \in \omega(\psi)$ satisfying $\Pi[w] = B^2$.

An immediate consequence of the two previous lemmas is the result.

Prop 4.3. Let Ω be a bounded subset of $\mathbb{R}^2$, $\psi \in H_0^1(\Omega)$ and $\kappa(s)$ be a function such that Lemma 6 and Lemma 7 are valid. If $u(x, t; \psi)$ is a global-in-time solution of (4.4) then

 (i) $\sup_{t \geq 0} \|\kappa(u(\cdot, t))\|_{H_0^1(\Omega)} < \infty$,

(ii) $\sup\limits_{t \geq \tau} \|u(\cdot, t)\|_\infty < \infty$ for any $\tau > 0$.

The main result of this section is the following.

Prop 4.4. Let $N = 1$, $\Omega = (-1, 1)$ or $N = 2$, $\Omega = B(0, 1)$ and $\psi(x) \in H_0^1(\Omega)$. If $\lambda > \lambda^*$ and $\kappa(s)$ is a function such that all the previous results are valid, then the solution $u(x, t; \psi)$ of (4.4) blows up in finite time.

Proof : If $u(x, t; \psi)$ is a global-in-time solution of problem (4.4) then according to Propositions 4.2 and 4.3 there holds $\sup\limits_{t \geq 0} \|\kappa(u(\cdot, t))\|_{H_0^1(\Omega)} < \infty$ and $\sup\limits_{t \geq \tau} \|u(\cdot, t)\|_\infty < \infty$ for $\tau > 0$. The latter implies the existence of a steady–state solution $w \in \omega(\psi)$ of (4.4), which is a contradiction since for $\lambda > \lambda^*$ there are no steady states. Hence there exists a sequence $(t_n)_{n \in \mathbb{N}}$ with $t_n \to t_{max}$ as $n \to \infty$ such that $\|\kappa(u(\cdot, t_n))\|_{H_0^1(\Omega)} \to \infty$ and $\|u(\cdot, t_n)\|_\infty \to \infty$ as $n \to \infty$. According to Lemma 1 we have that $t_{max} < \infty$, so the solution u blows up in finite time. This completes the proof.

References

1. D. G. Aronson, On the Green's function for second order parabolic equations with discontinuous coefficients, Bull. Amer. Math. Soc. 69 (1963), 841–847.
2. J. W. Bebernes and A. A. Lacey, Global existence and finite–time blow–up for a class of nonlocal parabolic problems, Adv. Diff. Eqns., 2 (1997), 927–953.
3. M. Fila, Boundeness of global solutions of nonlinear diffusion equations, J. Diff. Eqns. 98 (1992), 226–240.
4. M. Fila, Boundedness of global solutions of nonlocal parabolic equations. Proceedings of the Second World Congress of Nonlinear Analysts, Part 2 (Athens, 1996). Nonlinear Anal. 30 (1997), no. 2, 877–885.
5. G. M. Lieberman, Study of global solutions of parabolic equations via a priori estimates. Part II. Porous medium equations, Comm. Appl. Noul. Anal. 1, No. 3, (1994), 93–115.
6. D. E. Tzanetis and P. M. Vlamos, A nonlocal problem modelling ohmic heating with variable thermal conductivity, Nonlinear Anal. Real World Appl. 2 (2001), no. 4, 443–454.
7. D. E. Tzanetis and P. M. Vlamos, Some interesting special cases of a nonlocal problem modelling Ohmic heating with variable thermal conductivity, Proc. Edinb. Math. Soc. (2) 44 (2001), no. 3, 585–595.

HYSTERESIS MODELING AND APPLICATIONS

APHRODITE KTENA and DIMITRIOS I. FOTIADIS

Unit of Medical Technology and Intelligent Information Systems
Department of Computer Science, University of Ioannina
GR45110, Ioannina, Greece

CHRISTOS V. MASSALAS

Department of Materials Science, University of Ioannina,
GR45110, Ioannina, Greece

Hysteresis may be a desired of an undesired property of a material. In the first case, hysteresis modeling can be used as a predicting tool; in the second, as a compensator. Phenomenological models following the Preisach formalism are good candidates for both cases thanks to their efficient algorithms and ability to adjust to the material being modeled. In this work, we present the building blocks of the Preisach models and their application in ferromagnetic and magnetostrictive materials and shape memory alloys. The adaptability of the formalism is discussed for antiferromagnetically coupled magnetic recording media.

1. Introduction

1.1. *Hysteresis: blessing or curse*

The term of hysteresis, in the context discussed here, is used to describe those situations in which the response is lagging the excitation as a nonlinear function of the current and past excitations. It can be observed in a large number of input-output pairs in several systems and materials in nature, e.g. the magnetization, the magnetoimpedance or the strain may lag the applied magnetic field in magnetic and magnetostrictive materials; the strain may lag the stress in elastic materials, or the temperature in shape memory alloys; examples from systems include the turbidity in lakes with respect to the amount of the existing nutrients or the number of existing firms in an economy with respect to the exchange rate.

Hysteresis is the result of a complex network of interactions and competing energies among a material's particles, grains and phases or a system's constituent parts and their interaction with the external stimulus. It is related to energy dissipation, memory properties and metastability. From the stability point of view, there are more than one possible equilibrium states for a given input value. The state that is ultimately chosen by the system, as the one that minimizes its energy, depends on the history of the

314

system, i.e., on previous equilibrium states, hence the memory property. Alternatively, an external stimulus may elicit a partly or totally irreversible response; therefore, the original state cannot be recovered by levying the stimulus but further energy must be supplied in order to "assist" the system in returning back to the original state, hence the energy dissipation.

Hysteresis is a desired effect when we are interested in the stability of information or energy storage, as in the case of data storage media (tapes, disks) and permanent magnet applications [1]. Fig. 1 shows a typical major hysteresis loop of a ferromagnet where the magnetization M is plotted against the applied magnetic field, H. The loop is characteristic of a given material. It is symmetric, $M(H) = -M(-H)$, and the field at which the magnetization becomes zero is known as the coercivity or the coercive field, H_c. When H_c is negligible or zero the material exhibits no hysteresis and all process are reversible. A high coercivity suggests strong hysteresis and therefore higher storage stability, e.g. the information stored in a hard drive cannot be erased or altered due to stray magnetic fields as long as these are lower than H_c.

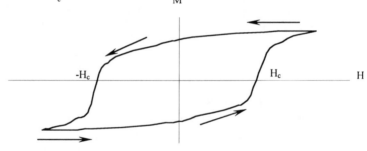

Fig. 1 – Hysteresis major loop of a ferromagnetic material

Hysteresis can be highly undesirable in sensing applications where the nonlinearity of the response adds to the uncertainty of the sensor [2]. Fig. 2 shows a typical hysteresis loop of a magnetostrictive material where the deformation λ is plotted against the applied magnetic field H. Magnetostrictive materials are a particular class of magnetic materials, used mainly as sensors. During the magnetization process, the magnetic dipoles, tending to align themselves with the applied field, apply stresses resulting in an elongation (positive magnetostriction) or shrinking (negative magnetostriction) of the material. Again, the higher the coercivity, the stronger the hysteresis is. This means that for a given elongation λ there are two possible field values causing it, depending on which branch the action is taking place. Therefore, in order to decrease the uncertainty in the control of a sensor, hysteresis must become negligible [2] or additional information is needed..

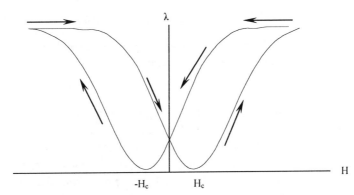

Fig. 2 – Hysteresis major loop of a magnetostrictive material

In either case, the modeling of hysteresis is necessary. Hysteresis models can be used as compensators, as prediction tools or as core models in simulations of a system's behavior. In all cases, they must be implementable by efficient algorithms and tunable to the systems or materials they model. Phenomenological models are, in this sense, better candidates than microscopic models based on the minimization of an energy equation consisting of the most prominent energy terms participating in the phenomenon. The Preisach formalism is probably the most popular class of phenomenological hysteresis models [3]. These models are abstract enough and therefore adjustable to a variety of materials; they account for the interactions in a statistical manner and therefore previous knowledge of the exact mechanism of hysteresis or energy terms involved is not necessary; they are efficient; they can be tuned to the material they model; when Mayergoyz' theorem [4] applies, their accuracy is excellent.

In the following section, we demonstrate how Preisach-type models can be applied to several types of materials and in section 4, we discuss how this type of modeling can be applied to a relatively new type of recording material, the antiferromagnetically coupled (AFC) media, in order to be used in recording simulations.

2. Building a Preisach-type model

The Preisach formalism postulates that hysteresis is the aggregate response of a distribution of elementary hysteresis operators. A hysteresis operator is the locus of the critical inputs at which irreversible switching (from one magnetization state to another, from one phase to another etc.) occurs. The classical Preisach model (CPM) predicts that, for an input *u(t)*, the output *f(t)* can be calculated as:

$$f(t) = \iint_{a \geq b} \rho(a,b)\gamma_{ab}u(t)dadb \tag{1}$$

where γ_{ab} is a local hysteresis operator with switching points at a and b and $\rho(a,b)$ is their probability density function.

2.1. Hysteresis operators γ_{ab}

The classical Preisach model (CPM) [4], introduced about 70 years ago for the modeling of ferromagnetic hysteresis, is based on the relay-type operator shown in Fig. 3a. The output switches between +1 and −1 at the respective upper and lower switching points (fields), (a,b). This pair of variables controls the width of the loop and its offset from the origin, thus incorporating the effect of interactions. The variations of this operator, shown in Figs. 3b and 3c, are mainly used in elastoplasticity. Notice, that the kp-operator (Fig. 3c) allows for a linear transition between the minimum and maximum values, and bi-directional horizontal movement at any point of the ascending or descending curve. These three operators are 1D allowing only for irreversible switching.

The phenomenon of hysteresis in ferromagnetism is, generally, vector in nature. The response to an applied magnetic field has a reversible part, due to small rotations, as well as an irreversible part due to switching. Figs. 3d and 3e depict two 2D operators allowing both rotation and switching. The Stoner-Wohlfarth astroid (Fig. 3d), a well-known model in micromagnetics [5], is described by the equation: $h_x^{2/3} + h_y^{2/3} = 1$, where h_x and h_y are the components of the applied field, normalized to the half-width of the astroid, along the easy and the hard axis of orientation of an anisotropic material. The normalized magnetization vector, **m**, is the tangent to the astroid passing through the tip of the applied field vector, **h**. Switching occurs when the magnetization vector, rotating from position $\mathbf{m}_k$ to a new position $\mathbf{m}_{k+1}$, crosses the astroid from the inside out. The vector operator of Fig.3e is the first order approximation of the asteroid and follows the same switching and rotation mechanism.

2.2. *The probability density function* $\rho(a,b)$

The pair of variables (a,b) defines a half plane, $a \geq b$, known as the Preisach plane. The distribution of a and b is obtained from a probability density function $\rho(a,b)$ defined over the Preisach plane, known as the Preisach or characteristic density. $\rho(a,b)$ is characteristic of the material being modeled and as a consequence the identification of the model consists in determining this density of a and b. In other words, the shape of a material's major hysteresis loop, or any other trajectory inside it, is related to the shape of the distribution, e.g. it is symmetrical in the cases of magnetic and magnetostrictive materials (Figs. 1-2), asymmetrical in the case of shape memory alloys (Fig. 6), and bimodal in AFC media (Fig. 8). It can be either reconstructed through detailed measurements of first order curves or as a weighted sum of bivariate normals whose parameters must be determined [6]. The first method is used in the case of the CPM [4]. It cannot be applied in the case of vector models or when these measurements cannot be carried out. The alternative method is based on the assumption that any pdf can be constructed as a sum of gaussians. It is more general because it needs only a major loop measurement and a curve-fitting procedure to determine the parameters of the pdf.

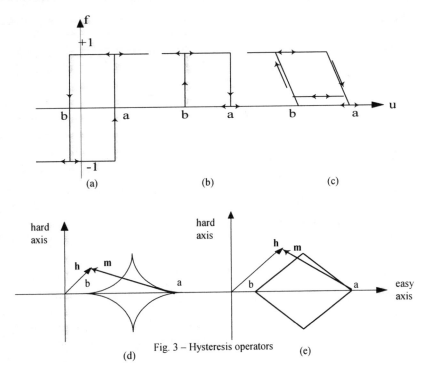

Fig. 3 – Hysteresis operators

318

2.3. *Vector model*

In the vector formulation, the 1D hysteresis operator is replaced by a vector one. An additional gaussian $\rho(\theta)$ is used for the angular dispersion around the material's easy axis [1]:

$$\mathbf{f}(t) = \int_{-\pi/2}^{\pi/2} \rho(\theta)d\theta \iint_{a \geq b} \rho(a,b)\gamma_{ab}\mathbf{u}(t)dadb \qquad (2)$$

2.4. *Examples*

2.4.1 Ferromagnets

In ferromagnets, hysteresis occurs as the material switches from positive to negative magnetization and vice versa. For an applied magnetic field, $H(t)$, the resulting magnetization, $M(t)$, is a function of the applied field as well as of an internal interaction field, which is in turn a function of the magnetization. Hence, the resulting magnetization state contains a positive feedback mechanism leading to hysteresis: $M(t) = M(H(t), M(t))$.

An appropriate hysteresis model may then be built using Eq. (2) along with a vector operator (Fig. 3e) and a bivariate gaussian. Fig. 4 shows the experimental vs. the calculated major loop of a SmFeN permanent magnet [6].

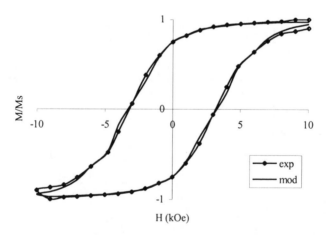

Fig. 4 – Experimental and calculated loop of a permanent magnet

2.4.2 Magnetostrictive materials

Under an externally applied field, the change in the Zeeman energy density, due to the external magnetic field, is counterbalanced by a change in the elastic energy of the bonds [2]. This may result in an increase (positive magnetostriction) or decrease (negative magnetostriction) of the sample length along the direction of the applied field, which because of the changes in microstructure and the ensuing interactions is hysteretic.

In order to generate the "butterfly" loop of Fig. 2, the model of Eq. (1) is used along with the operator of Fig. 3b and a density of the shape shown in Fig. 5.

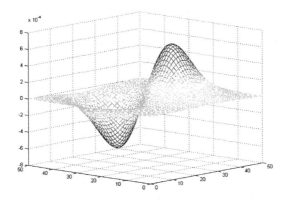

Fig. 5 – $\rho(a,b)$ used in the case of magnetostrictive materials

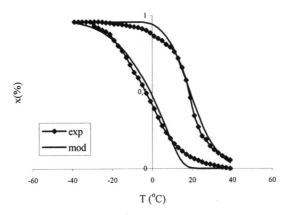

Fig. 6 – Experimental and calculated loop of a shape memory alloy

2.4.3 Shape memory alloys (SMAs)

Hysteresis in SMAs is observed as the material undergoes a phase transformation from the martensitic to the austenitic phase and vice versa. The input variable is temperature, $T(t)$, and the output is strain $x(t)$ [7].

In this case, Eq. (1) is used along with the operator of Fig. 3c and an asymmetrical bivariate gaussian [6]. Fig. 6 shows the calculated and experimental major loop of a NiTinol sample.

3. AFC recording media

In the quest for higher recording areal density, antiferromagnetically coupled (AFC) magnetic recording media have recently emerged as a promising technology in longitudinal recording. In conventional media, a recording density increase can be achieved by lowering the magnetic thickness parameter $M_r t$. M_r is the remanent magnetization after the recording field has been removed and t the film thickness. At the same time, however, decreasing the film thickness, the grain volume, V, decreases and eventually the media becomes thermally unstable as the superparamagnetic regime is approached. AFC media consist of an upper and lower ferromagnetic layer, of thicknesses t_U and t_L, antiferromagnetically coupled through a nonmagnetic layer (fig.7a). It turns out, that, for this structure, the effective magnetic thickness parameter $M_r t$ is the difference between the $M_r t$ values of each layer. As a result, higher recording densities can be achieved while keeping higher thermal stability [8]. The upper layer is a hard magnetic medium with coercivity H_{CU} of a few kOe and the lower layer is soft with coercivity H_{CL} of a few hundred Oe. For large applied fields, the magnetizations of the two layers are parallel to each other but, as the field decreases below the value $H_{EX}-H_{CL}$, the antiferromagnetic coupling mechanism is activated and the lower layer switches (Fig.7b). The lower layer magnetization remains antiparallel to that of the upper layer until H_{CU} is reached and the upper layer switches as well. The effective exchange field H_{EX} is a measure of the AF coupling strength depending, among others, on the thickness of the lower layer [8], i.e., the thinner the lower layer, the stronger the exchange bias and the resulting shift H_{EX}.

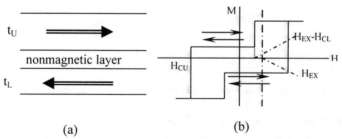

(a)   (b)

Fig. 7 – AFC media (a) layer structure (b) schematic of ideal hysteresis loop

A Preisach model for AFC media, appropriate for magnetic recording simulations [1] must account for the AF coupling effect on the hysteresis operator and/or on the distribution of effective switching fields required by the formalism, without compromising in efficiency.

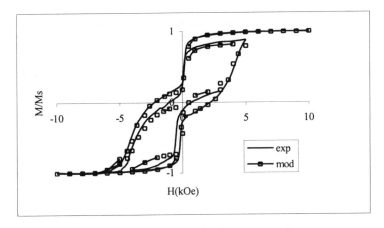

Fig. 8 – Experimental and calculated major and minor loops of an AFC sample

One approach is to use the vector model of Eq. (2) and the operator of Fig. 3e along with a mixture of two normal pdfs, one for each layer:

$$\rho(a,b) = w_L \rho_L(a,b) + w_U \rho_U(a,b), \quad w_U = 1 - w_L \quad (3)$$

where is shifted by $+H_{EX}$ for decreasing and $-H_{EX}$ for increasing input fields H(t). This formulation manages to reproduce the hysteresis characteristics of AFC media as shown in Fig. 8.

4. Conclusions

Hysteresis is undesired in magnetostrictive or magnetic sensors and actuators but it can or must be compensated by modeling. On the other hand, it is highly desired when the stability of information or energy storage is in question and can be accounted for by modeling for more accurate control or simulations in core transformers or magnetic recording. Models based on the Preisach formalism can be used to model hysteretic responses in several types of media regardless of the underlying microstructure or hysteresis mechanism. Depending on the major loop characteristic in question, the dimensionality (1D or 2D) of the model and the appropriate operator (scalar or vector) are first chosen. A bivariate normal pdf or a mixture of normal densities is used to generate the characteristic density and a least-squares fitting algorithm using data from an experimental major loop is used to identify the model for a given material. The results are satisfactory even in the case of AFC media where antiferromagnetic coupling occurs below a certain field at which point the hysteresis curve shows abrupt switching.

References

1. A. Ktena and S. H. Charap, Vector Preisach Modeling and Recording Applications, IEEE Trans. Magn., 29 (6), 3661-3663 (1993).
2. E. Hristoforou, Meas. Sci. & Technol., 14 R15 (2003).
3. S. H. Charap and A. Ktena, Vector Preisach modeling, J. Appl. Phys., 73, 5818-5823 (1993).
4. I. D. Mayergoyz, Mathematical models of hysteresis, Physical Review Letters, 56(15), 1518-1521 (1986).
5. E.C. Stoner and E.P. Wohlfarth, A mechanism of magnetic hysteresis in heterogeneous alloys, Phil. Trans. Roy. Soc., A240, 599-642 (1948).
6. A. Ktena, D.I. Fotiadis, P.D. Spanos, A. Berger and C.V. Massalas, Identification of 1D and 2D Preisach models for ferromagnets and shape memory alloys, Int. J. Eng. Sci., 40(20), 2235-2247 (2002).
7. Z. Bo, D. C. Lagoudas, Thermomechanical modeling of polycrystalline SMAs under cyclic loading, Part IV: modeling of minor hysteresis loops, Intl. J. Eng. Sci., 37, 1205-1249 (1999).
8. E.E. Fullerton, D.T. Margulies, N. Supper, Hoa Do, M. Schabes, A. Berger, and A. Moser, Antiferromagnetically coupled magnetic recording media, IEEE Trans. Mag., 39(2), 639-644 (2003).

MOMENTS' METHOD FOR INVERSE BOUNDARY VALUE PROBLEMS

Y. KURYLEV[*]

1. Introduction

This paper is devoted to the moments' method to solve certain types of inverse boundary problems (see [10], [9], [7], [8] for a more detailed exposition of the method for particular cases and its numerical implementations).There are currently two quite general methods to deal with fully-nonlinear inverse boundary problems:

i. The *complex geometric optics method* makes possible, starting from the Dirichlet-to-Neumann map, to evaluate some functionals of the unknown coefficient in the partial differential equation under consideration. For example,
in the case of a Schrdinger equation in a bounded domain $\Omega \in R^n, n \geq 2$ the Dirichlet-to-Neumann map, Λ_q, is given by

$$\Lambda_q(f) = \partial_n u^f \big|_{\partial\Omega}, \tag{1}$$

where u^f is the solution to the Dirichlet problem,

$$-\Delta u + qu = 0, \quad x \in \Omega, \quad u\big|_{\partial\Omega} = f. \tag{2}$$

It turns out that Λ_q uniquely determines the functionals of q of the form,

$$\Phi_q(\lambda, \mu) = \int_\Omega q(x)\psi(x,\lambda)e^{\mu \cdot x} dx, \tag{3}$$

for any $\lambda, \mu \in C$ with $\lambda^2 = \mu^2 = 0$, where $\psi(\cdot, \lambda)$ is the solution of the Schrodinger equation (2) in the whole R^n with the asymptotics $\psi(x,\lambda) = e^{\lambda \cdot x}(1 + O(|x|^{-1}))$, $x \to \infty$,
see [15], [13]. The knowledge of $\Phi_q(\lambda, \mu), \lambda^2 = \mu^2 = 0$ is then used to determine q due to the asymptotics $\psi(x,\lambda) = e^{\lambda \cdot x}(1 + (O|\lambda|^{-1}))$ valid for $|\lambda| \to \infty$ (see [15], [13] for $n \geq 3$, [12] for $n = 2$).

ii. The *boundary control (BC) method*, in its spectral version, e.g. [2], [6],is based on the possibility to determine, in terms of the boundary spectral data, the Fourier images in the spectral representation associated with the operator, of the eigenfunctions cut onto various domains of influences related to the points on $\partial\Omega$. For example, for the Schrodinger operator with Dirichlet boundary condition,

[*] .Department of Mathematical Sciences, Loughborough University, Loughborough, LE11 3TU, UK

$$A_q u = -\Delta u + qu, \quad u|_{\partial\Omega} = 0. \tag{4}$$

the boundary spectral data consist of all pairs $\{\sigma_k, \partial_n \phi_k |_{\partial\Omega}\}_{k=1}^{\infty}$, where σ_k are the eigenvalues and Φ_k the corresponding eigenfunctions of A_q. Assuming, for simplicity, Ω to be convex, the domains of Influence, $\Omega^{x,t}$ of the form,

$$\Omega^{x,t} = \Omega \cap B_t(x), \quad x \in \partial\Omega, \ t \succ 0 \tag{5}$$

with $B_t(x)$ being a ball of radius t centered in x. Then the boundary spectral data makes possible to find the Fourier coefficients $(\Phi_k^{x,t}, \Phi_l)$, where, for any function f,

$$f^{x,t} = \Xi^{x,t} f,$$

with $\Xi^{x,t}$ being the characteristic function of $\Omega^{x,t}$, $x \in \partial\Omega$, $t \succ 0$ and, eventually, the values $\Phi_k(x)$ for any $x \in \Omega$, k=1,2

In the case of an acoustic operator, $-\rho^{-1}(x)\Delta$, the BC-method makes use of the possibility, e.g. [2], to find the Fourier coefficients of $h^{x,t}$, where h is an arbitrary harmonic function. On the other hand, harmonic functions are the solutions of the equation,

$$A_\rho u := -\rho^{-1}\Delta u = 0, \tag{6}$$

which is the analog of (2) for the acoustic case, independently of a particular choice of ρ. This makes it natural to try to use some spectral ideas originated in the BC-method in order to find functionals of the form, similar to (3)

$$\Phi_\rho(h, g) = \int_\Omega \rho(x) h(x) g(x) dx \tag{7}$$

where h, g are harmonic functions, i.e. null-solutions of the "perturbed" acoustic equation, h, and "unperturbed" Laplace equation, g. Having found $\Phi_\rho(h, g)$, it is then natural to try to extract ρ from these data.

2. The boundary spectral data for the acoustic operator, A_ρ, is the set of eigenpairs $\{\sigma_k, \partial_n \Phi_k |_{\partial\Omega}\}_{k=1}^{\infty}$

$$-\rho^{-1}\Delta \phi_k = \sigma_k \phi_k, \quad \phi_k|_{\partial\Omega} = 0, \quad \int_\Omega \rho \phi_k \phi_k dx = \delta_{kl}.$$

If h is harmonic, then its Fourier coefficients with respect to the eigensystem ϕ_k are readily found in terms of the the boundary spectral data,

$$h = \sum h_k \phi_k, \quad h_k = \int_{\partial\Omega} \rho h \phi_k dx = -\frac{1}{\sigma_k} \int_{\partial\Omega} \partial_n \phi_k |_{\partial\Omega} h |_{\partial\Omega} d A \tag{8}$$

Therefore, we can evaluate all inner products of harmonic functions in terms

of $\{\sigma_k \partial_n \phi_k |_{\partial\Omega}\}$,

$$\int_\Omega \rho h g dx = \sum \frac{1}{\sigma_k^2} \left(\int_{\partial\Omega} \partial_n \phi_k |_{\partial\Omega} h |_{\partial\Omega} dA \right) \tag{9}$$

Concentrate now on the harmonic polynomials, p instead of arbitrary harmonicfunctions with HPm standing for the .nite-dimensional space of harmonic polynomials of degree less than m. The crucial observation is that any monomial, x^α, α being a multiindex, $\alpha = (\alpha 1, \ldots, \alpha n)$ can be represented in the form

$$x^\alpha = \sum p_j^1(x)p_j^2(x), \quad p_j^{1,2} \in HP^{|\alpha|} \tag{10}$$

where the sum in rhs of (9) is .nite. This is clear for $n = 2$, indeed,

$$x_1^{\alpha 1}x_2^{\alpha 2} = \left(\frac{z+\overline{z}}{2}\right)^{\alpha 1}\left(\frac{z-\overline{z}}{2i}\right)^{\alpha 2}, \quad z = x_1 + ix_2 \tag{11}$$

with rhs in (11) having the form (10) since z^l, $\overline{z}^l$ are harmonic polynomials. For $n > 2$, the proof of (10) is more involved using the notion of directional moments, [10]. It is based, roughly speaking, on the introduction of a complex structure on any two-dimensional plane in R^n so that a representation of type (11) is valid for multindeces α with only two components not equal to 0 with further increase of the dimensionality of α by means of some recurrent procedure.

Summarizing above considerations, the boundary spectral data determines all polynomial moments,

$$M_\alpha = \int_\Omega \rho(x)x^\alpha dx \tag{12}$$

Moreover, M_α can be directly evaluated from these data using (9) and the fact that the polynomials $p_j^{1,2}$ in (10) can be algorithmically found for any α, as seen from (11) for $n = 2$ (see [10] for $n > 2$).

It is well-known that the set of all $M\alpha$ uniquely determines ρ, due to the boundedness of Ω. Indeed, (12) may be read as,

$$M_\alpha = i^{|\alpha|}\partial_\xi^\alpha \hat{\rho}\big|_{\xi=0}, \quad \hat{\rho}(\xi) = \int_\Omega e^{i\xi \cdot x}\rho(x)dx \tag{13}$$

As $\hat{\rho}$ is analytic, $\partial_\xi^\alpha \hat{\rho}\big|_{\xi=0}$ determines ρ.

3. Having in mind numerical applications of the above procedure, we should take into account two obstacles:

i. In reality, instead of complete boundary spectral data, we possess only an incomplete set, $\{\sigma_k, \partial_n\phi_k\big|_{\partial\Omega}\}_{k=1}^K$ with $\sigma_k, \partial_n\phi_k\big|_{\partial\Omega}$. contaminated by some measurement's errors;

ii. Determination of $\hat{\rho}(\xi), \xi \in R^n$ from its derivatives, $\partial_\xi^\alpha \hat{\rho}\big|_{\xi=0}$, is an unstable procedure, this last observation being a manifestation of the ill-posedness of inverse problems.

A standard way to deal with the ill-posedness of inverse problems is by means of a "conditional stabilization", see e.g [1] for the pioneering work or [5] for further developments. It is achieved by assuming *a priori* that the unknown coefficient is bounded in some stronger functional space, *B* and using this fact in proving a continuous dependence of this coefficient on the inverse data in a weaker functional space, *X*, for the case when the embedding $B \to X$ is compact. The choice of *B* and *X* may be quite wide, for example, we can take $B = L^\infty(\Omega) \cap H^\delta(\Omega), \delta \succ 1/2$ and $X = L^2(\Omega)$.

This choice will make possible to apply our results to a practically important case when Ω consists of several components with ρ smoothly varying in each of these components and having jump discontinuities across the interfaces between different components. Thus we assume that, for some $c \succ 1$,

$$c^{-1} \leq \rho \leq c, \qquad \|\rho\|_{H^\delta} \leq c. \tag{14}$$

Condition (14) implies that if we cut $\hat{\rho}(\xi)$ onto a large ball of radius *R*, then
$$\|\hat{\rho} - \Xi_{R\hat{\rho}}\|_{L^2(R^n)} \to 0 \text{ when } R \to \infty.$$
This enables us to use a polynomial approximation, $\hat{\rho}_\alpha$ to $\hat{\rho}$ in *BR*,

$$\hat{\rho}_\alpha(\xi) = \Xi_R(\xi) \sum_{|\gamma| \leq N} \frac{1}{\gamma!} \partial_\xi^\gamma \hat{\rho}(0) \xi^\gamma. \tag{15}$$

When $\sigma_k, \partial_n \phi_k\big|_{\partial\Omega}$ are known only approximately and only for $k \leq K$, i.e.

we are given instead $\tilde{\sigma}_k, \tilde{\psi}_k\big|_{\partial\Omega}, k = 1,....,K$ and

$$|\sigma_k - \tilde{\sigma}_k| < \varepsilon, \quad \left\|\partial_n \phi_k\big|_{\partial\Omega} - \tilde{\psi}_k\big|_{\partial\Omega}\right\|L^2(\partial\Omega) < \varepsilon, \tag{16}$$

for some $\varepsilon > 0$, we know $\partial_\xi^\gamma \hat{\rho}(0)$ only approximately with the relative error growing with γ. Thus we should choose the parameter *N* and the radius of the cut-off ball *R* in (15) to optimize the error $\|\hat{\rho} - \hat{\rho}_\alpha\|$. Clearly, the choice of *N, R* will depend on ε, *K* and also δ, *c* in (14). A detailed analysis, see [10], shows that the reconstructed *pa* satisfies

$$\left\| \hat{\rho} - \hat{\rho}_\alpha \right\|_{L^2} \le C \left[\left(\log K \right)^{-\delta} + \varepsilon \left(\log K \right)^{n/2} K^{1/2 + \delta/n} + \varepsilon^2 \left(\log K \right)^{n/2} K^{1 + \delta/n} \right] \tag{17}$$

where the constant C depends on δ, c and the constant c in (14). This estimate shows that, for a given ε, the reconstruction error first decreases with K but then starts to increase. Optimization in (17) leads to the choice of K of the form

$$K = 0 \left(\varepsilon^{-2n/(n+2\delta)} \left| \log \varepsilon \right|^{-n} \right)$$

which gives rise to a logarithmic-type dependence of the reconstruction error on the measurements error, ε assuming that we have an optimal amount of data. As we have seen that approximations to $\partial_\xi^\gamma \hat{\rho}(0)$ can be evaluated using incomplete boundary spectral data with a finite, instead of infinite, sum in (9). Therefore, approximation polynomials $\hat{\rho}_\alpha \left(\xi \right)$ can be constructed to obtain an approximation ρ_α to ρ. In practice, we do not need to evaluate integrals to make the Fourier inversion since, if we cut $\hat{\rho}_\alpha \left(\xi \right)$ onto the qube Q_R rather than the ball B_R, we can use the fact that

$$\int_{-R}^{R} e^{i\xi x} \xi^p d\xi = 2 \left(\frac{1}{i} \frac{d}{dx} \right) \left(\frac{\sin R_x}{x} \right)$$

This algorithm was tested numerically in [9] in the two-dimensional case for several smooth densities ρ. We mainly assumed that the measurement's error is very small so that the accuracy in the reconstruction depends on the number of eigenpairs, K. Due to the logarithmic behavior of the reconstruction error on K, the convergence was painfully slow. To achieve an $L1$-error in the reconstruction of 5 per cent we should employ about 180 eigenpairs. The good side, however, is that, exactly for the same reason of a slow convergence, 16 eigenpairs recover ρ with 12-15 per cent error (see [9] for further details). This brings upon an idea to combine the moments' method with more classical methods based on regularization and optimization techniques so that the moments' method provides a "first approximation" to ρ and regularization techniques further improve the reconstruction (for the modern exposition of regularization methods for inverse problems see e.g. [4]).

4. It is clear from the above that we can apply the moments' method as soon as measured boundary data make possible to evaluate functionals $\Phi_\rho(h, g)$ given by (7). For example, although the Dirichlet-to-Neumann map Λq of form (2) does not supply integral (7), this map for non-zero energy $\lambda \ne 0$,

$$\Lambda_{q,\lambda}(f) = |\partial_n u|_{\partial\Omega}, \quad \text{where} \quad \rho^{-1}\Delta u = -\lambda u, u|_{\partial\Omega} = f \tag{18}$$

makes possible to find $\lambda \int ph(\lambda) g dx$, where $h(x, \lambda)$ is the solution to (18) with $f = h_{\partial\Omega}, h$ being a harmonic function. As $h(\lambda) \to h$ when $\lambda \to \infty$, we can determine $\Phi_\rho(h, g)$ and, therefore, ρ (cf. [14]). Dealing with incomplete data contaminated with error and assuming $\lambda \neq 0$ to be sufficiently small, one should analyse stability of such approach in various functional spaces. To our knowledge, this is not yet done.

Another way to use the moments method is to apply it to the inverse problem for the heat equation,

$$\rho u_t - \Delta u = 0 \text{ in } \Omega \times R_+, \quad u|_{t=0} = 0, \quad u|_{\partial\Omega \times R} = f \tag{19}$$

Inverse data may be the *parabolic response operator*,

$$R^\rho : f \to \partial_n u^f \big|_{\partial\Omega \times R_+}, \tag{20}$$

where u^f is the solution to (19). The fact that R^ρ determines ρ is wellknown, e.g. [5]. In the moments method it is possible to restrict the data to the set of boundary sources of the form

$$f = P|_{\partial\Omega} \cdot H(t), \tag{21}$$

where p is a harmonic polynomial and $H(t)$ is the Heaviside function. What is more, the moments method provides an algorithm to construct an approximation to ρ when data is incomplete and contaminated by errors. For the heat equation this means that we are given only $\partial_n u^\rho(t)$, where $\rho \in HP^m$, and for $0 \le t \le t_0$, or, more precisely, approximate values of the functionals

$$\Phi_\rho(P_1, P_2, t) = \int_\Omega \rho(x) u^{P_1}(x,t) P_2(x) dx, \tag{22}$$

which are related to $\partial_n u^{P1}$ via

$$\Phi_\rho(P1, P2, t) = \int_0^t \int_{\partial\Omega} P2 |\partial\Omega^{\partial_n}(u^{P1} - P2)|\partial\Omega^{dAdt'} \tag{23}$$

(see [7] for more details about these and further calculations). Since, clearly,

$$\Phi_\rho(P1, P2, t) \to \Phi_\rho(P1, P2), \quad \text{when } t \to \infty,$$

and $\Phi_\rho(P1, P2, t)$ is analytic for $\operatorname{Re} t > 0$, functionals $\Phi_\rho(P1, P2, t)$ known for all harmonic $P1, P2$ and $t1 < t < t2$ determine all moments $M\alpha$ of form (12) and, therefore, ρ. In order to apply the algorithm of an approximate reconstruction of ρ from a finite number of approximately found $M\alpha$ and to estimate the reconstruction error, as in sections 2., 3., it remains to find the way to determine $\Phi_\rho(P1, P2)$ from $\Phi_\rho(P1, P2, t)$ and analyze the resulting error. A naive but still numerically viable approach is based on the eigenfunction expansion

$$u^\rho(t) = \rho - \sum \pi_K e^{-\lambda_K t} \Phi_K, \quad \pi_K = \int_\Omega \rho P \Phi_K dx. \tag{24}$$

This implies that

$$\Phi_\rho(P1, P2) - \Phi_\rho(P1, P2, t) = o(e^{-\lambda_1 t}), \tag{25}$$

so that $\Phi_\rho(P1, P2, t_0)$ is itself a good approximation to $\Phi_\rho(P1, P2)$ at least for sufficiently large $t0$. It is, however, possible to achieve a better approximation by using both the exponential convergence of $\Phi_\rho(t)$ and its analyticity in t. Although the actual data $\tilde{\Phi}_\rho(P1, P2, t)$ is not analytical in t due to the measurements errors, we can employ the Carleman method of quasianalytical continuation [3], see also [11], to construct from $\Phi_\rho(P1, P2, t)$ an approximate functional $\Phi_\rho(P1, P2)$ such that

$$\left| \Phi_\rho(P1, P2) - \tilde{\Phi}_\rho(P1, P2) \right| \le C1 \left| \log \varepsilon \right|^{1/3} \exp\left\{ -C2 \left| \log \varepsilon \right|^{1/3} \right\} \|P1\| \|P2\|, \tag{26}$$

where constants $C1, C2$ depend on Ω, $t0$ and the constant c in (14). The parameter ε characterizes the measurements errors,

$$\left| \Phi_\rho(P1, P2, t) - \tilde{\Phi}_\rho(P1, P2, t) \right| \, \left| < \varepsilon \text{ for } \frac{1}{2} t_0 \le t \le t_0. \right. \tag{27}$$

$\tilde{\Phi}_\rho(P1, P2)$ is then defined by

$$\tilde{\Phi}_\rho(P1, P2) = \tag{28}$$

$$\frac{\exp\left\{ -\frac{1}{4} \left| \log \varepsilon \right|^{1/3} \right\}}{2\pi i} \int_{\pi/2}^{3\pi/2} \tilde{\Phi}_\rho(P1, P2, t(\omega)) \exp\left\{ -\frac{1}{4}(\omega - 1)^2 \right\} \frac{i\omega d\psi}{1 - \left| \log \varepsilon \right|^{-1/3} \omega},$$

where

$$\omega = e^{i\psi}, t = (\omega) = (1-\omega)^{-1}\sqrt{2(1+\omega^2)}t_0.$$

The above formulae indicate the possibility of a numerical algorithm to reconstruct ρ in (19) by means of the moments method. The only difference with the algorithm described in section **3.** is the use of formulae (23), with an approximate $\partial_n u^\rho$, and (28) instead of (9) to evaluate $\tilde{\Phi}_\rho(P1,P2)$. This was carried out numerically in [8]. It turned out, however, that the evaluation of the integral in rhs of (28) is very unstable due to a highly-oscillatory factor $\exp(-\log\varepsilon\frac{1}{4}(\omega-1)^2)$. A simpler approximation of $\Phi_\rho(P1,P2)$ used in [8] is based on an extrapolation, due to (24),

$$\Phi_\rho^*(P1,P2,t) \approx \Phi_\rho(P1,P2) - a_1 e^{-\lambda_1 t}, a_1 = \pi_1(P_1)\pi_1(P_2).$$

This leads to the following approximation to $\Phi\rho$,

$$\Phi_\rho^*(P_1,P_2) = \frac{\tilde{\Phi}_\rho(P_1,P_2,t_0) - \exp(-\lambda_1\tau)\tilde{\Phi}_\rho(P_1,P_2,t_0-\tau)}{1-\exp(-\lambda_1\tau)} \qquad (29)$$

where

$$\exp(-\lambda_1\tau) = \frac{\tilde{\Phi}_\rho(t_0) - \tilde{\Phi}_\rho(t_0-\tau)}{\tilde{\Phi}_\rho(t_0-\tau) - \tilde{\Phi}_\rho(t_0-2\tau)}. \qquad (30)$$

Assuming for simplicity that $\varepsilon = 0$ and the goal is to extrapolate $\Phi_\rho(P1,P2,t)$ to get $\Phi_\rho(P1,P2)$, it is shown in [8] that

$$\Phi_\rho(P_1,P_2) - \Phi_\rho^*(P_1,P_2) = 0(a_1\frac{\exp(-\lambda_2 t_0)}{\lambda_1^2\tau^2}) + 0(\frac{\exp(-\lambda_2 t_0)}{\lambda_1\tau}), \quad (31)$$

improving on the error estimate without extrapolation,

$$\Phi_\rho(P_1,P_2) - \Phi_\rho(P_1,P_2,t_0) = 0(a_1\exp(-\lambda_1 t_0)) + o(\exp(-\lambda_2 t_0)), (32)$$

at least for not too small, in comparison with $\Phi_\rho(P1,P2)$, $a1$. Numerical experiments in [8] for both smooth and piecewise smooth ρ's showed that theuse of the extrapolation (29), (30) makes $\Phi^*(P1,P2)$ a better approximation to $\Phi_\rho(P1,P2)$ than $\tilde{\Phi}_\rho(P1,P2,t_0)$ by a factor 6 – 8 for large $\Phi_\rho(P1,P2)$ most important in the reconstruction of ρ. As in [9], reconstruction in [8] was restricted to the two-dimensional case. Moreover, it used only the first 5 harmonic polynomial corresponding to $HP2$ to provide a good first guess to further use other reconstruction techniques like those mentioned in section **3.** We hope that such a combination of a fully-nonlinear moments method and regularization-type techniques would be carried out in the nearest future.

References

[1] Alessandrini G. Stable determination of conductivity by boundary measurements, Appl. Anal., **27** (1988), 153-172.

[2] Belishev M. An approach to multidimensional inverse problems for the wave equation (Russian), Dokl. Akad. Nauk SSSR **287** (1987), N3, 524-527.

[3] Carleman T. Les Fonctions Quasi-Analitiques, Gauthier-Villard, 1926.

[4] Engl H.W., Hanke M., Neubauer A. Regularization of Inverse Problems.Kluwer, 1996. 321pp.

[5] Isakov V. Inverse Problems for Partial Differential Equations. Springer,1998, 284pp.

[6] Katchalov A., Kurylev Y., Lassas M. Inverse Boundary Spectral Problems.Chapman/CRC, 2001, 290pp.

[7] Kawashita M., Kurylev Y., Soga H. Harmonic moments and an inverse problem for the heat equation, SIAM J. Math. Anal., **32** (2000), 522-537.

[8] KurylevY., Mandache N., Peat K. Hausdorff moments in an inverse problem for the heat equation: numerical experiment, Inverse Problems, **19** (2003), 253-264.

[9] Kurylev Y., Peat K. Hausdorff moments in two-dimensional inverse acoustic problems, Inverse Problems, **13** (1997), 1363-1377.

[10] Kurylev Y., Starkov A. Directional moments in the acoustic inverse problem. in: *Inv. Probl. in Wave Propag.*, IMA Vol. Math. Appl., **90** (1997), Springer, 295-323.

[11] Lavrent'ev M.M. Cauchy problem for the Laplace equation (Russian),Izv. Akad. Nauk SSSR, Ser. Matem., **20** (1956), 819-842.

[12] Nachman A. Global uniqueness for a two-dimensional inverse boundary value pronlem, Ann. Math., **143** (1996), 71-96.

[13] Nachman A., Sylvester J., Uhlmann G. An n-dimensional Borg-Levinson theorem. Comm. Math. Phys., **115** (1988), 595-605.

[14] Novikov R. Multidimensional inverse spectral problem for the equation:$\Delta \psi + (v(x) - Eu(x))\psi = 0$ (Russian), Funkt.Anal. Priloz., **22** (1988),11-22.

[15] Sylvester J., Uhlmann G. A global uniqueness theorem for an inverse boundary value problem. Ann. Math., **125** (1987), 153-169.

THE ELECTROMAGNETIC IMAGE METHOD

I.V. LINDELL

Electromagnetics Laboratory
Helsinki University of Technology
Otakaari 5A, 02015 Espoo, Finland
E-mail: ismo.lindell@hut.fi

The operational calculus originally introduced by Heaviside is shown to be applicable to forming image principles for various electromagnetic problems. The image expression is first reduced to the form of a pseudo-differential operator applied to the original source function (for a point source this is the delta function). The final step is its interpretation in terms of computable functions. For complicated operators an approximation can be done at this stage. The Heaviside method replaces previously applied, often heuristic, identification processes by a logical procedure. The method is first applied to one-dimensional transmission-line problems which arise naturally from three-dimensional physical problems. Examples show the use of the method for various electromagnetic boundary-value problems.

1. The Image Principle

The electromagnetic image principle allows one to solve problems involving material structures in a homogeneous space by replacing the structure by an equivalent source which is called the image of the original source [1,2]. The image principle was introduced to electricity by William Thomson (Kelvin) in 1849 [3]:

> The term *Electrical Images*, which will be applied to the imaginary electrical points or groups of electrical points, is suggested by the received language of Optics; and the close analogy of optical images will, it is hoped, be considered as a sufficient justification for the introduction of a new and extremely convenient mode of expression into the Theory of Electricity.

As an example of how to use the principle, let us consider a simple problem of electrostatics. A static charge distribution $\varrho_o(\mathbf{r})$ and a dielectric object of permittivity $\epsilon(\mathbf{r}) = \epsilon_o \epsilon_r(\mathbf{r})$ in air $z > 0$ are both above a layered planar dielectric structure ('ground') $z < 0$, Fig. 1. The poten-

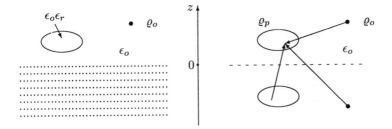

Figure 1. Original charge ϱ_o and a scatterer above a planar structure. The integral equation for the potential in the equivalent polarization charge ϱ_p is obtained with the aid of images of the original charge and the unknown polarization charge. The figure is drawn in simplified form, because, in general, the image of a point source is not a simple point source.

tial scattered by the object can be computed after solving the polarization distribution within the object. This can be done numerically through an integral equation for the equivalent polarization charge density $\varrho_p(\mathbf{r})$. If the Green function corresponding to the dielectric ground is known, there is no problem to build up the integral equation. However, finding such a Green function may be the greatest problem. Now, if we can find the image of a point charge in the presence of the ground, the integral equation can be formed using only the Free-space Green function. In fact, the potential inside the object is now due to four sources: (1) the original source, (2) the image of the original source, (3) the unknown polarization charge inside the object and (4) the image of the unknown polarization charge. Using the connection between the potential and the polarization in the material of the object, an integral equation for the unknown polarization can be formed. Similar arguments apply to problems involving time-harmonic sources with vector sources and dyadic Green functions. It is thus of basic interest to solve the problem of point source in front of different structures in terms of an image source which gives the same reflected field as the structure. Such a source is not unique because there exist nonradiating sources which can be added to the image without changing the reflected field [2].

2. Transmission-line theory

To solve a three-dimensional time-harmonic problem with suppressed time dependence $e^{j\omega t}$, it often helps if one can reduce the dimension from three to one, for example, by making a Fourier transformation in two dimensions. After solving the problem in one dimension, the solution can be transformed back to three dimensions.

2.1. *Transmission-line equations*

The one-dimensional model of the electromagnetic problem is a transmission line. The electromagnetic field vectors are replaced by voltage $U(z)$ and current $I(z)$ functions and the Maxwell equations become the transmission-line equations

$$\partial_z \begin{pmatrix} U(z) \\ I(z) \end{pmatrix} = -j\beta \begin{pmatrix} 0 & Z \\ Y & 0 \end{pmatrix} \begin{pmatrix} U(z) \\ I(z) \end{pmatrix} + \begin{pmatrix} u(z) \\ i(z) \end{pmatrix}. \tag{1}$$

This can be also interpreted as defining the operator giving the source of a voltage and current distribution on the line,

$$\begin{pmatrix} u(z) \\ i(z) \end{pmatrix} = \begin{pmatrix} \partial_z & j\beta Z \\ j\beta Y & \partial_z \end{pmatrix} \begin{pmatrix} U(z) \\ I(z) \end{pmatrix}. \tag{2}$$

Here, $u(z)$ and $i(z)$ denote the distributed voltage and current source functions. β is the (complex) propagation constant and $Z = 1/Y$ the line impedance and they are assumed constant, corresponding to a homogeneous 3D space. The second-order transmission-line equations are obtained after operating

$$\begin{pmatrix} \partial_z & -j\beta Z \\ -j\beta Y & \partial_z \end{pmatrix} \begin{pmatrix} u(z) \\ i(z) \end{pmatrix} = (\partial_z^2 + \beta^2) \begin{pmatrix} U(z) \\ I(z) \end{pmatrix}. \tag{3}$$

In some problems nonsymmetric transmission lines are encountered with different line parameters β, Z for waves traveling to opposite directions along the line. In such cases the matrix on the right-hand side of (1) has also diagonal elements. It can be shown that (2) must then be written as [4]

$$\begin{pmatrix} u(z) \\ i(z) \end{pmatrix} = \begin{pmatrix} \partial_z + j\frac{\beta_+ Z_+ - \beta_- Z_-}{Z_+ + Z_-} & j\frac{(\beta_+ + \beta_-)Z_+ Z_-}{Z_+ + Z_-} \\ j\frac{\beta_+ + \beta_-}{Z_+ + Z_-} & \partial_z + j\frac{\beta_+ Z_- - \beta_- Z_+}{Z_+ + Z_-} \end{pmatrix} \begin{pmatrix} U(z) \\ I(z) \end{pmatrix}, \tag{4}$$

where β_+, Z_+ and β_-, Z_- are the propagation constant and wave impedance for the two waves propagating to the respective positive and negative directions along the z axis.

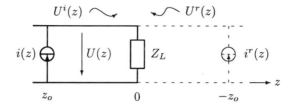

Figure 2. Reflection image i^r of a current source i at $z_o < 0$ lies on the extension of the transmission line at $-z_o$. It gives the voltage wave $U^r(z)$ reflected from the load impedance Z_L.

2.2. Reflection from load impedance

Let us consider a homogeneous transmission line $z < 0$ terminated by a load impedance Z_L at $z = 0$, Fig. 2. Assuming a combined voltage-current point source at $z = z_o$ on the transmission line

$$u^i(z) = U_o\delta(z - z_o), \qquad i^i(z) = I_o\delta(z - z_o), \tag{5}$$

it creates the incident voltage and current functions

$$\begin{pmatrix} U^i(z) \\ I^i(z) \end{pmatrix} = \frac{1}{2} \begin{pmatrix} \mathrm{sgn}(z - z_o) & Z \\ Y & \mathrm{sgn}(z - z_o) \end{pmatrix} \begin{pmatrix} U_o \\ I_o \end{pmatrix} e^{-j\beta|z - z_o|}. \tag{6}$$

Between the source and the termination, $0 < z < z_o$, the wave incident to the load is

$$\begin{pmatrix} U^i(z) \\ I^i(z) \end{pmatrix} = \frac{1}{2} \begin{pmatrix} -1 & Z \\ Y & -1 \end{pmatrix} \begin{pmatrix} U_o \\ I_o \end{pmatrix} e^{j\beta(z - z_o)} = \begin{pmatrix} U^i(0) \\ I^i(0) \end{pmatrix} e^{j\beta z}, \quad 0 < z < z_o. \tag{7}$$

the reflected voltage and current waves can be expressed as

$$\begin{pmatrix} U^r(z) \\ I^r(z) \end{pmatrix} = \begin{pmatrix} R & 0 \\ 0 & -R \end{pmatrix} \begin{pmatrix} U^i(0) \\ I^i(0) \end{pmatrix} e^{-j\beta z} = \begin{pmatrix} R & 0 \\ 0 & -R \end{pmatrix} \begin{pmatrix} U^i(-z) \\ I^i(-z) \end{pmatrix}. \tag{8}$$

The load impedance Z_L may represent a physical structure (an inhomogeneous transmission line, for example) existing in the region $z > 0$. In the general case, the reflection coefficient is a function of the propagation constant, $R = R(\beta)$. When operating on an exponential function we replace $R(\beta)$ by the Heaviside operator $R(-j\partial_z)$, whence the reflected wave becomes

$$\begin{pmatrix} U^r(z) \\ I^r(z) \end{pmatrix} = \begin{pmatrix} R(-j\partial_z) & 0 \\ 0 & -R(-j\partial_z) \end{pmatrix} \begin{pmatrix} U^i(-z) \\ I^i(-z) \end{pmatrix}. \tag{9}$$

If $R(\beta)$ is a polynomial function, $R(-j\partial_z)$ is a differential operator, otherwise it is a pseudo-differential operator. The form of the voltage reflection-coefficient function $R(\beta)$ is obtained from the boundary condition as

$$U^i(0) + U^r(0) = Z_L[I^i(0) + I^r(0)] \quad \Rightarrow \quad R(\beta) = \frac{Z_L(\beta) - Z}{Z_L(\beta) + Z}. \tag{10}$$

The generalization to nonsymmetric lines is [4]

$$R(\beta_-) = \frac{(Z_L(\beta_-) - Z_+)Z_-}{(Z_L(\beta_-) + Z_-)Z_+}, \tag{11}$$

which reduces to (10) for the symmetric line.

2.3. *Image sources*

Let us now replace the load impedance by image sources in the extension $z > 0$ of the homogeneous transmission line giving the same reflected wave (9) in the region $z < 0$, Fig. 2. The source of the reflected voltage $U^r(z)$ can be obtained through the operator in (2) and inserting (9),

$$\begin{pmatrix} u^r(z) \\ i^r(z) \end{pmatrix} = \begin{pmatrix} \partial_z & j\beta Z \\ j\beta Y & \partial_z \end{pmatrix} \begin{pmatrix} U^r(z) \\ I^r(z) \end{pmatrix}$$

$$= \begin{pmatrix} \partial_z & j\beta Z \\ j\beta Y & \partial_z \end{pmatrix} \begin{pmatrix} R(-j\partial_z) & 0 \\ 0 & -R(-j\partial_z) \end{pmatrix} \begin{pmatrix} U^i(-z) \\ I^i(-z) \end{pmatrix}$$

$$= -\begin{pmatrix} R(-j\partial_z) & 0 \\ 0 & -R(-j\partial_z) \end{pmatrix} \begin{pmatrix} -\partial_z & j\beta Z \\ j\beta Y & -\partial_z \end{pmatrix} \begin{pmatrix} U^i(-z) \\ I^i(-z) \end{pmatrix}$$

$$= -\begin{pmatrix} R(-j\partial_z) & 0 \\ 0 & -R(-j\partial_z) \end{pmatrix} \begin{pmatrix} u(-z) \\ i(-z) \end{pmatrix} = -\begin{pmatrix} U_o \\ -I_o \end{pmatrix} R(-j\partial_z)\delta(z + z_o), \tag{12}$$

from which we can identify the image sources as

$$u^r(z) = -R(-j\partial_z)U_o\delta(z + z_o), \quad i^r(z) = R(-j\partial_z)I_o\delta(z + z_o). \tag{13}$$

The expressions (13) form the basis of the image method discussed here. In the simplest case when the reflection coefficient R does not depend on β, the reflection image consists of voltage and current generators at the image point $z = -z_o$ with the respective amplitudes $-RU_o$, RI_o.

3. Time-harmonic planar problems

Let us consider problems involving a planar stratified structure, with material parameters depending on one coordinate z only. In this case, the one-dimensional image theory can be applied by making the Fourier transform in the xy plane, whence the problem becomes analogous to that of a transmission line [2].

3.1. *Transmission-line analogy*

The Fourier-transformed Maxwell equations are

$$\mathbf{u}_z \times \partial_z \mathbf{E} - j\mathbf{K} \times \mathbf{E} = -j\omega\mu\mathbf{H} - \mathbf{M}, \tag{14}$$

$$\mathbf{u}_z \times \partial_z \mathbf{H} - j\mathbf{K} \times \mathbf{H} = j\omega\epsilon\mathbf{E} + \mathbf{J}, \tag{15}$$

where the electric and magnetic fields $\mathbf{E}, \mathbf{H}$ and the electric and magnetic sources $\mathbf{J}, \mathbf{M}$ are functions of z and the two-dimensional Fourier variable $\mathbf{K}$. Vector equations (14), (15) represent a coupled pair of transmission lines defining two sets of voltage and current quantities as

$$U_e = \frac{1}{K^2}\mathbf{K} \cdot \mathbf{E}, \quad I_e = \frac{1}{K^2}\mathbf{u}_z \times \mathbf{K} \cdot \mathbf{H}, \tag{16}$$

$$U_m = \frac{1}{K^2}\mathbf{u}_z \times \mathbf{K} \cdot \mathbf{E}, \quad I_m = -\frac{1}{K^2}\mathbf{K} \cdot \mathbf{H}. \tag{17}$$

The field components transverse to $\mathbf{u}_z$ can now be expressed as

$$\mathbf{E}_t = KU_e + \mathbf{u}_z \times KU_m, \tag{18}$$

$$-\mathbf{u}_z \times \mathbf{H} = KI_e + \mathbf{u}_z \times KI_m. \tag{19}$$

The two transmission lines are uncoupled in homogeneous space and obey the equations

$$\begin{pmatrix} \partial_z & j\beta Z_{e,m} \\ j\beta Y_{e,m} & \partial_z \end{pmatrix} \begin{pmatrix} U_{e,m}(z) \\ I_{e,m}(z) \end{pmatrix} = \begin{pmatrix} u_{e,m}(z) \\ i_{e,m}(z) \end{pmatrix} \tag{20}$$

where the source functions are defined by

$$u_e(z) = -\frac{1}{K^2}\mathbf{u}_z \times \mathbf{K} \cdot \mathbf{M}(z) + \frac{1}{\omega\epsilon}\mathbf{u}_z \cdot \mathbf{J}(z), \quad i_e(z) = -\frac{1}{K^2}\mathbf{K} \cdot \mathbf{J}(z), \tag{21}$$

$$i_m(z) = -\frac{1}{K^2}\mathbf{u}_z \times \mathbf{K} \cdot \mathbf{J}(z) - \frac{1}{\omega\mu}\mathbf{u}_z \cdot \mathbf{M}(z), \quad u_m(z) = \frac{1}{K^2}\mathbf{K} \cdot \mathbf{M}(z). \tag{22}$$

The propagation factor is the same for both lines but the impedance is different:

$$\beta = \sqrt{k^2 - K^2}, \quad k = \omega\sqrt{\mu\epsilon}, \tag{23}$$

$$Z_e = \frac{\beta}{\omega\epsilon} = \frac{\beta}{k}\eta, \quad Z_m = \frac{\omega\mu}{\beta} = \frac{k}{\beta}\eta, \quad \eta = \sqrt{\frac{\mu}{\epsilon}}. \tag{24}$$

When there is an interface of two media at $z = 0$, coupling may arise between the two lines, which can be represented by a coupling two-port network. This means for example that, a source in the e line may create image sources in both e and m lines. However, for an interface between two isotropic media, there is no such coupling and the two lines can be understood as terminated by individual impedances Z_{Le} and Z_{Lm}.

3.2. Vertical magnetic dipole

As a simple example, let us consider a vertical magnetic current element (dipole) in the air (ϵ_o, μ_o) region $z z 0$, above a dielectric ($\epsilon_r \epsilon_o, \mu_o$) half space $z > 0$. Since all other sources except i_m vanish, only the m line exists and the subscript m will be omitted. The corerponding reflection coefficient at the interface is

$$R(\beta_o) = \frac{Z_L - Z}{Z_L + Z} = -\frac{\beta - \beta_o}{\beta + \beta_o}. \tag{25}$$

The relation between the two propagation factors β, β_o is

$$\beta = \sqrt{k^2 - K^2} = \sqrt{k^2 - k_o^2 + k_o^2 - K^2} = \sqrt{\beta_o^2 - B^2}, \tag{26}$$

with

$$B = \sqrt{k_o^2 - k^2} = k_o\sqrt{1 - \epsilon_r}. \tag{27}$$

For perfectly electrically conducting (PEC) or perfectly magnetically conducting (PMC) surfaces corresponding respectively to $\epsilon_r = \infty$ and $\epsilon_r = 0$, we have $R = -1$ and $R = +1$ corresponding to short and open circuit at $z = 0$.

The vertical magnetic dipole of moment M_oL corresponds to a current source in the transmission line:

$$\mathbf{M}(\mathbf{r}) = \mathbf{u}_z M_o L \delta(\mathbf{r} - \mathbf{u}_z z_o) \quad \Rightarrow \quad i(z) = I_o \delta(z - z_o), \quad I_o = -\frac{M_o L}{\omega\mu_o}. \tag{28}$$

For the reflection image we now have from (13) the simple formal solution

$$i^r(z) = R(-j\partial_z)I_o\delta(z + z_o), \tag{29}$$

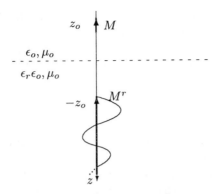

Figure 3. The dielectric half space $z > 0$ can be replaced by the reflection image M^r of the original magnetic dipole M. The image consists of a magnetic dipole and a magnetic line current obeying the Bessel function law.

or, more explicitly,

$$i^r(z) = -I_o \frac{\sqrt{\partial_z^2 + B^2} - \partial_z}{\sqrt{\partial_z^2 + B^2} + \partial_z} \delta(z + z_o). \tag{30}$$

Invoking (69) for $n = 2$ from the Appendix, the solution can be expressed as

$$i^r(z) = -2I_o \frac{J_2(Bz)}{z} \Theta(z). \tag{31}$$

Because the image does not depend on the Fourier variable $\mathbf{K}$, the corresponding vertical magnetic current source has $\delta(\boldsymbol{\rho}) = \delta(x)\delta(y)$ dependence,

$$\mathbf{M}^r(\mathbf{r}) = -\mathbf{u}_z \omega \mu_o i^r(z) \delta(\boldsymbol{\rho}) = -\mathbf{u}_z M_o L \frac{2}{z} J_2(Bz) \Theta(z + z_o) \delta(\boldsymbol{\rho}). \tag{32}$$

Thus, the image is a magnetic line source extending from $z = -z_o > 0$ to ∞ obeying the Bessel function $J_2(Bz)$, Fig.3. The result (32) was previously obtained in [2] through a much more complicated analysis.

It must be noted that the image expression works well for real $\epsilon_r < 1$, ϵ_r being the ratio of the two permittivities. If $\epsilon_r > 1$, the coefficient B is imaginary and the Bessel function $J_2(Bz)$ diverges for real z. The same happens for complex ϵ_r. By assuming the image along a line in complex space [2], a useful converging image source is, however, obtained. A source in complex space can be applied numerically to compute fields in the physical space with the same ease as a source in real space. Because the Green function decays exponentially when the source lies in complex space, some computation time can be saved. In, fact, such an image principle can show dramatic advantage in numerical computation when compared to classical Sommerfeld integrals [5].

4. The dielectric sphere

As another example, consider the electrostatic problem of the dielectric sphere $r < a$, $\epsilon = \epsilon_r \epsilon_o$ in air, $\epsilon = \epsilon_o$. The source is a point charge Q_o at the distance $r = r_o > a$ from the center of the sphere. The idea for an extended line image of the point charge was given first by C. Neumann in 1883 [6] in an appendix of a book on hydrodynamics. This principle lay dormant until it was rediscovered about a century later, [1].

4.1. Transmission-line analogy

The spherical geometry can be transformed to a transmission line by first transforming the source and fields through the spherical harmonic functions

$$Y_n^m(\theta, \varphi) = (-1)^m \sqrt{\frac{2n+1}{4\pi} \frac{(n-m)!}{(n+m)!}} P_n^m(\cos\theta) e^{jm\varphi}. \tag{33}$$

This corresponds to the Fourier transformation applied in the preceding example for the planar geometry. The functions of the spatial vector $\mathbf{r}$ are transformed to functions of the r coordinate and two indices n and m as [7]

$$f(\mathbf{r}) \Rightarrow f_n^m(r) = \int_{4\pi} f(\mathbf{r}) Y_n^{m*}(\theta, \varphi) d\Omega, \quad d\Omega = \sin\theta d\theta d\varphi. \tag{34}$$

The original function can be recovered from the sum

$$f(\mathbf{r}) = \sum_{n=0}^{\infty} \sum_{m=-n}^{n} f_n^m(r) Y_n^m(\theta, \varphi). \tag{35}$$

Because the Laplacian operator is transformed to [7]

$$\nabla^2 f(\mathbf{r}) \Rightarrow \frac{1}{r^2} [\partial_r (r^2 \partial_r f_n^m(r)) - n(n+1) f_n^m(r)] \tag{36}$$

and the charge density function corresponding to a radial line charge function $q(r)$ to

$$\varrho(\mathbf{r}) = \frac{q(r)}{r^2 \sin\theta} \delta(\theta - \theta_o) \delta(\varphi - \varphi_o) \Rightarrow \frac{q(r)}{r^2} Y_n^{m*}(\theta_o, \varphi_o), \tag{37}$$

the transformed Poisson equation for the scalar potential $\phi(\mathbf{r})$ becomes

$$\partial_r[\epsilon_r r^2 \partial_r \phi_n^m(r)] - n(n+1)\epsilon_r \phi_n^m(r) = -\frac{q(r)}{\epsilon_o} Y_n^{m*}(\theta_o, \varphi_o). \qquad (38)$$

After another transformation

$$r = ae^{-\tau}, \quad \tau = \ln(a/r), \qquad (39)$$

(38) for a layered sphere $\epsilon_r(r)$ becomes

$$\partial_\tau[\epsilon_r ae^{-\tau} \partial_\tau \phi_n^m(ae^{-\tau})] - n(n+1)\epsilon_r ae^{-\tau} \phi_n^m(ae^{-\tau})$$

$$= -\frac{q(ae^{-\tau})ae^{-\tau}}{\epsilon_o} Y_n^{m*}(\theta_o, \varphi_o). \qquad (40)$$

Let us now define the voltage and current quantities as

$$U(\tau) = ae^{-\tau}\phi(ae^{-\tau}), \quad I(\tau) = -\epsilon_r ae^{-\tau}\partial_\tau\phi(ae^{-\tau}). \qquad (41)$$

Because of the boundary conditions satisfied by the potential, these quantities are continuous at a junction of two different transmission lines. The following set of transmission-line equations is obtained from (40) and (41):

$$\partial_\tau \begin{pmatrix} U(\tau) \\ I(\tau) \end{pmatrix} = -\begin{pmatrix} 1 & 1/\epsilon_r \\ n(n+1)\epsilon_r & 0 \end{pmatrix} \begin{pmatrix} U(\tau) \\ I(\tau) \end{pmatrix} + \begin{pmatrix} 0 \\ i(\tau) \end{pmatrix}, \qquad (42)$$

with the source term defined as

$$i(\tau) = \frac{q(ae^{-\tau})ae^{-\tau}}{\epsilon_o} Y_n^{m*}(\theta_o\varphi_o). \qquad (43)$$

The equation pair (42) obviously corresponds to that of an asymmetric transmission line (4). The propagation constants and impedances for waves propagating in the opposite directions can be identified as

$$\alpha_+ = j\beta_+ = n+1, \quad \alpha_- = j\beta_- = n, \quad Z_+ = \frac{1}{n\epsilon_r}, \quad Z_- = \frac{1}{(n+1)\epsilon_r}. \qquad (44)$$

The second-order equation for the voltage

$$(\partial_\tau - (n+1))(\partial_\tau + n)U(\tau) = -\frac{1}{\epsilon_r}i(\tau) \qquad (45)$$

can be used to find the current source $i(\tau)$ corresponding to a voltage function $U(\tau)$.

4.2. Reflection image

Let us now consider the effect of the dielectric sphere on the potential from a point charge $q(r) = Q_o \delta(r - r_o)$. Because of

$$\delta(r - r_o) = \frac{\delta(\tau - \tau_o)}{ae^{-\tau}}, \qquad \tau_o = \ln(a/r_o), \tag{46}$$

the corresponding current generator on the transmission line is

$$i(\tau) = I_o \delta(\tau - \tau_o), \qquad I_o = \frac{Q_o}{\epsilon_o} Y_n^{m*}(\theta_o, \varphi_o), \tag{47}$$

The sphere (radius a and relative permittivity ϵ_r) in air is mapped to the transmission-line section $\tau > 0$ while the point charge Q_o is mapped from $r_o > a$ to $\tau_o < 0$. The impedances in the line $z < 0$ are $Z_+ = 1/n$, $Z_- = 1/(n+1)$. The impedance seen at $\tau = 0$ is $Z_L = 1/n\epsilon_r$, whence the reflection coefficient is

$$R(\alpha_-) = R(n) = \frac{(Z_L - Z_+)Z_-}{(Z_L + Z_-)Z_+} = \frac{(\frac{1}{n\epsilon_r} - \frac{1}{n})\frac{1}{n+1}}{(\frac{1}{n\epsilon_r} + \frac{1}{n+1})\frac{1}{n}} = \frac{(1 - \epsilon_r)n}{(1 + \epsilon_r)n + 1}. \tag{48}$$

The voltage incident to the junction $z = 0$ from the source is

$$U_+^i(\tau, \tau_o) = \frac{Z_+ Z_-}{Z_+ + Z_-} I_o e^{-\alpha_+(\tau - \tau_o)} = \frac{1}{2n+1} I_o e^{-(n+1)(\tau - \tau_o)}. \tag{49}$$

and the reflected voltage can be expressed as

$$U^r(\tau) = R(n)U_+^i(0, \tau_o)e^{\alpha - \tau} = e^{\tau_o} R(\partial_\tau)U_-^i(\tau, -\tau_o). \tag{50}$$

From (45) the reflection image $i^r(\tau)$, the source of the reflected voltage $U^r(\tau)$, is obtained as

$$(\partial_\tau - (n+1))(\partial_\tau + n)U^r(\tau) = -i^r(\tau). \tag{51}$$

Substituting (50), we have

$$i^r(\tau) = -e^{\tau_o} R(\partial_\tau)(\partial_\tau - (n+1))(\partial_\tau + n)U^i(\tau, -\tau_o), \tag{52}$$

and, finally, the image in the transform space is simply

$$i^r(\tau) = e^{\tau_o} R(\partial_\tau)I_o \delta(\tau + \tau_o). \tag{53}$$

Instead of transforming this to the physical space, let us interpret the image expression and transform it to a function of r. In the interpretation we expand (53) by applying (63) from the Appendix

$$R(\partial_\tau)\delta(\tau + \tau_o) = -\frac{\epsilon_r - 1}{\epsilon_r + 1}\partial_\tau \left(\frac{1}{\partial_\tau + \frac{1}{\epsilon_r+1}}\delta(\tau + \tau_o)\right)$$

$$= -\frac{\epsilon_r - 1}{\epsilon_r + 1}\partial_\tau \left(e^{-\frac{1}{\epsilon_r+1}(\tau+\tau_o)}\Theta(\tau+\tau_o)\right)$$

$$= -\frac{\epsilon_r - 1}{\epsilon_r + 1}\delta(\tau+\tau_o) + \frac{\epsilon_r - 1}{(\epsilon_r + 1)^2}e^{-\frac{1}{\epsilon_r+1}(\tau+\tau_o)}\Theta(\tau+\tau_o). \tag{54}$$

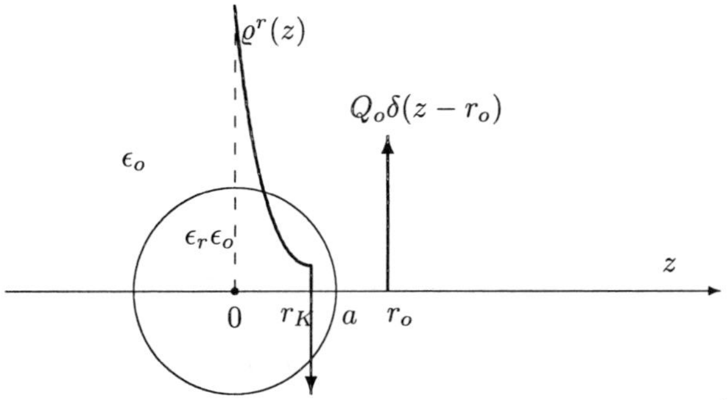

Figure 4. Image of a static point charge Q_o in a dielectric sphere consists of a point charge at the Kelvin point $r_K = a^2/r_o$ and a line charge with an integrable singularity at the origin.

The image on the transmission line is represented by the current generator

$$i^r(\tau) = -\frac{Q_o}{\epsilon_o}Y_n^{m*}(\theta_o,\varphi_o)\times$$

$$\left(\frac{\epsilon_r - 1}{\epsilon_r + 1}e^{\tau_o}\delta(\tau+\tau_o) - \frac{\epsilon_r - 1}{(\epsilon_r + 1)^2}e^{\tau_o}e^{-\frac{1}{\epsilon_r+1}(\tau+\tau_o)}\Theta(\tau+\tau_o)\right). \tag{55}$$

Noting that

$$\exp\left(-\frac{\tau+\tau_o}{\epsilon_r+1}\right) = \exp\left(-\frac{\ln(a^2/rr_o)}{\epsilon_r+1}\right) = \left(\frac{rr_o}{a^2}\right)^{\frac{1}{\epsilon_r+1}} \tag{56}$$

and

$$\delta(\tau+\tau_o) = \delta(\ln(a^2/rr_o)) = \delta(-\ln r + \ln(a^2/r_o)) = r\delta(r - \frac{a^2}{r_o}), \tag{57}$$

we can finally write the image in the physical space by comparing with (43):

$$q^r(r) = \frac{\epsilon_o i^r(\ln(a/r))}{rY_n^{m*}(\theta_o,\varphi_o)}$$

$$= -Q_o \left(\frac{\epsilon_r - 1}{\epsilon_r + 1} \frac{a}{r_o} \delta(r - \frac{a^2}{r_o}) - \frac{\epsilon_r - 1}{(\epsilon_r + 1)^2} \frac{1}{a} \left(\frac{r r_o}{a^2} \right)^{-\frac{\epsilon_r}{\epsilon_r + 1}} \Theta(-r + \frac{a^2}{r_o}) \right).$$

(58)

The image consist of a point charge at the Kelvin point $r = a^2/r_o$ plus a line charge with an integrable singularity at the origin. For $\epsilon_r \to \infty$ the continuous image reduces to a point charge at the center which corresponds to the classical Kelvin's image for a perfectly conducting sphere with zero total charge.

Appendix A. Heaviside operators

Oliver Heaviside (1850-1925) constructed the modern electromagnetic theory in compact form by introducing the vector notation simultaneously with J.W. Gibbs. In his analysis of electromagnetic equations he used a form of operational calculus, which later led to the Laplace transform technique. In the image theory we arrive at expressions of the form $F(\partial_z)\delta(z)$ where $F(\partial_z)$ is a function of the differential operator ∂_z. To interprete such an expression in terms of computable functions $f(z)$ we need operational rules. Such a rule is obtained through the Laplace transformation $F(s) \to f(t)$ as

$$F(\partial_z)\delta(z - z_o) = \int_0^\infty f(p) e^{-p\partial_z} dp \delta(z - z_o) = f(z - z_o)\Theta(z - z_o),$$

(59)

by applying the shifting rule $e^{-p\partial_z} f(z) = f(z - p)$. Here $\Theta(z)$ denotes the Heaviside unit step function

$$\Theta(z) = 0, \quad z < 0, \qquad \Theta(z) = 1, \quad z \geq 0.$$

(60)

Applying a standard table of Laplace transforms, a corresponding table for pseudo-differential operators $F(\partial_z)$ operating on the delta function can be compiled in the form $F(\partial_z)\delta(z) = f(z)\Theta(z)$. A short one is given below:

$$\partial_z^{-n}\delta(z) = \frac{z^{n-1}}{(n-1)!}\Theta(z),$$

(61)

$$\partial_z^{-\nu}\delta(z) = \frac{z^{\nu-1}}{\Gamma(\nu)}\Theta(z), \quad \nu > 0,$$

(62)

$$\frac{1}{\partial_z + B}\delta(z) = e^{-Bz}\Theta(z),$$

(63)

$$\frac{1}{(\partial_z + B)^2}\delta(z) = ze^{-Bz}\Theta(z), \tag{64}$$

$$\frac{1}{(\partial_z + B_1)(\partial_z + B_2)}\delta(z) = -\frac{e^{-B_1 z} - e^{-B_2 z}}{B_1 - B_2}\Theta(z), \tag{65}$$

$$\frac{1}{\partial_z^2 + B^2}\delta(z) = \frac{1}{B}\sin(Bz)\Theta(z), \tag{66}$$

$$\frac{1}{\sqrt{\partial_z^2 + B^2}}\delta(z) = J_0(Bz)\Theta(z), \tag{67}$$

$$\frac{(\sqrt{\partial_z^2 + B^2} - \partial_z)^\nu}{\sqrt{\partial_z^2 + B^2}}\delta(z) = B^\nu J_\nu(Bz)\Theta(z), \tag{68}$$

$$\left(\frac{\sqrt{\partial_z^2 + B^2} - \partial_z}{\sqrt{\partial_z^2 + B^2} + \partial_z}\right)^{n/2}\delta(z) = \frac{n}{z}J_n(Bz)\Theta(z), \tag{69}$$

References

1. I. V. Lindell, "Application of the image concept in electromagnetism", in *The Review of Radio Science* 1990-1992, W. Ross Stone, ed. Oxford University Press, 107–126 (1993).
2. I. V. Lindell, *Methods for Electromagnetic Field Analysis*, 2nd ed. Oxford University Press and IEEE Press (1995).
3. W. Thomson, *Reprint of Papers on Electrostatics and Magnetism*. Macmillan, London, 85, 144 (1872).
4. I. V. Lindell, M. E. Valtonen, A. H. Sihvola, "Theory of nonreciprocal and nonsymmetric uniform transmission lines", *IEEE Trans. Microwave Theory Tech.*, **42**, 291-297 (1994).
5. K. Sarabandi, M. D. Casciato, I. S. Koh, "Efficient calculation of the fields of a dipole radiating above an impedance surface", *IEEE Trans. Ant. Propagat.*, **50** 1222–1235 (2002).
6. C. Neumann, *Hydrodynamische Untersuchungen*, Leipzig: Teubner, 271–283 (1883).
7. G. Arfken, *Mathematical Methods for Physicists*, 3rd ed., San Diego: Academic Press (1985).
8. I. V. Lindell, "Electrostatic image theory for the dielectric sphere", *Radio Sci.* **27** 1–8 (1992).

ACOUSTIC-EMISSION SOURCE LOCATION USING THE LOWER-FREQUENCIES OF FLEXURAL WAVES

E. LYMPERTOS

FORTH-ICE/HT Institute of Chemical Engineering & High Temperature Chemical Processes, 26500 Patras, Hellas, e-mail: libertos@ee.upatras.gr

E. DERMATAS

Department of Electrical & Computer Engineering, University of Patras, 26500 Patras, Hellas, e-mail: dermatas@george.wcl2.ee.upatras.gr

In this paper a novel method for acoustic emission (AE) source location in dispersive media is presented by processing the low-frequency components of the flexural waves arrived in arbitrary positioning transducers. The accurate location of the source is obtained by minimizing the time-domain square error (TDSE) between a model-based AE source signals arrived at any sensor and the real signals. In the multi-variable optimization problem, the huge number of local minima is avoided using modifications in the definition of the TDSE-function and the use of a dedicated simulated annealing method. In the case of thin plates, a three sensors configuration can be used to estimate the AE source position. It has also been showed that the proposed method can be used to determine the AE source position in more complex surfaces even in the presence of additive Gaussian noise in extremely low signal-to-noise ratios.

1. Introduction

The location of a signal source is a wide applicable problem in target detection for radar, underwater, and seismic signals. In isotropic materials the location of the AE source is estimated using the differences in the time-of-arrival (DTA) from sensors, the distances between sensors and triangulation techniques [1,3,5-8]. In practice, the propagating waves are dispersive and noisy, introducing errors in the DTA estimation methods.

In [1] the DTA of the flexural wave in plates is estimated using the cross-correlation between the transducer signal and a single frequency component modulated by a Gaussian pulse. An improvement is presented in [5], where a wavelet transformation is applied to the signals and then the method of the cross-correlation is used to determine the DTA; the introduced noise is reducing and a better estimation is achieved. Toyama *et al* [3] evaluate the two-dimensional source location problem with only two AE sensors by using the wavelet transform (WT) for a thin CFRP plate. The authors use the two modes of the AE wave to achieve an estimation of the source position with two sensors. In [8], an experimental method has been developed for detecting flexural waves in plates. The recorded signals are analyzed by a wavelet transform to determine the waves arrival-time at different frequencies.

A method for reducing the number of sensors in AE source location is presented in [7]. The derived DTA equations are non linear and the solution is obtained via minimization. In this direction, a new family of exact solutions, for the case of four receivers located in a plane, is presented [6]. In [4] a different approach rather than calculating DTA's is introduced using a general regression neural network, which is trained from localization examples.

Taking into account that single frequency methods are very sensitive in the presence of additive and convoluted noise, we present and evaluate a model-based source location method by processing the lower-frequencies of the AE flexural waves. Therefore, the proposed method can be implemented using low-cost hardware. It is also shown that the proposed method is robust in case where the structure permits multi-path wave propagation from source to sensors (e.g. cylindrical tanks).

2. Simulated annealing method

Simulated annealing is a global optimization method that makes use of stochastic search to avoid terminating in local extremum. It is used in locating the global extremum of objective, or cost functions derived from complex nonlinear systems, where the number elements in the configuration space is factorially large, so that they cannot explored exhaustively.

Starting from an initial point, the algorithm takes a step and the function is evaluated. When minimizing a function, any downhill step is accepted and the process is repeating from this new point. An uphill step may be accepted. Thus, it can escape from local optima. As the optimization process proceeds, the length of the steps decline and the algorithm approaches the global optimum. Since the algorithm makes very few assumptions regarding the function to be optimized, it is quite robust with respect to non-quadratic surfaces. In fact, simulated annealing can be used as a local optimizer for difficult functions.

3. Source location

The location of an AE source is estimated using the TDSE function of a model-based wave signal for the flexural displacement and the real signal arrived in the transducer. The estimation of the AE source signal is achieved by solving the minimization problem using the simulated annealing method.

Two novel error functions were used to reduce some undesirable properties of the ordinary error function.

3.1. *Source signal*

According to Medick [2], the flexural displacement at distance r from source and time t is given by the equation:

$$w(r,t) = 2\left(\pi/2 - Si(r^2/4bt)\right) \tag{1}$$

$$b^2 = Eh^2/12p(1-v^2) \qquad Si(z) = \int_0^z \frac{\sin(t)}{t} dt \approx \pi/2 - \cos(z)/z \tag{2}$$

where, E is the modulus of elasticity, p is the material density, v is Poisson's ratio, and h is the material thickness.

The approximation for the Si(z) is valid only when the argument z is large enough (over 10). So (1) becomes:

$$w(r,t) = 2\cos(r^2/4bt)/(r^2/4bt) \tag{3}$$

3.2. *Error function*

The TDSE of the sensor signal $o(t)$ and the flexural displacement $w(r,t)$ is a function of two unknown variables, r and T:

$$E(r,T) = \sum_{t=1}^{L} \left(w(r,t+T) - o(t)\right)^2 \tag{4}$$

where, T is the time difference between AE and data acquisition starting time and, L is the length of the recorded signal.

In figure 1(a), the TDSE is plotted for known T and sensor distance $r=40$ m, when the sensor signal is replaced by the flexural displacement $w(r_o,t)$. In figure 1(b) the same function is plotted in the neighbor of the global minimum.

The derivation of r cannot be achieved in a close-form, and the well-known minimization methods based on derivatives fails due to the presence of a vast number of local minima. Moreover, the error function decreases continuously in the selected plotting scale. A more detail plot introduces a great number of minima in the neighbor of the global-minimum, as shown in figure 1(b).

In the proposed error function:

$$En(r,T) = r^2 \sum_{t=1}^{L} \left(w(r,t+T) - o(t)\right)^2 \tag{5}$$

the position of the global minimum remains the same to the ordinary TDSE function. If the sensor signal $o(t)$ is replaced by the flexural displacement $w(r_o,t)$, the global minimum can be found in distance r_o for all T and windows length L. Moreover, the proposed TDSE function increases as a function of r after the global minimum, as shown in figure 1(a), and the mean value of the TDSE function in the presence of additive noise at the global minimum is r_o^2 times the energy of the noise.

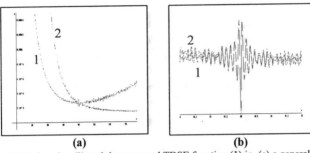

(a)   (b)

Figure 1. TDSE function (2) and the proposed TDSE function (1) in, (a) a general view, (b) a detailed view in the neighbor of the global minimum.

3.3. Source location using multiple sensors

In the case of using more than one sensor, the total error is the sum of the TDSE for all sensors and the same time frame. In this approach, the additional unknown parameters (r_i, T_i) for the ith sensor, can be significantly reduced using the sensors and the AE source Cartesian coordinates. Two types of individual unknown parameters are meet, the AE source sensor coordinates and the arrival-time of the AE wave to the nearest sensor.

In applications where each sensor has an individual trigger threshold, the TDSE function can be easily expressed as a function of the AE source sensor coordinates and the arrival-time of the AE wave to the nearest sensor, assuming known time differences between each individual triggering.

In the planar case, the total error depends on three parameters, the two Cartesian coordinates of the AE source (x,y) and the wave arrival-time to the nearest sensor T. If two sensors are placed in (x_1,y_1) and (x_2,y_2), the total TDSE of the two sensor problem becomes:

$$En(x,y,T) = \sum_i En_i(x,y,T)$$

$$En_i(x,y,T) = r_i^2 \sum_{t=1}^{L} \left(2\cos\left(r_i^2/4b(t+T)\right)/\left(r_i^2/4b(t+T)\right) - o_i(t)\right)^2$$

where, $r_i^2 = (x_i-x)^2 + (y_i-y)^2$ (6)

In figure 2, a contour plot of the total error in x-y plane is showed in case where three sensors are placed at $S1(50,0)$, $S2(85,50)$, $S3(120,0)$ and the AE source are placed at *(60, 20)*. In the gray scale plot (the dark area denotes low total TDSE), three circles with centers at *S1, S2* and *S3* and radius the distance between each center and the AE source appears. In these circles the TDSE function for each pair sensor-receiver is minimized. The AE source is placed in the common area of three circles. In a typical simulated annealing implementation the AE source is initially located at a position on the circles perimeter. Consequently, in the neighbor of the circle perimeter a second search reduces the total TDSE.

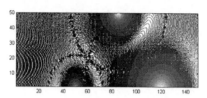

Figure 2. A contour plot of the total Error function in the x-y plane in case of three sensors

The efficiency of the second search is improved significantly by introducing a normalization procedure for each partial TDSE:

$$E_i = \frac{\sum_{t=1}^{L}\left(2\cos\left(r_i^2/4b(t+T)\right)/\left(r_i^2/4b(t+T)\right)-w_i(t,r_o,T)\right)^2}{\sqrt{\sum_{t=1}^{L}\left(2\cos\left(r_i^2/4b(t+T)\right)/\left(r_i^2/4b(t+T)\right)\right)^2\sum_{t=1}^{L}w_i^2(t,r_o,T)}} \tag{7}$$

The normalization expression produces TDSE values in the same range in the neighbor of the global minimum, for all L, r_o, and T.

3.4. *Source location in cylindrical surfaces*

In surfaces where the signal can follow different paths from source to receiver, the AE location problem becomes more complex. In some applications the multi-path propagation phenomenon can be studied as related noise. If the secondary paths distances are close to the shortest path, the interference strongly distorts the direct AE signal.

The multi-path effect can be simulated by the use of virtual sensors using the well-known image method. In this study two assumptions were used: The flexural waves are absorbed completely on the cylinder boundaries, and two virtual sensors are adequate to model the multi-path effect. Using the above assumptions the total TDSE function in the case of two sensors becomes:

$$E(x,y,T) = \sum_j E_j(x,y,T)$$

$$E_j(x,y,T) = \left(\min(r_{ji})\right)^2 \sum_{t=1}^{L}\left(\sum_{i=1}^{3}\left(2\cos\left(r_{ji}^2/4b(t+T)\right)/\left(r_{ji}^2/4b(t+T)\right)\right)-o_j(t)\right)^2$$

where $r_{ji}^2 = (x_j + (i-2)P - x)^2 + (y_j - y)^2$, and P is the perimeter of the cylinder.

3.5. *Robust simulated annealing for AE source location*

A typical implementation of the simulated annealing method in the AE source localization problem is failed even in case of using the normalized TDSE error. As shown in figure 2, the presence of a vast number of local minima in an environment where a great number of valleys are met in the total TDSE function, the probability of the simulated annealing method to converge in the global minimum is reduced significantly.

The probability of reaching the global minimum is improved significantly in the proposed modified simulated annealing method, consisting of three individual searching procedures. Initially, the simulated annealing is used to derive an estimation of *x, y, T*. In the second procedure, the searching is restricted in the neighbor of a circle perimeter, giving a better estimation of the source position (x',y'). The center of this circle is the sensor position with the minimum normalized TDSE (eq. 7) and the radius is equal to the distance between the estimated AE position (x,y) and the sensor position. In the final procedure, the simulated annealing method is search for the global minimum in the neighbor of (x',y') for fine tuning of source position, by performing full scale search for the parameter *T*.

4. Experimental results

Three sensors have been used for the location of an AE source placed in five positions in a plane surface. The sensors coordinates were $S_1(50,0)$, $S_2(120,0)$, $S_3(85,50)$ while the source is placed at $R_1(85,0)$, $R_2(85,40)$, $R_3(51,2)$, $R_4(60,20)$, $R_5(35,10)$. The material properties were, $E=70$ *GPA*, $p=2750$ Kg/m^3, and $v=0.33$ and the thickness was *5 mm*. A sampling frequency of *1MHz* was used to simulate a digital representation of the flexural wave. The experiments have been set-up in different signal to noise ratios *(-20, -10, 0, 10, 20dB SNR)*. The SNR has been calculated in the frequency range of *50-150 KHz*. The additive noise was white Gaussian with zero mean value. In this configuration the frequency components of the digital signal varies from *10 KHz* to *500 KHz*.

In all experiments the source location is estimated by applying the modified simulated annealing method to the derived minimization problem for the total proposed TDSE function. The experimental results are compared to the AE source positions estimated by the conventional solution using DTA and triangular methods. The results are showed in Tables 1 and 2.

In Tables 1, 3 and 4 three error-values are given for each AE source location experiment and each noise level. The left value is the Euclidean distance of the estimated and the real position of the AE source in meters, the right upper value is the total TDSE estimated by eq. 6, and the right lower value is the sum of the noise energy, normalized by the distance of each individual sensor with the estimated AE source position. A reliable estimation of the source position is obtained in the case where the absolute difference between the upper and lower values is minimized.

Table 1. Estimation Error using the modified simulated annealing method.

	20dB		10dB		0dB		-10dB		-20dB	
R_1	4.61	37.6	5.69	38.7	5.38	49.3	5.26	128	12.4	885
		0.09		0.88		8.69		88.4		838.7
R_2	0.73	22.4	0.26	16.4	2.13	39.6	2.42	128	2.04	1018
		0.12		1.01		9.95		99.6		992
R_3	$7\ 10^{-5}$	0.12	0.10	10.1	0.635	35.4	$3\ 10^{-5}$	115	1.91	1191
		0.12		1.15		11.7		115		1155
R_4	2.36	35.1	3.72	34.2	1.75	43.3	8.60	128	14.3	927
		0.10		0.97		10.1		90.1		879.1
R_5	0.0003	0.17	0.095	22.0	0.996	54.4	24.07	159	30.6	811.6
		0.14		1.37		14.2		77.9		720.8

Table 2. Estimation Error using DTA's and triangular methods.

AE position	20dB	10dB	0dB	-10dB	-20dB
R_1	0.14	0.25	0.29	0.61	25.05
R_2	0.12	0.07	0.20	52.80	36.39
R_3	0.03	0.26	0.48	0.70	46.01
R_4	0.19	0.46	0.08	42.35	30.39
R_5	0.49	1.47	5.83	1.27	81.32

Table 3. Estimation Error using 100 KHz sampling frequency on a plane surface.

	20dB		10dB		0dB		-10dB		-20dB	
R_1	0.15	0.04 / 0.04	0.65	0.51 / 0.36	2.97	9.17 / 3.37	1.17	40.3 / 35.5	1.51	369 / 369
R_2	29.11	6.57 / 0.03	33.33	6.62 / 0.24	33.99	8.43 / 2.26	39.3	28.8 / 23.1	39.67	237 / 233
R_3	$8\ 10^{-4}$	0.05 / 0.05	$7\ 10^{-4}$	0.53 / 0.53	$1\ 10^{-3}$	4.79 / 4.79	1.20	52.2 / 52.2	3.83	499 / 499
R_4	1.51	2.88 / 0.05	17.94	6.43 / 0.36	15.13	9.67 / 3.66	21.8	43.2 / 34.9	22.58	361 / 354
R_5	3.30	2.34 / 0.07	$2\ 10^{-4}$	0.67 / 0.67	4.80	13.6 / 5.82	28.7	43.0 / 30.7	30.58	328 / 319

A second experiment has been set-up to evaluate the capability of the proposed method to locate the position of an AE source by processing the low-frequency components of the flexural waves on cylindrical surfaces. Using only the two sensors (S_1, S_2) from the previous experiment and the same material, the five positions of the AE source are re-estimated. The sampling frequency is reduced to *100 KHz* and the SNR has been calculated in the frequency range of *10-20KHz*. The frequency components of the signal, which is used to derive the AE source position, were restricted to the range of *5-50 KHz*. In the simulation experiments have been assumed that the sensors are placed on a plane (Table 3) or a cylindrical surface (Table 4). In case of a cylinder surface a random point at the bottom of the cylinder is assumed to be the *(0,0)* point.

Table 4. Estimation Error in five AE source positions using 100 KHz sampling frequency on a cylindrical surface.

	20dB		10dB		0dB		-10dB		-20dB	
R_1	0.57	0.20 / 0.04	0.49	0.47 / 0.35	1.79	9.08 / 3.45	0.68	41.1 / 35.8	0.35	369.7 / 363.8
R_2	33.4	6.52 / 0.03	32.82	6.73 / 0.31	32.9	8.88 / 2.53	37.26	30.7 / 24.2	39.3	248.5 / 244.0
R_3	0.08	0.43 / 0.08	$7\ 10^{-4}$	0.79 / 0.79	0.47	9.38 / 7.63	0.94	79.5 / 77.4	3.37	788.1 / 778.1
R_4	30.4	6.40 / 0.06	$6\ 10^{-5}$	0.61 / 0.61	29.1	10.5 / 5.82	22.95	49.9 / 40.7	24.7	413.9 / 407.6
R_5	25.7	7.44 / 0.05	23.02	8.28 / 0.53	25.9	13.3 / 5.27	35.03	54.3 / 44.2	38.8	445.8 / 439.0

References

1. S. Ziola and Michael R.Gorman, "Source location in thin plates using cross-correlation", *J. Acoust. Soc. Am.* **90(5)**, 2551-2556 (1991).

2. M. A. Medick, "On Classical Plate Theory and Wave Propagation", *J. Appl. Mech.* **28**, 223-228 (1961).

3. N. Toyama, J. H. Koo, R. Oishi, M. Enoki, and T. Kishi, "Two-dimensional AE source location with two sensors in thin CFRP plates", *J. of Materials Science Letters* **20(19)**, 1823-1825 (2001).

4. T. Kosel, I. Grabec and Franc Kosel, "Intelligent location of simultaneously active acoustic emission sources: Part I", *Aircraft Engineering and Aerospace Technology* **75(1)**, 11-17 (2003).

5. Q. Wang and F. Chu, "Experimental determination of the rubbing location by means of acoustic emission and wavelet transform", *J. of Sound and Vibration* **248(1)**, 91-103 (2001).

6. R. Duraiswami, D. Zotkin, and L. Davis, "Exact solutions for the problem of source location from measured time differences of arrival", *138th Meeting: Acoustical Society of America*, (1999).

7. M. Surgeon, and M. Wevers, "One sensor linear location of acoustic emission events using plate wave theories", *Materials Science and Engineering* **265(1-2)**, 254-261 (1999).

8. L. Gaul and S. Hurlebaus, "Identification of the impact location on a plate using wavelets", *Mechanical Systems and Signal Processing* **12(6)**, 783-795 (1997).

TIME-ESTIMATES OF BURNT FOOD FOR A NONLOCAL REACTIVE-CONVECTIVE PROBLEM FROM THE FOOD INDUSTRY

C.V. NIKOLOPOULOS

Department of Mathematics, University of the Aegean, Samos, Greece
E-mail: cnikolo@aegean.gr

D.E. TZANETIS[*]

Department of Mathematics, Faculty of Applied Sciences, National Technical
University of Athens, Zografou Campus, 157 80 Athens, Greece
E-mail: dtzan@math.ntua.gr

We estimate the blow-up time for a non-local hyperbolic problem of Ohmic type, $u_t + u_x = \lambda f(u)/(\int_0^1 f(u)\,dx)^2$. It is known, that for $f(s)$, $-f'(s)$ positive and $\int_0^\infty f(s)ds < \infty$, there exists a critical value of the parameter $\lambda > 0$, say λ^*, such that for $\lambda > \lambda^*$ there is no stationary solution and the solution $u(x,t)$ blows up globally in finite time t^*, while for $\lambda \leq \lambda^*$ there exist stationary solutions. Estimates for t^* were found for λ greater than the critical value λ^* and fixed initial data $u_0(x) \geq 0$. The estimates are obtained by comparison and by numerical methods.

1. Introduction

We consider the non-local initial boundary value problem,

$$u_t(x,t) + u_x(x,t) = \lambda \frac{f(u(x,t))}{\left(\int_0^1 f(u(x,t))\,dx\right)^2}, \qquad 0 < x < 1, \quad t > 0, \quad (1.1a)$$

$$u(0,t) = 0, \ t > 0, \qquad\qquad\qquad (1.1b)$$

$$u(x,0) = u_0(x) \geq 0, \qquad 0 < x < 1, \qquad (1.1c)$$

---

[*]An extended version of this work will appear in a Journal. The talk was given by D.E. Tzanetis.

where $\lambda > 0$. The function $u(x,t)$ represents the dimensionless temperature when an electric current flows through a conductor (e.g. food) with
temperature dependent on electrical resistivity $f(u) > 0$, subject to a fixed
potential difference $V > 0$. In the following we assume f to satisfy

$$f(s) > 0, \quad f'(s) < 0, \quad s \geq 0, \qquad (1.2a)$$

$$\int_0^\infty f(s)\, ds < \infty, \qquad (1.2b)$$

for instance either $f(s) = e^{-s}$ or $f(s) = (1+s)^{-p}, \; p > 1$, satisfy (1.2).
For the initial data it is required that $u_0(x)$, $u_0'(x)$ be bounded and
$u_0(x) \geq 0$ in $[0,1]$ (the last requirement is a consequence of the fact that for
any initial data the solution u becomes non-negative over $(0,1]$ sometime,
[4,6]).

Also we emphasize that for $\lambda > \lambda^*$ and for all $x \in (0,1]$ we have:

$$F(u) = \frac{f(u)}{(\int_0^1 f(u)dx)^2} \to \infty \text{ as } t \to t^* - < \infty, \qquad (1.3a)$$

$$u(x,t;\lambda) \to \infty \text{ as } t \to t^* - < \infty, \qquad (1.3b)$$

the latter means that $u(x,t;\lambda)$ blows up globally, see Refs. 4, 5, 6.

In the present work, we find estimates for the non-local problem (1.1),
for $f(s) = e^{-s}$ and for general $f(s)$ which satisfies (1.2).

2. Comparison methods: upper and lower bounds of t^* for $\lambda > \lambda^*$

If the function f satisfies (1.2), one can prove (see appendix in Ref. 6)
that a maximum principle holds for (1.1) (here is where we need f to be
decreasing). Then we may, in the usual way, define upper and lower so-
lutions of (1.1): an upper (lower) solution $\bar{u}$ ($\underline{u}$) is defined as a function
which satisfies (1.1) if we substitute $\geq$ ($\leq$) for $=$, see Refs. 4, 5, 6. In the
following work, we use similar ideas and techniques as in Ref. 1.

An upper bound for t^*: We now wish to find an upper bound for the blow-up time t^*. For simplicity, we assume that $0 \leq u_0 < w^*$. Firstly, we write the steady-state problem in a slightly different way,

$$w' = \mu f(w) = \frac{\lambda f(w)}{\left(\int_0^1 f(w)\,dx\right)^2} = \lambda F(w), \quad 0 < x < 1, \quad w(0) = 0, \text{(2.1)}$$

where $F(\cdot) = f(\cdot) / \left(\int_0^1 f(\cdot)dx\right)^2$ and λ is a positive parameter (eigenvalue). Then the related linearized eigenvalue problem of (2.1) is:

$$\phi' - \lambda\, \delta F(w; \phi) = -\rho(w, \lambda)\, \phi, \quad 0 < x < 1, \quad \phi(0) = \phi_0 = 0, \quad \text{(2.2)}$$

where $\phi = \phi(x; \lambda)$ and $\delta F(w; \phi)$ is the first variation (or Gâteaux derivative) of F at w in the direction of ϕ, $(F(w; \phi) := F(w + \epsilon\phi) = J(\epsilon)$ and $\delta F(w; \phi) = J'(0) = \lim_{\epsilon \to 0} \frac{F(w+\epsilon\phi)-F(w)}{\epsilon})$.

As regards the first variation $\delta F(w; \phi)$ we have,

$$\delta F(w; \phi) = \frac{f'(w)\phi}{(\int_0^1 f(w)\,dx)^2} - \frac{2f(w)\int_0^1 f'(w)\phi\,dx}{(\int_0^1 f(w)\,dx)^3}.$$

Now we have the following Lemmas concerning the eigenpair of problem (2.2).

Lemma 1. *Problem (2.2) has the eigenpair (ρ, ϕ) where $\phi(x) > 0$ in $(0, 1]$ and its spectrum is a continuum of eigenvalues in $\mathbb{R}$, generated by $\rho = \rho(w, \lambda)$ for every $\lambda \in (0, \lambda^*]$. The function $\rho(w, \lambda)$ is continuous with respect to λ.*

Lemma 2. *Let w_1, w_2 with $w_1 < w_2$ be the solutions of (2.1) at $\lambda < \lambda^*$, then $\rho_1 = \rho(w_1, \lambda) < 0$, $\rho_2 = \rho(w_2, \lambda) > 0$ and $\rho^* = \rho(w^*, \lambda^*) = 0$ where ρ represents the eigenvalues of problem (2.2).*

Hence problem (2.2) at $\lambda = \lambda^*$, with $\phi^*(x) > 0$, becomes

$$\phi^{*'} - \lambda^* \,\delta F(w^*; \phi^*) = 0, \quad 0 < x < 1, \quad \phi^*(0) = \phi_0^* = 0. \quad \text{(2.3)}$$

Now, in order to find an upper bound for t^*, we take the difference,

$$v = v(x, t) = v(x, t; \lambda) = u(x, t; \lambda) - w^*(x) = u - w^*. \quad \text{(2.4)}$$

Since w^* is bounded, v blows up at the same time as u does and in the same manner i.e. globally. Hence $t^* = t^*(u) = t^*(v)$ ($t^*(u)$ is the blow-up time for u) and $v(x, t) \to \infty$ as $t \to t^*-$ for all $x \in (0, 1]$. In the following, we find an A-problem, where $A = A(t)$ blows up and is such that: $t^*(u) = t^*(v) \leq T^* = T^*(A)$, for some T^*, thus we find an upper bound T^* for $t^*(u)$.

Therefore we obtain

$$v_t = u_t = -u_x + \lambda F(u) = -u_x + \lambda F(u) - \lambda^* F(w^*) + w^{*\prime}$$
$$= -v_x + (\lambda - \lambda^*)F(u) + \lambda^* (F(u) - F(w^*)). \quad (2.5)$$

We set $v = u - w^* = \theta \hat{v}$, for $0 < \theta = \lambda - \lambda^* \ll 1$. This is simplified to

$$\hat{v}_t + \hat{v}_x = F(u) + \lambda^* \, \delta F(w^*; \hat{v}) + \frac{\lambda^*}{2}\theta \hat{J}''(\xi), \quad 0 < x < 1, \quad t > t_1, \quad (2.6)$$

where $J(\xi) = \theta^2 \hat{J}(\xi) = \theta^2 \delta^2 F(z; \hat{v})$ is the second Gâteaux derivative. Now we find a lower solution ψ for the $\hat{v}$-problem. Therefore we require $\psi = \psi(x, t) = A(t) \phi^*(x)$ to satisfy

$$\psi_t + \psi_x \leq F(u) + \lambda^* \, \delta F(w^*; \psi) + \frac{\lambda^*}{2}\theta \delta^2 F(z; \psi). \quad (2.7)$$

Taking c small enough so that $\theta \leq \frac{c}{A(t)}$, ($c$ is about the time that u is smaller than order one) we have that $A(t) \leq c_1 e^{\beta_1 t} \leq \frac{c}{\theta}$ with $c_1 = A(\tau_1) e^{-\beta_1 \tau_1}$ and this holds for time $t = \tau = \frac{1}{\beta_1} \ln(\frac{c}{\theta c_1})$. Now we can obtain an upper estimate T_u^* for $t^*(u)$ which is $t^*(u) < T_u^* = \tau + t_1^*$, where t_1^* is the blow-up time of the problem:

$$u_t(x, t) + u_x(x, t) = \lambda \frac{f(u(x, t))}{\left(\int_0^1 f(u(x, t)) \, dx\right)^2}, \quad 0 < x < 1, \quad t > \tau, \quad (2.8a)$$

$$u(0, t) = 0, \quad t > \tau, \quad (2.8b)$$

$$u(x, \tau) = w^* + c\phi^* \geq 0, \quad 0 < x < 1, \quad (2.8c)$$

and $t_1^* \ll \tau$.

A lower bound for t^*: We take $u_0(x)$ such that $u_0(x) < w^*(x)$ for $0 < x < 1$ and $u_0(0) = w^*(0)$. Let $u^* = u^*(x, t) = u(x, t; \lambda^*)$ be the solution to (1.1) with $u_0^* = u_0$.

In the following we use a similar concept to those in Refs. 1, 2. Therefore we set $u = u^* + u_1 \leq u^* + \psi_1 = w^* - \hat{u} + \psi_1 \leq w^* - \psi + \psi_1$, where $\hat{u}$ is given by $\hat{u} = w^* - u^* > 0$ and satisfies (2.9), u_1 solves (2.11), ψ_1 is an upper solution to the u_1-problem and ψ is lower solution to $\hat{u}$-problem i.e. $\psi_1 \geq u_1$ and $\psi \leq \hat{u}$. The $\hat{u}$-problem is defined by

$$\hat{u}_t = -u_t^* = u_x^* - \lambda^* F(u^*) - w^{*\prime} + \lambda^* F(w^*)$$
$$= -\hat{u}_x - \lambda^* (F(u^*) - F(w^*)), \quad 0 < x < 1, \quad 0 < t < T, \quad (2.9a)$$

$$\hat{u}(0,t) = w^*(0) - u^*(0,t) = 0, \ 0 < t < T, \tag{2.9b}$$

$$\hat{u}(x,0) = \hat{u}_0(x) = w^*(x) - u_0^*(x), \ 0 < x < 1, \tag{2.9c}$$

with $\hat{u}_0 > 0$, hence $\hat{u} > 0$, for $0 < x < 1$, $0 < t < T < t^*$ and for some $T > 0$.

Thus equation (2.9a) gives:

$$L(\hat{u}) := \hat{u}_t + \hat{u}_x - \lambda^* \delta F(w^*; \hat{u}) + \frac{\lambda^*}{2} \delta^2 F(z; \hat{u}) = 0. \tag{2.10}$$

Now we introduce the function $\psi = \psi(x,t) = \frac{c\phi^*(x)}{t+t_0} + \frac{u_2(x)}{(t+t_0)^2}$, where c, t_0 (positive constants), $u_2 = u_2(x) \geq 0$ are to be determined and $\phi^* = \phi^*(x)$ satisfies problem (2.3). For $u^* = w^* - \psi - r = w^* - \frac{c\phi^*(x)}{t+t_0} - \frac{u_2(x)}{(t+t_0)^2} - r \geq 0$, where $r = r(x,t) = \frac{u_3(x)}{(t+t_0)^3} + \frac{u_4(x)}{(t+t_0)^4} + ...$, $(r \in \mathbb{R})$, since $u^* \to w^*-$ or $(\psi + r) \to 0+$ as $t \to \infty$ $(u_0^*(x) < w^*(x))$, actually $u^* < w^*$ and $\psi + r > 0$ for all $t \geq 0$, [6]. Substituting now ψ, for $\hat{u}$ and taking into the account ϕ^*-problem, we obtain,

$$L(\psi) = \frac{c}{t+t_0} (\phi^{*\prime} - \lambda^* \delta F(w^*; \phi^*)) + \frac{1}{(t+t_0)^2} (u_2' - \lambda^* \delta F(w^*; u_2) -$$

$$c\phi^* + \frac{\lambda^*}{2} c^2 \delta^2 F(z; \phi^*) + O\left(\frac{1}{(t+t_0)^3}\right) \leq 0.$$

Then we have that ψ is a lower solution to $\hat{u}$-problem and thus $\psi \leq \hat{u}$.

We write $u = u^* + u_1 \leq w^*$ and find an upper solution to u_1-problem. Thus the u_1-problem now becomes

$$u_{1t} = -u_{1x} + (\lambda - \lambda^*)F(w^*) + \lambda^* \delta F(w^*; v) + Q(w^*, z, v)$$

$$+ \ \lambda^* \delta F(w^*; \hat{u}) - \frac{\lambda^*}{2} \delta^2 F(\zeta; \hat{u}), \ 0 < x < 1, \ t > 0, \tag{2.11a}$$

$$u_1(0,t) = 0, \quad t > 0, \tag{2.11b}$$

$$u_1(x,0) = u_0(x) - u_0^*(x) = 0, \quad 0 < x < 1, \tag{2.11c}$$

where $0 < z < w^*$, $0 < \zeta < w^*$, $0 < \hat{u} < w^*$, $u < u_1 < w^*$ as far as $u < w^*$, so that $Q(w^*, z, v)$, $J_2''(\xi_2)$ are bounded from above and below.

Now we introduce $\psi_1(x,t) = [(\lambda - \lambda^*)\Lambda(t + t_0)]\phi^*(x)$, where Λ is a constant which is determined, so that ψ_1 to be an upper solution to u_1-problem. Here ϕ^* again satisfies problem (2.3).

Then, ψ_1 is a "restricted" upper solution for u_1-problem in the interval

$[\gamma, 1]$, [3]. Here we have to notice that u blows up globally, see relation (1.3b), this means that $u(x,t)$ is bounded for $(x,t) \in [0,\gamma] \times [0,T]$ for some $T < t^*$. In other words u is bounded in $[\gamma, 1] \times [0, T]$, (actually we require $u \leq w^*$, see below) and hence a lower bound for t^* can be found working in the restricted interval $[\gamma, 1]$.

Hence, $u \leq w^*$ as far $\psi - \psi_1 \geq 0$, thus we have

$$u = u_1 + u^* = w^* - \frac{c\phi^*}{t + t_0} - \frac{u_2}{(t + t_0)^2} + (\lambda - \lambda^*)\Lambda\,(t + t_0)\,\phi^*.$$

The right-hand side of the above relation is no greater than w^*, as long as $\psi \geq \psi_1$ or

$$\frac{c\phi^*}{(t + t_0)} + \frac{u_2}{(t + t_0)^2} \geq (\lambda - \lambda^*)\,\Lambda\,\phi^*\,(t + t_0), \quad x \in [\gamma, 1].$$

Hence, $u \leq w^*$ as far $\psi - \psi_1 \geq 0$; it is enough $u \leq w^*$ for any $x \in [\gamma, 1]$, for some $\gamma > 0$, due to the fact that u blows up globally.

Therefore, for simplicity, it is enough to choose

$$\frac{c\phi^*}{t + t_0} \geq (\lambda - \lambda^*)\,\Lambda\,\phi^*\,(t + t_0), \quad x \in [\gamma, 1],$$

so that $\quad c \geq (\lambda - \lambda^*)\Lambda(t + t_0)^2.$

Thus we get

$$t \leq c^{1/2}\,\Lambda^{-1/2}\,(\lambda - \lambda^*)^{-1/2} - t_0,$$

which for λ sufficiently close to λ^* $(\lambda > \lambda^*)$ gives

$$t \lesssim c^{1/2}\,\Lambda^{-1/2}\,(\lambda - \lambda^*)^{-1/2} = t_l(\lambda - \lambda^*)^{-1/2},$$

Hence, as long as $u = u(x,t) < w^*$ at $t = t_l(\lambda - \lambda^*)^{-1/2}$, we deduce that $t^* > t_l(\lambda - \lambda^*)^{-1/2}$ and $t_l(\lambda - \lambda^*)^{-1/2}$ is a lower bound for t^* with $t_l = c^{1/2}\Lambda^{-1/2}$.

3. Numerical Solutions

We solve problem (1.1) by using a two-step up-wind scheme. For the linear terms we apply the usual form of the scheme:

$$v_j^{n+1} = u_j^n - r\left(u_j^n - u_{j-1}^n\right) + \lambda F(u_j^n),$$

where u_j^n is the temperature at the nth time level and at the jth space grid, $r = \frac{\delta t}{\delta x}$ and the non-local term $F(u_j^n)$ is evaluated at the nth time step. For this term we have

$$F(u_j^n) = \frac{f(u_j^n)}{\left(\int_0^1 f(u_j^n)\, dx\right)^2}.$$

Finally u at the $(n+1)$th time step is approximated by

$$u_j^{n+1} = \frac{1}{2}(v_j^{n+1} + w_j^{n+1}).$$

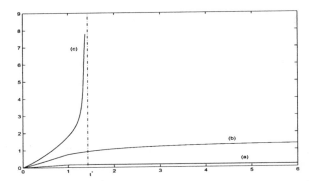

Figure 1. *Numerical solution of problem (1.1). We plot the $\max_x(u(x,t)) = u(1,t) = M(t)$, for x in $[0,1]$ against time for $\delta x = 0.033$, $\delta t = 0.002$ (the upper curve, (c), corresponds to $\lambda = 1.1476 > \lambda^* = 0.6476$, the intermediate, (b), to $\lambda = \lambda^*$ and the lower one, (a), to $\lambda = 0.1476 < \lambda^*$). Also the dash-dotted axis corresponds to the asymptotic estimate of the blow up time $t^* \sim 1.3367$ for $\lambda = 1.1476$. To obtain this estimate we calculate numerically w^* by an iteration scheme and then we solve the equation for ϕ^* using the appropriate normalization.*

In Figure 1 we use this scheme to solve the problem numerically for $f(u) = e^{-u}$ and taking $u(x,0) = 0$. We see that for $\lambda < \lambda^*$ the solution u tends to a steady state, for $\lambda = \lambda^*$ the behaviour is similar, and for $\lambda > \lambda^*$ the solution blows up (the decay is faster for $\lambda < \lambda^*$ than it is for $\lambda = \lambda^*$). More precisely, in Figure 1 the maximum of solutions are plotted against time.

In Figure 2, we plot the numerical solution of u for $\lambda = 0.5476$.

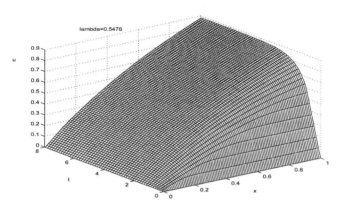

Figure 2. *Numerical solution to problem (1.1) for $\lambda = 0.5476 < \lambda^*$.*

References

1. N.I. Kavallaris, C.V. Nikolopoulos & D.E. Tzanetis, Estimates of blow-up time for a non-local problem modelling an Ohmic heating process, *Euro. Jl. Appl. Maths.* vol. **13**, (2002), pp. 337–351.

2. A.A. Lacey, Mathematical analysis of thermal runaway for spartially inhomogeneous reactions, *SIAM J. Appl. Math.*, **43**, (1983), 1350–1366.

3. A.A. Lacey & D. Tzanetis, Global existence and convergence to a singular steady state for a semilinear heat equation, *Proc. Royal Soc. Edin.*, **105A**, (1986), 289–305.

4. A.A. Lacey, Thermal runaway in a non-local problem modelling Ohmic heating. Part I : Model derivation and some special cases, *Euro. Jl. Appl. Maths.*, **6**, (1995a), 127–144.

5. A.A. Lacey, Thermal runaway in a non-local problem modelling Ohmic heating. Part II : General proof of blow-up and asymptotics of runaway, *Euro. Jl. Appl. Maths.*, **6**, (1995b), 201–224.

6. A.A. Lacey, D.E. Tzanetis & P.M. Vlamos, Behaviour of a nonlocal reactive convective problem modelling Ohmic heating of foods, *Quart. Jl. Mech. Apll. Maths.* **5**, No 4, (1999), 623–644.

7. J. Lopez-Gomez, On the structure and stability of the set of solutions of a non-local problem modelling Ohmic heating, *J. Dynam. Diff. Eqns* **10**, 4, (1998), 537–559.

THE $(F^*F)^{1/4}$-METHOD FOR 2D PENETRABLE ELASTIC BODIES

G. PELEKANOS

Department of Mathematics and Statistics, Southern Illinois University,
Edwardsville, IL 62026, USA
E-mail: gpeleka@siue.edu

V. SEVROGLOU

Department of Mathematics, University of Ioannina,
45110 Ioannina, Greece
E-mail: bsevro@cc.uoi.gr

In this paper we discuss an inversion algorithm for the determination of the shape of a penetrable object from knowledge of the scattered elastic field generated by a given P or SV incident wave. In particular Kirsch's improved variant of the linear sampling method (LSM), the so called $(F^* F)^{1/4}$-method is extended to the two dimensional elastic case.

1. Introduction

The linear sampling method (LSM) was developed by Colton and Kirsch [3] and mathematically clarified in [4], provides an effective way to tackle inverse scattering problems. Two of its attractive features are that no low- or high frequency approximation is needed and that a priori knowledge of the boundary condition(s) is not required. Its main drawback however is that it requires complete far-field data.

Recently Kirsch [6] improved the original version of the linear sampling method, leading to the so-called $(F^*F)^{1/4}$ -method.

This method along with its previous version is known to give only an explicit characterization of the scattering obstacle (i.e. it only determines the support of the refractive index) and it is also very easy to implement, since it just involves the solution of a linear integral equation whose kernel consists of the measured data.

The mathematical analysis presented will be based on an interior transmission problem where a weak-type solution is introduced. After appropriate regularization of the linear integral equation described above, the boundary of the scatterer will be found by noting that the L^2- norm of its Herglotz kernel is not bounded as the source point of the fundamental solution approaches the boundary from inside.

Finally, numerical results from P and SV wave excitations will be presented showing the robustness of the $(F^*F)^{1/4}$-method with respect to noise.

2. Notation and Problem Statement

Consider a bounded, simply connected domain D located in an unbounded homogeneous background medium with constant Lamé coefficients λ and μ. Let a scattering object (or objects) B, whose location and contrast are unknown, embedded in D and assume that B is made by an inhomogeneous elastic material, also characterized by the constant Lamé coefficients λ and μ. The vector $\boldsymbol{x}$ denotes the vectorial position in $\mathbb{R}^2$. We irradiate the object by a number of known incident fields of one single frequency, $\boldsymbol{u}^{inc}(\boldsymbol{x})$. It has been shown, see [10], that the total field in D satisfies the following domain integral equation

$$\boldsymbol{u}(\boldsymbol{x}) = \boldsymbol{u}^{inc}(\boldsymbol{x}) + (k^S)^2 \mu \int_D \boldsymbol{\Gamma}(\boldsymbol{x}, \boldsymbol{x}')\chi(\boldsymbol{x}')\boldsymbol{u}(\boldsymbol{x}')\mathrm{d}v(\boldsymbol{x}'), \quad \boldsymbol{x} \in D, \quad (1)$$

where

$$\chi(\boldsymbol{x}') = \frac{\rho(\boldsymbol{x}') - \rho_e}{\rho_e}, \quad (2)$$

is the normalized contrast between the mass density $\rho = \rho(\boldsymbol{x})$ of the scattering object and the mass density ρ_e of the homogeneous embedding. The Green's displacement tensor $\boldsymbol{\Gamma}(\boldsymbol{x}, \boldsymbol{x}')$ is given by Morse and Feshbach [11] as

$$\boldsymbol{\Gamma}(\boldsymbol{x}, \boldsymbol{x}') = \frac{i}{4}\{\frac{1}{\mu}\boldsymbol{I}H_0^{(1)}(k^S R) - \frac{1}{\omega^2 \rho_e}\boldsymbol{\nabla}\boldsymbol{\nabla}\left[H_0^{(1)}(k^P R) - H_0^{(1)}(k^S R)\right]\}, (3)$$

where $R = |\boldsymbol{x} - \boldsymbol{x}'|$, while k^P and k^S are the wave numbers for the P- and the SV-wave respectively, and $\boldsymbol{I}$ is a unit matrix.

If $\boldsymbol{u}$ solves the equation above, then the scattered field is obtained from the integral representation

$$\boldsymbol{u}^{sct}(\boldsymbol{x}) = (k^S)^2 \mu \int_D \boldsymbol{\Gamma}(\boldsymbol{x}, \boldsymbol{x}')\chi(\boldsymbol{x}')\boldsymbol{u}(\boldsymbol{x}')\mathrm{d}v(\boldsymbol{x}'), \quad \boldsymbol{x} \in S, \quad (4)$$

where S is the measurement domain located outside of and surrounding D.

Our goal in this paper is to present reconstructions based on the above scattered field $\boldsymbol{u}^{sct}$ from P and SV -wave incidence.

3. The Simple Method

We begin our discussion by introducing the linear integral operator $F:[L^2(\Omega)]^2 \to [L^2(\Omega)]^2$, where $\Omega = \{x \in \mathbb{R}^2, |\mathbf{x}| = 1\}$ as follows

$$(F\widetilde{\mathbf{g}})(\hat{\mathbf{x}}) = \int_\Omega \mathbf{u}_\infty(\hat{\mathbf{x}}, \mathbf{d}, \mathbf{d}) \, \mathbf{g}_p(\mathbf{d}) \, ds(\mathbf{d}) + \int_\Omega \mathbf{u}_\infty(\hat{\mathbf{x}}, \mathbf{d}, \mathbf{d}^\perp) \, \mathbf{g}_s(\mathbf{d}) \, ds(\mathbf{d}) \quad (5)$$

where the kernel $\widetilde{\mathbf{g}}$ has columns the $L^2(\Omega)$ vectors $\mathbf{g}_p$, $\mathbf{g}_s$ whereas $\mathbf{d}$, $\hat{\mathbf{x}} \in \Omega$ denote the incident and observation directions respectively. In addition, $\mathbf{u}_\infty$ represents the far field pattern of $\mathbf{u}^{sct}$ with

$$\mathbf{u}_\infty(\cdot, \mathbf{d}, \mathbf{d}) = (u_\infty^P(\cdot, \mathbf{d}, \mathbf{d}), u_\infty^S(\cdot, \mathbf{d}, \mathbf{d})) \quad (6)$$

$$\mathbf{u}_\infty(\cdot, \mathbf{d}, \mathbf{d}^\perp) = (u_\infty^P(\cdot, \mathbf{d}, \mathbf{d}^\perp), u_\infty^S(\cdot, \mathbf{d}, \mathbf{d}^\perp)) \quad (7)$$

For any $\mathbf{y} \in \mathbb{R}^2$ and any direction $\mathbf{p} \in \Omega$, an elastic point source in $\mathbf{y}$ with polarization $\mathbf{p}$ is given by [1]

$$\mathbf{u}(\mathbf{x}) = \mathbf{\Gamma}(\mathbf{x}, \mathbf{y}) \cdot \mathbf{p}, \quad \mathbf{x} \in \mathbb{R}^2 - \{\mathbf{y}\}$$

The far-field pattern $\mathbf{\Gamma}_\infty(\cdot, \mathbf{y}; \mathbf{p})$ of this point source is given by

$$\Gamma_\infty^P(\hat{\mathbf{x}}, \mathbf{y}; \mathbf{p}) = \frac{1}{2\mu + \lambda} \frac{i+1}{4\sqrt{\pi k^P}} e^{-ik^P \hat{\mathbf{x}} \cdot \mathbf{y}} \hat{\mathbf{x}} \cdot \mathbf{p} \quad (8)$$

$$\Gamma_\infty^S(\hat{\mathbf{x}}, \mathbf{y}; \mathbf{p}) = \frac{1}{\mu} \frac{i+1}{4\sqrt{\pi k^S}} e^{-ik^S \hat{\mathbf{x}} \cdot \mathbf{y}} \hat{\mathbf{x}}^\perp \cdot \mathbf{p} \quad (9)$$

The basic idea of the method is to determine a

$$\mathbf{g}(\cdot, \mathbf{y}_o; \mathbf{p}) = (g_p(\cdot, \mathbf{y}_o; \mathbf{p}), g_s(\cdot, \mathbf{y}_o; \mathbf{p}))$$

that solves the far field equations

$$((F^\star F)^{1/4} \mathbf{g})_p = \Gamma_\infty^P(\hat{\mathbf{x}}, \mathbf{y}_o; \mathbf{p}) \quad (10)$$

and

$$((F^\star F)^{1/4} \mathbf{g})_s = \Gamma_\infty^S(\hat{\mathbf{x}}, \mathbf{y}_o; \mathbf{p}) \quad (11)$$

or in compact form

$$(F^\star F)^{1/4} \mathbf{g}(\hat{\mathbf{x}}, \mathbf{y}_o; \mathbf{p}) = \mathbf{\Gamma}_\infty(\hat{\mathbf{x}}, \mathbf{y}_o; \mathbf{p}) \quad (12)$$

where $\mathbf{y}_o$ is a point in the interior of the scattering object, and $\mathbf{p} \in \Omega$ denotes the polarization of the point source.

This 'improved' version of the simple method was proposed by Kirsch [6], [7] in order to clarify what happens when $\mathbf{y}_o \notin B$.

It is also important to note here [8], that equation (12) has a solution if and only if the following interior transmission problem

$$\Delta^* \mathbf{w} + \rho_i\, \omega^2 \mathbf{w} = 0, \quad \Delta^* \mathbf{v} + \rho_e\, \omega^2 \mathbf{v} = 0 \qquad in\ \ B, \qquad (13)$$

$$\mathbf{w} - \mathbf{v} = \boldsymbol{\Gamma}, \quad \mathbf{Tw} - \mathbf{Tv} = \mathbf{T\Gamma} \qquad on\ \ \partial B, \qquad (14)$$

with $\mathbf{y}_o \in B$, where the traction operator $\mathbf{T}$ is given by

$$\mathbf{T} = 2\mu\widehat{\mathbf{n}} \cdot \nabla + \lambda\widehat{\mathbf{n}}\nabla \cdot +\mu\widehat{\mathbf{n}} \times \nabla\times, \qquad (15)$$

whereas Δ^* by

$$\Delta^* = \mu\nabla^2 + (\lambda + \mu)\nabla\nabla\cdot \qquad (16)$$

has a solution $\mathbf{v}, \mathbf{w} \in [C^2(B)]^2 \cap [C^1(\bar{B})]^2$, such that $\mathbf{v}$ is a Herglotz function [9] with kernel $\mathbf{g}$ given by

$$\mathbf{v}(\mathbf{x}) = \int_\Omega e^{ik^p\mathbf{x}\cdot\mathbf{d}}\, g_p(\mathbf{d})\, ds(\mathbf{d}) + \int_\Omega e^{ik^S\mathbf{x}\cdot\mathbf{d}}\, g_s(\mathbf{d})\, ds(\mathbf{d}), \ \ \mathbf{x} \in \mathbb{R}^2 \qquad (17)$$

As it is pointed out in [3] the above is true only in special cases. However, it has been shown that we can always find an approximate solution of the far field equation (12) that has the important property

$$\|\mathbf{g}(\cdot, \mathbf{y}_o)\|_{[L^2(\Omega)]^2} \to \infty, \quad \|\mathbf{v}\|_{[L^2(B)]^2} \to \infty. \qquad (18)$$

as $\mathbf{y}_o \to \partial B$. The existence of such functions is shown in the following theorem, provided that ω^2 is not a transmission eigenvalue.

Theorem 3.1. *For every $\epsilon > 0$ and $\mathbf{y_o} \in B$ there is a solution $\mathbf{g}(:, \mathbf{y_o}; \mathbf{p}) \in [L^2(\Omega)]^2$ of the inequality*

$$\left\|(F^\star F)^{1/4}\mathbf{g}(\cdot, \mathbf{y_o}; \mathbf{p}) - \boldsymbol{\Gamma}_\infty(\cdot, \mathbf{y_o}; \mathbf{p})\right\|_{[L^2(\Omega)]^2} \le \epsilon$$

where $\boldsymbol{\Gamma}_\infty(\cdot, \mathbf{y_o}; \mathbf{p}) = (\Gamma_\infty^P(\cdot, \mathbf{y_o}; \mathbf{p}), \Gamma_\infty^S(\cdot, \mathbf{y_o}; \mathbf{p}))$, such that

$$\|\mathbf{g}(\cdot, \mathbf{y_o})\|_{[L^2(\Omega)]^2} \to \infty, \quad \|\mathbf{v}\|_{[L^2(B)]^2} \to \infty \qquad (19)$$

as $\mathbf{y_o} \to \partial B$, where $\mathbf{v}$ is the Herglotz function with kernel $\mathbf{g}$.

The proof can be found in [4] and hence is omitted for brevity.

4. Regularization

Clearly compactness of the far field operator F, given by equation (5), constitutes the linear inverse problem of determining the function $\mathbf{g}(:, \mathbf{y_o}; \mathbf{p})$ ill-posed in the sense of Hadamard. Thus appropriate regularization is required.

In general we let F_δ to be a version of F which is affected by a measurement error of the order of magnitude of δ, i.e.

$$\|F_\delta - F\|_{[L^2(\Omega)]^2} \leq \delta \tag{20}$$

We now compute $\mathbf{g}$ by minimizing the Tikhonov functional [5]

$$\left\|(F_\delta^\star F_\delta)^{1/4}\mathbf{g}(\cdot, \mathbf{y_o}; \mathbf{p}) - \mathbf{\Gamma}_\infty(\cdot, \mathbf{y_o}; \mathbf{p})\right\|^2 + \gamma \|\mathbf{g}(\cdot, \mathbf{y_o}; \mathbf{p})\|^2 \tag{21}$$

where the regularization parameter γ is chosen by a generalized Morozov's discrepancy principle [4], in such a way as

$$\left\|(F_\delta^\star F_\delta)^{1/4}\mathbf{g}(\cdot, \mathbf{y_o}; \mathbf{p}) - \mathbf{\Gamma}_\infty(\cdot, \mathbf{y_o}; \mathbf{p})\right\| = \delta \|\mathbf{g}(\cdot, \mathbf{y_o}; \mathbf{p})\| \tag{22}$$

Minimizing the Tikhonov functional is equivalent to solving the normal equation

$$\gamma\mathbf{g} + (F_\delta^\star F_\delta)^{1/2}\mathbf{g} = (F_\delta^\star F_\delta)^{1/4}\mathbf{\Gamma}_\infty(\cdot, \mathbf{y_o}; \mathbf{p}) \tag{23}$$

We are now ready to proceed with some numerical applications.

5. Numerical Results

In our first numerical example the scatterer was taken to be a circular cylinder of radius $a = 0.35$ m and density $\rho = 1.6$, while the outer medium's density was $\rho_e = 1$. Hence, the contrast is $\chi = 0.6$. The circular cylinder is centered at $(0, 0)$ m. The scatterer is located in a test domain. This test domain was divided into 21×21 subsquares of 0.1×0.1 m^2. The wavenumbers of the P-waves and the SV-waves are $k^P = 3$ and $k^S = 6$, respectively.

The synthetic data in our numerical experiments were generated by using the weak form of the conjugate gradient FFT method [12]. We compute the far-field patterns $\mathbf{u}_\infty(\hat{\mathbf{x}}, \mathbf{d}, \mathbf{d})$ and $\mathbf{u}_\infty(\hat{\mathbf{x}}, \mathbf{d}, \mathbf{d}^\perp)$ at 29 equidistantly distributed observation points and 29 uniformly distributed directions. Furthermore, a 5% Gaussian noise is added to each element of the far-field matrix.

Figure 1 shows the contours for the circles of radius .35 m generated by P-wave incidence (left) and SV- wave incidence (right).

368

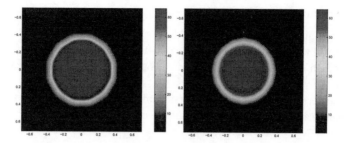

Figure 1. Reconstruction of a circular cylinder from P-wave incidence (left), and SV-wave incidence (right).

We now repeat the above experiment for two square scatterers with side .41 m, that their centers are located at locations $(-.4, 0)$ and $(.4, 0)$ respectively on the test square. The test domain is subdivided again into 21×21 subsquares, and 5% Gaussian noise is added to the data. The reconstruction results are given in Figure 2 where we clearly observe the presence of two square objects. Note however that the reconstruction obtained from S-wave incidence (right) is better comparing to the one obtained via P-wave incidence.

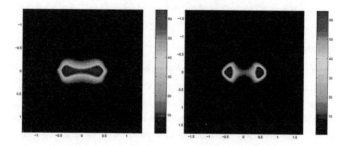

Figure 2. Reconstruction of two square cylinders from P-wave incidence (left), and SV-wave incidence (right).

References

1. T. Arens, Linear Sampling Methods for 2D Inverse Elastic Wave Scattering, Inverse Problems **17** pp. 1445-1464 (2001).
2. D. Colton and R. Kress, Inverse Acoustic and Electromagnetic Scattering Theory, Berlin: Springer (1992).

3. D. Colton and A. Kirsch, A Simple Method for Solving Inverse Scattering Problems in the Resonance Region, Inverse Problems **12**, pp. 383-393 (1996).

4. D. Colton, M. Piana and R. Potthast, A Simple Method using Morozov's Discrepancy Principle for Solving Inverse Scattering Problems, Inverse Problems **13**, pp. 1477-1493 (1997).

5. C. W. Groetsch, The Theory of Tikhonov Regularization for Fredholm Equations of the First Kind, (Pitman, Boston) (1984).

6. A. Kirsch, Characterization of the Shape of a Scattering Obstacle using the Spectral Data of the Far Field Operator, Inverse Problems **14**, pp. 489-512 (1998).

7. A. Kirsch, Factorization of the Far Field Operator for the Inhomogeneous Medium Case and an Application in Inverse Scattering Theory, Inverse Problems **15**, pp. 413-429 (1999).

8. A. Kirsch, An Introduction to the Mathematical Theory of Inverse Problems, (Berlin: Springer) (1996).

9. K. Kiriaki and V. Sevroglou, On Herglotz Functions in Two-Dimensional Linear Elasticity, Scattering Theory and Biomedical Engineering Modeling and Applications, Proceedings of the 4th International Workshop, pp. 151-158 (1999).

10. V. D. Kupradze, Potential Methods in the Theory of Elasticity, (Israel Program for Scientific Translations, Jerusalem) (1965).

11. P. M. Morse and H. Feshbach, Methods of Theoretical Physics, Vols. I, II., McGraw-Hill, New York (1953).

12. G. Pelekanos, R. E. Kleinman and P. M. Van den Berg, A Weak Form of the Conjugate Gradient FFT Method for Two-Dimensional Elastodynamics, Journal of Comp. Phys. **160**, pp. 597-611 (2000).

ECHOES FROM TIME DEPENDENT PERTURBATIONS

G. F. ROACH

Department of Mathematics, University of Strathclyde
Richmond Street, Glasgow. G 1 1XH, Scotland. UK

1. Introduction

The use of various types of wave energy as a probe is an increasingly promising non-destructive means of identifying objects and of diagnosing the properties of quite complicated materials.

In this note the central problem with which we shall be concerned can be simply stated as follows.

A system consists of a medium containing a transmitter and a receiver. The transmitter emits a signal which is eventually detected at the receiver, possibly after it has been perturbed, that is scattered, by some inhomogeneity in the medium. We are interested in the manner in which the emitted signal evolves through the medium and the form that it assumes at the receiver. The direction of arrival, time delays and frequency shifts associated with the scattered or echo signal can then be used to estimate the properties of inhomogeneities in the medium. When the media involved are either stationary or possess time-independent characteristics, the so-called autonomous problems (AP), then the mathematical analysis of the associated scattering effects is now quite well developed. However, the investigations of corresponding scattering phenomena when the media are either moving or have time-dependent characteristics, the so-called non-autonomous problems (NAP), have not reached such a well developed state.

The study of NAP is motivated by problems arising in such areas as, for example, radar, sonar, non-destructive testing and ultrasonic medical diagnosis. In all these areas a powerful diagnostic is the dynamical response of the media to the emitted signal. Mathematically, many of these problems can be conveniently modelled in terms of an initial boundary value problem (IBVP). To fix ideas we shall confine attention here to acoustic problems and to IBVP for the classical wave equation. Specifically, we shall be interested in an IBVP that has the following typical form.
Let

$$Q = \left\{ (x, t) \in \mathbf{R}^n \times \mathbf{R} \right\}$$

$$\Omega(t) = \{x \in \mathbf{R}^n : (x,t) \in Q\} \quad , \quad B(t) = \{x \in \mathbf{R}^n : (x,t) \notin Q\}$$

The region Q is assumed to be open in $(\mathbf{R}^n \times \mathbf{R})$ and $\Omega(t)$ denotes the exterior, at time t, of a scattering target $B(t)$. For each value of t the domain $\Omega(t)$ is open in $(\mathbf{R}^n \times \mathbf{R})$ and is assumed to have a smooth boundary $\partial\Omega(t)$. The lateral surface of Q, denoted ∂Q, is defined by $\partial \mathcal{U} := \underset{t \in I}{\cup} \partial\Omega(t)$ where for $T \geq 0$ fixed $I := \{t \in \mathbf{R} : 0 \leq t \leq T\}$. We shall assume throughout that $B(t)$ remains in a fixed ball in $\mathbf{R}^n$ and that ∂Q is time like in the sense that the boundary speed is less than the speed of propagation.

2. Mathematical Model

The basic mathematical problem we investigate is the following.
Determine a quantity $u(x,t)$ satisfying the IBVP

$$\{\partial_t^2 + L(x,t)\}u(x,t) = f(x,t) \quad , \quad (x,t) \in Q \tag{1}$$

$$u(x,s) = \varphi(x,s) \quad , \quad u_t(x,s) = \psi(x,s) \quad , \quad x \in \Omega(s) \quad , \quad s \in \mathbf{R} \tag{2}$$

$$u(x,t) \in (bc)(t) \quad , \quad (x,t) \in \partial\Omega(t) \times \mathbf{R} \tag{3}$$

where
$L(x,t)$ is a differential expression characterising the wave field
$f, \varphi(.,s), \psi(.,s)$ are given data functions
$s \in \mathbf{R}$ is a fixed initial time
$(bc)(t)$ indicates boundary condition to be satisfied by $u(.,.)$.
We remark that in (1) the inhomogeneous term f characterises the transmitter and the signals which it emits.
To fix ideas we shall confine attention here to an acoustic Dirichlet problem for which we take

$$L(x,t) = -\Delta \quad , \quad \text{for all } (x,t) \tag{4}$$

$$(bc)(t) = \{u(.,.): u(x,t) = 0 , (x,t) \in \partial\Omega(t) \times \mathbf{R})\} \tag{5}$$

Our aim here is to discuss and determine solutions of the acoustic Dirichlet problem defined above. This is most conveniently done by representing the problem in an energy space setting.

3. Energy Space Settings and Solution Concepts

We shall use the notation

$$\mathbf{f} = \begin{bmatrix} f_1 \\ f_2 \end{bmatrix} = <f_1, f_2> \tag{6}$$

We introduce the energy norm

$$\|\mathbf{f}\|^2 = \|<f_1, f_2>\|^2 = \tfrac{1}{2}\int_{\mathbf{R}^n}\{|\nabla f_1(x)|^2 + |f_2(x)|^2\}dx \tag{7}$$

where we assume $f_1, f_2 \in C_0^\infty(\mathbf{R}^n)$.

Associated with the norm in (7) is the inner product

$$(\mathbf{f}, \mathbf{g}) = (\nabla f_1, \nabla g_1) + (f_2, g_2) \tag{8}$$

where on the right hand side of (8) the notation $(.,.)$ denotes the usual $L_2(\mathbf{R}^n)$ inner product.

We shall write $H_0 = H_0(\mathbf{R}^n)$ to denote the completion of $C_0^\infty(\mathbf{R}^n) \times C_0^\infty(\mathbf{R}^n)$ with respect to the energy norm (7) and introduce

$$H(t) = \{f \in H_0 : \mathbf{f} = 0 \text{ on } B(t)\} \tag{9}$$

$$H^{loc}(t) = \{\mathbf{f} =<f_1, f_2>: \zeta \mathbf{f} \in H(t) \ \forall \zeta \in C_0^\infty(\mathbf{R}^n)\} \tag{10}$$

We notice that

(i) the first component of $\mathbf{f} \in H^{loc}(t)$ must vanish on $\partial\Omega(t)$

(ii) the first component of $\mathbf{f} \in H^{loc}(t)$ must vanish on $B(t)$.

It is convenient, at this stage, to emphasise the following notation.

(iii) $h = h(.,.) : (x,t) \to h(x,t) \in \mathbf{K} = \mathbf{R}$ or $\mathbf{C}$

(iv) $h = h(.,.) : t \to h(.,t) =: h(t) \in H_0$

In this latter case we refer to h as an H_0-valued function of t.

We are now in a position to say what we mean by a solution of our problem.

Definition 3.1

A function $u = u(.,.)$ is a solution of locally finite energy of the IBVP (1)-(5) if

(i) $\mathbf{u} =<u_1, .u_2> \in C(\mathbf{R}, H^{loc}(t))$

(ii) $\{\partial_t^2 + L(x,t)\}u(x,t) = f(x,t)$, $(x,t) \in Q$

(in the sense of distributions).

We say that u is a solution of finite energy if $\mathbf{u} \in C(\mathbf{R}, H_0)$ and $\mathbf{u}(t) \in H(t)$ for each t. A function u is a free solution of finite energy if $\mathbf{u} \in C(\mathbf{R}, H_0)$.

If $\mathbf{u} \in H_0$ then u has finite energy and we write

$$\|\mathbf{u}(t)\|^2 = \|u(t)\|^2 = \tfrac{1}{2}\int_{\mathbf{R}^n}\{|\nabla u_1(x,t)|^2 + |u_2(x,t)|^2\}dx \tag{11}$$

for the total energy of u at time t.

The wave energy in a sphere is obtained from (11) by restricting the range of integration appropriately.

4. Reduction to a First Order System

By giving the IBVP (1)-(5) an energy space setting we are led to a consideration of an initial value problem, (IVP), of the following form

$$\{\partial_t + N(t)\}\mathbf{u}(x,t) \;=\; \mathbf{F}(x,t) \quad, \quad (x,t) \in Q \tag{12}$$

$$\mathbf{u}(x,s) \;=\; <\varphi(.,s), \psi(.,s)>(x) \quad, \quad x \in \Omega(s) \tag{13}$$

where

$$\mathbf{u}(x,t) \;=\; <u, u_t>(x,t) \quad, \quad \mathbf{F}(x,t) \;=\; <0, f>(x,t)$$

$$N(t) \;:=\; \begin{bmatrix} 0 & -I \\ A(t) & 0 \end{bmatrix}$$

and denoting by $L_2^D(\Omega(t))$ the completion of $C_0^\infty(\Omega(t))$ in $L_2(\Omega(t))$ we have introduced

$$A(t) : \; L_2^D(\Omega(t)) \to L_2^D(\Omega(t))$$

$$A(t)u(.,t) \;=\; L(.,t)u(.,t) \quad, \quad u(.,t) \in D(A(t))$$

$$D(A(t)) \;=\; \{u \in L_2^D(\Omega(t)) : \; L(.,t)u(.,t) \in L_2^D(\Omega(t))\}$$

Throughout we assume that the receiver, R_x, and the transmitter T_x are in the far field of $B(t)$ and, furthermore, that supp $f \subset \{|(x,t)|, \; t_0 \le t \le T,$ $|x - x_0| \le \delta_0\}$ where x_0 denotes the position of T_x and t_0, T, δ_0 are constants.

If we now introduce

$$G(t) : \; H(t) \to H(t) \tag{14}$$

$$G(t)\mathbf{u}(t) \;=\; -iN(t)\mathbf{u}(t) \quad, \quad \mathbf{u}(t) \in D(G(t))$$

$$D(G(t)) \;=\; \{\mathbf{u}(t) \in H(t) : \; N(t)\mathbf{u}(t) \in H(t)\}$$

then the IVP (12) can now be realised as a first order system in $H(t)$ in the form

$$\{d_t + iG(t)\}\mathbf{u}(t) \quad = \quad \mathbf{F}(t) \quad , \quad t \in \mathbf{R}$$
$$\mathbf{u}(s) \qquad = \quad \mathbf{u}_s \tag{15}$$

A general theory of IVPs of the form (15) has been developed by a number of authors and in this connection we would cite [1, 2, 6, 7]. These authors introduced the notion of a fundamental solution or propagator, denoted $U(t,s)$, for the IVP (14), which was required to satisfy

(i) $U(t,s)$ has values in $\mathbf{B}(H_0(\mathbf{R}^n))$, the set of bounded linear operators in $H_0(\mathbf{R}^n)$, for $0 \le s \le t \le T$ for some fixed T

(ii) $U(t,r)U(r,s) = U(t,s)$, $U(s,s) = I$

(iii) $\partial_t U(t,s) =: \partial_1 U(t,s) = -iG(t)U(t,s)$

(iv) $\partial_s U(t,s) =: \partial_2 U(t,s) = iU(t,s)G(s)$

The quantity $U(t,s)$ is also referred to as an evolution operator for the family $\{iG(t)\}_{0 \le t \le T}$. The evolution operator is the operator $\mathbf{u}(s) \to \mathbf{u}(t)$ which maps $H(s)$ onto $H(t)$. The abstract theory provides conditions under which the family $\{iG(t)\}_{0 \le t \le T}$ has at most one associated propagator. In the present case a simple integrating factor technique applied to (15) yields

$$\mathbf{u}(t) \quad = \quad U(t,s)\mathbf{u}_s + \int_s^t U(t,\tau)\mathbf{F}(\tau)d\tau \tag{16}$$

where

$$U(t,s) = \exp\left\{-i\int_s^t G(\eta)d\eta\right\} \tag{17}$$

The relation (16) is the familiar variation of parameter formula and the integral involved is a corresponding Duhamel type integral.

5. Scattering Aspects and Echo Analysis

Scattering theory is concerned with the (asymptotic) comparison of two systems; these systems being characterised by the families of operators $\{iG_j(t)\}$, $j = 1,2$ respectively.

We shall assume

(i) $\{iG_j(t)\}$, $j = 1,2$ are self adjoint operators defined on suitable Hilbert space(s)

(ii) The families $\{iG_j(t)\}$, $j = 1,2$ satisfy conditions which ensure that

$$\{d_t + iG_j(t)\}\mathbf{u}_j(t) \quad = \quad \mathbf{F}(t) \quad , \quad \mathbf{u}_j(s) \quad = \quad \mathbf{u}_{sj} \quad , \quad j = 1,2 \tag{18}$$

have unique solutions.

(iii) $U_j(t,s)$, $j = 1,2$ denote the associated propagators.

We can now introduce, in much the same manner as for AP, [see 5, 7]
Wave Operators (WO) :

$$W_{s^{\pm}}(G_1, G_2) := s - \lim U_1(s,t)U(t,s) \tag{19}$$

Scattering Operator (SO):

$$S_s(G_1, G_2) = W_{s^+}(G_1, G_2)W_{s^-}(G_1, G_2)^* \tag{20}$$

The various properties of these two items are much the same as for similar
IVP in the AP case. These are indicated in detail in [5, 7, 4].
We remark that in most practical applications one or other of G_1 and G_2
can often be taken as being independent of t.
In developing an echo analysis for the IBVP (1)-(5) we shall assume that
the medium is initially at rest and concern ourselves with the IVP

$$\{\{d_t + iG_j(t)\}\mathbf{u}_j(t)\} = \mathbf{F}(t) \quad , \quad \mathbf{u}_j(s) = 0 \quad , \quad j = 1,2 \tag{21}$$

The Free (unperturbed) Problem (FP) obtains when $j = 1$ and
$\Omega(t) = \Omega = \mathbf{R}^n$.
The Perturbed Problem (PP) obtains when $j = 2$ and $\Omega(t) \subset \mathbf{R}^n$.
If we bear in mind the various properties of the propagators and, in
particular, the fact that $U(t,s): H(s) \to H(t)$ then the variation of
parameters formula indicates that solutions of (21) can be written,
for $j = 1,2$, in the form

$$\mathbf{u}_j(t) = U_j(t,s)\int_{t_0}^{t_0+T} U_j(s,\tau)\mathbf{F}_j(\tau)d\tau =: U_j(t,s)\mathbf{h}_j(s) \tag{22}$$

Scattering phenomena involve three fundamental items; the incident, the
total and the scattered wave fields. In the present case we have

Incident field $\quad \mathbf{u}_1(t) = U_1(t,s)\mathbf{h}_1(s)$
Total field $\qquad \mathbf{u}_2(t) = U_2(t,s)\mathbf{h}_2(s)$
Scattered field $\quad \mathbf{u}^s(t) = \mathbf{u}_2(t) - \mathbf{u}_1(t)$

The definition of the WO and SO enables us to write

$$\mathbf{u}_2(t) = U_1(t,s)W_{s^+}(G_1, G_2)\mathbf{h}_2(s) + \sigma_t(1) \quad \text{as} \quad t \to \infty$$

where $\sigma_t(1)$ is an $L_2(\mathbf{R}^n)$ function of t with $\sigma_t(1) \to 0$ as $t \to \infty$.
Similarly we can obtain

$$\mathbf{u}^s(t) = U_1(t,s)\{W_{s^+}(G_1, G_2)\mathbf{h}_2(s) - \mathbf{h}_1(s)\} + \sigma_t(1) \tag{23}$$

For $t << 0$ there is no scattered field. Hence $\mathbf{u}_2(t) = \mathbf{u}_1(t)$ that is, $U_2(t,s)\mathbf{h}_2(s) = U_1(t,s)\mathbf{h}_1(s)$. Now, the asymptotic equality of the FP and the PP requires, for example [3],

$$\mathbf{h}_1(s) = W_{s^-}(G_1, G_2)\mathbf{h}_2(s) + \sigma_{x_0}(1) \quad \text{as } |x_0| \to \infty$$

where $\sigma_{x_0}(1)$ is an $L_2(\mathbf{R}^n)$ function of x with $\sigma_{x_0}(1) \to 0$ as $|x_0| \to \infty$. Consequently, operating on both sides of this result with S_s and recalling (20) we obtain

$$
\begin{aligned}
S_s(G_1, G_2)\mathbf{h}_1(s) &= S_s(G_1, G_2)W_{s^-}\mathbf{h}_2(s) + \sigma_{x_0}(1) \\
&= W_{s^+}\mathbf{h}_2(s) + \sigma_{x_0}(1)
\end{aligned}
\tag{24}
$$

If we now substitute (24) in (23) then we obtain

$$\mathbf{u}^s(t) = U_1(t,s)\{S_s(G_1, G_2) - I\}\mathbf{h}_1(s) + \sigma_t(1) + \sigma_{x_0}(1) \tag{25}$$

Thus we see that the scattered (echo) field is determined in the far field by the SO and the far field data.

6. On the Construction of the Propagators

For ease of presentation we shall temporarily drop the subscript j, and $\mathbf{u}$ will represent either $\mathbf{u}_1$ or $\mathbf{u}_2$ as appropriate.

To determine the propagator, $U(t,s)$, we write it as a perturbation of the semigroup generated by $(-iG(s))$, s fixed and $iG(s)$ self-adjoint on $H(s)$. Stone's Theorem then indicates that the first term in

$$U(t,s) = \exp\{-i(t-s)G(s)\} + K(t,s) =: U(t-s) + K(t,s) \tag{26}$$

is well defined. We then obtain [1, 4]

$$U(t,s) = U(t-s) + \int_s^t U(t-\tau)R(\tau,s)d\tau \tag{27}$$

where

$$
\begin{aligned}
R(t,s) &= \sum_{m=1}^{\infty} R_m(t,s) \\
R_m(t,s) &= \int_s^t R_{m-1}(t,\tau)R_1(\tau,s)d\tau \quad , \quad m \geq 2 \quad , \quad 0 \leq s \leq t \leq T \\
R_m(t,s) &= -i\{G(t) - G(s)\}U(t-s)
\end{aligned}
$$

and

$$\|R(t,s)\| \leq c(t-s)^{b-1}$$

where

$$iG(.) \in C^b([0,T],\mathbf{B}(H(.)))$$

It then follows that an exact representation of the required solution is

$$\mathbf{u}(t) = \{U(t-s) + \sum_{m=1}^{\infty} \int_s^t U(t-\tau)R_m(\tau,s)d\tau\}\mathbf{h}(s) \tag{28}$$

The results (27) and (28) provide a firm base from which to develop approximations for $W_{s^{\pm}}$, S_s and $\mathbf{u}^s$.

The required solution of problems of the form (1)-(5) are given by the first component of $\mathbf{u}$ in (28). This is obtained by making use of the result [3]

$$U(t) = \exp(-itG(s))$$

$$= \begin{bmatrix} \cos tA^{\frac{1}{2}}(s) & A^{-\frac{1}{2}}(s)\sin tA^{\frac{1}{2}}(s) \\ -A^{\frac{1}{2}}(s)\sin tA^{\frac{1}{2}}(s) & \cos tA^{\frac{1}{2}}(s) \end{bmatrix}$$

Since s is fixed in time then the various entries in the above matrix can be interpreted by means of either the familiar spectral theorem or by some other particularly appropriate generalised eigenfunction expansion theorem.

References

1. H. Amann, "Ordinary Differential Equations, An Introduction to Nonlinear Analysis", W. de Gruyter, Berlin (1990).
2. T. Kato, "Perturbation Theory for Linear Operators", Springer, New York (1966).
3. G. F. Roach, "An Introduction to Linear and Nonlinear Scattering Theory", Pitman Monographs and Surveys in Pure and Applied Mathematics 78, Longman, Essex, UK (1995).
4. G. F. Roach, "Wave scattering by time dependent perturbations, Fractional Calculus and Appl. Anal.4 (2),209-236 (2001).
5. E. J. P. Schmidt, "On scattering by time dependent potentials, Indiana Univ. Math. Jour. 24 (10), 925-934 (1975).
6. P. E. Sobolevski, "Equations of parabolic type in a Banach space", Amer. Math. Soc.Transl .49, 1-62 (1996).
7. H. Tanabe, "Evolution Equations", Pitman Monographs and Studies in Mathematics, 6, Pitman, London (1979).

A HYBRID LBIE/BEM METHODOLOGY FOR SOLVING FREQUENCY DOMAIN ELASTODYNAMIC PROBLEMS

E. J. SELLOYNTOS AND D. POLYZOS*

Department of Mechanical and Aeronautical Engineering.
University of Patras
Greece,
E-mail: polyzos@mech.upatras.gr

This work addresses a coupling of a Local Boundary Integral Equation (LBIE) method, recently proposed by the authors for elastodynamic problems, and the Boundary Element Method (BEM). Because both methods conclude to a final system of linear equations expressed in terms of nodal displacement and tractions, their combination is accomplished directly with no further transformations as it is happens in other meshless/BEM formulations as well as in typical hybrid Finite Element Method/BEM schemes. The coupling approach is demonstrated for a frequency domain elastodynamic problem.

1. Introduction

The goal of any numerical hybrid scheme appearing to date in the literature is to combine properly two numerical methods in order to exploit the potential of each method and to confine their deficiencies. The BEM is a well-known and robust numerical tool, successfully used to solve various types of engineering elastic problems (Beskos[1,2]). The main advantage offered by the BEM as compared to FEM is the reduction of the dimensionality of the problem by one, which means that two- and three- dimensional problems are accurately solved by discretizing only the surfaces surrounding the domain of interest. Since a boundary element formulation takes automatically into account conditions at infinity, this advantage becomes more pronounced in problems where infinite or semi-infinite domains are considered. Nevertheless, the requirement of using the fundamental solution of the differential equation or system of differential equations that describe

---

*Work partially supported by grant of the Institute of Chemical Engng. and High Temperature Process.

the problem of interest renders the BEM less attractive than FEM when non-linear, non-homogeneous and anisotropic elastic problems are considered. Also, the final system of linear equations taken by a BEM formulation leads to unsymmetric and full-populated matrices the numerical treatment of which is in general computationally expensive. Recently, Zhu, Zhang and Atluri [3,4,5] proposed a meshless methodology called Local Boundary Integral Equation (LBIE) method that seems to circumvent the two aforementioned problems associated with a conventional boundary element formulation. Their methodology is characterized as "meshless" since a cloud of properly distributed nodal points covering the domain of interest as well as the surrounding global boundary is employed instead of any boundary or finite element discretization. The combination of these two methods appears to be the solution for many engineering problems the treatment of which cannot be accomplished by the use of either the BEM or the LBIE method. Although some hybrid meshless/BEM methodologies have been reported in the literature so far (Liu and Gu[6,7], Sellountos and Polyzos[8], Li, Paulino and Aluru[9]) there is not a work dealing with the combination of BEM and LBIE method. Very recently, a new LBIE method for solving frequency domain elastodynamic problems has been proposed by the authors of the present paper (Sellountos and Polyzos[8]). The goal of the present paper is the combination of this LBIE method with the BEM.

2. BEM formulation

Consider a two-dimensional linear elastic domain V surrounded by a surface S part of which is subjected to an exterior harmonic excitation. The developed displacement field $\mathbf{u}$ satisfies the Navier-Cauchy differential equation:

$$\mu\nabla^2\mathbf{u} + (\lambda + \mu)\nabla\nabla\cdot\mathbf{u} + \rho\omega^2\mathbf{u} = 0 \tag{1}$$

where λ,μ and ρ stand for the Lame constants and the mass density, respectively, ∇ is the gradient operator and ω the excitation frequency. Considering the fundamental solutions of the above differential equation (Polyzos et al[10]) and employing the well-known Betti's reciprocal identity, one can obtain the integral equation (Manolis and Beskos[11], Dominguez[12])

$$\alpha\mathbf{u}(\mathbf{x}) + \int_S \tilde{\mathbf{t}}^*(\mathbf{x},\mathbf{y})\cdot\mathbf{u}(\mathbf{y})\,dS_y = \int_S \bar{\mathbf{u}}^*(\mathbf{x},\mathbf{y})\cdot\mathbf{t}(\mathbf{y})\,dS_y \tag{2}$$

where $\bar{\mathbf{u}}^*$, $\tilde{\mathbf{t}}^*$ are the dynamic fundamental displacement and the corresponding traction tensor (Polyzos et al[10]), respectively, and α is a jump coefficient taking the value 1 for interior field points $\mathbf{x}$ and the value $1/2$

when boundary points are considered.

In order to solve numerically Eq. (2), the boundary S is discretized into isoparametric quadratic line elements. For smooth boundaries full continuous elements are employed, while a combination of continuous-discontinuous or partially discontinuous elements are used in order to treat boundaries with corners and discontinuous boundary conditions. Collocating the discretized integral Eq. (2) at each node, one obtains a system of linear algebraic equations having the form

$$[\mathbf{H}] \cdot \{\mathbf{u}\} = [\mathbf{G}] \cdot \{\mathbf{t}\} \tag{3}$$

where the vectors $\{\mathbf{u}\}$, $\{\mathbf{t}\}$ contain all the nodal components of the displacement and traction vectors, respectively, and $[\mathbf{H}]$, $[\mathbf{G}]$ are full populated matrices with complex elements each of which is a function of frequency, material properties and structure's geometry.

3. The LBIE methodology

The departure point for the local boundary integral representation of the problem is the Eq. (2). Since both $\tilde{\mathbf{u}}^*$ and $\tilde{\mathbf{t}}^*$ become singular only when $\mathbf{y}$ approaches $\mathbf{x}$, it is easy to find one that the integral Eq. (2) can also be wriiten in the form

$$\mathbf{u}(\mathbf{x}) + \int_{\partial\Omega_S} \tilde{\mathbf{t}}^*(\mathbf{x}, \mathbf{y}) \cdot \mathbf{u}(\mathbf{y}) \, dS_y = \int_{\partial\Omega_S} \tilde{\mathbf{u}}^*(\mathbf{x}, \mathbf{y}) \cdot \mathbf{t}(\mathbf{y}) \, dS_y \tag{4}$$

where $\partial\Omega_S$ is the boundary of an arbitrarily small circle Ω_S centered at the field point $\mathbf{x}$. In case where the field point $\mathbf{x}$ locates either near or on the global boundary S so that the corresponding local domain Ω_S intersects the global boundary S, Eq. (4) obtains the form

$$\alpha\mathbf{u}(\mathbf{x}) + \int_{\partial\Omega_S \cup \Gamma_s} \tilde{\mathbf{t}}^*(\mathbf{x}, \mathbf{y}) \cdot \mathbf{u}(\mathbf{y}) \, dS_y = \int_{\partial\Omega_S \cup \Gamma_s} \tilde{\mathbf{u}}^*(\mathbf{x}, \mathbf{y}) \cdot \mathbf{t}(\mathbf{y}) \, dS_y \tag{5}$$

with Γ_s being the part of S intersected by the sub-domain Ω_S and $\partial\Omega_S$ the boundary of Ω_S belonging in the interior space V (Fig. 1). In order to get rid of the unknown tractions on the circular boundaries $\partial\Omega_S$, the elastodynamic companion solution $\tilde{\mathbf{U}}^c$, derived in Sellountos and Polyzos[8] is employed. By invoking Betti's reciprocal identity for the fields $\mathbf{u}, \tilde{\mathbf{U}}^c$ over the sub-domain Ω_s the local integral representations of Eqs (4) and (5) obtain eventually the form

$$\mathbf{u}(\mathbf{x}) + \int_{\partial\Omega_s} \left[\tilde{\mathbf{t}}^*(\mathbf{x}, \mathbf{y}) - \tilde{\mathbf{T}}^c(\mathbf{x}, \mathbf{y}) \right] \cdot \mathbf{u}(\mathbf{y}) \, dS_y = \mathbf{0} \tag{6}$$

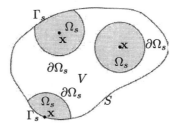

Figure 1. Local domains and local boundaries used for the integral representation of displacements at point **x**

and

$$\alpha \mathbf{u}\left(\mathbf{x}\right) + \int_{\partial\Omega_s \cup \Gamma_s} \left[\tilde{\mathbf{t}}^*\left(\mathbf{x}, \mathbf{y}\right) - \tilde{\mathbf{T}}^c\left(\mathbf{x}, \mathbf{y}\right)\right] \cdot \mathbf{u}\left(\mathbf{y}\right) dS_y =$$
$$\int_{\Gamma_s} \left[\tilde{\mathbf{u}}^*\left(\mathbf{x}, \mathbf{y}\right) - \tilde{\mathbf{U}}^c\left(\mathbf{x}, \mathbf{y}\right)\right] \cdot \mathbf{t}\left(\mathbf{y}\right) dS_y \qquad (7)$$

where both $\tilde{\mathbf{U}}^c$ and $\tilde{\mathbf{T}}^c$ can be found in Sellountos and Polyzos[8] In almost all the LBIE methodologies appearing to date in the literature, the assembly of the local integral equations valid for each nodal point **x** is accomplished with the MLS local interpolation scheme. Thus displacements and boundary traction fields are written in the following interpolation form

$$\mathbf{u}^h\left(\mathbf{x}\right) = \sum_{j=1}^{n} \phi_j\left(\mathbf{x}, \mathbf{x}^{(j)}\right) \hat{\mathbf{u}}^{(j)} \qquad (8)$$

$$\mathbf{t}^h\left(\mathbf{x}\right) = \sum_{j=1}^{n} \phi_j\left(\mathbf{x}, \mathbf{x}^{(j)}\right) \hat{\mathbf{t}}\left(\mathbf{x}^{(j)}\right) \qquad (9)$$

where $\hat{\mathbf{u}}$ and $\hat{\mathbf{t}}$ are the fictitious nodal values of displacement and traction, respectively, and ϕ_j are the MLS interpolants explained in Sellountos and Polyzos[8]. Inserting Eq. (8) and Eq. (9) into (6) and (7) and collocating at all internal and boundary nodes one obtains the following final system of algebraic equations

$$\tilde{\mathbf{K}} \cdot \hat{\mathbf{u}} + \tilde{\mathbf{R}} \cdot \hat{\mathbf{t}} = 0 \qquad (10)$$

where the vector $\hat{\mathbf{u}}$ comprises all the components of the fictitious displacement vectors corresponding to N internal and L boundary nodes, while the vector $\hat{\mathbf{t}}$ consists of the L fictitious boundary traction vectors. The matrices $\tilde{\mathbf{K}}$ and $\tilde{\mathbf{R}}$ contain integrals, the evaluation of which is reported in Sellountos and Polyzos[8].

4. Hybrid BEM/LBIEM

The elastic domain V considered in the two previous sections is subdivided now into two regions V_1 and V_2. The external boundary S_1 and the interfacial surface S_i of the region V_1 are discretized into quadratic line elements, while the region V_2 as well as the external boundary S_2 and the interfacial surface S_i are covered by a cloud of properly distributed points as it is shown in Fig. (2). The region V_1 is treated by the standard BEM and according to the Eq. (3), one obtains the following linear system of algebraic equations

$$\begin{bmatrix} \mathbf{H}^1 & \mathbf{H}^{12} \end{bmatrix} \cdot \left\{ \begin{array}{c} \mathbf{u}^1 \\ \mathbf{u}^{12} \end{array} \right\} = \begin{bmatrix} \mathbf{G}^1 & \mathbf{G}^{12} \end{bmatrix} \cdot \left\{ \begin{array}{c} \mathbf{t}^1 \\ \mathbf{t}^{12} \end{array} \right\} \tag{11}$$

where $\left(\mathbf{u}^1, \mathbf{t}^1\right)$ and $\left(\mathbf{u}^{12}, \mathbf{t}^{12}\right)$ are displacements and tractions defined at the nodes of the external boundary S_1 and the interfacial surface S_{12}, respectively. In the sequel, applying the LBIE procedure explained in section 3 for

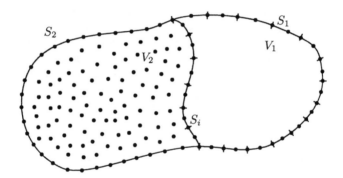

Figure 2.   Division of the domain to one BEM region V_1 and one LBIE region V_2.

the region V_2, one concludes to a final system of linear algebraic equations that according to Eq. (10) can be written in the form

$$\begin{bmatrix} \tilde{\mathbf{K}}^2 & \tilde{\mathbf{K}}^{21} & \tilde{\mathbf{K}}^i \end{bmatrix} \cdot \left\{ \begin{array}{c} \hat{\mathbf{u}}^2 \\ \hat{\mathbf{u}}^{21} \\ \hat{\mathbf{u}}^i \end{array} \right\} + \begin{bmatrix} \tilde{\mathbf{R}}^2 & \tilde{\mathbf{R}}^{21} \end{bmatrix} \cdot \left\{ \begin{array}{c} \hat{\mathbf{t}}^2 \\ \hat{\mathbf{t}}^{21} \end{array} \right\} = \mathbf{0} \tag{12}$$

where the vectors $\hat{\mathbf{u}}^2, \hat{\mathbf{t}}^2$ and $\hat{\mathbf{u}}^{21}, \hat{\mathbf{t}}^{21}$ represent fictitious displacement and traction nodal values defined at the external boundary S_2 and the interfacial surface S_i, respectively, while the vector $\mathbf{u}^i$ comprises all the components of the fictitious displacement vectors defined through the MLS approximation at the interior nodes. At the interfacial surface S_i, the continuity conditions must be satisfied. However, observing the system (12) one easily realizes that it is expressed in terms of the fictitious displacements and tractions $\hat{\mathbf{u}}^{21}, \hat{\mathbf{t}}^{21}$, which in general are not identical to the corresponding nodal ones $\mathbf{u}^{21}, \mathbf{t}^{21}$ due to the lack of Kronecker delta property of the MLS interpolants ϕ_j. Thus, the question here is how to enforce one in Eq. (12) the displacement and traction nodal values instead of the fictitious ones, in order the continuity conditions as well as the essential boundary conditions valid for the boundary S_2 to be satisfied. In the present work is the direct procedure of Gosz and Liu [13] where evenly distributed points along to the global boundary are considered and the essential boundary conditions are directly imposed on the fictitious values of displacements and tractions of Eq. (12). Although Gosz and Liu claim that this procedure works only for piece- wise linear global boundaries, numerical experiments performed in Sellountos and Polyzos [8] shown that a Kronecker delta behavior of the MLS approximation of boundary displacements and tractions is also possible for problems with curved boundaries.

5. A numerical example

In order to demonstrate the accuracy of the proposed hybrid BEM/LBIE methodology a representative example is solved. Consider a cantilever beam (Fig. (3)) under a harmonic excitation with dimensions $D = 1m$, $L = 8m$

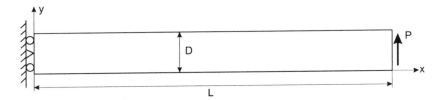

Figure 3.   Cantilever beam

and material properties $E = 2.4e^6 N/m^2$, $\nu = 0.20$ and $\rho = 1.0Kg/m^3$. The discretized model consists of one BEM region and one LBIE region as it is shown in Fig. (4). The problem is solved for the range of frequencies $4 - 2000 rad/s$ and the derived results are compared with those obtained

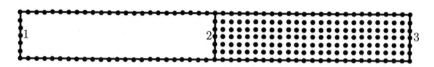

Figure 4.   The discretized model of the cantilever beam

from a BEM program the accuracy of which is reported in Polyzos et al[10].

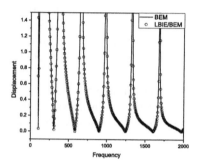

Figure 5.   Amplitude of displacement at point 3

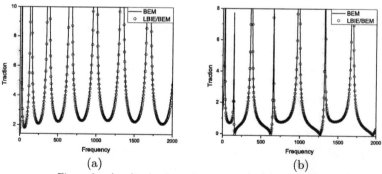

(a)   (b)

Figure 6.   Amplitude of traction at points 1(a) and 2(b)

References

1. D.E. Beskos, Boundary element methods in dynamic analysis, Appl. Mech. Rev. 40 (1987) 1-23.
2. D.E. Beskos, Boundary element methods in dynamic analysis. Part II, Appl. Mech. Rev. 50 (1997) 149-197.
3. T. Zhu, J.D. Zhang, S.N. Atluri, A local boundary integral equation (LBIE) method in computational mechanics and a meshless discretization approach, Comp. Mech. 21 (1998) 223-235.
4. T. Zhu, J.D. Zhang, S.N. Atluri, A meshless local boundary integral equation (LBIE) for solving nonlinear problems, Comp. Mech. 22 (1998) 174-186.
5. T. Zhu, J.D. Zhang, S.N. Atluri, A meshless numerical method based on the local boundary integral equation (LBIE) to solve linear and non-linear boundary value problems, Engng. Anal. Boundary Elem 23 (1999) 375-389.
6. G.R. Liu, Y.T. Gu, Meshless local Petrov-Galerkin (MLPG) method in combination with finite element and boundary element approaches, Comput. Mech. 26 (2000) 536-546.
7. G.R. Liu, Y.T. Gu, Coupling of element free Galerkin and hybrid boundary element methods using modified variational formulation, Comput. Mech. 26 (2000) 166-173.
8. E.J. Sellountos, D. Polyzos A MLPG (LBIE) method for solving frequency domain elastic problems Comp. Model. in Eng. Sci. Accepted for publication.
9. G. Li, G.H. Paulino, N.R. Aluru, Coupling of the mesh-free finite cloud method with the boundary element method: a collocation approach, Comput. Methods Appl. Mech. Engng. 192 (2003) 2355-2375.
10. D. Polyzos, S.V. Tsinopoulos, D.E. Beskos, Static and dynamic boundary element analysis in incompressible linear elasticity, European J. Mech. A/Solids. 17 (1987) 515-536.
11. G.D. Manolis, D.E. Beskos Boundary Element Methods in Elastodynamics, Unwin Hyman, London.
12. J. Dominguez Boundary Elements in Dynamics, CMP, Southampton and Elsevier Applied Science, London.
13. S. Gosz, W.K. Liu, Admissible approximations for essential boundary conditions in the reproducing kernel particle method, Comp. Mech. 19 (1996) 120-135.

SEQUENTIALLY FITTING GAUSSIAN MIXTURES USING AN OUTLIER COMPONENT

M. K. TITSIAS AND C. K. I. WILLIAMS

School of Informatics,
University of Edinburgh,
Edinburgh EH1 2QL, UK

We describe a method for training mixture models by learning one model at a time and thus building the mixture model in a sequential manner. We do this by incorporating an auxiliary outlier component (a uniform density to any of the data points) into the mixture model which allows us to fit just one data cluster by "ignoring" the rest of the clusters. Once a model is fitted we remove from consideration in a probabilistic fashion all the data explained by this model and then repeat the operation. This process can be viewed as fitting a mixture model using a constrained EM algorithm. A natural stopping criterion is to stop adding components once the outlier component fits no data (or just real background clutter). We also apply the algorithm to train J-component mixtures of Gaussians and show results real data.

1. Introduction

We address the problem of learning a mixture density model with J components

$$P(\mathbf{x}) = \sum_{j=1}^{J} \pi_j P_j(\mathbf{x}|\theta_j) \qquad (1)$$

where $P_j(\mathbf{x}|\theta_j)$ is the jth component having parameters θ_j and π_j the mixing coefficient. Mixture models have been widely used in statistical modelling as density estimation methods [5]. Given a set of i.i.d data $X = \{\mathbf{x}^1, \ldots, \mathbf{x}^N\}$ we wish to estimate the underlying density of $\mathbf{x}$ by a mixture model of the form (1). The component densities $P_j(\mathbf{x}|\theta_j)$ can be chosen from some parametric family such as the exponential family. Most of the presentation in the rest of the paper assumes that $P_j(\mathbf{x}|\theta_j)$ can be any distribution while in out experiments we specify this to be a multivariate Gaussian in the case of continuous data and a multinomial for discrete-valued data.

In this paper we describe a method for training mixture models by learning one component model at a time and thus building the mixture in a sequential manner. The key idea in doing this is that we incorporate an auxiliary outlier component (a uniform density to any possible data point) into the mixture model which allows us to fit one cluster of the data by "ignoring" the rest of the clusters. Data points that are fitted by a model are then "removed from consideration" in a probabilistic fashion so that at next stage a new model can fit to a unexplored region of the data space and so on. Intuitively the algorithm begins by considering all data as outliers (noise) and at each stage successively refines this belief by searching for clusters (structure) in all data that previously was labelled as outliers. Such a method can be also viewed as a kind of boosting algorithm for density estimation. An additional feature is that the algorithm can indicate when to stop adding new components; when the outlier component fits no data (or fits only background clutter data) we have potentially reached the desirable number of components and we can stop. We show that this can be used to find the number of components in some simple clustering problems.

The sequential algorithm can be useful for improving parameter initialization of EM algorithm when it is applied for training mixture models. This is because at each stage it provides a way to initialize a density model to data regions that are not well explained by the already fitted models, e.g. for Gaussian mixtures this can be more effective than simultaneously initializing the centres by randomly selecting data points.

The structure of the remainder of the paper is as follows: In section 2 we describe the sequential algorithm for fitting a mixture model assuming any form for the component density $P_j(\mathbf{x}|\theta_j)$. In section 3 we show some experiments using Gaussian mixtures and we make a comparison with the regular EM training. We conclude with a discussion in section 4.

2. Sequential algorithm for Mixture Models

2.1. *Fitting one density model together with an outlier component*

We wish to learn a density model $P(\mathbf{x}|\theta)$ together with a uniform distribution $U(\mathbf{x})$, called the outlier component so that

$$P(\mathbf{x}) = \alpha P(\mathbf{x}|\theta) + (1 - \alpha)U(\mathbf{x}). \tag{2}$$

For clarity assume at the moment that the model $P(\mathbf{x}|\theta)$ is a multivariate Gaussian with parameters $\theta = \{\boldsymbol{\mu}, \Sigma\}$. Selecting first a value for α we can learn the parameters by maximizing the log likelihood $L = \sum_{n=1}^{N} \log P(\mathbf{x}^n)$ using the EM algorithm.

Notice that if $\alpha = 1$ the outlier component is neglected and the parameter estimate for the Gaussian is the maximum likelihood solution. As α decreases the Gaussian becomes more and more focused on some population of the data and as α approaches zero the Gaussian fits very few data points ending up with fitting just one data point[a].

An obvious use of this model is to robustify the Gaussian estimate by choosing a high value for α (say 0.9) which can be useful in situations where one data cluster is embedded in background clutter. However, what is less obvious is the fact that by choosing properly the value of α the robust model of equation (2) can be used for learning just one data cluster by ignoring any other clusters of the data.

If we can fit just one cluster of the data by the model described above we can then, by repeating the process, fit all the data clusters sequentially. This motivates the sequential algorithm for fitting mixture models described in next section.

2.2. *Fitting mixture models sequentially*

In this section we discuss how we can use an outlier component to fit a mixture model rather than a single density model. Such a mixture should have the form

$$P(\mathbf{x}) = \sum_{j=1}^{J} \pi_j P_j(\mathbf{x}|\theta_j) + (1 - \sum_{j=1}^{J} \pi_j) U(\mathbf{x}) \qquad (3)$$

where generally $\sum_{j=1}^{J} \pi_j \leq 1$. We wish to train this mixture model sequentially by learning only one density model $P_j(\mathbf{x}|\theta_j)$ at each stage. An intuitive way of thinking about this is that we start by assuming that the mixing coefficients π_j, $j, \ldots, J$ are set to zero, so that the outlier component has all the probability. At j^{th} stage the mixing coefficient π_j is set free to get a positive value and the corresponding component model $P_j(\mathbf{x}|\theta_j)$ is allowed to fit some part of the data. At each stage the mixing coefficient of the outlier component always decreases which implies that the probability of what is considered as outlying data (noise) decreases sequentially.

---

[a]If we know the smallest distance between any two training points we can easily work out a lower bound of α that below of that value the Gaussian fits exactly one data point.

[1] has shown how mixture models can be fitted using variational Bayesian methods. It is interesting to note that if one component of the mixture at a time is updated repeatedly while keeping the other components fixed then a scheme quite similar to ours would emerge, as the as-yet-unfitted components would tend to have vague distributions not dissimilar to our uniform component[b].

We now describe our algorithm in detail, starting with training the first component $P_1(\mathbf{x}|\theta_1)$. By allowing the coefficient π_1 to obtain a positive value we have the mixture

$$P(\mathbf{x}) = \pi_1 P_1(\mathbf{x}|\theta_1) + (1 - \pi_1)U(\mathbf{x}) \tag{4}$$

which is exactly the model discussed in section 2.1 and thus learning the parameters $\{\theta_1, \pi_1\}$ can be done by maximizing the log likelihood using EM.

Suppose now that we have fitted a model $P_1(\mathbf{x}|\theta_1)$ to the data. What we wish to do next is to train the second mixture component $P_2(\mathbf{x}|\theta_2)$ by considering the mixture

$$P(\mathbf{x}) = \pi_1 P_1(\mathbf{x}|\theta_1) + \pi_2 P_2(\mathbf{x}|\theta_2) + (1 - \pi_1 - \pi_2)U(\mathbf{x}). \tag{5}$$

This case now becomes a little more complicated in the sense that we wish the second model not to fit data that are already well explained by the first model. Generally the new model $P_2(\mathbf{x}|\theta_2)$ should fit a subset of the data that is reasonably different from all the data fitted by $P_1(\mathbf{x}|\theta_1)$. Such a constraint can be efficiently taken into account by applying a constrained EM algorithm where instead of the log likelihood we maximize a lower bound of the log likelihood [7]. Particularly, we compute the responsibilities of the uniform component for each data point $\mathbf{x}^n$ by

$$z_1^n = \frac{(1 - \pi_1)U(\mathbf{x}^n)}{\pi_1 P_1(\mathbf{x}^n|\theta_1) + (1 - \pi_1)U(\mathbf{x}^n)}, \tag{6}$$

which are computed by having only trained the first component and then we express a lower bound of the log likelihood of the model (5)

$$F = \sum_{n=1}^{N}(1 - z_1^n) \log \pi_1 P_1(\mathbf{x}^n|\theta_1)$$
$$+ \sum_{n=1}^{N} z_1^n \log \{\pi_2 P_2(\mathbf{x}^n|\theta_2) + (1 - \pi_1 - \pi_2)U(\mathbf{x}^n)\} + H(\{z_1^n\}) \tag{7}$$

---

[b]We thank Steve Roberts for a helpful discussion on this point.

where $H(\{z_1^n\})$ denotes an entropic term independent of $\{\pi_2, \theta_2\}$. Since the parameters of the first model are fixed, maximizing the above bound simplifies to maximizing only the second term in the above sum under the constrain that π_2 should receive a value smaller or equal to $1 - \pi_1$. According to the form of the above objective function maximizing with respect to $\{\theta_2, \pi_2\}$ favours solutions where the model fits data that previously was explained mainly by the outlier component. To make this more obvious note that the weights $\{z_1^n\}$ are close to zero for all data explained by $P_1(\mathbf{x}|\theta_1)$ and close to one for all data explained by the outlier component. So the objective function (7) effectively removes from consideration in a probabilistic fashion data fitted by the first model. This process of fitting the mixture components to the data can be performed sequentially. The algorithm is summarised below

(1) Set $j = 0$. Initialize: $z_0^n = 1$ for all n.
(2) Set $j = j + 1$. Initialize θ_j and $\pi_j = \alpha(1 - \sum_{i=1}^{j-1} \pi_i)$, where $\alpha < 1$ (we use $\alpha = 0.5$).
(3) Optimize the parameters $\{\theta_j, \pi_j\}$ by running EM and maximizing:

$$\sum_{n=1}^{N} z_{j-1}^n \log\{\pi_j P_j(\mathbf{x}^n|\theta_j) + (1 - \sum_{i=1}^{j-1} \pi_i - \pi_j)U(\mathbf{x}^n)\} \qquad (8)$$

where π_j is sparsely updated by EM[c] (every 10 iterations).
(4) Update the log likelihood weights

$$z_j^n = \frac{(1 - \sum_{i=1}^{j} \pi_i)U(\mathbf{x}^n)}{\sum_{i=1}^{j} \pi_i P_i(\mathbf{x}^n|\theta_i) + (1 - \sum_{i=1}^{j} \pi_i)U(\mathbf{x}^n)}. \qquad (9)$$

(5) Go to step 2 or output the mixture $P_j(\mathbf{x}) = \sum_{i=1}^{j} \pi_i P_i(\mathbf{x}|\theta_i) + (1 - \sum_{i=1}^{j} \pi_i)U(\mathbf{x})$.

At each stage of the above algorithm a new model is trained (step 3) by maximizing a weighted log likelihood where the weights $\{z_j^n\}$ mask out (probabilistically) data fitted by the previously learned models. At step 4 the $\{z_j^n\}$ values are updated so that at the next stage we can fit a new model most probably to a different data subset. The sequential process can be considered as a kind of boosting algorithm for density estimation as the data points are reweighted on each iteration.

Note that the mixing coefficient of the outlier component $1 - \sum_{i=1}^{j} \pi_i$ can only decrease at each stage and naturally the learning process stops

---

[c]The constraint that $\pi_j \in [0, 1 - \sum_{i=1}^{j-1} \pi_i]$ is automatically satisfied in the EM update.

once this coefficient becomes very close to zero. In section 3 we describe a simple stopping criterion based on this idea and we use it to find the number of clusters in some simple clustering problems.

We can also use the outcome of the sequential algorithm to initialize a mixture model. In such case the outlier component is discarded and the coefficients π_j, $j = 1, \ldots, J$ are normalized to sum to one. The parameters of the resulting mixture can be refined by applying EM and maximizing the likelihood.

3. Experiments

In our experimentx we use the Brodatz textures images following an experimental setup used by [11]. The task is to cluster a set of $16 \times 16 = 256$ patches taken from 256×256 pixel Brodatz texture images. We consider the number of clusters (textures) from which patches are extracted to be $J = \{3, 5, 7, 9\}$. For a specific J we randomly choose J textures from the 37 available textures, create a set of $900J$ patches and then keep the half $(450J)$ for training and the rest for testing. We repeat this experiment 50 times. Each data set was also projected from the 256 to 50 dimensions using PCA in order to speed up the experiment. For each of the 50 datasets of a certain J we train a mixture model with J components using (i) k-means initialized EM[d] ($k$means), (ii) the sequential algorithm with refinement (Ref) and with (iii) no refinement (NoRef) steps. Table 1 displays the t-statistic values of the difference of the average log likelihoods. Note that when we consider the differences in log likelihoods of the method A and B $(A - B$ in the notation in the Table 2) and the t-statistic is larger than 2.01 the method A is significantly better than B at level 5% $(t_{0.025,50} = 2.01)$. When the t-statistic is less than -2.01 the method A is significantly worse. Observe that for $J \geq 5$ the sequential fitting algorithm with refinement beats EM initiated with k-means and that these differences are significant for $J = 7$, 9. Also note that using refinement always improves the results. We have also run the algorithm for mixture fitting proposed by [2] on this data using their code (available from http://www.lx.it.pt/~mtf/mixturecode.zip). However, the pruning strategy they use means that one cannot guarantee to get J components in the final model, and when fewer than J components are selected the test set log likelihoods are low leading to poor performance in comparison to the methods reported in Table 1.

---

[d]We used the NETLAB implementation available from http://www.ncrg.aston.ac.uk/netlab.

J	Ref - $k$means	Ref - NoRef	NoRef - $k$means
3	-0.61	1.8	-1.88
5	0.97	0.7	-0.05
7	**2.25**	1.68	0.2
9	**2.6**	**2.04**	1.2

4. Discussion

Our sequential algorithm deals with two issues (i) fitting a model with J components, and (ii) model selection. Of course there has been a huge amount of work concerning both of these topics. For (i) in addition to the standard EM algorithm with various initialization strategies there have also been proposed. [2] and others have demonstrated a backwards selection method, starting with many components and pruning some away using a prior that favours sparsity. There is another forward sequential (greedy) algorithm for Gaussian mixtures proposed by [12] and [11]. In their method they initialize the Gaussians one after the other by comparing at each stage a set of candidate initializations (ideally all the training points). One important difference with our method is that we use the outlier component which allows the Gaussians at each stage to fit some part of the data, while in [12,11] the Gaussians at each stage fit all the data. By avoiding using a set of candidate initializations our algorithm is faster. However, our method could probably be significantly improved by combining it with the idea of using a small set of candidate initializations.

In terms of model selection, Bayesian methods using the marginal likelihood as a selection criterion or approximations such as BIC penalties are most common. While our method is unlikely to be able to compete with sophisticated Bayesian methods such as reversible jump MCMC [4] on densities whose components are not well separated, it does provide a much more rapid answer.

The sequential algorithm can be regarded as a boosting density estimation algorithm. There has been some recent work [10,9] on extending boosting from the supervised learning problem to the density estimation problem. Our sequential formulation of fitting process is reminiscent of these boosting algorithms, however one attractive feature of our scheme is that the boosting view derives from a constrained EM formulation of the problem which derives the weightings in a particular way. After we had developed our idea we learned of the work of [8] who have shown that a sequential fitting approach to the mixture of experts architecture gave a boosting-like algorithm for supervised learning.

References

1. Attias, H. (2000). A variational Bayesian framework for graphical models. In *Advances in Neural Information Processing Systems 12*. MIT Press.
2. Figueiredo, M. A. T. and Jain, A. K. (2002). Unsupervised learning of finite mixture models. *PAMI*, 24(3):381–396.
3. Frey, B., Hinton, G., and Dayan, P. (1996). Does the wake-sleep algorithm produce good density estimators. In Touretsky, D., Mozer, M., and Hasselmo, M., editors, *Advances in Neural Information Processing Systems 8*. MIT Press.
4. Green, P. J. (1995). Reversible Jump Markov chain Monte Carlo computation and Bayesian model determination. *Biometrika*, 82(4):711–732.
5. McLachlan, B. G. and Peel, D. (2000). *Finite Mixture Models*. Wiley, New York.
6. Meila, M. and Heckerman, D. (2001). An experimental comparison of model-based clustering methods. *Machine Learning*, 42:9–29.
7. Neal, R. and Hinton, G. (1998). A view of the EM algorithm that justifies incremental, sparse and other variants. In Jordan, M., editor, *Learning in Graphical Models*, pages 355–368. Kluwer Academic Publishers, Dordrecht, The Netherlands.
8. Neal, R. and Mackay, D. (1998). Likelihood–based boosting. Unpublished paper available at http://www.inference.phy.cam.ac.uk/mackay/BayesICA.html.
9. Rosset, S. and Segal, E. (2003). Boosting density estimation. In Becker, S., Thrun, S., and Obermayer, K., editors, *Advances in Neural Information Processing Systems 15*. MIT Press.
10. Thollard, F., Sebban, M., and Ezequel, P. (2002). Boosting density function estimators. In *13th European Conference on Machine Learning*, pages 431–443.
11. Verbeek, J., Vlassis, N., and Krose, B. (2003). Efficient greedy learning of Gaussian mixture models. *Neural Computation*, 15:469–485.
12. Vlassis, N. and Likas, A. (2002). A greedy EM for Gaussian mixture learning. *Neural Processing Letters*, 15:77–87.

EXPLICIT ACCUMULATION MODEL FOR GRANULAR MATERIALS UNDER MULTIAXIAL CYCLIC LOADING

THEODOR TRIANTAFYLLIDIS
TORSTEN WICHTMANN
ANDRZEJ NIEMUNIS

Institute of Soil Mechanics and Foundation Engineering,
Ruhr-University Bochum, Germany

The prediction of cyclic-driven accumulation of stress or strain in granular materials is difficult due to a number of subtle effects in the soil structure. Also from the numerical point of view such prediction turns out to be troublesome because even small systematic errors of the general-purpose constitutive models are quickly accumulated. A remedy could be a so called explicit model that treats accumulation as a kind of creep process especially for engineering problems like compaction or liquefaction. However, for a good assessment of accumulation a detailed definition of strain amplitude is required. Consideration of the polarization and the openness of strain cycles on one hand and the degree of adaptation of the fabric on the other hand is crucial. A novel "back polarization" tensor is introduced to memorize the history of cyclic deformation. Multiaxial strain amplitude is defined considering the shape of the strain loop and rotation of principal directions of strain tensor. Some experimental evidence for the assumptions made is provided. Finally attempts to correlate the in-situ degree of adaptation with dynamic soil properties are reported.

1 Introduction

A considerable displacement of structures may be caused by an accumulation of the irreversible deformation of soil with load cycles. Even relatively small amplitudes may significantly contribute. This can endanger the long-term serviceability of structures which have large cyclic load contribution and small displacement tolerance (e.g. magnetic levitation train). Under undrained conditions similar phenomena may lead to an accumulation of pore water pressure, to soil liquefaction and eventually to a loss of overall stability.

From a physical point of view, displacements due to cyclic loading are rather difficult to describe. They depend strongly on several subtle properties of state (distribution of grain contact normals, arrangement of grains) which cannot be expressed by the customary state variables (stress $\mathbf{T}$ and void ratio e) used in geotechnical engineering. From a numerical point of view, two computational strategies can be considered: an implicit and an explicit one (time integration is not meant here).

Explicit or *N-type* models are similar to creep laws in which in place of time the number of cycles N is used. Therefore rates are understood in terms of the number of cycles, i.e. $\dot{\sqcup} = \partial \sqcup / \partial N$. Generally, explicit models can be seen as special-purpose constitutive relations that are thought to predict the accumulation due to a bunch of cycles at a time. The recoverable (resilient) part of the deformation is calculated in a conventional way (using many strain increments per cycle) in order to estimate the amplitude. Having the amplitude we assume that it remains constant over a number of the following cycles. The permanent (residual) deformation due to packages of cycles is calculated with special empirical formulas, that can be deducted from laboratory tests. In 'semi-explicit' models the cyclic creep procedure is interrupted by so-called *control cycles* calculated incrementally. Such cycles are useful to check the admissibility of the stress state, the overall stability (which may be lost if large pore pressures are generated) and, if necessary, to modify the amplitude (it may change due to a stress redistribution).

Implicit constitutive models are general-purpose relations which reproduce each single load cycle by many small strain increments. The accumulation of stress or strain appears as a by-product of this calculation, resulting from the fact that the loops are not perfectly closed. Implicit strategies require much computation time and magnify systematic errors. This requires a constitutive model of unreachable perfection. These as well as some other numerical problems discussed by Niemunis (2000) speak for the application of an explicit strategy, especially if the number of cycles is large.

2 Material model

For an arbitrary state variable $\sqcup$ we define its average value $\sqcup^{av}$ upon a cycle in such way that $\sqcup^{av}$ is the center of the smallest sphere that encompasses all states $\sqcup^{(i)}$ of a given cycle. The average value $\sqcup^{av}$ should not be mixed up with the *mean* value $\frac{1}{n} \sum_{i=1}^{n} \sqcup^{(i)}$. The amplitude is defined as $\sqcup^{ampl} = \max \| \sqcup^{(i)} - \sqcup^{av} \|$. A more elaborated definition of multiaxial amplitude is given further in this text.

Performing triaxial tests we use the Roscoe variables $p = -tr(\mathbf{T})/3$ and $q = \sqrt{3/2}\,\|\mathbf{T}^*\|$ and the conjugated strain rates D_v and D_q with the scalar product $p\,D_v + q\,D_q = -\mathbf{T} : \mathbf{D}$. $\mathbf{D}$ is the strain rate and $\sqcup^*$ denotes the deviatoric part of $\sqcup$. In the triaxial case strain and stress are axial-symmetric. We denote the axial components of stress (only effective stresses are considered) and strain with the index $\sqcup_1$ and the lateral components with $\sqcup_2$ and $\sqcup_3$ using the principal stresses $p = -(T_1 + 2T_3)/3$ and $q = -(T_1 - T_3)$ and the conjugated strain rates $D_v = -(D_1 + 2D_3)$ and $D_q = -2/3\,(D_1 - D_3)$. Stress ratio is expressed by $\eta = q/p$ or $\bar{Y} = (Y - 9)/(Y_c - 9)$ with $Y = -I_1 I_2 / I_3$, $Y_c = (9 - \sin^2\varphi)/(1 - \sin^2\varphi)$ and I_i being the invariants of $\mathbf{T}$. In the cyclic triaxial tests presented below the vertical stress component was cyclically varied with an amplitude T_1^{ampl} at a constant average stress level $\mathbf{T}^{\mathrm{av}}$. Also $\zeta = T_1^{\mathrm{ampl}}/p^{\mathrm{av}}$ is used.

The strains under cyclic loading can be decomposed into a residual and a resilient part denoted by $\sqcup^{\mathrm{acc}}$ and $\sqcup^{\mathrm{ampl}}$, respectively. The accumulated strain $\varepsilon^{\mathrm{acc}} = \|\varepsilon^{\mathrm{acc}}\| = \|\int \mathbf{D}^{\mathrm{acc}} dN\|$ and its ratio $\omega = \varepsilon_v^{\mathrm{acc}}/\varepsilon_q^{\mathrm{acc}}$ are usually measured. The shear strain amplitude $\gamma^{\mathrm{ampl}} = (\varepsilon_1 - \varepsilon_3)^{\mathrm{ampl}} = \sqrt{3/2}\,\|(\varepsilon^*)^{\mathrm{ampl}}\|$ is used in the evaluation of cyclic triaxial tests.

The general stress-strain relation has the form

$$\mathring{\mathbf{T}} = \mathsf{E} : (\mathbf{D} - \mathbf{D}^{\mathrm{acc}}) \tag{1}$$

wherein E denotes the elastic stiffness. The rate of strain accumulation $\mathbf{D}^{\mathrm{acc}}$ is proposed to be

$$\mathbf{D}^{\mathrm{acc}} = D^{\mathrm{acc}}\,\mathbf{m} = f_{\mathrm{ampl}}\,\dot{f}_N\,f_p\,f_Y\,f_e\,f_\pi\,\mathbf{m} \tag{2}$$

with the direction expressed by the unit tensor $\mathbf{m}$ (= flow rule) and with the intensity D^{acc} given by six partial functions f. For triaxial tests $\omega = \sqrt{\frac{3}{2}}\,\mathrm{tr}(\mathbf{m})/\|\mathbf{m}^*\|$ holds. The intensity of strain accumulation D^{acc} depends on the strain amplitude γ^{ampl} (function f_{ampl}), the number of cycles N (function $\dot{f}_N$), the average stress p^{av} (function f_p), $\bar{Y}^{\mathrm{av}}$ (function f_Y), the void ratio e (function f_e), the cyclic strain history π (function f_π) and the shape of the strain loop (via function f_{ampl}). The detailed forms of the functions are given in Sec. 3.

The first so-called *irregular* cycle generates, at least for fresh pluviated samples, a much larger residual deformation than the so-called *regular* (second and subsequent) cycles. In the explicit method the irregular and the first regular cycle are calculated implicitly (Fig. 1a). This is necessary for representative information about one full regular cycle to estimate the strain amplitude, its polarization etc. The irregular cycle is not suitable for this purpose. Note that if the irregular cycle happened to be similar to the subsequent ones (to be actually regular) the above mentioned precaution in the finite element (FE) calculation is redundant but safe. Equation (2) describes only the accumulation during the regular cycles. Also all diagrams presented in Sec. 3 show the residual strain due to the regular cycles only.

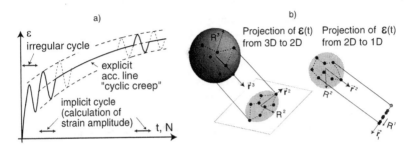

Figure 1   a) Calculation procedure in explicit models, b) Reduction steps (projections) from the 3-dimensional path to the 1-dimensional path

An important factor influencing $\mathbf{D}^{acc}$ is the shape (openness) of the strain cycle. For example, a twirled multiaxial strain loop that encloses some volume in the strain space causes larger accumulation than a 1-D cycle of the same *scalar* amplitude. A more general definition of an amplitude is required, see also fatigue problem for metals, Ekberg (2000) and Papadopoulos (1994). In order to consider the openness and the complexity of the stress loop a novel definition is proposed here. The amplitude is assumed to be a fourth rank dyadic tensor. Experiments presented in Sec. 3 show that change in strain polarization increases the rate of accumulation. This effect is described in the following using a so-called "back polarization" tensor $\boldsymbol{\pi}$, which memorizes the cyclic strain history.

The multiaxial amplitude A_ϵ is defined as a combination of several specially chosen projections of the strain path (loop) weighted by their perimeters. Suppose that we are given a single strain loop in form of the strain path consisting of a sequence of discrete strain points $\epsilon(t_i)$, $i = 1, \dots, N$ (this loop need not be closed). These strain states lie in a 6-D strain space and need not be coaxial. The following flow chart demonstrates how to calculate the subsequent projections and their perimeters. The upper index indicate the number of dimensions of the strain space which need to be considered.

(1) Calculate the deviatoric projection $\mathbf{e}^5(t_i) = \boldsymbol{\epsilon}(t_i) - \frac{1}{3}\mathbf{1}\mathrm{tr}(\boldsymbol{\epsilon}(t_i))$ of the strain points $\boldsymbol{\epsilon}(t_i)$. For soils it was observed that only the deviatoric part of strain amplitude influences the process of cyclic relaxation or cyclic creep. The space in which the $\mathbf{e}^5(t_i)$-path could be drawn has 5 dimensions because the deviator of symmetric 3×3 tensor has 5 independent components.

(2) Calculate the perimeter $P_5 = \sum_{i=1}^{N} \|\mathbf{e}^5(t_i) - \mathbf{e}^5(t_{i-1})\|$ of the loop (define $\mathbf{e}^5(t_0) = \mathbf{e}^5(t_N)$ to close the loop) wherein N denotes the number of strain states (points) used to record the loop.

(3) Find the average point $\mathbf{e}_{av}^5$ and the radius R_5 of the smallest 5-d sphere $R_5 = \|\mathbf{e}^5 - \mathbf{e}_{av}^5\|$ that encompasses the loop (we may do it numerically using $\mathbf{e}_g = \frac{1}{N}\sum_{i=1}^{N}\mathbf{e}^5(t_i)$ as the first approximation of $\mathbf{e}_{av}^5$).

(4) Calculate the unit tensor $\mathbf{r}^5$ along the line that connects the average strain $\mathbf{e}_{av}^5$ with the most distant point $\mathbf{e}^5(t_i)$ of the loop. Usually there are two equally distant points (antipodes). In case of more than two equally distant points choose anyone of antipodes.

(5) Project the loop onto the plane perpendicular to $\mathbf{r}^5$ calculating $\mathbf{e}^4(t_i) = \mathbf{e}^5(t_i) - \mathbf{r}^5 : \mathbf{e}^5(t_i)\mathbf{r}^5$.

(6) Analogously find $P_4, R_4, \mathbf{r}^4$ and then $P_3, R_3, \mathbf{r}^3$, $P_2, R_2, \mathbf{r}^2$ and $P_1, R_1, \mathbf{r}^1$.

Reduction steps from the 3-D to the 1-D are shown in Fig. 1b. For any D-dimensional sub-space $\mathbf{e}^D$ is preserved in full tensorial (3×3) form. Doing this, the conventional definition of the distance, e.g. $R = \sqrt{[e_{ij} - e_{ij}^{av}][e_{ij} - e_{ij}^{av}]}$, remains insensible to the choice of the coordinate system.

After a series of projections a list of radii R_D, perimeters P_D and orientations $\mathbf{r}^D$ is calculated with dimensions $D = 1\ldots5$. The orientations are all mutually perpendicular, $\mathbf{r}^i : \mathbf{r}^j = \delta_{ij}$. The sense of the orientation $\mathbf{r}^D$ must not enter the definition of the amplitude. Therefore the dyadic products $\mathbf{r}^D\mathbf{r}^D$ are used and their weighted sum.

$$\mathsf{A}_\epsilon = \frac{1}{4}\sum_{D=1}^{5} P_D\, \mathbf{r}^D\mathbf{r}^D \qquad (3)$$

is proposed to be the definition of the amplitude. The unit amplitude $\vec{\mathsf{A}}_\epsilon = \mathsf{A}_\epsilon/\|\mathsf{A}_\epsilon\|$ is further called *polarization*.

3 Experiments

The functions f_i in Eq. (2) are determined empirically on the basis of experimental data from cyclic triaxial and cyclic multiaxial direct simple shear (CMDSS) tests. The main experimental results are briefly summarized in the following. In all tests medium coarse sand (mean diameter $d_{50} = 0.5$ mm, uniformity index $U = d_{60}/d_{10} = 1.8$, maximum and minimum void ratios $e_{max} = 0.874$, $e_{min} = 0.577$) was used. The soil's density is described by $I_D = (e_{max} - e)/(e_{max} - e_{min})$. The subscript $\sqcup_0$ denotes the initial value before cyclic loading.

3.1 Cyclic triaxial tests

Details of the cyclic triaxial apparatus, the specimen preparation procedure and the test results are given by Wichtmann et al. (2004a). Tests with varying amplitude, average stress and density were performed. All tests exhibit an increase of residual strain ε^{acc} with the number of cycles N with an accompanying decrease of the accumulation rate $\dot{\varepsilon}^{acc}$.

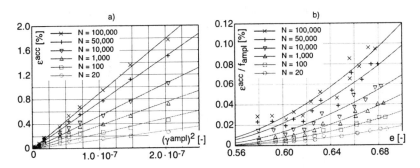

Figure 2   Rate of accumulation influenced by a) amplitude and b) void ratio

Tests with identical average stress $\mathbf{T}^{av}$ ($p^{av} = 200$ kPa, $\eta^{av} = 0.75$) and similar initial density $I_{D0} = 0.55$–0.64 but different amplitudes T_1^{ampl} (10 kPa–90 kPa) exhibit a proportionality $\varepsilon^{acc} \sim \left(\gamma^{ampl}\right)^2$ independent on the number of applied cycles (Fig. 2a). Thus, f_{ampl} was proposed as

$$f_{ampl} = \left(\gamma^{ampl}/\gamma^{ampl}_{ref}\right)^2 \tag{4}$$

with a reference amplitude $\gamma^{ampl}_{ref} = 10^{-4}$.

In tests with different initial densities ($I_{D0} = 0.63$–0.99) but identical average and cyclic stress ($p^{av} = 200$ kPa, $\eta^{av} = 0.75$, $\zeta = 0.3$) an increase of accumulation with void ratio e was observed (Fig. 2b). The partial function f_e is proposed to be

$$f_e = \frac{(C_e - e)^2}{1 + e} \frac{1 + e_{ref}}{(C_e - e_{ref})^2} \tag{5}$$

with the material constant C_e and the reference void ratio $e_{ref} = e_{max}$.

Tests with varying average stress $\mathbf{T}^{av}$ ($p^{av} = 50$–300 kPa, $\eta^{av} = 0.25$–1.375) but identical amplitude ratio $\zeta = 0.3$ and similar initial densities ($I_{D0} = 0.57$–0.69) show a faster accumulation with decreasing average mean pressure (Fig. 3a) and increasing stress anisotropy (Fig. 3b). For the purpose of purging the influence of slightly different shear strain amplitudes due to the dependence of sand stiffness on $\mathbf{T}$ the accumulated strain in Figs. 3a and 3b is divided by f_{ampl}. The exponential functions

$$f_p = \exp\left[-C_p \left(\frac{p^{av}}{p_{ref}} - 1\right)\right] \qquad f_Y = \exp\left(C_Y \, \bar{Y}^{av}\right) \tag{6}$$

with C_p and C_Y being material constants and the reference pressure $p_{ref} = p_{atm} = 100$ kPa were found to describe the experimental results .

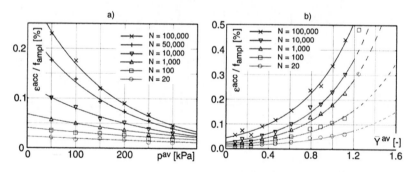

Figure 3   Rate of accumulation influenced by a) average mean pressure and b) stress anisotropy

Figure 4a presents the accumulated strain ε^{acc} in all test series adressed above divided by the partial functions f_p, f_Y, f_{ampl}, f_e and f_π ($f_\pi =$ const $= 1$ in the cyclic triaxial case) as a function of the number of cycles. The curves in Fig. 4a exhibit an increase of accumulated strain faster than the logarithm of N. The current description of the dependence of the accumulation rate on the number of cycles is

$$\dot{f}_N = C_{N1} \left[\frac{C_{N2}}{1 + C_{N2}\ N} + C_{N3} \right] \tag{7}$$

and drawn as the solid line in Fig. 4a. A more sophisticated model using a compaction tensor and passing on the function $\dot{f}_N$ is under way.

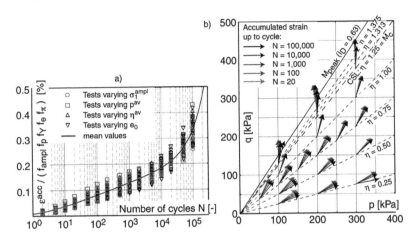

Figure 4   a) Rate of accumulation influenced by the number of cycles, b) flow rule

The flow rule **m** was found to be independent on density and amplitude but strongly dependent on $\mathbf{T}^{av}$ as presented in Fig. 4b. In Fig. 4b the critical state line (CSL) is presented with an inclination $M_c = (6 \sin\varphi)/(3 - \sin\varphi) = 1.25$ in the p - q - space where $\varphi = 31.2°$ is the critical friction angle determined in static tests. At an average stress $\mathbf{T}^{av}$ below the CSL cyclic loading leads to compaction whereas dilation is observed for $\eta^{av} > M_c$. A slight increase of the volumetric portion of ε^{acc} with N was observed but currently this effect is not incorporated in our cyclic accumulation model. The well known flow rules of constitute models for monotonic loading (e.g. Cam-clay and hypoplastic models) are sufficient.

3.2   Cyclic multiaxial direct simple shear (CMDSS) tests

A special simple shear device was constructed to study the material behavior under cyclic multiaxial shearing. The apparatus allows to compare the accumulation under one-dimensional cyclic shear with the residual strain due to an application of circular strain paths. Furthermore the influence of strain polarization of cyclic shear can be studied in this device since the direction of shear can be orthogonally changed after a given number of cycles. A detailed description of the apparatus, the procedure of specimen preparation and the test series is given by Wichtmann *et al.* (2004b).

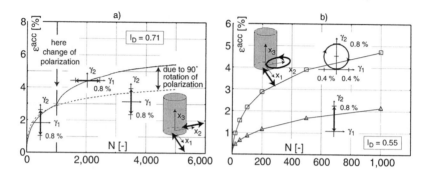

Figure 5 Rate of accumulation influenced by a) rapid change of strain polarization, b) shape of strain loop

Figure 5a presents the accumulation of strain measured in two tests with $\gamma^{\text{ampl}} = 8 \cdot 10^{-3}$. In the first test 5,000 cycles of one-directional cyclic shear were applied whereas in the second test the direction of shear was changed for 90° after 1,000 cycles and 4,000 cycles in the new direction followed. From Fig. 5a it is obvious that the change of strain polarization leads to a significant increase in accumulation rate. In order to mathematically describe these experimental findings the partial function f_π and the evolution of π, respectively were proposed as

$$f_\pi = 1 + C_{\pi 1}\left[1 - \left(\vec{\mathbf{A}}_\epsilon :: \pi\right)^{C_{\pi 2}}\right] \qquad \overset{\circ}{\pi} = C_{\pi 3}\left(\vec{\mathbf{A}}_\epsilon - \pi\right)\|\mathbf{A}_\epsilon\|^2 \qquad (8)$$

with $C_{\pi 1}, C_{\pi 2}$ and $C_{\pi 3}$ being material constants. Figure 5b contains a comparison of strain paths with circular and one-directional cyclic shear having identical amplitudes $\gamma^{\text{ampl}} = 8 \cdot 10^{-3}$ in one direction. The circular strain path leads to a twice larger accumulation in comparison with the one-directional cyclic shearing. This effect is captured by the definition of the amplitude $\mathbf{A}_\epsilon$.

4 Problem with N_0

Experimental observations indicate a strong dependence of the accumulation rate on the preloading history. Despite identical void ratios and stresses the accumulation rates $\dot{e}$ of two specimens may be quite different depending on their cyclic preloading history. A volume of sand *in situ* is less compactable than a freshly pluviated laboratory specimen. The preloading history (i.e. the number of cycles, the amplitudes and the polarization of the cycles) has to be determined. For the present the cyclic history is lumped together into a single scalar variable N_0 that enters the equation for $\dot{f}_N$ as the *initial number of cycles*.

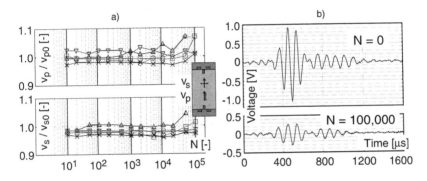

Figure 6   Impact of preloading history on a) v_p and v_s and b) signal intensity

Cyclic triaxial tests were performed in a test apparatus instrumented with piezoelectric elements for measuring compression (v_p) and shear (v_s) wave velocities. Details of the apparatus and the test series can be found in Wichtmann and Triantafyllidis (2004b). The cyclic loading was interrupted after given cycle numbers and v_p and v_s were measured at $\mathbf{T}^{av}$. The development of the wave velocities with the number of cycles is presented in Fig. 6b for tests with varying average stress (p^{av} = 100–200 kPa, q^{av} = 100–200 kPa) but identical amplitude T_1^{ampl} = 60 kPa and similar initial densities (I_{D0} = 0.57–0.59). No significant changes of v_p and v_s and thus small strain stiffness could be detected. However, in the case of the elements that send and receive shear waves the intensity of the received signal tended to abate with the number of cycles, probably caused by an increased material damping. Further tests will check if N_0 can be correlated with damping.

Several undrained cyclic triaxial tests with drained preloading history exhibit a correlation between cyclic undrained strength (i.e. the stress amplitude T_1^{ampl} needed to cause a definite strain amplitude ε_1^{ampl} in a given number of cycles) and N_0.

As an alternative for the determination of preloading history via indirect measurements N_0 could be detected by a back analysis of settlements caused by a strong vibration *in situ* (e.g. applied by a vibrator placed on the ground surface with accompanying measurements of the time history of nearby settlements).

5 Finite element (FE) calculation

The cyclic accumulation model was used to calculate a model test which was performed in the geotechnical centrifuge at our institute (Helm *et al.* 2000). In the model test (acceleration level 20g) a strip foundation (width 1 m) on a fine sand was loaded with an oscillating stress 104 kPa ± 69 kPa. After 70,000 cycles a settlement of 6.8 cm was observed below the middle of the foundation. The accumulation model predicts a settlement of 7.8 cm after 70,000 cycles (Fig. 7, Hammami 2003), which satisfactory agrees with the observations of the model test.

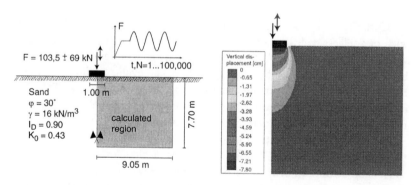

Figure 7   FE calculation of the settlement of a strip foundation under cyclic loading

Acknowledgements

The authors are grateful to German Research Council (DFG, Project A8 / SFB 398) for the financial support.

References

Ekberg, A. (2000). Rolling contact fatigue of railway wheels, Ph.D. thesis, Chalmers University of Technology, Solid Mechanics

Hammami, M. (2003). Numerische Fehler bei der expliziten FE-Berechnung der Verdichtbarkeit von Sand infolge zyklischer Belastung, Diploma thesis at Institute of Soil Mechanics and Foundation Engineering, Ruhr-University Bochum.

Helm, J., Laue, J. and Triantafyllidis, T. (2000). Untersuchungen an der RUB zur Verformungsentwicklung von Böden unter zyklischen Beanspruchungen, *Beiträge zum Workshop "Boden unter fast zyklischer Belastung: Erfahrungen und Forschungsergebnisse"*, Rep. No. 32, pp. 201–222.

Papadopoulos, I. (1994). A new criterion of fatigue strength for out-of-phase bending and torsion of hard metals. *International Journal of Fatigue*, **16**, pp. 377–384.

Niemunis, A. (2000). Akkumulation der Verformung infolge zyklischer Belastung - numerische Strategien, *Beiträge zum Workshop "Boden unter fast zyklischer Belastung: Erfahrungen und Forschungsergebnisse"*, Rep. No. 32, pp. 1–20.

Niemunis, A. (2003). Extended hypoplastic models for soils, *Habilitation*, Rep. No. 34, Institute of Soil Mechanics and Foundation Engineering, Ruhr-University Bochum.

Wichtmann, T. and Triantafyllidis, T. (2004a). Influence of a cyclic and dynamic loading history on dynamic properties of dry sand, part I: cyclic and dynamic torsional prestraining, *Soil Dynamics and Earthquake Engineering*, **24**, pp. 127–147.

Wichtmann, T. and Triantafyllidis, T. (2004b). Influence of a cyclic and dynamic loading history on dynamic properties of dry sand, part II: cyclic axial preloading, *Soil Dynamics and Earthquake Engineering (accepted)*.

Wichtmann, T., Niemunis, A. and Triantafyllidis, T. (2004a): Strain accumulation in sand due to drained uniaxial cyclic loading. *Cyclic Behaviour of Soils and Liquefaction Phenomena*, Proc. of CBS04, Bochum, March/April 2004, Balkema, pp. 233–246.

Wichtmann, T., Niemunis, A. and Triantafyllidis, T. (2004b): The effect of volumetric and out-of-phase cyclic loading on strain accumulation. *Cyclic Behaviour of Soils and Liquefaction Phenomena*, Proc. of CBS04, Bochum, March/April 2004, Balkema, pp. 247–256.

ON AN INVERSE PROBLEM OF $\nabla^2\psi = -\phi\psi^*$

G.A. TSIHRINTZIS AND P.S. LAMPROPOULOU

Department of Informatics, University of Piraeus, Piraeus 18534, GREECE

E-mail: {geoatsi,vlamp}@unipi.gr

We address an inverse problem associated with the partial differential equation $\nabla^2\psi = -\phi\psi$. A translation property is derived relating the fields scattered from a known object placed at different locations. The result is used in the derivation of the optimum (in the maximum likelihood sense) algorithm for detecting a target object, estimating its location, and classifying it from noisy scattered field measurements.

1. Introduction

Significant research activity has taken place over the past thirty years on the problem of quantitative determination of the structure of an unknown object by computerized processing of measurements of the wavefields diffracted by the object in a set of scattering experiments [1,2,3]. This activity is founded on the mathematical theories of Inverse Scattering [4,5] and is applicable in a number of seemingly different scientific disciplines, such as crystal structure determination [6], medical ultrasound tomography [7], acoustic and electromagnetic underground surveying [8,9,10,11], optical and coherent x-ray microscopy [12], and elastic wave inverse scattering [13]. Related activities also form the well-known technologies of Computerized Tomography [14] and, in fact, x-ray Computerized Tomography is the special case of the general discipline of Inverse Scattering obtained as the wavelength of the probing radiation approaches zero. Finally, electrocardiology is yet another discipline that may benefit from the theories of Inverse Scattering, as was recently shown in [15].

The reconstruction of the spatial distribution of the (complex-valued) index of refraction inside a scattering object requires the inversion of the (nonlinear) mathematical mapping that relates the probing wave, the ob-

---

*Partial support for this work was provided by the University of Piraeus Research Center.

ject's refraction index, and the measurable total wave, a task that is non-trivial to achieve and remains to date an open problem of active research. An alternative approach that has often met with success is based on parametric modeling of the unknown object and the corresponding scattered wave data and the use of well-known techniques of estimation theory for estimation of appropriate values for the unknown parameters. This approach was followed within the domain of Computerized [16] and linearized Diffraction [17] Tomography and was later extended into the domain of exact (nonlinear) scattering theory [18] and availability of intensity-only data [19]. More specifically, in [18] it was shown that the (log) likelihood function for detection of the presence of a known object and estimation of its location from scattered wave data could be computed via a *filtered backpropagation algorithm* [1] that could be efficiently implemented with fast Fourier transform algorithms.

In the present paper, we address an inverse problem associated with the partial differential equation $\nabla^2 \psi = -\phi\psi$, which arises in electrocardiology [15] and low frequency diffraction tomography. We consider the "*object function*" ϕ as a spatially-shifted version of an otherwise completely known object function ϕ_0, i.e., $\phi(x) = \phi_0(x - x_0)$, where x_0 is the unknown location of the object. More specifically, the paper is organized as follows: In Section 2, we derive a translation property relating the fields scattered by the object ϕ_0 placed at different locations. In Section 3, we use the translation property of Section 2 to derive an algorithm for the computation of the maximum likelihood estimate of the object location from noisy scattered field measurements. Finally, in Section 4, we summarize the paper, draw conclusions and point to related future work.

2. Scattered Field Translation Property

A. Measurement Configuration and Relevant Equations
For simplicity and without loss of generality, we limit the discussion to the two-dimensional case, where the object is independent of one coordinate axis. We consider the configuration in Fig. 1, in which we defined a fixed Cartesian coordinate system with axes x_1 and x_2 and a second system that is centered at the origin of the (x_1, x_2) system and has axes (t, s) with unit vectors u and v, respectively. The (t, s) system is allowed to rotate around the common origin, with the t axis forming an angle θ with the positive x_1 axis.

An object, described by the object function ϕ and fully contained in the

408

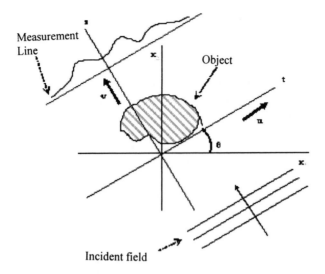

Figure 1.   The data measurement configuration

finite volume **V**, is probed with the exponentially-decaying incident field[a] $\psi_\theta^{in}(tu + sv) = e^{-\kappa s} = e^{-\kappa <v,x>}$, $\kappa > 0$. The interaction of the incident field with the object results in the formation of a total field ψ satisfying the differential equation

$$\nabla^2 \psi_\theta = -\phi \psi_\theta, \qquad (1)$$

where ϕ is the, so-called, object function that quantifies the object properties and structure.

The solution to Eq.(1) can be formally expressed in integral form[b] with use of Green function techniques [20] as

$$\psi_\theta(x) = \psi_\theta^{in}(x) + \psi_\theta^s(x), \qquad (2)$$

where we have defined the *scattered* field ψ_θ^s via

$$\psi_\theta^s(x) = -\int d^2x' \, G(x - x')\phi(x')\psi_\theta(x'). \qquad (3)$$

---

[a]The index θ parameterizes the direction of the unit vector v along which the incident field decays exponentially.

[b]Unless explicitly denoted otherwise, all integrations are over the support volume **V** of the object.

Therefore, Eq.(2) becomes

$$\psi_\theta(x) = e^{-\kappa<v,x>} - \int d^2x'\, G(x-x')\phi(x')[e^{-\kappa<v,x'>} + \psi_\theta^s(x')]. \quad (4)$$

The appropriate Green function G is given by [20]

$$G(tu+sv) = \ln\frac{1}{t^2+s^2} = -\frac{1}{2\pi}\int_{-\infty}^{\infty} dp\, e^{ipt}\frac{e^{-p|s|}}{p}. \quad (5)$$

The rightmost side in Eq.(3) is known as the Weyl expansion [20] to the Green function and will be used in the following portion of the paper.

B. *The Induced Source*

At this point, it is useful to define the "*induced source*" via the equation

$$\rho_\theta(x) = \phi(x)\psi_\theta(x). \quad (6)$$

Clearly, Eq.(3) gives

$$\psi_\theta^s(x) = -\int d^2x'\, G(x-x')\rho_\theta(x'). \quad (7)$$

The relation in Eq.(7) admits an interesting interpretation when the scattered field is evaluated at points on a straight line $x = tu + sv$, $t \in R$, s: fixed. Indeed, using the Weyl expansion for the Green function in Eq.(3) and for points in the shadow region of the object (i.e., $s > s'$), we have

$$\psi_\theta^s(x = tu + sv) = \frac{1}{2\pi}\int dp\, e^{ipt}\frac{e^{-ps}}{p}\int\int dt'ds'\, e^{-ipt'} e^{ps'}\rho_\theta(t'u + s'v)$$

$$= \frac{1}{2\pi}\int dp\, e^{ipt}\frac{e^{-ps}}{p}\tilde{\rho}_\theta(p,-p), \quad (8)$$

where $\tilde{\rho}_\theta(p,-p)$ indicates the Fourier (w.r.t. the t' variable)/Laplace (w.r.t. the s' variable) transform of $\rho_\theta(t'u + s'v)$. That is, *the one-dimensional Fourier transform $\tilde{\psi}_\theta(pu + sv)$ of the scattered field w.r.t. the straight line position variable t relates through the factor $\frac{e^{-ps}}{p}$ to the Fourier/Laplace transform $\tilde{\rho}_\theta(p,-p)$ of the induced source.*

C. *Translation Property*

We consider now an object function $\phi(x) = \phi_0(x - x_0)$, i.e., an object function that arises from a spatial shift by x_0 of the object function ϕ_0. Substituting this expression into Eq.(4) and multiplying both sides with the factor $e^{\kappa<v,x_0>}$, we obtain

$$
\begin{aligned}
e^{\kappa<v,x_0>}\psi_\theta(x;x_0) &= e^{-\kappa<v,(x-x_0)>} \\
&\quad - \int d^2x'\, G(x-x')\phi_0(x'-x_0)[e^{-\kappa<v,(x'-x_0)>} \\
&\quad\quad + e^{\kappa<v,x_0>}\psi_\theta^s(x';x_0)], \\
&= e^{-\kappa<v,(x-x_0)>} \\
&\quad - \int d^2x'\, G(x-x')\phi_0(x'-x_0)e^{\kappa<v,x_0>}\psi_\theta(x';x_0)
\end{aligned}
\tag{9}
$$

in which we indicate explicitly the dependence of the fields on the object location x_0. Next, we effect the integration variable substitution $\chi = x' - x_0$ and obtain

$$
e^{\kappa<v,x_0>}\psi_\theta(x;x_0) = e^{-\kappa<v,(x-x_0)>}
\tag{10}
$$
$$
- \int d^2\chi\, G(x-x_0-\chi)\phi_0(\chi)[e^{-\kappa<v,\chi>} + e^{\kappa<v,x_0>}\psi_\theta^s(\chi+x_0;x_0)].
$$

On the other hand, Eq.(4) needs to be satisfied by a total field $\psi_\theta(x;0)$ which corresponds to the object located at $x_0 = 0$, i.e.,

$$
\psi_\theta(x;0) = e^{-\kappa<v,x>} - \int d^2x'\, G(x-x')\phi_0(x')[e^{-\kappa<v,x'>} + \psi_\theta^s(x';0)]. \tag{11}
$$

Eq.(11) gives

$$
\begin{aligned}
\psi_\theta(x-x_0;0) &= e^{-\kappa<v,x-x_0>} \\
&\quad - \int d^2x'\, G(x-x_0-x')\phi_0(x')[e^{-\kappa<v,x'>} + \psi_\theta^s(x';0)] \\
&= e^{-\kappa<v,x-x_0>} \\
&\quad - \int d^2x'\, G(x-x_0-x')\phi_0(x')\psi_\theta(x',0).
\end{aligned}
\tag{12}
$$

Since $\psi_\theta(x';0) = \psi_\theta(x'+x_0-x_0;0)$, we observe that the quantities $e^{\kappa<v,x_0>}\psi_\theta(x;x_0)$ and $\psi_\theta(x-x_0;0)$ satisfy the same equation. From the uniqueness of the solution of the equation [20], we are readily led to the conclusion that

$$
e^{\kappa<v,x_0>}\psi_\theta(x;x_0) = \psi_\theta(x-x_0;0),
\tag{13}
$$

or equivalently,

$$
\psi_\theta(x;x_0) = e^{-\kappa<v,x_0>}\psi_\theta(x-x_0;0).
\tag{14}
$$

Eq.(14) defines *a translation property that relates the fields scattered from a known object placed at different locations, when the object is probed with the same incident field.*

A similar translation property can be derived for the corresponding induced sources. Indeed, Eqs.(14) and (6) give

$$\rho_\theta(x; x_0) = e^{-\kappa<v,x_0>}\rho_\theta(x - x_0; 0) \qquad (15)$$

or, equivalently,

$$\tilde{\rho}_\theta(p, -p; x_0) = e^{-ip<u,x_0>}e^{(p-\kappa)<v,x_0>}\tilde{\rho}_\theta(p, -p; 0), \qquad (16)$$

in which we have used well-known properties of the Fourier and Laplace transforms.

3. Object Location Estimation

We consider again the configuration in Fig. 1, where the object $\phi(x) = \phi_0(x - x_0)$ is probed with known fields $e^{-\kappa<v,x>}$, $\kappa > 0$, which decay exponentially in the direction of the positive t axis and corresponding scattered field data are measured along the straight line $<v, x> = s = s_0$, where s_0 is a fixed measurement offset from the coordinate axis origin. We assume that the data are measured for a number of probing directions θ in some finite set Θ and are modeled for each probing direction as

$$d(t, \theta) = \alpha_\theta(t; x_0) + n_\theta(t) \equiv r_\theta(t) *_t \psi_\theta^s(tu + s_0v; x_0) + n_\theta(t), \qquad (17)$$

where r_θ and n_θ are a convolutional filter and a stochastic process modeling filtering effects and noise addition that may be present in the measurement process.

For the noise process, we assume that it is a zero-mean Gaussian process which is white with respect to both the probing direction parameter θ and the measurement position t, i.e.,

$$E\{\overline{n_\theta}(t)n_{\theta'}(t')\} = R_\theta(t)\delta_{\theta,\theta'}\delta(t - t'), \qquad (18)$$

with the overbar indicating the complex conjugate. The assumption of whiteness w.r.t. the t variable is not crucial and can be removed with use of a corresponding whitening filter which can be incorporated into the measurement filter.

Assuming the object function ϕ_0 completely known, our goal is to *estimate its unknown location x_0 from the data $d(t, \theta)$, $t \in R$, $\theta \in \Theta$.* For that, we compute the corresponding log likelihood function, which under the given noise assumptions attains the form [21]

$$L(x_c) = \Re \sum_{\theta \in \Theta} \left\{ \int_{-\infty}^{\infty} dt\, d(t,\theta) \overline{\alpha_\theta}(t; x_c) - \frac{1}{2} \int_{-\infty}^{\infty} dt\, |\alpha(t; x_c)|^2 \right\}, \qquad (19)$$

where $x_c = t_c u + s_c v$ is a test location and $\Re$ indicates the real part. After computation of the log likelihood function, the estimate of the unknown location x_0 is taken to be the point X_c of global maximum of the function, i.e. $X_c = \mathrm{argmax}_{x_c} L$.

The second term in the log likelihood function in Eq.(19) is independent of the data and, therefore, can be precomputed and stored for a number of test locations x_c. The first term in the same equation can be computed with the use of efficient fast Fourier transform-based algorithms. Indeed, we the help of Eqs.(7), (16) and Parseval's equality we can show that

$$\int_{-\infty}^{\infty} dt\, d(t,\theta) \overline{\alpha_\theta}(t; x_c) = \int_{-\infty}^{\infty} dp\, e^{-ip<u,x_c>} e^{(p-\kappa)<v,x_c>} \overline{\tilde{\alpha}_\theta}(p; 0) \tilde{d}(p, \theta),$$

$$(20)$$

where the overbar indicates the complex conjugate. Therefore, we see that *the first term in the log likelihood function in Eq.(19) can be computed via a tomographic procedure in which, for each probing direction, the measurements are convolutionally filtered and* backpropagated *with the backpropagation kernel $e^{(p-\kappa)<v,x_c>}$ and then coherently summed. The convolutional filter is the filter matched to the scattered field produced by the given object ϕ_0 when located at the origin of the coordinate system.*

4. Summary, Conclusions, and Future Research

In this paper, we addressed an inverse problem associated with the partial differential equation $\nabla^2 \psi = -\phi\psi$ and derived a translation property relating the fields scattered from a known object placed at different locations. The result was used to show that the maximum likelihood algorithm for detecting a target object, estimating its location, and classifying it from noisy scattered field measurement attains a form that can be efficiently implemented with fast Fourier transforms.

In the future, a detailed evaluation will be conducted of the performance of the proposed algorithm. Future relevant research may also follow the avenues of linear and nonlinear tomographic inversion of the partial differential equation $\nabla^2 \psi = -\phi\psi$ for estimation of the object function ϕ from measurements of the scattered fields ψ_θ. These and other research avenues are currently being explored and the findings will be reported elsewhere.

References

1. Devaney,A. J. (1982). A filtered backpropagation algorithm for diffraction tomography, Ultrasonic Imaging,Vol.1, p.336
2. Devaney,A. J. (1986). Reconstructive tomography with diffracting wavefields, Inverse Problems,Vol.2, p.161
3. Wolf,E.(1996). Principles and development of diffraction tomography, Trends in Optics,A Consortini, Academic Press, San Diego
4. Chadan,K. and Colton,D. and Paivarinta,L. and Rundell,W.(1997) An Introduction to Inverse Scattering and Inverse Spectral Problems, SIAM , Philadelphia
5. Colton,D. and Kress,R.(1998) Inverse Acoustic and Electromagnetic Scattering, Springer-Verlag,Berlin, 2nd edition
6. Lipson,H. and Cochran,W.(1966). The Determination of Crystal Structures, Cornell University Press,Ithaca, New York
7. Greenleaf,J. F.(1983). Computerized tomography with ultrasound,Proc. IEEE, Vol.71, p.330
8. Devaney,A.J.(1984). Geophysical diffraction tomography,IEEE Trans., Geosci. and Remote Sensing, Vol.GE-22, p.3
9. Witten,A. and Long,E.(1986) Shallow applications of geophysical diffraction tomography,Geosci. and Remote Sensing, Vol.GE-24, p.654
10. Witten,A. and Tuggle,J and Waag,J.C.(1988) A practical approach to ultrasonic imaging using diffraction tomography, J. Acoust. Soc. Am., Vol83, p.1645
11. Witten,A. and King,W.C.(1990) Acoustical imaging of subsurface features,J. Envir. Eng., Vol.116, p.166
12. Maleki,M.H and Devaney,A.J and Schatzberg,A. (1992) Tomographic reconstruction from optical scattered intensities, J. Opt. Soc. Am. A , Vol.9, pp.1356-1363
13. Devaney,A.J. (1990) Elastic Wave Inverse Scattering, Elastic Waves and Ultrasonic Nondestructive Evaluation, Elsevier Science Publishers,New York,S.K. Datta and J.D. Achenbach and Y.S. Rajapakse
14. Kak,A.C. and Slaney,M.(1988). Principles of Computerized Tomographic Imaging, IEEE Press, New York
15. Macleod,R.S. and Brooks,D.H.(1998). Recent Progress in Inverse Problems in Electrocardiology, IEEE Engineering Medicine and Biology, pp.83
16. Rossi,D.J. and Willsky, A.S.(1984). Reconstruction from Projections Based on Detection and Estimation of Objects, Parts I and II: Performance Analysis and Robustness Analysis, IEEE Transactions on Acoustics, Speech, and Signal Processing, Vol. ASSP-32, p.886
17. Devaney,A.J. and Tsihrintzis,G.A.(1991).Maximum likelihood estimation of object location in diffraction tomography, IEEE Transactions on Signal Processing, Vol. SP-39, p.672
18. Tsihrintzis,G.A. and Devaney,A.J. (1991).Maximum likelihood estimation of object location in diffraction tomography, Part II: Strongly scattering objects, IEEE Transactions on Signal Processing, Vol. SP-39, p.1466
19. Tsihrintzis,G.A. and Devaney,A.J. (1991).Estimation of object location from diffraction tomographic intensity data, IEEE Transactions on Signal Processing, Vol. SP-39, p.2136
20. Hansen, T.B. and Yaghjian, A.D.(1999). Plane-Wave Theory of Time-Domain Fields: Near-Field Scanning Applications, IEEE Press, Piscataway, NJ, USA
21. Van Trees, H.L.(1968) Detection, Estimation, and Modulation Theory, Part I, Wiley, New York

ELECTROMAGNETIC ANALYSIS OF A TWO DIMENSIONAL DIFFRACTION GRATING REFLECTOR

N. L. TSITSAS AND N. K. UZUNOGLU

Microwaves and Fiber Optics Laboratory,
School of Electrical and Computer Engineering,
National Technical University of Athens
Heroon Polytechniou 9, GR-15773 Zografou, Athens
GREECE
e-mail: ntsitsas@central.ntua.gr

The diffraction phenomena taking place in a reflector with grooved grating are analyzed by using integral equation techniques. The grating geometry is assumed to be two dimensional, consisting of an external surface of circular shape. The electric current flowing on the surface of the diffraction grating reflector is the unknown quantity and the free space Green's function for the two dimensional electromagnetic fields is utilized to develop an integral equation formulation. This integral equation is solved by using a *method of moments Galerkin technique*. The current distributions on the facets of the diffraction grating are described, by using linear functions fulfilling the edge conditions involved. Numerical computations have been carried out to quantify the diffraction phenomena and the associated dispersion characteristics are investigated. All the integrations involved are carried out analytically, thus reducing the numerical computations to a significant degree. Various diffraction grating geometries have been analyzed and will be presented.

1. Introduction

Diffraction gratings traditionally were constructed on polished metallic surfaces by mechanical etching and used in spectral analysis of optical radiation in optical engineering. Recently, because of the growing interest in wavelength division multiplexing in optical communications and the necessity of using grating diffraction phenomena in a highly controllable way, new technologies based on holographic techniques and chemical etching methods were developed. In traditional optical engineering, grating phenomena were analyzed by using geometrical optics methods. However, in gratings used in optical communications, information on the details concerning depolarization and higher order dispersion phenomena is required. Then wave analysis based on electromagnetic field theory is required to understand these second order important phenomena and to design grating structures.

In this paper *a method of moments Galerkin technique* is utilized to analyze two dimensional grating structures by solving numerically the resulting integral

414

equation. An $e^{j\omega t}$ time convention is assumed and suppressed throughout the paper.

2. Formulation of the problem

The diffraction grating under consideration is a metallic periodic structure, which consists of N identical parts (see Fig. 1). Each part is composed of two indentations with corresponding arcs lying on concentric circles and consists of four areas (I, II, III, IV), two areas of constant radius (I, III) and two areas of constant angle (II, IV). The i-th part of the grating is determined by the angle φ_i and has angular span $2\delta\varphi$. The grating is determined by its inner radius a, its outer radius b, the angle $\delta\varphi$ and the number N.

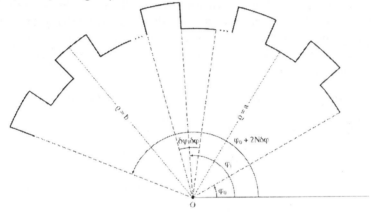

Figure 1. Geometry of the two dimensional diffraction grating.

The structure is illuminated by a plane electromagnetic wave impinging from a direction, determined by polar angle θ. According to *Huygens* principle, surface currents are induced on the metallic surface of the grating and every point of it acts as a source of secondary radiation. As a result, the incident electromagnetic wave is scattered to all directions. The total electromagnetic field at any point in space is the vector sum of the incident and the scattered fields.

The goal is to determine the surface currents, which are induced in every part of the grating and use them to compute the far electromagnetic field for various incident angles θ. For the description of these surface currents the edge radiation conditions are used. The surface current for each area of the grating's i-th part is determined by computing an unknown coefficient, which depends on the specific area. Thus, four coefficients a^i, b^i, c^i, d^i need to be computed for the i-th part and $4N$ coefficients totally.

The continuity of the tangential components of the total electric field (incident and scattered) to each one of the four areas of the grating's j-th part is described by the integral equation:

$$E_0 + (-j\omega\mu_0) \cdot \sum_{i=1}^{N} [\iint_S G(\bar{r}_j, \bar{r}') J^i(\bar{r}')d\bar{r}'] = 0 \tag{1}$$

where $E_0 = e^{-jk_0\rho\cos(\phi-\theta)}$ is the incident plane electric field, $G(\bar{r},\bar{r}') = (-j/4) \cdot H_0^{(2)}(k_0 |\bar{r}-\bar{r}'|)$ is the Green's function for the two dimensional problem (k_0 is the free space's wavenumber), J^i the induced surface currents to the i-th part of the grating's surface, $\bar{r}_j$ the observation vector (considered on the j-th part of the conductor's surface), $\bar{r}'$ the vector of the current sources and S is the grating's surface. Thus, for the j-th part four equations are derived (one for each area). All current sources, lying at the whole length of the conductor, have a contribution to each one of these equations, consequently the summation is extended to every part of the surface ($1 \leq i \leq N$). Thus, integral equation (1) becomes:

$$-j\omega\mu_0 \sum_{i=1}^{N} [\int_{\phi'=\phi_i-\delta\phi}^{\phi_i} d\phi' a(\frac{-j}{4})H_0^{(2)}(k_0\sqrt{(x-a\cos\phi')^2+(y-a\sin\phi')^2})J_z^i(a,\phi') +$$

$$\int_{\rho'=a}^{b} d\rho'(\frac{-j}{4})H_0^{(2)}(k_0\sqrt{(x-\rho'\cos\phi_i)^2+(y-\rho'\sin\phi_i)^2})J_z^i(\rho',\phi_i) + \tag{2}$$

$$\int_{\phi'=\phi_i}^{\phi_i+\delta\phi} d\phi' b(\frac{-j}{4})H_0^{(2)}(k_0\sqrt{(x-b\cos\phi')^2+(y-b\sin\phi')^2})J_z^i(b,\phi') +$$

$$\int_{\rho'=b}^{a} d\rho'(\frac{-j}{4})H_0^{(2)}(k_0\sqrt{(x-\rho'\cos(\phi_i+\delta\phi))^2+(y-\rho'(\sin(\phi_i+\delta\phi))^2})J_z^i(\rho',\phi_i+\delta\phi)] +$$

$$+e^{-jk_0\rho\cos(\phi-\theta)} = 0$$

The surface current distributions are chosen in such a way that the boundary edge conditions are fulfilled, which for the electromagnetic fields on the edges suggest [3, 9.2 (8)]:

$$E_\rho, E_\phi, H_\rho, H_\phi \propto \rho^{\frac{\beta-\pi}{2\pi-\beta}}, E_z \propto \rho^{\frac{\pi}{2\pi-\beta}}, H_z \propto constant \tag{3}$$

where angle β is shown in Fig. 2

By relation (3), the parallel component H_z of the magnetic field on the edges must be constant, while for the transverse components H_ϕ and H_ρ the following relations hold:

$$\beta = \frac{\pi}{2} \Rightarrow H_\phi, H_\rho \propto \rho^{\frac{\frac{\pi}{2}-\pi}{2\pi-\frac{\pi}{2}}} = \rho^{-\frac{1}{3}}, \beta = \frac{3\pi}{2} \Rightarrow H_\phi, H_\rho \propto \rho^{\frac{\frac{3\pi}{2}-\pi}{2\pi-\frac{3\pi}{2}}} = \rho^1 \tag{4}$$

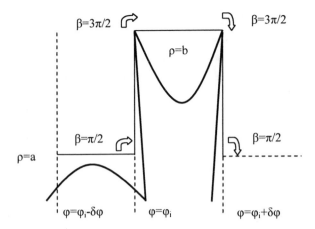

Figure 2. Current distributions on the four areas of the i-th part.

In order to fulfill conditions (4), the following current distributions are chosen for the grating's i-th part ($1 \leq i \leq N$):

Area I $\phi_i - \delta\phi < \phi < \phi_i$, $\rho = a$: $\qquad J_z^i = b' \cosh(\ l(\phi - (\phi_i - \frac{\delta\phi}{2})))$ (5)

Area II $\phi = \phi_i$, $\alpha < \rho < b$: $\qquad\qquad\qquad\qquad J_z^i = c'(\rho - b)$ (6)

Area III $\phi_i < \phi < \phi_i + \delta\phi$, $\rho = b$: $J_z^i = a' \cos(\ \frac{\pi}{2} \frac{\phi - (\phi_i + \frac{\delta\phi}{2})}{\frac{\delta\phi}{2}})$ (7)

Area IV $\phi = \phi_i + \delta\phi$, $\alpha < \rho < b$: $\qquad\qquad J_z^i = d'(\rho - b)$ (8)

Now, by combining (2), (5), (6), (7), (8) and using the addition theorem for Hankel functions [5, 1323]:

$$H_0^{(2)}(k_0 |\vec{r} - \vec{r}'|) = \sum_{m=-\infty}^{+\infty} H_m^{(2)}(k_0 (max(|\vec{r}|, |\vec{r}'|)))J_m(k_0 (min(|\vec{r}|, |\vec{r}'|)))e^{jm(\phi - \phi')} \quad (9)$$

the integral equations for the four areas of the j-th part of the conductor's surface (where the observation vector lies) are obtained.

3. Solution of the integral equations and far field computation

Using integration of power series term by term, the integrals of the resulting series are reduced to series of integrals. A method of moments Galerkin technique is then applied to the four equations of the j-th part. First, each equation is multiplied by the corresponding current distribution and then integration on the corresponding interval is carried out. For the areas of constant radius, integration is carried out with respect to the polar angle φ and for the areas of constant angle with respect to the radius ρ. After computing the resulting integrals, a linear system of four equations of the j-th part with $4N$ unknown current coefficients is fully determined.

1) Observation vector in the area: $\phi_j - \delta\phi < \phi < \phi_j$, $\rho = a$, of the j-th part

Multiplication of the equation by $cosh(\ l(\phi - (\phi_j - \dfrac{\delta\phi}{2})))$ and integration with respect to φ on the interval $[\varphi_j-\delta\varphi,\varphi_j]$ yields:

$$-\frac{\omega\mu_0}{4}\sum_{i=1}^{N}[b^i\sum_{m=-\infty}^{+\infty}aJ_m(k_0a)H_m^{(2)}(k_0a)\int_{\phi=\phi_j-\delta\phi}^{\phi_j}d\phi e^{jm\phi}\ cosh(l(\phi-(\phi_j-\frac{\delta\phi}{2})))$$

$$\int_{\phi'=\phi_j-\delta\phi}^{\phi_j}d\phi' e^{-jm\phi'}\ cosh(l(\phi'-(\phi_j-\frac{\delta\phi}{2})))+ \tag{10}$$

$$c^i\sum_{m=-\infty}^{+\infty}e^{-jm\phi_j}J_m(k_0a)\int_{\phi=\phi_j-\delta\phi}^{\phi_j}d\phi e^{jm\phi}\ cosh(l(\phi-(\phi_j-\frac{\delta\phi}{2})))\int_{\rho'=a}^{b}d\rho'\,H_m^{(2)}(k_0\rho')(\rho'-b)+$$

$$a^i\sum_{m=-\infty}^{+\infty}bJ_m(k_0a)H_m^{(2)}(k_0b)\int_{\phi=\phi_j-\delta\phi}^{\phi_j}d\phi e^{jm\phi}\ cosh(l(\phi-(\phi_j-\frac{\delta\phi}{2})))\int_{\phi'=\phi_j}^{\phi_j+\delta\phi}d\phi' e^{-jm\phi'}\ cos(\frac{\pi}{2}\frac{\phi'-(\phi_j+\frac{\delta\phi}{2})}{\delta\phi})+$$

$$d^i\sum_{m=-\infty}^{+\infty}e^{-jm(\phi_j+\delta\phi)}J_m(k_0a)\int_{\phi=\phi_j-\delta\phi}^{\phi_j}d\phi e^{jm\phi}\ cosh(l(\phi-(\phi_j-\frac{\delta\phi}{2})))\int_{\rho'=a}^{b}d\rho'\,H_m^{(2)}(k_0\rho')(\rho'-b)]=$$

$$-\int_{\phi_j-\delta\phi}^{\phi_j}d\phi e^{-jk_0a\cos(\phi-\theta)}\ cosh(l(\phi-(\phi_j-\frac{\delta\phi}{2})))$$

2) Observation vector in the area: $\phi = \phi_j$, $a < \rho < b$, of the j-th part

Multiplication of the equation by $(\rho-b)$ and integration with respect to ρ on the interval $[a,b]$ yields:

$$-\frac{\omega\mu_0}{4}\sum_{i=1}^{N}[b^i\sum_{m=-\infty}^{+\infty}ae^{jm\phi_i}J_m(k_0a)\int_a^b d\rho(\rho-b)H_m^{(2)}(k_0\rho)\int_{\phi'=\phi_i-\delta\phi}^{\phi_i}d\phi'e^{-jm\phi'}\cosh(l(\phi'-(\phi_i-\frac{\delta\phi}{2})))$$

$$c^i\sum_{m=-\infty}^{+\infty}e^{jm(\phi_j-\phi_i)}\int_a^b d\rho\{(\rho-b)H_m^{(2)}(k_0\rho)\int_{\rho'=a}^{\rho}d\rho'J_m(k_0\rho')(\rho'-b)\}+$$

$$c^i\sum_{m=-\infty}^{+\infty}e^{jm(\phi_j-\phi_i)}\int_a^b d\rho\{(\rho-b)J_m(k_0\rho)\int_{\rho'=\rho}^{b}d\rho'H_m^{(2)}(k_0\rho')(\rho'-b)\}+$$

$$a^i\sum_{m=-\infty}^{+\infty}be^{jm\phi_i}H_m^{(2)}(k_0b)\int_a^b d\rho(\rho-b)J_m(k_0\rho)\int_{\phi'=\phi_i}^{\phi_i+\delta\phi}d\phi'e^{-jm\phi'}\cos(\frac{\pi}{2}\frac{\phi'-(\phi_i+\frac{\delta\phi}{2})}{\frac{\delta\phi}{2}})+$$

$$d^i\sum_{m=-\infty}^{+\infty}e^{jm(\phi_i-(\phi_i+\delta\phi))}\int_a^b d\rho\{(\rho-b)J_m(k_0\rho)\int_{\rho'=b}^{\rho}d\rho'H_m^{(2)}(k_0\rho')(\rho'-b)\}+$$

$$d^i\sum_{m=-\infty}^{+\infty}e^{jm(\phi_i-(\phi_i+\delta\phi))}\int_a^b d\rho\{(\rho-b)H_m^{(2)}(k_0\rho)\int_{\rho'=\rho}^{a}d\rho'J_m(k_0\rho')(\rho'-b)\}]=$$

$$-\int_a^b d\rho e^{-jk_0\rho\cos(\phi_i-\theta)}(\rho-b)\tag{11}$$

3) Observation vector in the area: $\phi_j<\phi<\phi_j+\delta\phi$, $\rho=b$, of the j-th part

Multiplication of the equation by $\cos(\frac{\pi}{2}\frac{\phi-(\phi_j+\frac{\delta\phi}{2})}{\frac{\delta\phi}{2}})$ and integration with respect to φ on the interval $[\varphi_j,\varphi_j+\delta\varphi]$ yields:

$$-\frac{\omega\mu_0}{4}\sum_{i=1}^{N}[b^i\sum_{m=-\infty}^{+\infty}aJ_m(k_0a)H_m^{(2)}(k_0b)\int_{\phi=\phi_j}^{\phi_j+\delta\phi}d\phi e^{jm\phi}\cos(\frac{\pi}{2}\frac{\phi-(\phi_j+\frac{\delta\phi}{2})}{\frac{\delta\phi}{2}})\int_{\phi'=\phi_i-\delta\phi}^{\phi_i}d\phi'e^{-jm\phi'}\cosh(l(\phi'-(\phi_i-\frac{\delta\phi}{2})))+$$

$$c^i\sum_{m=-\infty}^{+\infty}e^{-jm\phi_i}H_m^{(2)}(k_0b)\int_{\phi=\phi_j}^{\phi_j+\delta\phi}d\phi e^{jm\phi}\cos(\frac{\pi}{2}\frac{\phi-(\phi_j+\frac{\delta\phi}{2})}{\frac{\delta\phi}{2}})\int_{\rho'=a}^{b}d\rho'J_m(k_0\rho')(\rho'-b)+$$

$$a^i\sum_{m=-\infty}^{+\infty}bJ_m(k_0b)H_m^{(2)}(k_0b)\int_{\phi=\phi_j}^{\phi_j+\delta\phi}d\phi e^{jm\phi}\cos(\frac{\pi}{2}\frac{\phi-(\phi_j+\frac{\delta\phi}{2})}{\frac{\delta\phi}{2}})\int_{\phi'=\phi_i}^{\phi_i+\delta\phi}d\phi'e^{-jm\phi'}\cos(\frac{\pi}{2}\frac{\phi'-(\phi_i+\frac{\delta\phi}{2})}{\frac{\delta\phi}{2}})+$$

$$d^i\sum_{m=-\infty}^{+\infty}e^{-jm(\phi_i+\delta\phi)}H_m^{(2)}(k_0b)\int_{\phi=\phi_j}^{\phi_j+\delta\phi}d\phi e^{jm\phi}\cos(\frac{\pi}{2}\frac{\phi-(\phi_j+\frac{\delta\phi}{2})}{\frac{\delta\phi}{2}})\int_{\rho'=b}^{a}d\rho'J_m(k_0\rho')(\rho'-b)]=$$

$$-\int_{\phi_j}^{\phi_j+\delta\phi}d\phi e^{-jk_0b\cos(\phi-\theta)}\cos(\frac{\pi}{2}\frac{\phi-(\phi_j+\frac{\delta\phi}{2})}{\frac{\delta\phi}{2}})\tag{12}$$

4) Observation vector in the area: $\phi=\phi_j+\delta\phi$, $a<\rho<b$, of the j-th part

Multiplication of the equation by $(\rho-b)$ and integration with respect to ρ from b to a yields:

$$-\frac{\omega\mu_0}{4}\sum_{i=1}^{N}[b^i\sum_{m=-\infty}^{+\infty}ae^{jm(\phi_i+\delta\phi)}J_m(k_0a)\int_b^a d\rho(\rho-b)H_m^{(2)}(k_0\rho)\int_{\phi'=\phi_i-\delta\phi}^{\phi_i}d\phi'e^{-jm\phi'}\cosh(l(\phi'-(\phi_i-\frac{\delta\phi}{2})))$$

$$c^i\sum_{m=-\infty}^{+\infty}e^{jm(\phi_i+\delta\phi-\phi_i)}\int_b^a d\rho\{(\rho-b)H_m^{(2)}(k_0\rho)\int_{\rho'=a}^{\rho}d\rho'J_m(k_0\rho')(\rho'-b)\}+$$

$$c^i\sum_{m=-\infty}^{+\infty}e^{jm(\phi_i+\delta\phi-\phi_i)}\int_a^b d\rho\{(\rho-b)J_m(k_0\rho)\int_{\rho'=\rho}^{b}d\rho'H_m^{(2)}(k_0\rho')(\rho'-b)\}+$$

$$a^i\sum_{m=-\infty}^{+\infty}be^{jm(\phi_i+\delta\phi)}H_m^{(2)}(k_0b)\int_b^a d\rho(\rho-b)J_m(k_0\rho)\int_{\phi'=\phi_i}^{\phi_i+\delta\phi}d\phi'e^{-jm\phi'}\cos(\frac{\pi}{2}\frac{\phi'-(\phi_i+\frac{\delta\phi}{2})}{\frac{\delta\phi}{2}})+$$

$$d^i\sum_{m=-\infty}^{+\infty}e^{jm((\phi_i+\delta\phi)-(\phi_i+\delta\phi))}\int_b^a d\rho\{(\rho-b)J_m(k_0\rho)\int_{\rho'=b}^{\rho}d\rho'H_m^{(2)}(k_0\rho')(\rho'-b)\}+$$

$$d^i\sum_{m=-\infty}^{+\infty}e^{jm((\phi_i+\delta\phi)-(\phi_i+\delta\phi))}\int_b^a d\rho\{(\rho-b)H_m^{(2)}(k_0\rho)\int_{\rho'=\rho}^{a}d\rho'J_m(k_0\rho')(\rho'-b)\}]=$$

$$-\int_b^a d\rho e^{-jk_0\rho\cos(\phi_j+\delta\phi-\theta)}(\rho-b) \tag{13}$$

Thus, letting j vary from 1 to N (so that the observation vector lies at the whole surface of the grating) a linear system of $4N$ equations with $4N$ unknown current coefficients is obtained. The current coefficients are computed by solving the $4N\cdot4N$ resulting linear system, thus the current distributions on the grating's surface are determined. Then taking into account the asymptotic form of the Green's function [1, 9.2.4]:

$$G(|\bar{r}|\to\infty,\bar{r}')\sim\sqrt{\frac{2}{\pi k_0}}\frac{e^{-jk_0\rho}}{\sqrt{\rho}}e^{jk_0(x'\cos\phi+y'\sin\phi)},\rho=\sqrt{x^2+y^2} \tag{14}$$

the far field is expressed as follows:

$$E(\bar{r})=\frac{-\omega\mu_0}{4}\frac{e^{-jk_0\rho}}{\sqrt{\rho}}\sum_{i=0}^{N}[b^ia\int_{\phi'=\phi_i-\delta\phi}^{\phi_i}d\phi'e^{jk_0a\cos(\phi'-\phi)}\cosh(l(\phi'-(\phi_i-\frac{\delta\phi}{2})))+$$

$$c^i\int_{\rho'=a}^{b}d\rho'e^{jk_0\rho'\cos(\phi_i-\phi)}(\rho'-b)+a^ib\int_{\phi'=\phi_i}^{\phi_i+\delta\phi}d\phi'e^{jk_0b\cos(\phi'-\phi)}\cos(\frac{\pi}{2}\frac{\phi'-(\phi_i+\frac{\delta\phi}{2})}{\frac{\delta\phi}{2}})+ \tag{15}$$

$$d^i\int_{\rho'=b}^{a}d\rho'e^{jk_0\rho'\cos(\phi_i+\delta\phi-\phi)}(\rho'-b)]+e^{-jk_0\rho\cos(\phi-\theta)}$$

4. Numerical results and discussion

The scattered far field is computed by using the approximate form of the Green's function for large values of the observation distance and the computed current coefficients. By letting the polar angle vary from 0 to 360 degrees, the total far electric field is computed for each angle and the radiation diagram is plotted.

The radiation diagram indicates the directions of intense and weak scattering by the distribution of the intensity of the far electric field.

Fig. 3-6 present the intensity of the far electric field in dB versus the polar angle φ, for different incident angles θ. The current distributions along the entire grating's length are shown in Fig. 7, while Fig. 8 indicates the current distributions along the grating's first part.

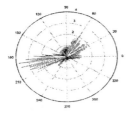

Figure 3. Intensity of the far electric field in dB as a function of φ for $\varphi_0=0^0$, $N=10$, $\delta\varphi=\pi/40$. $\theta=10^0$. $k_0=1$. $a=10$-π. $b=10$

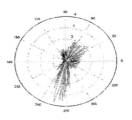

Figure 4. Intensity of the far electric field in dB as a function of φ for $\varphi_0=0^0$, $N=10$, $\delta\varphi=\pi/40$. $\theta=80^0$. $k_0=1$. $a=10$-π. $b=10$

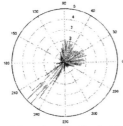

Figure 5. Intensity of the far electric field in dB as a function of φ for $\varphi_0=0^0$, $N=10$, $\delta\varphi=\pi/40$, $\theta=45^0$, $k_0=1$, $a=10$-π, $b=10$

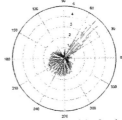

Figure 6. Intensity of the far electric field in dB as a function of φ for $\varphi_0=0^0$, $N=10$, $\delta\varphi=\pi/40$, $\theta=225^0$, $k_0=1$, $a=10$-π, $b=10$

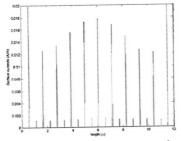

Figure 7. Surface currents versus the grating's length for $\varphi_0=0^0$, $N=11$, $\delta\varphi=\pi/88$, $\theta=22.5^0$, $k_0=1$, $a=10$-π, $b=10$

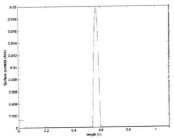

Figure 8. Surface currents of the grating's first part versus its length for $\varphi_0=0^0$, $N=11$, $\delta\varphi=\pi/88$. $\theta=22.5^0$. $k_0=1$. $a=10$-π. $b=10$

422

As shown in Fig. 3-6 the two directions of intense scattering are the direction of reflection and the direction of transmission. In particular, the maximum of the intensity of the scattered field lies in the transmission direction. This is due to the fact that the grating's dimensions are comparable to the wavelength and consequently a great percentage of the electromagnetic wave passes through the grating. The largest surface currents are induced on areas I and III and especially on the exterior circular grating's surface (area III) (see Fig. 7 and 8).

References

1. M. Abramowitz and I. A. Stegun, *Handbook of Mathematical Functions*, Dover Publications, Inc., New York, 1970.
2. I. S. Gradshteyn and I. M. Ryzhik, *Table of Integrals, Series, and Products*, Academic Press, 2000.
3. D. S. Jones, *Theory of Electromagnetism*, Pergamon Press, 1964.
4. G. N. Watson, *A Treatise on the Theory of Bessel Functions*, Cambridge University Press & MacMillan Company, 1944.
5. P. M .Morse and H. Feshbach, *Methods of Theoretical Physics, Parts I & II*, McGraw-Hill Book Company, Inc., New York, 1953.

WAVE PROPAGATION IN PLATES WITH MICROSTRUCTURE

V. VAVOURAKIS, K. G. TSEPOURA, D. POLYZOS

*Department of Mechanical Engineers and Aeronautics, University of Patras
Patras, TK 26500, Greece and Institute of Chemical Engineering and High
Temperature Processes (FORTH/ICE-HT)
Patras, TK 26504, Greece*

In this work, the Boundary Element Method is employed to solve three dimensional wave propagation problems in plates characterized by microstructure. The microstructural effects are taken into account with the aid of a simple gradient elastic theory, which is the simplest case of the general gradient elastic theory proposed by Mindlin. A representative numerical example dealing with the propagation of a transient, narrow band pulse is numerically solved through a boundary element methodology recently proposed by the last two authors for axisymmetric gradient elastic bodies. Comparisons made between the obtained results and the corresponding classical elastic solutions demonstrate the microstructural effects on the propagated waves.

1. Introduction

The effect of microstructure on the macroscopic description of the mechanical behavior of a linear elastic material can be adequately taken into account with the aid of higher-order strain gradient theories. Among those who have developed such theories one can mention the generalized gradient elastic theory of Mindlin [1], the couple stresses theory of Cosserat brothers [2], the non-local elastic theory of Eringen [3], the simple gradient elastic theory of Aifantis [4] and the gradient elastic theory with surface energy proposed by Vardoulakis and Sulem [5]. Although Mindlin's theory can be considered as the most general and comprehensive gradient elastic theory appearing to date in the literature, the simpler theory of Aifantis has been successfully used in the past to eliminate singularities or discontinuities of classical elasticity theory and to demonstrate their ability to capture size and edge effects, necking in bars, nano-structured materials behavior and wave dispersion in cases where this was not possible in the classical elasticity framework. However, use of the gradient elastic theory in boundary value problems increases considerably the solution difficulties in comparison with the case of the classical elasticity. For this reason, the need of using numerical methods for the treatment of those problems is apparent. Shu *et al.* [6] and Amanatidou and Aravas [7] have used the Finite Element Method

(FEM) for solving two-dimensional elastostatic problems in the framework of the general theories of Mindlin. On the other hand, Tsepoura and Polyzos [8], Polyzos *et al.* [9] and Tsepoura *et al.* [10] employing the Boundary Element Method (BEM) have solved three dimensional elastostatic and harmonic problems in the context of the simple strain-gradient elastic theories of Aifantis and Vardoulakis and Sulem Finally, Tang *et al.* [11] have proposed a meshless local Petrov-Galerkin methodology for the treatment of materials with strain gradient effects.

In the present work, the Boundary Element Method proposed by Tsepoura and Polyzos [8] is employed to solve three dimensional wave propagation problems in plates characterized by microstructure. The microstructural effects are taken into account with the aid of the simple gradient elastic theory of Aifantis, which is the simplest case of the general gradient elastic theory proposed by Mindlin. A representative numerical example dealing with the propagation of a transient, narrow band pulse is numerically solved through a boundary element methodology recently proposed by the last two authors for axisymmetric gradient elastic bodies. Comparisons made between the obtained results and the corresponding classical theory of elasticity solutions demonstrate the microstructural effects on the propagated waves.

2. Integral representations used in BEM solutions of elastodynamic and gradient elastodynamic problems

In this section, the equations of motion as well as the integral representations of classical elastic and gradient elastic problems are presented in brief. The BEM procedure adopted here for the solution of the considered elastic and gradient elastic wave propagation problems are described in detail in [12] and [8], respectively.

Consider a 3D linear isotropic elastic body, of volume Ω surrounding by a surface Γ. The dynamic behavior of this body is described by the well-known Navier-Cauchy equation

$$\mu\Delta\mathbf{u} + (\lambda + \mu)\nabla\nabla\cdot\mathbf{u} = \rho_m\ddot{\mathbf{u}} \tag{1}$$

where μ, λ are the Lamé constants, ρ_m is the mass density, while $\mathbf{u}$ is the displacement field vector. In order to have one a well posed mathematical problem Eq. (1) is complemented by the boundary conditions: $\mathbf{u}(\mathbf{x}) = \mathbf{u}_0$ and $\mathbf{p}(\mathbf{x}) = \tilde{\boldsymbol{\sigma}}\cdot\hat{\mathbf{n}} = \mathbf{p}_0$, where $\tilde{\boldsymbol{\sigma}}$ is the stress tensor, $\mathbf{p}$ is the surface traction vector, $\hat{\mathbf{n}}$ the normal to the surface vector and $\mathbf{u}_0, \mathbf{p}_0$ prescribed values at Γ.

Assuming harmonic dependence in time, the integral representation of the just described boundary value problem is [12]

$$\tilde{c}(x) \cdot u(x;\omega) + \int_\Gamma \tilde{P}^*(x,y;\omega) \cdot u(y;\omega) d\Gamma_y = \int_\Gamma \tilde{U}^*(x,y;\omega) \cdot p(y;\omega) d\Gamma_y \qquad (2)$$

where c is the jump tensor, P^* and U^* are the fundamental traction and displacement tensors, respectively, and ω is the excited frequency.

Considering now a 3D linear elastic body with microstructure, the equation of motion (1) is replaced by the following one [8]:

$$\left(1 - g^2\Delta\right)\left(\mu\Delta u + (\lambda+\mu)\nabla\nabla\cdot u\right) = \rho_m \ddot{u} \qquad (3)$$

where g is the volumetric energy strain-gradient coefficient, which correlates the microstructure with the macrostructure of the analyzed materials. Eq. (3) is accompanied by the classical boundary conditions: $u(x) = u_0$ and/or $p(x) = \tilde{\sigma} \cdot \hat{n} = p_0$, and the non-classical ones: $q(x) = \partial u(x)/\partial n = q_0$ and/or $r(x) = \hat{n} \cdot \tilde{\mu} \cdot \hat{n} = r_0$, where $\tilde{\mu}$ is Mindlin's double stress tensor.

The evaluation of all unknown fields $u(x)$, $p(x)$, $\partial u(x)/\partial n = q(x)$ and $r(x)$ requires the existence of two integral equations, as explained in [8], having the form:

$$\tilde{c}(x) \cdot u(x;\omega) + \int_\Gamma \tilde{P}^*(x,y;\omega) \cdot u(y;\omega) d\Gamma_y + \int_\Gamma \tilde{R}^*(x,y;\omega) \cdot q(y;\omega) d\Gamma_y =$$
$$\int_\Gamma \tilde{U}^*(x,y;\omega) \cdot p(y;\omega) d\Gamma_y + \int_\Gamma \tilde{Q}^*(x,y;\omega) \cdot r(y;\omega) d\Gamma_y \qquad (4)$$

$$\tilde{c}(x) \cdot q(x;\omega) + \int_\Gamma \partial\tilde{P}^*(x,y;\omega)/\partial n_x \cdot u(y;\omega) d\Gamma_y +$$
$$\int_\Gamma \partial\tilde{R}^*(x,y;\omega)/\partial n_x \cdot q(y;\omega) d\Gamma_y = \int_\Gamma \partial\tilde{U}^*(x,y;\omega)/\partial n_x \cdot p(y;\omega) d\Gamma_y + \qquad (5)$$
$$\int_\Gamma \partial\tilde{Q}^*(x,y;\omega)/\partial n_x \cdot r(y;\omega) d\Gamma_y$$

where all kernels appearing in the above integral equations are given explicitly in [8].

3. BEM solution of a wave propagation problem in plate with microstructure

Consider an isotropic steel disk-like plate having dimensions: radius $r=50mm$ and thickness $h=2mm$, while its material properties are: Young modulus of elasticity $E=200GPa$, Poisson ratio $v=0.29$, mass density $\rho_m=7850Kgr/m^3$ and negligible damping. In the case of the gradient elastic BEM case we considered for the volumetric energy strain-gradient coefficient two values: $g=0.5$ and 1.0.

A traction-free or/and a double traction-free disk is considered as boundary condition for either the conventional or for the gradient elastic BEM, respectively, subjected to a normal to the down surface time-depended displacement loading, applied on the center of it. The time function was chosen to be a 5-cycle Hanning window tone burst, as shown in Fig 1(a). A matter of great importance in time sampling is the time-window definition as well as the number of time samples. In our case, we choose a sufficiently large window ($T=9.6\mu s$, samples $N=128$) so as the P-wave could manage to travel from the center till the free end of the disk ($\approx 8.5\mu s$). The frequency spectrum of the transient loading is shown in Fig 1(b), where it can be easily seen its narrow band of excited modes in the wave-guide. Due to the fact that only 15 frequency harmonics (between 0.625 and 2.08MHz) have approximately non-zero Fourier coefficients, the time-harmonic problem is solved numerically only for those harmonics while neglecting all others.

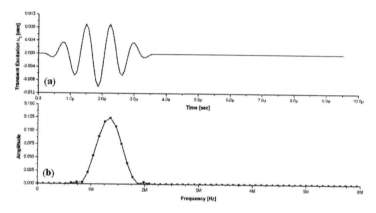

Figure 1. (a) The 5-cycle Hanning window tone burst time function and
(b) its bell-like frequency spectrum.

Apart from time sampling, special care must be taken on the spatial sampling of the BEM model. For good result convergence, one can follow the rule that requires at least ten nodal points per the minimum wavelength propagating in the medium. Due to the fact that only three-node linear elements are used, five elements per the minimum wavelength (λ_{min}) prove to be adequate enough. Thus, if the maximum frequency of the propagating mode in the disk is $f_{max}=2.08MHz$ and the phase velocity of the S-wave is $c_S=3142m/s$ then the element length for the conventional elastic BEM model must be:

$$l_e = \lambda_{min} / 5 = c_S / (5 f_{max}) \approx 0.3mm \qquad (6)$$

It must be noted here that for the gradient theory presented herein the dispersion relation is significantly different from the one of the classical elastic theory:

$$c = \frac{\omega}{k\sqrt{k^2 g^2 + 1}} \qquad (7)$$

where for g equal to zero we conclude to the well known dispersion relation: $c = \omega/k$. This means that for the BEM models described by the gradient elastic BEM the element length is as large as g increases. For both input values of the volumetric strain-gradient coefficient the element length is equal to disk's thickness. Consequently, lesser elements (52 linear elements) are needed to discretize the BEM model, in contrast with the conventional elastic BEM models (510 linear elements).

Finally, after solving numerically the time-harmonic problem through the BEM for the 15 harmonics of the transient excitation in the frequency domain, nodal displacements are transformed into the time domain through an Inverse Fast Fourier algorithm.

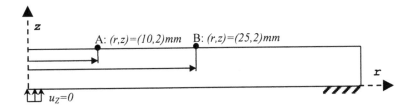

Figure 2. Schematic of the axisymmetric disk in the cylindrical coordinate system (r, z).

428

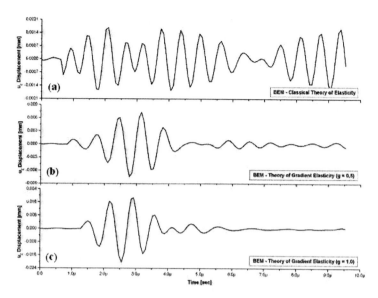

Figure 3. Time-history of u_z displacement of the boundary point: $(r, z)=(10, 2)mm$ according to: (a) the conventional elastic BEM and (b) the gradient elastic BEM for $g=0.5$ or (c) for $g=1.0$.

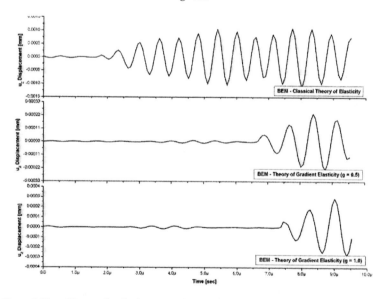

Figure 4. Time-history of u_z displacement of the boundary point: $(r, z)=(25, 2)mm$ according to: (a) the conventional elastic BEM and (b) the gradient elastic BEM for $g=0.5$ or (c) for $g=1.0$.

In Figs. 3 and 4 the displacements of the boundary points A and B (Fig. 2) evaluated by the gradient elastic BEM for various values of the g coefficient, are compared with the corresponding ones taken by the conventional elastic BEM (Fig. 2(a)). From these figures it can be easily seen that as g increases the time signal delays enough, resulting to less reflections from the free surfaces of the disk. This delay phenomenon can be explained by the dispersion Eq. (7), where as g increases the phase velocity c of each propagating mode decreases. In addition, the propagating pulse seems not to be affected by the geometrical dispersion observed in an elastic plate. This reveals that the material dispersion of the gradient elastic medium cancels the dispersion imposed by the boundaries of the plate.

4. Conclusions

In this work, a BEM proposed in [8] has been used to solve three dimensional wave propagation problems in plates characterized by microstructure. The obtained results have shown that as the volumetric gradient elastic coefficient increases the velocity of the propagating pulses in the plate decreases. Also the propagating pulses seem not to be affected by the geometrical dispersion observed in a classical elastic plate. Thus, one can say that the material dispersion of the gradient elastic medium cancels the dispersion imposed by the boundaries of the plate.

Acknowledgments

The authors would like to thank the Greek Secretariat of Research and Technology for the support at this work, in the framework of the PENED 01 EΔ 420 program.

References

1. Mindlin R.D., "Microstructure in linear elasticity", *Arch. Rat. Mech. Anal.* **10**, pp. 51-78 (1964).
2. Cosserat E. et F., "Theorie des Corps Deformables", *A. Hermann et Cie*, Paris (1909).
3. Eringen, A.C., "Vistas of Non-local Continuum Physics", *Int. J. Eng. Sc.* **30** (10), 1551-1565 (1992).
4. Aifantis E.C., "On the microstructural origin of certain inelastic models", *ASME J. Engng. Mat. Tech.* **106**, pp. 326-330 (1984).
5. Vardoulakis I., Sulem, J., *Bifurcation Analysis in Geomechanics*, Blackie/Chapman and Hall, London (1995).

6. Shu J.Y., King W.E., Fleck, N.A., "Finite elements for materials with strain gradient effects", *Int. J. Num. Meth. Engng.* **44**, pp. 373-391 (1999).

7. Amanatidou E., Aravas N., "Mixed finite element formulations of strain-gradient elasticity problems", *Comp. Meth. Appl. Mech. Eng.* **191**, pp. 1723-1751 (2002).

8. Tsepoura K.G., Polyzos D., "Axisymmetric BEM solutions of static and frequency domain gradient elastic problems", *Comp. Mech.* **32**, pp. 89-103, (2003).

9. Polyzos D., Tsepoura K.G., Tsinopoulos S.V., Beskos D.E., "A Boundary Element Method for solving 2-D and 3-D Static Gradient Elastic problems. Part I: Integral Formulation", *Comp. Meth. Appl. Mech. Eng.*, (in press) (2003).

10. Tsepoura K.G., Tsinopoulos S.V., Polyzos D., Beskos D.E., A Boundary Element Method for solving 2-D and 3-D Static Gradient Elastic problems. Part II: Numerical Implementation, *Comp. Meth. Appl. Mech. Eng.*, (in press) (2003).

11. Tang Z., Shen S., Atluri S.N., "Analysis of materials with strain-gradient effects: A meshless local Petrov-Galerkin (MLPG) approach, with nodal displacements only", *CMES: Comp. Modeling in Eng. & Sc.*, pp.177-196 (2003).

12. Polyzos D, Tsinopoulos SV, Beskos DE, "Staic and dynamic boundary elements analysis in incompressible linear elasticity", *European J. Mech. A/Solids* **17** (3), pp. 515-536 (1998).

A HYBRID METHOD FOR NEURAL–NETWORK TRAINING

C. VOGLIS AND I. E. LAGARIS

Department of Computer Science, University of Ioannina
IOANNINA - GREECE 45110

We present a special purpose method suitable for supervised training of feed-forward artificial neural networks, with one hidden layer and sigmoidal activation functions. The resulting Sum-of-Squares objective function is minimized using a hybrid technique that switches between the Gauss–Newton approach in the small residual case, and Newton's method in case where large residuals are detected. This is done in the spirit of Fletcher[1] where instead of Newton's method, a variable metric method (BFGS) was preferred in order to avoid the calculation of the Hessian matrix, which in the general case is both costly and cumbersome. In the special case that we consider here, the Hessian matrix can be expressed analytically and calculated efficiently by taking advantage of the properties of the sigmoidal activation function and its derivatives.

1. Introduction

Artificial Neural Network (ANN) training is a subject of central interest due to the widespread involvement of ANNs in a variety of scientific as well as practical tasks, such as data fitting, modelling, classification, pattern recognition, solution of differential equations etc. The "back–propagation" technique, that has been widely used mainly due to the simplicity of its implementation, is far from being satisfactory. Its main shortcomings, i.e. the oscillatory behavior and the sensitivity to round–off that causes premature termination, were early recognized. As a consequence, several alternative approaches employing efficient and robust optimization methods have been tried out. Among the various optimization techniques, Newton's method is the one with the most desirable properties. However it requires the computation of the Hessian matrix which may be inaccurate and very expensive if performed numerically, or very complicated to be expressed analytically. Hence due to this extra burden, Newton's method is not the preferred method in a host of practical applications. Training ANNs is an optimization problem, where the objective function can be cast as a sum of squared terms. This structure can be exploited to devise efficient approaches. In cases where the value of the objective function at the minimum is close to zero, an excellent approximation for the Hessian matrix exists that does not involve second derivatives. This is the well known "Gauss–Newton" approach. When however the value at the minimum is far from zero then the Hessian is not well approximated with severe consequences on the convergence of the approach. Taking the above into account, a hybrid method has been developed by Fletcher[1], which at run time detects, according to a simple criterion, the problem category, and switches appropriately to either the Gauss-Newton or to the BFGS Quasi–Newton method. Quasi–Newton methods do not use second order derivatives; instead they maintain at each iteration a positive definite approximation for the Hessian via an updating scheme, using gradient information only. Nowadays Quasi–Newton are considered the most succesful general purpose optimization methods and are being widely used. In this article we focus on the special problem of ANN training and we use a modified Newton instead of BFGS in the proposed framework of Fletcher[1]. We derive analytical closed-form expressions for the Hessian of this problem and present the sigmoidal properties that can be exploited to make the implementation efficient.

2. Description

Let $N(x, p)$ denote an ANN with input vector x and weights p. In our case this will be a perceptron with one hidden layer with sigmoidal units and linear output activation, i.e.

$$N(x, p) = \sum_{i=1}^{h} p_{i(n+2)-(n+1)} \sigma \left(\sum_{k=1}^{n} p_{i(n+2)-(n+1)+k} x_k + p_{i(n+2)} \right) \quad (1)$$

where:

- x_i, $\forall i = 1, \cdots, n$ are the components of the input vector $x \in R^{(n)}$.
- p_i, $\forall i = 1, \cdots, h(n+2)$ are the components of the weight vector p.
- h, denotes the number of hidden units.
- $\sigma(z) \equiv (1 + exp(-z))^{-1}$ is the sigmoid used as activation.

The training of the ANN to existing data is performed by minimizing the following "Error function":

$$f(p) = \frac{1}{2} \sum_{K=1}^{M} r_K^2 \equiv \frac{1}{2} \sum_{K=1}^{M} [N(x_K, p) - y_K]^2 \quad (2)$$

Useful expressions are its gradient:

$$g \equiv \nabla_p f(p) = \sum_{K=1}^{M} r_K \nabla_p r_K = J^T r \tag{3}$$

and the Hessian:

$$G \equiv \nabla_p^2 f(p) = J^T J + \sum_{K=1}^{M} r_K \nabla_p^2 r_K \tag{4}$$

Where J is the Jacobian given by: $J_{K,j} = \frac{\partial r_K}{\partial p_j}$

3. Methodology

Nonlinear least-squares problems are among the most commonly occurring and important applications, such as neural network training. Let $r : \Re^n \to \Re^m$, with $m \geq n$ be a nonlinear mapping. The problem is:

Find a local minimum x^* of

$$f(x) = \frac{1}{2} \sum_{i=1}^{m} [r_i(x)]^2 = \frac{1}{2} r(x)^T r(x)$$

Assuming that $r_i(x)$ is twice differentiable function then the derivatives of f are given by:

$$\mathbf{g}(x) = \nabla f(x) = J(x)\mathbf{r}(x) \tag{5}$$

$$G(x) = \nabla^2 f(x) = J^T(x)J(x) + \sum_{i=1}^{m} r_i(x)\nabla^2 r_i(x) \tag{6}$$

where $J(x)$ is the Jacobian matrix with elements $J_{ij} = \frac{\partial r_i(x)}{x_j}$ and $G(x)$ is the Hessian matrix, with elements $G_{ij} = \frac{\partial^2 f}{\partial x_i \partial x_j}$. Throughout this paper we will use the notation J_k, G_k, g_k and f_k for $J(x_k)$, $G(x_k)$, $g(x_k)$ and $f(x_k)$ respectively.

Many methods have been suggested for solving such problems. In this work we only consider Newton-like methods with line search. These methods have the form of the minimization algorithm MA1.

Within this framework, different methods correspond to different choices for the matrix B_k. Two well known methods which have been extensively studied and constitute the basis for several others are the damped *modified Newton* method for general nonlinear optimization and the damped *Gauss-Newton* for nonlinear least squares problems.

Algorithm MA 1

Let x_k be the current estimate of x^*

S1. [*Test for convergence*] If the conditions for convergence are satisfied, the algorithm terminates with $x^* = x_k$ as solution

S2. [*Compute search direction*] Compute a non-zero vector p_k, by solving

$$B_k p_k = -g_k \tag{7}$$

where B_k is some positive definite approximation of the hessian $\nabla^2 f_k$. This property ensures that p_k is a descent direction.

S3. [*Compute step length*] Compute a positive scalar a_k, called step length, which satisfies the two conditions

$$f(x_k + a_k p_k) \leq f_k + \rho a_k g_k^T p_k \tag{8}$$

$$|g(x_k + a_k s_k)^T s_k| \leq -\sigma g_k^T s_k \tag{9}$$

$\rho \in (0, 1)$ and $\sigma \in (\rho, 1)$ known as Wolfe conditions.

S4. [*Update the estimate of the minimum*] Set $x_{k+1} \leftarrow x_k + a_k p_k, k \leftarrow k + 1$ and go back to step S1.

3.1. *Modified Newton*

In the case of modified Newton method we solve the Newton equation (7) in order to obtain the search direction p_k using for B_k a modification of the Hessian matrix so that the resulting B is positive definite.

A variety of modifications have been proposed form simply adding a multiple of the identity matrix (poor results, fairly cheap) to modify the eigenvalues of the Hessian (accurate results, costly method).

We use the $L^T DL$ decomposition $\nabla^2 f_k \approx L_k^T D_k L_k$ and maintain the D_{ii} elements positive.

3.2. *Gauss - Newton*

The the Gauss-Newton method takes advantage of the special structure of least-squares problems in the following way. Since r is being minimized in the least-square sense, it is often the case that the components r_i are small, which suggests that the second term of the Hessian matrix (6) $\sum_{i=1}^{m} r_i(x) \nabla^2 r_i(x)$ may be ignored and

$$B_k \approx J_k^T J_k \tag{10}$$

is a reasonable approximation to $\nabla^2 f_k$. This choice of the Hessian gives the Gauss-Newton method. If $r(x^*) = 0$, then $B_k \to \nabla^2 f(x_k)$ and the order of convergence is similar to the modified Newton, superlinear and usually quadratic. When however $r(x^*) \neq 0$ then the order of convergence may be linear and in fact depends on the size of the residuals $r_i(x)$. The main drawbacks that disable the method from being of general purpose are that it may converge to a nonstationary point when J_k looses rank in the limit, and that the search direction p_k may become orthogonal to the gradient vector g_k while x_k is remote from x^*

3.3. *Combining the methods*

In comparing GN and Newton methods, the GN is generally preferred for zero residual problem (ZRP) that is when $\mathbf{r}(x^*) = 0$, whereas Newton-like methods are preferred for large residual problems (LRP) or when J_k looses rank.

Usually is not known beforehand whether a problem will turn out to have small or large residuals at the solution. It seems reasonable, therefore, to consider *hybrid algorithms*, which would behave like Gauss-Newton if the residuals turn out to be small (and take advantage of the cost savings associated with these methods) but switch to Newton like steps if the residuals at the solution are large (with the cost of approximating or computing second order derivatives).

We use Fletcher's criterion to switch between the GN approximation $(J_k^T J_k)$ and a positive definite modification of the full Hessian.

In this way the method will asymptotically take Newton steps for a LRP and GN steps for ZRP.

Following Fletcher, the quantity

$$\lim_{k \to \infty} \frac{f_k - f_{k+1}}{f_k} = \begin{cases} 0 & \text{for the LRP,} \\ 1 & \text{for the ZRP.} \end{cases}$$

Therefore this quantity defines a straightforward criterion that can be used to switch the minimizing procedure from GN to modified Newton.

We replace the second step of the minimization algorithm with the following

---

Set $B_k = \begin{cases} \nabla^2 f_k & \text{if } f_{k-1} - f_k/f_{k-1} < \epsilon, \\ J_k^T J_k & \text{otherwise.} \end{cases}$

Solve $B_k p_k = -g_k$ to get the search direction p_k

---

4. Hessian Calculation

Using the mapping $i = l(n+2) - (n+1) + m$ and $j = r(n+2) - (n+1) + s$ where $l, r = 1 \ldots h$ and $m, s = 0 \ldots n+1$ we can write $\frac{\partial^2 N(x_K, p)}{\partial p_i \partial p_j}$ in a form

$$\frac{\partial^2 N(x_K, p)}{\partial p_{l(n+2)-(n+1)+m} \partial p_{r(n+2)-(n+1)+s}} \tag{11}$$

For simplicity we denote $Y_j = \sum_{k=1}^{n} p_{j(n+2)-(n+1)+k} x_k + p_{j(n+2)}$. Using the above notation we can derive an analytic formula for the Hessian matrix of a feedforward artificial neural network $N(x, p)$. The resulting formula is displayed in Table 1.

Table 1. Analytic Hessian calculation

$l = r$	$m = 0$	$s = 0$	0
		$s = 1 \ldots n$	$\sigma'(Y_j) x_s$
		$s = n+1$	$\sigma'(Y_j)$
	$m = 1 \ldots n$	$s = 0$	$\sigma'(Y_j) x_m$
		$s = 1 \ldots n$	$p_{l(n+2)-(n+1)} x_m x_s \sigma''(Y_j)$
		$s = n+1$	$p_{l(n+2)-(n+1)} x_m \sigma''(Y_j)$
	$m = n+1$	$s = 0$	$\sigma'(Y_j)$
		$s = 1 \ldots n$	$p_{l(n+2)-(n+1)} x_s \sigma''(Y_j)$
		$s = n+1$	$p_{l(n+2)-(n+1)} \sigma''(Y_j)$
$l \neq r$	$m = 0 \ldots n+1$	$s = 0 \ldots n+1$	0

5. Experimental Results

In order to compare the convergence speed of our hybrid method, to other well known algorithms we have contacted a series of experiments using the Merlin Optimization environment[6]. The Merlin testbed provides a set of powerful and robust minimization routines, that guarantee its effectiveness.

We have tested our method against five other minimization procedures namely the Quasi-Newton (Tolmin[7]), the Gauss-Newton with line search, the Hybrid BFGS–Gauss-Newton[a],the damped modified Newton and the conjugate gradients with Polack–Ribiere updates.

The strategy that we followed was to start the minimization in the neighborhood of a local minimum and calculate the iterations and function calls that each method performed in order to reach it. In this way we have a clear view of the convergence rate.

---

[a]As it was presented in the original paper of Fletcher

In the tables (2) and (3) we present the results for a training problem with input dimension $n = 2$ hidden nodes $h = 5$ and training data $M = 100$. This problem falls into the LRP category because the minimum reached is non-zero. Each table displays the results for two different minima of the same training problem.

Table 2. LRP: Minimum No 1

Method	Iterations	Function calls
Hybrid Newton	126	357
Hybrid BFGS	335	652
Newton	719	1000
Gauss-Newton	1000	3000
Tolmin	174	252
Conjugate Gradient	1480	6000
Minimum value		18.486

Table 3. LRP: Minimum No 2

Method	Iterations	Function calls
Hybrid Newton	20	64
Hybrid BFGS	33	100
Newton	35	57
Gauss-Newton	150	301
Tolmin	74	107
Conjugate Gradient	77	302
Minimum value		19.266

On the other hand, the results for a ZRP case are shown in the tables (4) and (5)This training problem has input dimension $n = 10$ hidden nodes $h = 10$ and training data $M = 100$. Again we present two results for different local minima.

Table 4. ZRP: Minimum No 1

Method	Iterations	Function calls	Minimum reached
Hybrid Newton	115	275	0
Hybrid BFGS	363	487	0
Newton	316	364	0
Gauss-Newton	257	296	0
Tolmin	513	697	0
Conjugate Gradient	2318	10000	1.0768
Minimum value			0

Table 5. ZRP: Minimum No 2

Method	Iterations	Function calls	Minimum reached
Hybrid Newton	599	1000	0.0961
Hybrid BFGS	623	713	0.0101
Newton	759	1000	2.3282
Gauss-Newton	734	1000	0.4374
Tolmin	710	1000	0.0733
Conjugate Gradient	965	4000	1.2228
Minimum value			0

6. Conclusion

We have presented a novel hybrid method focused in Artificial Neural Network training. Our proposal combines the hybrid ideas of Fletcher[1] and an efficient way to compute analytically second order derivatives. Our preliminary results are promising, however more extensive experimentation should be contacted.

We are currently investigating online schemes for exact Hessian calculation, in order to increase the speed of our method. We also search for a better way to distinguish between LRP and SRP. Another extension is to calculate derivatives for alternative ANN architectures and use the proposed algorithm for their training process.

References

1. Fletcher R. and Xu C. *Hybrid Methods for Nonlinear Least Squares*, IMA Journal on Numerical Analysis 7 (1987) 371-389.
2. Hornik K., Stinchcombe M., and White H.,Neural Networks 2 (1989) 359.
3. Cybenko G.,*Approximation by superpositions of a sigmoidal function*, Mathematics of Control Signals and Systems 2 (1989) 303-314.
4. Bishop C., *Neural Networks for Pattern recognition*, Oxford University Press, 1995.
5. Fletcher R., *A new approach to variable metric algorithms*, Computer Journal 13 (1970) 317-322
6. Lagaris I. E., Papageorgiou D. G. and Demetropoulos I. N. *MERLIN-3.0 A multidimensional optimization environment*, Comput. Phys. Commun. 109 (1998) 227-249
7. Powell M.J.D., *TOLMIN: a Fortran package for linearly constrained C optimization calculations*, Technical Report NA2,DAMTP,University of Cambridge, Cambridge, 1998

CLEANING ASTRONOMICAL DATABASES USING HOUGH TRANSFORMS AND RENEWAL STRINGS

C. K. I. WILLIAMS AND A. J. STORKEY

School of Informatics, University of Edinburgh,
5 Forrest Hill, Edinburgh EH1 2QL

N. C. HAMBLY AND R. G. MANN*

Institute for Astronomy, University of Edinburgh,
Blackford Hill, Edinburgh, EH9 3HJ

In this paper we are concerned with detecting artefactual linear features such as satellite or aeroplane tracks and scratches in astronomical images. The standard approach to line detection is the Hough transform (HT). By combining the HT with renewal processes and hidden Markov models we obtain a probabilistic model (renewal strings) that is a highly effective method for removing these artefacts.

1. Introduction

Very large scientific datasets are now being created, and the challenge of scientific data mining is to extract scientific knowledge out of this data. Sometimes standard data mining/machine learning methods such as predictive and descriptive modelling can be applied [1], but in other cases we may need to build models that incorporate domain knowledge.

In this study we focus on the detection of spurious objects in astronomical catalogues. Spurious objects can arise by a number of causes including satellite or aeroplane tracks, scratches, fibres and other linear phenomena introduced to the plate, circular haloes around bright stars due to internal reflections within the telescope and diffraction spikes near to bright stars. In this paper we focus on linear phenomena; full details of the entire system and how all the different types of spurious object are detected can be found in [2]. A standard approach to the detection of lines is to use the *Hough transform* [3]. However, we will see that the Hough transform does not

---

*Also at National E-Science Centre, South College Street, Edinburgh EH8 9AA.

perform adequately. The *renewal strings* method was developed in order to solve these problems. It is a probabilistic technique which combines the Hough transform, renewal processes and hidden Markov models and has proved highly effective in this context.

The structure of the remainder of the paper is as follows: In section 2 we describe the astronomical data available and the kinds of spurious objects that occur. Section 3 explains the Hough transform approach to linear feature detection and some of the problems with this method. Section 4 describes the renewal strings method and its application to real data. We conclude with a discussion in section 5.

2. Astronomical Databases and Spurious Objects

Digital sky surveys are a fundamental research resource in astronomy (see e.g. [4]). Surveys are carried out in all wavelength ranges, from high energy gamma rays to the longest wavelength radio atlases. Despite this diversity, there are certain features common to most digital surveys: pixel images at a given spatial and spectral resolution are processed using a pixel analysis engine to generate lists of object detections containing parameters describing each detection. In most cases, the object detection algorithm has to be capable of finding a heterogeneous family of objects, for example point–like sources (stars, quasars); resolved sources (e.g. galaxies) and diffuse, low surface–brightness, extended objects (e.g. nebulae). Object parameters describing each detection typically include positions, intensities and shapes.

This paper looks at a class of problems which are the most significant sources of unwanted records in the SuperCOSMOS Sky Survey (SSS) data. The SSS is described in a series of papers ([5] and references therein). Briefly, the SSS consists of Schmidt photographic plates scanned using the fast, high precision microdensitometer SuperCOSMOS. The survey is made from 894 overlapping fields in each of three colours (blue, red and near–infrared denoted by the labels J, R and I respectively), one colour (R) is available at two epochs to provide additional temporal information. Each image contains approximately 10^9 2-byte pixels. The pixel data from each photograph in each colour and in each field are processed into a file of object detections; each object record contains parameters describing that object. The number of object detections on a plate varies from about half a million to ten million records. Presently, the entire southern hemisphere is covered, primarily using plates from the UK Schmidt Telescope. Data and many more details are available online at http://www-wfau.roe.ac.uk/sss.

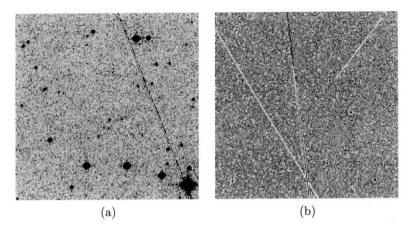

(a)   (b)

Figure 1. (a) A satellite track with spurious objects elongated along the track. The 'blocky' appearance of the sky pixels is a result of them having passed through a Haar-transform compression algorithm. (b) A number of aeroplane tracks in field UKJ413.

The focus of this paper is on locating artefactual objects in an astronomical dataset derived from or affected by satellite tracks, aeroplane tracks or scratches, and how can we accurately distinguish them from true astronomical objects. Because much work has usually already been done deriving the object data from images, because in many cases original image data may not be available, and because of the huge size of the images involved, we are not considering working with the images directly, only with the derived datasets. Examples of these kinds of linear features are shown in Figure 1.

Spurious objects introduce errors in statistical results derived from the data, and make locating particular classes of objects harder. Many tracks result in spurious, elliptical objects with low surface brightness that resemble galaxies, contaminating the respective object catalogue. Also we may be interested in real objects which might be in one dataset but not in an other, such as objects which are evident at one wavelength but not another. Fast moving stars will also be in different places in catalogues derived from observations at different times, meaning that they will not have exact positional matches across the datasets, see e.g. [6]. Unfortunately satellite track artefacts have the same characteristics; they will only ever appear (in the same place) in one dataset, and not in any other. Searches on non-matching objects will bring up all the objects of interest *plus* all of these artefacts. When searching for rare objects the spurious records can be overwhelming.

Removing spurious objects, then, is of broad importance in astronomy.

3. The Hough Transform

The most obvious way to locate lines of objects in two dimensional data utilises the Hough transform. Indeed this approach was followed in [7, 8]. The Hough transform [3] is a standard image processing method from which other related approaches have been developed. In its standard form it is generally used in low dimensional situations to find lines containing a high density of points hidden amongst a large number of other points distributed widely across the whole space. Commonly it is used for line detection in images.

The Hough transform works by moving from the space of points to the Hough space, that is the space of lines. Every point (d, θ) in Hough space corresponds to a line in the original space which is at a perpendicular distance d from the origin in the data space and inclined at angle θ from the vertical.

One method of implementing the Hough transform searches through a finite number of line angles θ. For each angle all the data points are considered. For each data point we find the (perpendicular) distance from the origin of the straight line *through* that point at the relevant angle. This distance is then discretised, and the count in an accumulator corresponding to this discretised distance and angle pair is increased by one[a]. Neglecting dependencies between the accumulators at different angles and assuming, as a null hypothesis, a uniform scattering of points[b] in the data space, we know the distribution of the count in a given Hough accumulator will be Poisson with a mean proportional to the length of the corresponding line. If on the other hand there is also a line of high density points in amongst the uniform scattering, then this Poisson distribution will not be the correct model for the Hough accumulator corresponding to this line. In fact the count will be significantly higher than that expected under the null hypothesis. Hence looking at the probability of the actual count under this null hypothesis, and ideally comparing this to an alternative hypothesis based on some prior model of line counts for satellite tracks, will indicate how likely it is that this accumulator corresponds to a satellite track. A surprisingly large number of papers on, and applications of, the Hough transform focus

---

[a] A concise tutorial/demo of the Hough transform can be found at http://www.storkey.org/hough.html

[b] More formally assuming that points are sampled from a homogeneous Poisson process.

on finding large absolute values contained within the Hough accumulators rather than comparing them with the null distribution. Needless to say that approach is significantly less accurate and powerful and is not to be recommended.

For an SSS dataset derived from field UKJ005, Figure 2a illustrates the Hough transform of the data. In this figure lighter regions correspond to higher accumulator counts. The large scale variation in light and dark regions comes from the square shape of the plate: lines through the centre along a diagonal are longer than off centre or off diagonal lines, and hence will generally contain more stars. It is also possible to see some sinusoidal lines of slightly increased intensity. These are caused by a local cluster of large numbers of objects—either a galaxy or artefacts surrounding a bright star.

The points in Figure 2a which have been circled correspond to points which have an accumulator count much higher than that which would be suggested by the null Poisson model. In fact one of these Hough accumulators combined with its highest count nearest neighbours corresponds to the data illustrated in Figure 2b. In this figure the data have been rotated so that the horizontal axis shows the length along the line of the Hough box (the region in data space corresponding to a given accumulator), and the vertical axis corresponds to the much smaller combined width of the two neighbouring Hough boxes. The representation shows that this Hough box does indeed contain (part of) a satellite track, in fact the most prominent track on the plate. Part of this track is illustrated in Figure 1(a).

The curved shape of the track gives some hints of the problems which will be encountered when working with satellite tracks. Many real stars and galaxies lie within the the smallest Hough box which could contain the satellite track. Hence flagging everything within the Hough box as possibly spurious will not suffice. Reducing the size of the Hough boxes means that the data from a single track will be split across a number of boxes, and the data within each box might begin to be swamped by the general variations in underlying star and galaxy distribution. This, combined with the fact that some of the tracks and scratches we are interested in locating are very short segments means that the data from the line can be swamped by the random variations in sample density of all the other points along the line. Thus fitting a nonlinear regression function in the Hough box using a robust error model would not be sufficient to extract the track. Add to this the problems of dashed aeroplane tracks and the variable curvature of scratches and it becomes clear that an approach is needed that is more flexible than

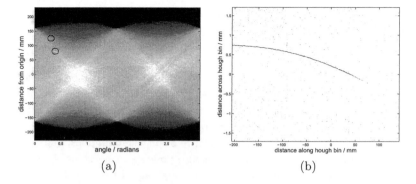

Figure 2. (a) Hough transform of data from field UKJ005. The vertical axis gives the distance from an origin in the centre of the plate (400 bins), the horizontal axis gives the angle of orientation (200 bins). Lighter colours are higher accumulator counts. The circled points are Hough accumulators with a significantly high count, and which correspond to satellite tracks on the plate. The original data which was accumulated in the 3 Hough accumulators at points (0.38,[79 81]) of (a), that is points in the lower circle, is shown in (b). Note the different scales of the two axes. The curvature of the track is obvious from this plot.

the Hough transform. Comparisons of the results obtained by the methods developed in this paper and a Hough approach can be found in section 4.3.

4. Renewal Strings

Renewal strings are a probabilistic data mining tool for finding subsets of records following unknown line segments in data space which are hidden within large amounts of other data. The method was developed specifically to address the problem of this paper and first published in [9]. Renewal strings combine a model for two dimensional background data and a set of models for small numbers of data points lying on one dimensional manifolds within the two dimensional space. The design of the model allows efficient line based techniques to be used for separating the background data from the different one dimensional manifolds.

The key point in detecting linear features is that even if we have a narrow strip of image which we believe contains a linear feature, there will still be objects in that strip which will not be from the linear feature but from the background. This happens mainly because the linear feature may not extend the whole length of the strip. Thus we need to *label* each object in the strip as either being derived from a linear feature (label 1)[c] or the background (label 0). To do this we measure the distance along the strip of each object, sort these distances and then compute the inter-point distances $\{\Delta t_i\}$. The key insight is that the distribution of these inter-point distances will be different in regions of pure background, and regions where there is a linear feature, and this drives the labelling process. We use strips that are narrow so that in regions of the strip where a linear feature is present the number of background objects projected onto the line will be small. Figure 2b gives a misleading impression here as the the number of Hough boxes used in the actual experiments reported below is considerably larger than the number used to create this figure, in particular the width is much smaller.

The two key probabilistic models we use in renewal strings are the *renewal process* and *hidden Markov models*; these are described in section 4.1 below. In section 4.2 we describe how these are put together to make a practical algorithm. Results of experiments are given in section 4.3.

4.1. *The model*

A **renewal process** is a model for event times obtained by defining a probability distribution for the time between events (commonly termed the inter-arrival time). The time at which event i occurs is dependent only on the time of the previous event $i - 1$, that is it has the Markov property. In this paper we will be looking at the distance between points on a line rather than the time between events, and hence this distance will be called the inter-point distance. A Poisson process on the line gives an exponential distribution of inter-point distances.

To tie together inferences over the whole set of inter-point distances along the line we use a **hidden Markov model** (HMM). HMMs are a ubiquitous tool, seen in many different applications such as speech recognition, gene sequence analysis, time series prediction and natural language processing. A standard introduction to hidden Markov models can be found in [10].

Suppose we are given an inter-point distance Δt_i at a point i and renewal process models for the linear feature and background processes. Then given a prior probability $P(X_i)$ of point i having label X_i and the inter-point

---

[c]It would be possible to use a number of different labels for different kinds of linear features such as aeroplane tracks, satellite tracks, scratches etc but for simplicity we have not subdivided the class of linear features.

distance Δt_i, we can obtain the posterior probability

$$P(X_i|\Delta t_i) = \frac{P(\Delta t_i|X_i)P(X_i)}{\sum_{X_i} P(\Delta t_i|X_i)P(X_i)}. \qquad (1)$$

However, the prior probability of a point being part of a satellite track will be highly dependent on whether the last point in the line was part of the same type of satellite track or not. Hence we need some prior model for satellite track continuity. This is most easily defined using a Markov model for the track labels X_i. Because X_i are not observable the whole model is called a hidden Markov model. This Markov model is defined by the conditional transition probabilities $P(X_i|X_{i-1})$ for the change in label between object $i-1$ and object i along the line, and the initialisation probability $P(X_0)$ of starting in the background or track states. A big advantage of the HMM model is that there is an efficient algorithm (called the forward-backward algorithm) which is linear in the number of points along the line for computing $P(X_i|D)$, the probability distribution over the label X_i given all of the data D along the line.

The combination of renewal processes and hidden Markov models is not new within temporal settings. It has been used for (amongst other things) modelling the pecking behaviour of pigeons [11]! Also, in the case that the renewal processes are all Poisson processes, there is a direct relationship between the renewal process hidden Markov model and the Markov modulated Poisson process [12].

One way to visualise the complete renewal string model involves building a background image of the stars and galaxies and then superimposing the linear features. We have used an inhomogeneous Poisson process as the background model. Then having decided on the number of the linear features, and the starting point and direction for each one, we thread beads onto a string for each satellite track, where the distances between the beads are defined by the hidden Markov renewal process, stopping when we get $X_i = 0$ in the hidden Markov model. We then place the beads down on to a background image, keeping the string tight. The final data consists of the positions of the stars and galaxies in the background model, combined with the positions of the beads.

4.2. The algorithm

We start by obtaining an estimate for the density $\rho(r)$ of background objects local to each point r. In practice this is estimated from a local region of size $s \times s$ where s is chosen so that the contributions from the satellite tracks to the total number of points, and the variation in background star/galaxy density are both negligible. For the SSS data ρ was estimated by gridding the plate into 40,000 regions.

To initialise the algorithm, we set the line width w based on the expected maximum width of the lines to be found. We then define the inter-point distance distribution $P(\Delta t | X, \rho)$ for each class; this can depend on the background object density at that point. Let Θ denote the set of angles to be considered (from 0 to 180 degrees), and $L = L(\theta)$ the set of lines at angle θ. Then the algorithm is:

For each angle θ from the set Θ

(a) For every point in the dataset, find all the lines L' of width w in L which contain the point. Store the position t *along* each line in L' in a bin corresponding to that line.

(b) For each line in L, sort all the distances in its bin. Use these distances as the data for an HMM with emission probabilities $P(\Delta t | X, \rho)$ and transition probabilities $P(X_i | X_{i-1})$. Run the usual forward-backward inference to get marginal posterior class probabilities for each point. Flag any points which have a low probability of being background objects.

End for

At the end of this process, the flagged points are the points suspected to be part of a track or scratch. The associated probability gives extra information regarding the certainty of this classification. The inference process is exact for one line but does not consider lines that cross; however, the number of objects affected by this situation is very small.

In this work the tracks are modelled as Poisson processes (a specific form of renewal process with an exponential inter-point distance). The fundamental reason for this is that along the line of a satellite track there will also be objects corresponding to stars and galaxies. The point density along a track from a satellite moving in front of a dense distribution of stars will be higher than one passing in front of a relatively sparse region of sky, and hence the line of objects along each track is a superposition. Poisson processes have the advantage that the superposition of two Poisson processes is a Poisson process. The equivalent statement is not true for more general forms of renewal processes.

4.3. *Experimental results*

The renewal process model was tested on plate datasets within the SSS. 1000 different angle settings were used, and 18000 different bins for the distance from the origin. Each data point was put in two bins (i.e. the line width was twice the distance separation). These values were obtained from simple geometric arguments. The number of bins for the distance from the origin was chosen based on the largest widths of the tracks which we were trying to detect. Then the angular variation was then chosen such that any significant length of any track will not be missed between two different angles.

Tuning of the HMM parameters could be done with the usual expectation maximisation algorithm [10]. However, as they are physically interpretable we were able to set them by analysis of the data directly. The inter-point distribution for the satellite track was set to be an exponential distribution using the empirical mean from a training set including 30 different satellite tracks from low density plates (the resulting mean inter-point distance was 360 microns on the plate, corresponding to 24 arcsec on the sky). As stars and galaxies also appear along satellite tracks, this empirical mean was added to the mean of the background process to model the density along a satellite track in different circumstances. The transition probabilities were set using approximate prior knowledge about the number of satellite tracks etc. on the training plates, the number of objects in total and the number of objects per satellite track. This resulted in the transition matrix $P(X_t|X_{t-1})$ for $X = \{background,\ track\}$ of

$$\begin{pmatrix} 0.999998 & 0.04 \\ 2 \times 10^{-6} & 0.96 \end{pmatrix}$$

The initial probabilities $P(X_0)$ were assumed to be the equilibrium probabilities of a Markov chain with these transition probabilities.

To evaluate the success of the renewal strings method the detections were evaluated by an astronomer (NCH), who looked through a printed version of the plate data for a whole plate (UKR001). The plate was split into 36 regions, each region being printed on an A3 sheet. These A3 sheets were examined closely for false negative and false positive detections, and the astronomer also commented on other aspects of the detection he felt notable. In this analysis features corresponding to small fibres were ignored. As the measured characteristics of true stars or galaxies along or very near a satellite track will be affected by the track, these objects should also be flagged.

Out of 429238 objects on the plate 8539 objects were flagged as belonging to linear features. 60 of these (0.7%) were false positives, while there were 14 (0.0033%) of false negatives. All the major satellite tracks were found and the ends of the tracks were generally accurately delineated. All of the small scratches were properly identified, although one of them involved a significant bend. Some of the objects along the bend were improperly classified as real objects. A small number of small false positive linear detections were made. Some objects due to fibres on the plate were also picked up, although as expected the method was not designed for and is not ideally suited to their detection.

It is also interesting to compare the performance of the renewal strings method to the Hough transform. This is not entirely straightforward as the Hough transform is designed for a slightly different problem (detecting lines which traverse the whole plate). However, we can assess how well the Hough transform can find strips which contain linear features. For a useful comparison with the renewal string results, we look at the significance level which would be needed to detect each track that was detected with the renewal string method. We also look at how many other false positive tracks would also be detected for given significance levels. The number of angles and line widths considered were set to match the renewal string settings.

Table 1. The number of the 35 tracks/scratches on UKR002 which would have been detected using the Hough transform.

SIGLEV	DET	TOT	THEOR
0.5	31	3.18×10^6	4.4×10^6
0.9	21	8.96×10^5	8.7×10^5
0.95	15	5.34×10^5	4.4×10^5
0.99	9	1.62×10^5	8.7×10^4
0.999	7	30147	8712
$1 - 10^{-5}$	4	1158	87.12
$1 - 10^{-7}$	2	71	0.87
$1 - 10^{-9}$	1	16	0.0087

This procedure was carried out on plate UKR002, which has no satellite tracks that traverse the whole plate, but does have some smaller to medium size (a quarter of plate width) tracks and scratches. Each track was located in a semi-automated way, and most diffraction spikes were ignored by removing all tracks within 1.5 degrees of the horizontal or vertical. The position and angle of each track was noted, and included in a track list.

In general each track was noted once. However, where there was a large curvature to a track, more than one reference could have been included in the list. The points corresponding to the plate notes in the bottom left of the plate, and detections relating to a halo about a bright star were removed by hand. This left 35 tracks or scratches in the reference list.

Results for plate UKR002 are shown in Table 1. SIGLEV gives the significance level used. DET gives the number of the tracks which would have been flagged at that significance level, TOT the total number of lines flagged as significant by the Hough transform, and THEOR the theoretical number of false positives for a homogeneous Poisson distribution. A significance level of $1 - 10^{-7}$ is needed to reduce the theoretical false positive detection rate to a suitably low level. Then only two of the tracks would have been detected, and in practice there would have been many false positives flagged.

For comparison purposes, we looked at all the listed tracks and calculated what significance level would be needed in order to detect the line containing that track with the Hough transform. The table shows the significance level required to detect the tracks along with the total number of tracks (true and false) which would have been detected by the Hough transform at various significance levels. These counts once again exclude Hough accumulators corresponding to lines within 1.5 degrees of the horizontal or vertical. The result was that a total of 968 different angles were considered. Accumulators with an expected count less than 12 were discarded as these are easily affected by isolated points.

Many of the tracks are picked up by the Hough transform for high significance levels. However some of the tracks are not even detectable at significance levels of 0.5 and smaller. Hence the renewal string approach is certainly increasing the detection rate compared with using the Hough transform alone. Furthermore the Hough transform produces large numbers of false positives even when only choosing very significant lines. The number of false positives on this plate is much greater than the theoretical number that should be found at the high significance levels. Some of these will be contributions from accumulators mapping to lines overlapping a track at a slight angle. However a dominant reason for the discrepancy is that global approaches like the Hough transform do not easily deal with variations in the background density; there is an assumption of homogeneity. If many stars are clustered in one location, then they can cause a significant contribution to a single Hough accumulator.

If only the more significant detections are wanted then the Hough transform can be used to select candidate search lines, and then the renewal string approach allows the exact points in the track to be found along that line (if there are any). This can be a significant speed up over running a hidden Markov renewal process along every line. How many tracks would be missed depends on the significance level used, and in this circumstance can be estimated from Table 1. The lower the significance level, the more lines that would have to be checked, and hence the greater the computational cost.

5. Discussion

Renewal Strings have certainly aided the process of detection of spurious objects in astronomical data: given very large amounts of data only a small number of detections were made, most of which were correct. The form of the model allows the use of the hidden Markov models and renewal processes, resulting in a model that is efficient even for huge datasets. It has been run all the plates of the SSS data (over 3000 in total), providing a valuable resource to astronomers.

The renewal string approach has also been adapted for diffraction spike detection, and shows promising results. The current method is providing accurate detection results, and enables the recognition of the vast majority of diffraction spike objects with relatively few false positives.

The renewal string approach shows clear benefits over Hough approaches, and has proven a highly effective method for detection of spurious data in the SuperCOSMOS Sky Survey. The results of the method will reduce the problem of spurious data in these surveys to insignificant levels. Furthermore the technique is general and can be adapted for use in future sky surveys. The techniques will also be useful in fully digital sky surveys. These techniques are particularly suitable for detection of the shorter satellite and aeroplane tracks which can be found in many digital surveys.

Acknowledgements

This work is part of a project funded by the University of Edinburgh. The authors also thank IBM for the generous provision of the P-Series machine Blue Dwarf to the School of Informatics, Edinburgh through the Shared University Research Programme. This machine was used for some of the runs on the SuperCOSMOS Sky Survey data. AJS would like to thank Microsoft Research for fellowship funding for the final stages of this work.

References

1. D. J. Hand, H. Mannila, and P. Smyth. *Principles of Data Mining.* MIT Press, 2001.
2. A. J. Storkey, N. C. Hambly, C. K. I. Williams, and R. G. Mann. Cleaning Sky Survey Databases using Hough Transform and Renewal String Approaches. Accepted for publication in Monthly Notices of the Royal Astronomical Society, 2003.
3. P.V.C. Hough. Machine analysis of bubble chamber pictures. In *International Conference on High Energy Accelerators and Instrumentation*, 1959.
4. H.T. MacGillivray and E.B. Thomson. Proceedings of conference on digitised optical sky surveys. *Astrophysics and Space Science Library*, 174, 1992.
5. N. C. Hambly, H. T. MacGillivray, et al. The SuperCOSMOS sky survey - I. Introduction and description. *Monthly Notices of the Royal Astronomical Society*, 326(4):1279–1294, 2001.
6. B.R. Oppenheimer, N.C. Hambly, A.P. Digby, S.T. Hodgkin, and D. Saumon. Direct detection of galactic halo dark matter. *Science*, 292:698, 2001.
7. M. Cheselka. Automatic detection of linear features in astronomical images. In *Astronomical Data Analysis Software and Systems VIII, ASP Conference Series, Vol. 172*, page 349, 1999.
8. B. Vandame. Fast Hough transform for robust detection of satellite tracks. In *Mining the Sky, Proceedings of the MPA/ESO/MPE Workshop*, page 595, 2001.
9. A. J. Storkey, N. C. Hambly, C. K. I. Williams, and R. G. Mann. Renewal Strings for Cleaning Astronomical Databases. In *Uncertainty in Artificial Intelligence: Proceedings of the Nineteenth Conference (UAI-2003)*, pages 559–566, 2003.
10. L. R. Rabiner. A tutorial on hidden Markov models and selected applications in speech recognition. *Proceedings of the IEEE*, 77:257–285, 1989.
11. J. Otterpohl, J. Haynes, F. Emmert-Streib, and G. Vetter. Extracting the Dynamics of Perceptual Switching from 'Noisy' Behavior. *Journal of Physiology*, 94:555–567, 2000.
12. S. L. Scott and P. Smyth. The Markov modulated Poisson process and Markov Poisson cascade with applications to web traffic modeling. To appear in Bayesian Statistics 7, 2003.

MESH MODELING AND ITS APPLICATIONS IN IMAGE PROCESSING*

YONGYI YANG

Department of Electrical and Computer Engineering, Illinois Institute of Technology
Chicago, IL 60616, USA

In this work we explore the use of a content-adaptive mesh model (CAMM) for solution of image inverse problems. In the proposed framework we first model the image to be determined by an efficient mesh representation. A CAMM can be viewed as a form of image representation using non-uniform samples, in which the mesh nodes (i.e., image samples) are adaptively placed according to the local content of the image. The image is then determined through estimating the model parameters (i.e., mesh nodal values) from the data. A CAMM provides a spatially-adaptive regularization framework. We present experiments in image restoration and tomographic image reconstruction to demonstrate the proposed approach.

1. Introduction

Mesh modeling of an image involves partitioning the domain of the image into a collection of non-overlapping (generally polygonal) patches called mesh elements. The image function is then determined over each element by interpolation from the values at the vertices, called mesh nodes, of the element. Mesh modeling is an efficient and compact method for image representation and is an effective tool for both rigid and non-rigid motion tracking in image sequences. Consequently, it has recently found many important applications in image processing, including image compression [1], motion tracking and compensation [2], and medical image analysis [3].

In our previous work we developed a fast and accurate approach for image representation using a content-adaptive mesh model (CAMM) [4]. The CAMM is essentially a form of image representation based on non-uniform sampling. Specifically, in a CAMM smaller mesh elements (hence more samples) are placed in regions of an image containing high frequency features, while larger elements are placed in regions containing predominantly low frequency components.

In this paper we explore the use of mesh modeling for solving image inverse problems. In this approach we first represent the image by a

* This work was supported by the Whitaker Foundation and by the National Institutes of Health under grant HL65425.

CAMM, and then determine the image by estimating the values of the mesh nodes from the observed data. Our goal is to establish a basic framework for the proposed mesh-modeling approach for solving image inverse problems, and to investigate the feasibility and benefits of this approach.

The use of a CAMM may have several potential benefits. First, a CAMM is a compact image representation, i.e., an image can be represented with far fewer mesh nodes than pixels, which can help alleviate the underdetermined nature of an inverse problem and result in numerically efficient solution algorithms. Second, a CAMM provides a built-in, spatially-adaptive smoothness mechanism.

2. Mesh Model for Image Representation

Let $f(\mathbf{x})$ denote an image function defined over a domain D, which in our problem can be two-dimensional (2D) or three-dimensional (3D). In a mesh model the domain D is partitioned into M non-overlapping mesh elements, denoted by D_m, $m = 1, 2, \cdots, M$, so that $D = \bigcup_{m=1}^{M} D_m$. In practice, polygonal elements (such as triangles or quadrangles) are usually used in mesh models because of the geometric simplicity and ease of manipulation of these shapes.

In a mesh model the function $f(\mathbf{x})$ is represented by interpolation over each element D_m from the values of its nodes. Specifically, we have

$$f(\mathbf{x}) = \sum_{n=1}^{N} f(\mathbf{x}_n) \phi_n(\mathbf{x}) + e(\mathbf{x}), \qquad (1)$$

where $\mathbf{x}_n$ denotes the location of the nth mesh node, $n = 1, 2, \cdots, N$, $\phi_n(\mathbf{x})$ is the interpolation basis function associated with $\mathbf{x}_n$, and $e(\mathbf{x})$ is the interpolation error. Note that the support of $\phi_n(\mathbf{x})$ is strictly limited to those elements attached to $\mathbf{x}_n$.

It is apparent from (1) that the mesh representation is an image description based on non-uniform sampling, wherein the mesh nodes are the sample points. Therefore, the mesh elements (and thus the nodes) should be placed strategically according to the local content of the image, with samples placed most densely in areas having the most image detail. Such placement results in a compact representation of the image determined by a relatively small number of mesh nodes.

In our previous work [4] we proposed a fast algorithm, based on a simple half-toning procedure, that can generate a very accurate CAMM representation of an image. For a given image $f(\mathbf{x})$, the mesh is generated by the following steps: 1) generate a feature map that represents the spatial distribution of the largest magnitude of the 2nd directional directives of

$f(\mathbf{x})$; 2) apply an error-diffusion algorithm—based on the well-known Floyd-Steinberg algorithm—to distribute the mesh nodes in the image domain according to the feature map; and 3) use Delaunay triangulation to connect the mesh nodes. The resulting mesh structure consists of triangular elements that are automatically adapted to the content of the image. The theoretical development of this algorithm and its justification are detailed in [4].

To demonstrate the idea, we show in Fig. 1(a) a 128×128 section cropped from the original 256×256 image "Lena"; in Fig. 1(b) we show a mesh structure obtained using the above procedure for the image in (a).

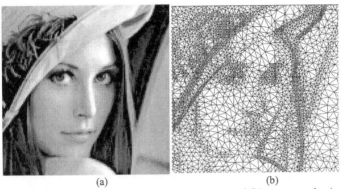

(a)   (b)

Figure 1.   (a) A 128×128 section of the original Lena image; and (b) a content-adaptive mesh generated for representation of the image in (a).

3.   MESH MODEL APPROACH FOR INVERSE PROBLEMS

3.1.   *Mesh Domain Data Acquisition Model*

Now we describe the data acquisition model for an inverse problem in terms of a mesh representation.   Generally speaking, the mean of the observed noisy image data can be modeled by

$$E[g_i] = \int_D h_i(\mathbf{x})f(\mathbf{x})d\mathbf{x} , \ i = 1, 2, \ldots, S , \qquad (2)$$

where g_i denotes the ith measurement, $h_i(\mathbf{x})$ denotes the response of measurement i to an impulse at location $\mathbf{x}$, and $E[\cdot]$ is the expectation operator. For example, in tomographic imaging (2) represents the projection measurement at the ith bin; in image restoration $h_i(\mathbf{x})$ corresponds to the point spread function (PSF) of the imaging system.

Our goal is to use a mesh model as a basis for estimation of $f(\mathbf{x})$ from the noisy measurements.   Substituting (1) into (2) we obtain:

$$E[g_i] = \sum_{n=1}^{N} \left[f(\mathbf{x}_n) \int_D h_i(\mathbf{x})\phi_n(\mathbf{x})d\mathbf{x} \right] + \int_D h_i(\mathbf{x})e(\mathbf{x})d\mathbf{x} . \qquad (3)$$

Defining

$$a_{i,n} = \int_D h_i(\mathbf{x})\phi_n(\mathbf{x})d\mathbf{x} \text{ , and } \hat{e}_i = \int_D h_i(\mathbf{x})e(\mathbf{x})d\mathbf{x} \text{ ,} \tag{4}$$

we rewrite (3) as

$$E[g_i] = \sum_{n=1}^{N} a_{i,n} f(\mathbf{x}_n) + \hat{e}_i \text{ .} \tag{5}$$

A few remarks are in order. First, it is evident from (4) that the term $a_{i,n}$ represents the noise-free response of measurement i to an impulse placed at the nth node in the mesh model. Second, also from (4) it is clear that $\hat{e}_i$ is simply the response of measurement i to the mesh modeling error.

Now we construct vectors $\mathbf{g} \equiv [g_1, g_2, \cdots, g_S]^T$ and $\hat{\mathbf{e}} \equiv [\hat{e}_1, \hat{e}_2, \cdots, \hat{e}_S]^T$ containing all the measured data and interpolation errors, respectively, and matrix $\mathbf{A} \equiv [a_{i,n}]_{S \times N}$ consisting of all the coefficients in (4). Furthermore, let $\mathbf{f}_m \equiv [f(\mathbf{x}_1), f(\mathbf{x}_2), \cdots, f(\mathbf{x}_N)]^T$.

Then, the mesh-domain imaging model becomes

$$E[\mathbf{g}] = \mathbf{A}\mathbf{f}_m + \hat{\mathbf{e}} \text{ .} \tag{6}$$

The inverse problem becomes that of estimating $\mathbf{f}_m$ from the observed data $\mathbf{g}$ through the system matrix $\mathbf{A}$. As demonstrated in our previous work [4], a CAMM can provide a very accurate image representation; therefore the error term $\hat{\mathbf{e}}$ in (6) is often negligible compared to the imaging noise. That is, the noise statistics in (6) is dominated by that of the imaging noise.

3.2. *Image Estimation Algorithms*

Note that the mesh-domain imaging model (6) has precisely the same form as the conventional pixel-domain imaging model, the only difference being in the form of their basis functions. Therefore, existing estimation algorithms for pixel-based image inverse problems can be used directly to solve (6). In this study we consider both maximum-likelihood (ML) and maximum *a posteriori* (MAP) methods [5].

The ML estimation is based on solution of the following problem

$$\hat{\mathbf{f}}_m = \arg\max_{\mathbf{f}_m} \left\{ \log p(\mathbf{g}; \mathbf{f}_m) \right\}, \tag{7}$$

where $p(\mathbf{g}; \mathbf{f}_m)$ is the likelihood function of $\mathbf{g}$ parameterized by $\mathbf{f}_m$.

On the other hand, the MAP estimate is obtained as:

$$\hat{\mathbf{f}}_m = \arg\max_{\mathbf{f}_m} \left\{ \log p(\mathbf{g}; \mathbf{f}_m) + \log p(\mathbf{f}_m) \right\}, \tag{8}$$

where $p(\mathbf{f}_m)$ is a prior on the unknown nodal values $\mathbf{f}_m$. In this study we assume a Gibbs prior for $p(\mathbf{f}_m)$ [5], i.e.,

$$p(\mathbf{f}_m) \sim \exp\left[-\beta\, U(\mathbf{f}_m)\right], \tag{9}$$

where β is a scalar weighting parameter, and the potential function $U(\mathbf{f}_m)$ is defined as:

$$U(\mathbf{f}_m) = \sum_{n=1}^{N} \sum_{j \in \Re_n} \left[f(\mathbf{x}_j) - f(\mathbf{x}_n)\right]^2. \tag{10}$$

In (10), $\Re_n$ denotes the index set of nodes connected to node n.

4. Experimental Results

In this section we present some results to demonstrate the proposed mesh based approach for applications in the following problems: (i) image noise filtering, (ii) image restoration, and (iii) tomographic image reconstruction.

4.1. *Filtering of Noisy Images*

Shown in Fig. 2(a) is the "Lena" image (Fig. 1(a)) corrupted with additive white Gaussian noise (SNR=25.91 dB). Shown in Fig. 2(b) is the filtered image using the mesh domain ML estimate in (7). The image in (a) was first lowpass processed (bandwidth=0.8π rad/pixel) before used to generate the mesh structure; N=4,042 mesh nodes were used. For Gaussian noise, the ML estimate in (7) simply corresponds to the least-squares solution of (6).

As can be seen, the mesh representation can effectively reduce the noise in the image. For comparison, we show in Figs. 2(c) and (d) images obtained using adaptive Wiener filtering [6] and median filtering of the noisy image, respectively. Note that most of the high-frequency features (such as edges and textures) are better preserved in the mesh representation in (b), even though the adaptive Wiener filtering produces a result with a slightly higher PSNR.

4.2. *Image Restoration*

Shown in Fig. 3(a) is the Lena image first blurred by a 2D Gaussian kernel (std=2 pixels), followed by additive white Gaussian noise (SNR=29.90 dB). Shown in Fig. 3(b) is the restored image using the mesh domain MAP estimate in (8). The noisy image in (a) was lowpass filtered (bandwidth=0.8π rad/pixel) to obtain the mesh structure, in which 6,500 mesh nodes were used. In (9) $\beta = 0.05$ was used. For Gaussian noise, the MAP estimate in (8) simply corresponds to a regularized least-squares solution of (6).

For comparison purpose, we show in Fig. 3(c) an image obtained with a Wiener inverse filter, where the signal spectrum was estimated from the image data. In addition, we show in Fig. 3(d) an image obtained using a

458

pixel based regularized restoration approach [7], for which the Laplacian operator was used in the regularization term and the regularization parameter was selected using a trial-and-error process to find the optimal value. As can be seen, the mesh-based image does not exhibit as much patterned noise as the pixel-based images.

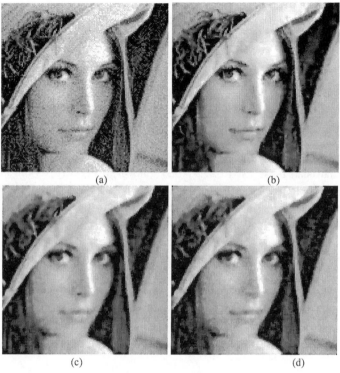

(a) (b)

(c) (d)

Figure 2. (a) The image in Fig. 1(a) corrupted with additive white Gaussian noise, SNR=25.91 dB; (b) image restored from mesh ML estimation, N=4,042 mesh nodes used; (c) image obtained using Wiener filtering, PSNR= 29.14dB; (d) image obtained using median filtering, PSNR= 27.61dB.

4.3. Tomographic Image Reconstruction

The proposed CAMM-based algorithms were also tested for tomographic image reconstruction using the 4D gated mathematical cardiac-torso (gMCAT) D1.01 phantom [8], which is a time sequence of 16 3D images. The field of view (FOV) was 28.8 cm. Poisson noise, at a level of 4 million total counts for the whole sequence, was introduced into the projections to simulate a clinical Tc^{99m} gated cardiac-perfusion single-photon emission computed tomography (SPECT) study. Our experiments were based on a single slice (No.35) of the phantom, which has approximately 2400 counts per frame. For each frame, the projections consisted of 64 bins at 64 views over $360°$, yielding a total of $S = 4096$ bins. The system had a blur of approximately 9 mm full width at half-maximum (FWHM) at the center of FOV.

In Fig. 4(a) we show a reconstructed image of frame 1 (slice #35) by the mesh domain ML estimate in (7). The image used for mesh generation was obtained by a filtered-backprojection reconstruction of a sum of the 16 frames of data, followed by lowpass filtering (bandwidth= 0.15 cycles/pixel). The number of mesh nodes used was 585. For comparison, we show in Fig. 4(b) a reconstructed image using the conventional pixel based ML method. As can be seen, the mesh reconstructed image appears to be less noisy than the pixel one.

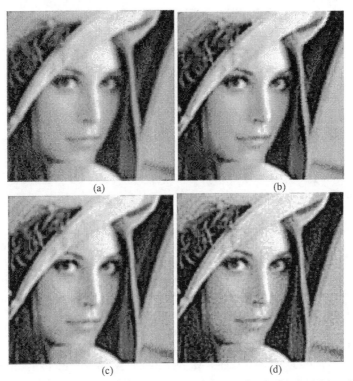

(a) (b)

(c) (d)

Figure 3. (a) Lena image degraded with a 2D Gaussian kernel (std =2) and additive white Gaussian noise at a level of SNR=29.90 dB; (b) image obtained with mesh based restoration, PSNR=28.77 dB; (c) image obtained using Wiener inverse filter with the image spectrum estimated from the noisy data, PSNR=28.15dB; and (d) image obtained using pixel based restoration, PSNR=28.52 dB.

460

5. Conclusions

In this paper we proposed a mesh modeling approach for image inverse problems. In this approach we first model the image to be restored by a compact mesh representation. The problem of image restoration then becomes that of estimating the parameters of this model. A key feature in this mesh model is that it is customized for the image, having an essential form of non-uniform sampling where samples are placed most densely in areas that contain significant detail. The imaging model was then derived based on this mesh representation, and the estimation algorithms were derived based on both ML and MAP methods. Our experiments in several image restoration and reconstruction problems demonstrate that the proposed approach can yield superior results to that of pixel-based approaches.

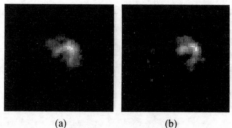

(a)　　　　　　　(b)

Figure 4.　(a) Reconstructed image of frame 1 using mesh based ML estimate, PSNR=28.5 dB; (b) reconstructed image using pixel based ML estimate PSNR=27.6 dB.

References

1.　K. Aizawa and T. S. Huang, "Model-based image coding: advanced video coding techniques for very low bit-rate applications," *Proc. of IEEE*, vol. 83, no.2, pp. 259-271, Feb. 1995.

2.　Y. Wang and O. Lee, "Active mesh--a feature seeking and tracking image sequence representation scheme," *IEEE Trans. Imag. Proc.,* vol. 3, no. 5, pp. 610-624, 1994.

3.　A. Singh, D. Goldgof, and D. Terzopoulos, ed., *Deformable Models in Medical Image Analysis*, IEEE Computer Society Press, 1998.

4.　Y. Yang, J. Brankov, and M. Wernick, "A computationally efficient approach for accurate content-adaptive mesh generation," *IEEE Trans. on Image Processing,* vol. 12, no. 8, pp. 866-881, 2003.

5.　S. Geman and D. Geman, "Stochastic relaxation, Gibbs distribution, and the Bayesian restoration of images," *IEEE Trans. Pattern Analysis and Machine Intelligence*, vol. 6, pp. 228-238, 1984.

6. J. S. Lim, *Two-Dimensional Signal and Image Processing*, Prentice-Hall, Inc., pp.536-540, 1990.
7. N. P. Galatsanos and A. K. Katsaggelos, "Methods for choosing the regularization parameter...," *IEEE Trans. on Image Processing*, vol. 1, no. 3, pp. 322-336, 1992.
8. P. H. Pretorius, M. A. King, *et al*, "Evaluation of right and left ventricular volume...," *J. of Nucl. Med,* vol. 38, pp. 1528-1534, 1997.

BIOMEDICAL ENGINEERING

AUTOREGRESSIVE SPECTRAL ANALYSIS OF PHRENIC NEUROGRAM BEFORE AND AFTER VAGOTOMY IN THE PIGLET

SHANNON AGNER, METIN AKAY

Thayer School of Engineering, Dartmouth College, Hanover, NH USA

Thayer School of Engineering,Dartmouth College

Introduction

The cardio-respiratory system is a complicated system of inputs and outputs. Many of the nerves that send signals from the pre-Botzinger complex in the brainstem act in conjunction with nerves that give feedback to the brainstem about the current state of the system. Two of the nerves involved in input and output from the brainstem are the phrenic nerve and the vagus nerve, respectively. The phrenic nerve carries signals from the brainstem to the diaphragm. Although not much is known about the effects of the vagus nerve on breathing, it is known that the vagus nerve innervates the stretch receptors in the lungs, preventing overinflation of the lungs by what is known as the Hering-Breuer reflex (Seeley, et al., 1998).

Previous studies in various animal models have shown that respiratory premotor and motor neurons undergo rapid changes in biochemical and bioelectrical properties during the first month of postnatal life (Scheibel and Scheibel 1971; Cameron et al. 1990). Early in postnatal life, there is an increase in the complexity of the morphology of the dendritic tree of respiratory neurons as it changes from a bipolar to a multipolar morphology. It is likely that the phrenic neurogram, the output of the respiratory neural networks, may provide useful information about the dynamics of the respiratory pattern generator.

Our group at Dartmouth has been working on the influence of afferents on the respiratory patterns during maturation to gain some insight into the mechanisms of sudden infant death syndrome (SIDS). Our previous studies suggested that the phrenic neurogram of a decerebrated piglet has a characteristic pattern during inspiration followed by a lack of activity in the phrenic neurogram during expiration. This characteristic pattern may vary depending on the age of the pig (Akay et al 2002). A sudden burst of activity is characteristic of a young (3-6 days) piglet while a ramp-like pattern is characteristic of older piglets (>16 days) (Akay et al 2002). These varying patterns are thought to be a result of varying patterns of neuron recruitment during respiration.

Previous studies have shown that spectral analyses, specifically Fast Fourier Transform (FFT) and autoregressive (AR) analysis, are useful tools in characterizing various breathing states such as hypoxia in piglets, kittens and puppies and have shown that there is a significant increase in the high frequency component of inspiratory activity during the first weeks of postnatal life (Suthers et al. 1995; Cohen et al.1987; Sica et al., 1988).

We have previously shown that gasping can also be differentiated from eupnea, in the frequency domain. Our results show that hypoxic gasping is characterized by a loss of power in the medium-frequency range (30-60 Hz), the appearance of significant power at frequencies < 30 Hz, and an inconsistent increase in power in high frequency oscillation (HFO) peaks but little change in the frequency where the HFO peak is located. However, the power spectra of the phrenic activity during hypercapnia and carotid sinus nerve (CSN) stimulation did not alter the frequencies of the power peaks observed during hyperoxic eupnea. Therefore, we suggested that the shift in the AR spectra during hypoxic gasping is a unique characteristic of this respiratory pattern (Akay et al, 1996).

In this study, we have examined the AR spectrum of the phrenic neurogram during development of vagus nerves in piglets during eupnea. The purpose of this study is to determine the influence of vagus nerves on the respiratory patterns during maturation in piglets. We have used the autoregressive (AR) method to analyze the phrenic neurogram signals before and after vagotomy since it provides more reliable spectral estimates for the short data segment.

Methods

General Preparation
Piglets (either sex) were provided by a USDA approved local swine breeder (Lucus Farm). The piglets were housed within the animal facility of the Dartmouth Medical School located in the Borwell Research Building in Lebanon, NH. All studies were conformed to animal care standards promulgated by the Council of the American Physiological Society and appropriate federal laws and regulations (Akay et al 2002).

Experimental Protocol
In these experiments, piglets were decerebrated to allow study of the phrenic neurogram during eupnea. After a control period during which piglets were ventilated with 40% O2 in N2, the phrenic neurograms were

recorded from 14 piglets at 2 postnatal age groups: 3-7 days (n=6) and 16-34 days (n=8) for 60 min. Then, the vagi were cut bilaterally and the phrenic neurogram was recorded for 60 min again while pressure and end-tidal CO2 were held constant throughout all studies (Akay et al 2002).

The phrenic nerve was isolated in the neck at the level of the C5 rootlet. The nerve was cut and placed on a bipolar electrode for neuronal recording. We have previously found that it is not necessary to desheath the phrenic nerve in the piglet to obtain excellent recordings of neural activity. The phrenic neurogram was amplified and bandpass filtered (10 - 300 Hz). Phrenic nerve activity was quantified from the moving average of the phrenic signal. Inspiratory activity was measured as the peak amplitude (PA) of the averaged phrenic neurogram. In all studies, physiological data was simultaneously recorded, digitized and stored on magnetic media for subsequent analysis. Data analyses were performed after completion of the study using a combination of commercially and locally-written computer software.

Data Analysis

After recording the phrenic neurograms of each subject, five bursts selected at random from each subject were analyzed using both FFT and AR methods. The power spectral density using both methods was then compared. Here, we would like to summarize the fundamental formulation of the AR process.

The AR process can be written in the time domain as follows:

$$e(n) = \sum_{m=0}^{x} a_m(n)y(n-m) \qquad (1)$$

where $e(n)$ is the estimation error.

Using AR methods, each sample of the phrenic neurogram is expressed as a linear combination of a previous sample and the error signal $e(n)$. The input $y(n)$ can be predicted from the linear combination of previous samples such that:

$$\hat{y}(n) = -\sum_{m=1}^{M} a_m(n)y(n-m) \qquad (2)$$

where $\hat{y}(n)$ represents the signal to be modeled, a_m is the AR coefficient and the mth stage, and M represents the AR model order. The PSD, S_{AR}, is then calculated at follows:

$$S_{AR}(w) = \frac{\sigma_e^2}{\left|1 + \sum_{m=1}^{M} a_m e^{-jwm\Delta t}\right|^2} \tag{3}$$

where σ_e^2 is the constant noise power, ω is the frequency, and Δt is the sampling interval. Since σ_e^2 is a constant, only the values that are necessary for calculating the shape of the PSD function are the prediction coefficients, a_m (Akay et al., 1996). For our purposes we found that the appropriate model order was 10.

Results

Figure 1a and b show a typical phrenic neurogram burst from a 4-day-old piglet before and after vagotomy. Note that the duration and shape of the phrenic neurogram burst were more or less the same. It appears that the vagotomy did not alter the respiratory patterns for the young animal. The respiratory patterns were more sudden and synchronized and short in duration as we observed in our previous studies (Akay et al., 1996). Figures 2a and b show the corresponding AR spectra of the phrenic bursts shown in Fig 1a and b. These spectra suggest that there is a slight increase in the total energy after the vagotomy.

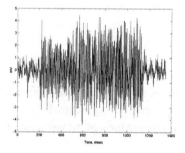 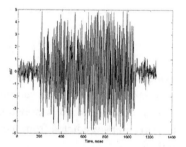

Figure 1a   Figure 1b

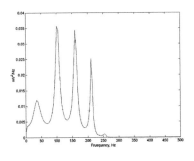

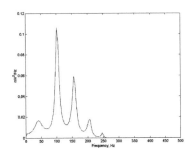

Figure 2a

Figure 2b

Figures 3a and b show the phrenic neurogram bursts for a typical 30 day old piglet before and after vagotomy. Fig 4a and b show the corresponding AR spectra of the phrenic bursts shown in Fig 3a and b. Note that the duration of the phrenic burst was much shorter after the vagotomy in the time domain compared to that before vagotomy. In addition, the shape of the phrenic burst became more synchronized and the early slowly firing neural activities became more silenced. This synchrony is reflected in the power spectra with more sharp spectral components.

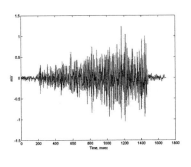

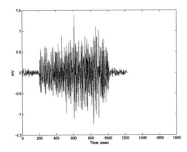

Figure 3a

Figure 3b

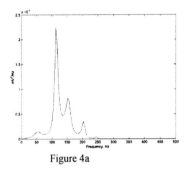

Figure 4a

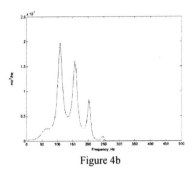

Figure 4b

After the spectral analysis, we quantified the AR spectra by diving into 3 spectral frequency bands: 10-60; 61-140; 141-240 Hz. The choice of these ranges was based on the distribution of the spectral components in the AR spectra. We estimated the mean and variance of the energy in these frequency bands before and after vagotomy at two postnatal age groups. For each condition, we analyzed and quantified 5 consecutive phrenic neurograms. Tables 1 and 2 show the mean and variance of the spectral energy and total energy in three different frequency ranges. The results showed that the mean spectral energy of the phrenic neurograms at these 3 spectral ranges did not change after vagotomy for the 3-7 days old group. However, the spectral energy drastically decreased after vagotomy for the 16-34 days old group.

To investigate the influence of the vagus on the respiratory patterns, the spectral energies at the three different frequency ranges were statistically analyzed using an analysis of variance (ANOVA). The mean values of the spectral components at these three frequency ranges were statistically different before and after vagotomy for the 3-7 days old group. However, those of the 16-34 days old group were significantly different before and after vagotomy ($p < 0.001$).

Table 1: Mean and Variance of the AR Spectral Energy for Young Subjects before and after Vagotomy						
young						
subject #	age		10-60	61-140	141-240	total
1	4	before	0.0012	0.0072	0.0063	0.0149
		after	0.0031	0.0176	0.0195	0.0409
2	4	before	0.0023	0.0097	0.0052	0.0175
		after	0.0012	0.0041	0.0026	0.0079
3	4	before	0.0041	0.0104	0.0084	0.0235
		after	0.0049	0.0218	0.0142	0.0416
4	5	before	0.0019	0.0073	0.0104	0.0200
		after	0.0037	0.0174	0.0159	0.0377
5	6	before	0.0014	0.0119	0.0169	0.0308
		after	0.0011	0.0074	0.0064	0.0150
6	7	before	0.0063	0.0232	0.0211	0.0516
		after	0.0043	0.0160	0.0122	0.0330
average+/ - SE		before	.002880+/ -.0003612	.011630+/ -.0010689	.011380+/ -.0011522	.026387+/ -.0023899
		after	.003057+/ -.0003029	.014043+/ -.0012511	.011783+/ -.0011480	.029367+/ -.0025781

Note: Mean values before and after vagotomy do not change much for the young subjects (see Table 1 above). This shows that vagotomy has little or no effect on respiratory patterns for young subjects.

Table 2: Mean and Variance of the AR Spectral Energy for Old Subjects before and after Vagotomy						
old						
subject #	age		10-60	61-140	141-240	total
7	16	before	0.0088	0.0242	0.0218	0.0562
		after	0.0026	0.0119	0.0131	0.0302
8	29	before	0.0164	0.1345	0.0831	0.2365
		after	0.0012	0.0073	0.0056	0.0143
9	30	before	0.0048	0.0115	0.0159	0.0329
		after	0.0054	0.0106	0.0172	0.0340
10	30	before	0.0372	0.2762	0.2645	0.5842
		after	0.0010	0.0046	0.0045	0.0103
11	30	before	0.0034	0.0374	0.0237	0.0651
		after	0.0056	0.0701	0.0502	0.1272
12	30	before	0.0088	0.0115	0.0144	0.0356
		after	0.0055	0.0125	0.0110	0.0297
13	30	before	0.0172	0.0545	0.0394	0.1130
		after	0.0027	0.0057	0.0082	0.0169
14	34	before	0.0092	0.1168	0.0508	0.1779
		after	0.0062	0.0442	0.0257	0.0768
average +/- SE		before	.013242 +/- .0016669	.083338 +/- .0138892	.064190 +/- .0126727	.162660+/- .0279326
		after	.003758 +/- .0003442	.020873 +/- .0036034	.016933 +/- .0022953	.042415+/- .0060685

Note that the energy in the 141-240 Hz range decreases on average. This shows that the influence of the vagus nerve appears after approximately 7 days.

Discussion

Previous studies suggest that there are drastic changes in the respiratory patterns during early maturation as well as biochemical and bioelectrical changes in the respiratory neurons. Several studies used the spectral analysis methods to analyze the respiratory patterns in kittens, piglets, rats and adult cats during eupnea and hypercapnia and the results suggested that the spectral features can be used to quantify the maturational changes in the phrenic neurograms. These encouraged us to examine the influence of the vagus nerves on the respiratory patters in the piglet since little is known about the effects of the vagus nerve on the phrenic neurograms.

In this study, we use the AR analysis method to quantify changes in spectral patterns of respiratory activity during early development before and after vagotomy. The data suggests that feedback to the medulla may affect the phrenic neurogram for the first week of postnatal maturation. Although we do not know the mechanism of such results, we suspect that the vagus nerve may not be mature in early development (between 3-7 days) and vagotomy therefore does not have an effect on phrenic nerve activity in young (3-7 days) subjects. However, as maturation proceeds, the phrenic neurogram signal drastically changes in both time and frequency domains after vagotomy. We speculate that this change could be attributed to a lack of feedback to the respiratory neural networks of the medulla.

Acknowledgement: This work was support by NIH grant (HL 65732). The authors thank Dr. Karen Moodie for her technical assistance in recording the phrenic neurograms.

References

1. Scheibel ME, Scheibel AB. (1971) Developmental relationship between spinal motoneuron dendrite bundles and patterned activity in the forelimb of cats. *Experimental Neurology* 30: 367-373.

2. Sica AL, Steele AM, Gandhi MR, Donnely DF, and Prasad N (1988) Power spectral analyses of inspiratory activities in neonatal pigs. *Brain Research* 440: 370-374.

3. Suthers GK, Henderson-Smart DJ, Read DJC (1977) Postnatal changes in the rate of high frequency bursts of inspiratory activity in cats and dogs. *Brain Research* 132: 537-540.

4. Cameron WE, Brozanski BS, Guthrie RD (1990) Postnatal development of phrenic motoneurons in the cat. *Brain Research, Developmental Brain Research* 51:142-145.

5. Cohen HL, Gootman PM, Steele AM, Eberle LP, Rao PP (1987) Age-related changes in power spectra of efferent phrenic activity in the piglet. *Brain Research* 426: 179-182.

6. Akay M, Lipping T, Moodie K, Hoopes PJ (2002) Effects of Hypoxia on the Complexity of Respiratory Patterns during Maturation, *Early Human Development*, 70:55-71.

7. Akay M, Melton JE, Welkowitz W, Edelman NH, Neubauer JA (1996) AR spectral Analysis of Phrenic Neurogram during Eupnea and Gasping, *Journal of Applied Physiology*, 81:530-540.

8. Akay, M.*Biomedical Signal Processing*. New York: Academic Press, 1994.

9. Seeley RR, Stephens TD, Tate P. *Anatomy and Physiology*, 4th ed. Boston: WCB McGraw-Hill, 1998.

MATHEMATICAL TREATMENT OF TUMOR GROWTH

IOANNIS ARAHOVITIS
Department of Mathematics, University of Athens,
Panepistimiopolis 15784
Athens,15784, Greece

In *Exploring Complexity* (cf. [10]) ,Gr. Nicolis and I. Prigogine propose the following dynamical system concerning the growth of tumors

$$dx / dt = (1 - \theta x)x - \beta x / (1 + x), \tag{0}$$

where x(t) is the population density of the proliferating malignant cells, the free cytotoxic cells acting aspredators .

In what follows we are proposed to examine

$$dx / dt = rx(1 - x / k) - p(x)$$

instead, where x(t) is the population density of the normal cells, the malignant cells acting as predators. This is in seeking to be more effective, since actually the free cytotoxic cells, in the rôle of predators, are proven to be inefficient, due possibly to chaos that we diagnosed. To this end we discuss various types of predator functions p(x) including cases of chaotic and catastrophic (in the sense of R.Thom) behaviour. Thus, after experimentation with different p(x), one can proceed in each case, by directly selecting the more appropriate treatment: surgery, chemotherapy, radiation, or a combination of the above (cf. Glattre and Nygård in [8]).

1. Introduction

We quote from [5], p. 262 that:
"Just as the research literature concerning cancer is immense, so too are there many different kinds of mathematical models of cancer". For example, emerging from (ibid. p. 266), the following linear dynamical system

$$dIC_1 / dt = a_{11}IC_1 + a_{14} NC$$
$$dIC_2 / dt = a_{21} IC_1 - a_{22}IC_2 \tag{1}$$
$$dMC / dt = a_{32}IC_2 + a_{33}MC$$
$$dNC / dt = 0 \text{ (i.e. constant NC)}$$

can serve as a multistage model for colon cancer, where $IC_1(t)$ is the population of intermediate cells 1, $IC_2(t)$ the population of intermediate cells 2, MC(t) that of malignant cells and NC(t) that of normal colon cells.

Using [4] one then proceeds to obtain the following solution of the system (cf. [5], p. 268):

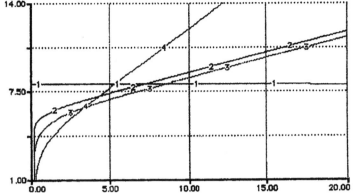

Figure 1. 1: Log of NCC, 2: Log of IC1, 3: Log of IC2, 4: Log of Malignant Cells.

It is clear that the malignant cells divide exponentially, with catastrophic results. At this point let us note that another model describing the population growth of the malignant cells, is the self-limiting one of Gomperzt, where the growth rate is proportional to the current value, but the proportionality factor decreases exponentially in time, i.e. we have:

$$dx/dt = kx\exp(-at), \text{ from which } x(t) = x(0)\exp\{k[1-\exp(-at)]/a\}.$$

This means that

$$\lim x(t) = x(0)\exp(k/a), \text{ for } t \to \infty$$

x(t) being the population of malignant cells. Certain experimental data agree surprisingly well with this model (cf. [9], p. 164 and also [11], p. 39, 2.3.3). Therefore the rate at which tumor cells divide is despite their reputation, not so fearsome (cf. [5], p. 267).

Searching for modifications in (1) it would be necessary e.g. to add more intermediate stages. But an axiom of Mathematical Modeling is 'keep it simple". Also, since for the tumor to progress, additional mutations must occur and since mutations are random events (cf. [5], p. 268), would it be worthwhile to introduce a mutation rate that is stochastic? This question may cause disorientation. In fact, in the early stages of tumor development, it is very likely that the acquired mutations are not random, but must affect specific genes involved in growth control, the *tumor suppressor genes* (c f . i b i d . p . 2 6 3) . Therefore, concerning mutations, instead of the transition

not randomness → randomness,

it is more likely to accept the transition

not randomness → chaos,

inasmuch as chaos is unpredictable, but not so unpredictable (cf. [4], p. 222) as a truly random process, a fact that frequently causes confusions . In this connection let us also note that in contrast to the normal cell, a tumor cell is theoretically able to divide indefinitely (cf. [3], p. 135) and this is one of the ways leading to chaos. On the other hand another question is (cf. [5], p. 269): what is the best way to model the effect of tumor suppressor genes? Is it necessary to lose both copies of the gene, or is there some loss of growth control if only one of the two chromosomes is modified? It seems, as we will see, that the second case is possible, as when in a system the resulting bifurcation leads to a cusp catastrophe.

Now, let us consider a general frame.

2. The Predator – Prey Model

A predator – prey mathematical model has the following form

$$dP/dt = R(P)P – p(P) \qquad (2)$$

where the rate of change for the population P is expressed as the difference of the *predation function* $p(P)$ from the *reproduction rule* $R(P)P$ (cf. [6], p. 17, (1.3.1.)), $R(P)$ being the *intrinsic growth rate* (or *net birth rate*).

Denoting by $N(t)$ the population density of the normal cells, we use an intrinsic growth rate of the form $R[1 – (N / K)]$, where the constant K is the *carrying capacity* and the constant R controls the intrinsic growth rate $R(N) = R[1 – (N / K)]$. This is of the same form used by [10], p. 225 in their model of tumor growth, in which the rôle of the prey and of the predator are interchanged, due to the fact that the organism sends killer cells to counteract the action of the malignant cells. An appropriate predation function (controlling the death rate) which has proven to be adequate in many situations, assumes the specific form $p(N)=BN^2 /(A^2+N^2)$, where A, B > 0 are constants. This can be explained because of the fact that the corresponding map $P_{n+1} = r P^2_n / [1+(P_n /K)^2]$ is a predator satiation model (cf. [6], p. 20) which shows a portion of a quite similar behaviour to that of the map $P_{n+1} = rP^2_n \exp(-P_n / K)$ as the following diagram 2 (obtained by means of [1]) shows, where the two functions $y = x^2 / (1+x^2)$ and $y = x^2 \exp(-x)$ are depicted.

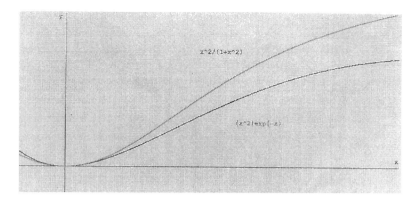

Figure 2 . The two Functions show a Similar Behaviour

Thus, (2) reads

$$dN(t) / dt = R[\ 1 - N(t) / K\ ] - BN^2(t) / [A^2 + N^2(t)] \qquad (3)$$

We proceed now to a nondimensionalization of (3). By letting $x = N / A$ and dividing by B, yields

$$(A/B)(dx / dt) = (RA / B)x[1 - x(A/K)] - x^2/(1+x^2).$$

Introducing $\tau = Bt/A$, $r = RA/B$ and $k = K/A$, (3) takes the final dimensionless form

$$dx/d\tau = rx(1-x/k) - x^2/ (1 + x^2) \qquad (4)$$

From (4) it is clear that the reproduction rule for the normal cells is a quite natural one, being of the logistic type. This however is not a natural assumption for the reproduction rule of the malignant cells considered in 6.5 of [10], p.225 (cf. equation (0) above), since these cells are not the "slow" variable: They tend to invade the organism by rapid proliferation which is better described by a predation function of the type considered above (cf. also diagram 1 and 2). In fact, Rescigno and De Lisi in 1977, as it is described in a survey by Swan (cf. [12], pp. 91-179), observe that: "Not all tumor cells are susceptible to attack and destruction by lymphocytes" and moreover that only a proportion of free tumor cell – free lymphocyte interactions result in binding".

Concerning the above model (4) let us remark the following:

i) The malignant cells, the "troublemakers" in the terminology of [10], are a function of the normal cells N, some of which have lost their physiological function for reasons examined by Molecular Biology.

ii) The dynamics of the phenomenon of tumor growth is not globally investigated by using the dynamical system 6.5, p. 225 of [10], since there is no mention in it of the normal cells' growth, which continues until the malignant cells to invade the organism.

iii) The presence and the effectiveness of the phagocytes the organism sends to destroy the tumor cells, in model (4) is incorporated into the form of the predation function p(N) we will choose, according to the case under consideration.

The dimensionless model considered in the above survey of Swan is the following:

$$dx/dt = -\lambda_1 x + a_1 xy^{2/3}(1-x/c) / (1+x)$$
$$dy/dt = \lambda_2 y - a_2 xy^{2/3} / (1+x) \tag{5}$$

where $x = K_2 L$, $c = K_2 L_M$, $y = K_2 C$, L stands for the free lymphocytes on the tumor spherical surface, C stands for the tumor cells in and on the tumor, K_1, K_2 are constants, λ_1, λ_2, a_1, a_2 are positive parameters and L_M is a maximum population bounding from above L. Here too the growth of the normal cells is ignored and the study of (5) makes apparent that uncontrolled tumor growth dominates the behaviour of the above family. At this point let us note that the family:

$$dx/dt = - (\mu_1 + y^2)$$
$$dy/dt = - (x + \mu_2 y + y^2) \tag{6}$$

investigated by Takens in 1974, shows a local phase portrait, characterized by a pair of separatrices which form a cuspoidal curve at the fixed point and consequently the resulting bifurcation is called a *cusp bifurcation*. Reference to (6) is made because it can be shown that (5) and (6) are qualitatively equivalent families, in the sense that all bifurcational behaviour exhibited by one must also occur in the other.

We now proceed to show that a cusp catastrophe can occur in the proposed model (4) too. The fixed points of (4) are given by the solutions of

$$rx(1 - x / k) = x^2 / (1+x^2) .$$

Thus (4) has an unstable fixed point at $x_0 = 0$, which means that predation is extremely weak for small x. Then the normal population grows exponentially. The other fixed points are the solutions of

$$r(1 - x / k) = x / (1+x^2),$$

resulting as the points of intersection of the straight line $y = r(1 - x / k)$ with the curve $y = x / (1 + x^2)$, as the following diagram shows, where, depending on the values of r and k, we can have one, two, or three such points.

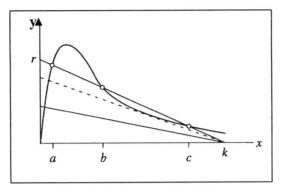

Figure 3. The three types of Fixed Points

In the case of three intersections with k fixed, the line rotates about k counter-clockwise and in the tangent position $x_2 = b$ and $x_3 = c$ collide, giving a staddle-node bifurcation. Then, the only remaining fixed points are $x_0 = 0$ and $x_1 = a$. In an analogous way, by increasing r, x_1 and x_2 can collide.

The following diagram depicts the stability type of three positive fixed points when r and k are in the appropriate range. Thus we have x_1 stable, x_2 unstable, x_3 stable:

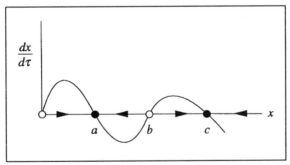

Figure 4. The Stability type of the Fixed Points

The stable fixed point a determines the *refuse level* and the other one c the *outbreak level* for the normal cells population. The fate of the system depends on the initial condition $x(0)$: If $b < x(0)$ we have an outbreak, so that x_2 plays the rôle of a *threshold*. Trying to keep the population at a, is the most satisfactory realistic situation.

480

In case the above straight line is tangent to the curve, i.e. in the case of a saddle-node bifurcation, we must have both

$r(1-x/k) = x/(1+x^2)$ and also $d[r(1-x/k)] / dx = d[x/ (1+ x^2)] /dx$.

From the last one we obtain

$-r / k = (1 - x^2) / (1 + x^2)^2$ and then from the first one $r = 2x^3 / (1 + x^2)^2$ which yields

$$k = 2x^3 / (x^2 -1).$$

In the parameter space these two curves in parametric form are depicted in the following diagram, where in order k to be > 0, we must have x > 1.

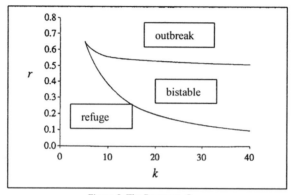

Figure 5. The Parameter Space

The above two curves determine the different regions according to the existing stable fixed points: low r – refuge a, large r – outbreak c. The above diagram can

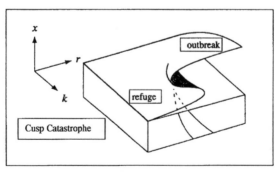

Figure 6. The Cusp Catastrophe

be regarded as the projection onto the parameter space of a cusp catastrophe surface as illustrated bellow:

Let us note that as the population of normal cells increases, in a model like (4) but where $x(\tau)$ stands for the population of malignant cells, the point (k,r) drifts upward in parameter space toward the outbreak region. This too supports our point of view, not to neglect the dynamics of the normal cells.

Now we turn our attention to the case where chaos is involved.

In equation (0), where malignant cells play the rôle of prey, the predation function used is of the form

$$p(P) = rP / (1 + P / K),$$

with corresponding discrete generation model the Verhulst map

$$P_{n+1} = rP_n / (1 + P_n / K). \tag{7}$$

As simple as the model (7) appears, there lie very complicated issues, since it is one of these producing chaos. In fact, by applying Feigenbaum theory (cf. e.g. [11]) we obtain, for $K = 1$, the first of the following diagrams

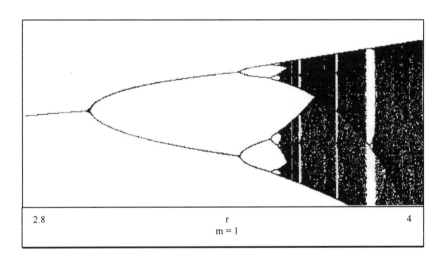

2.8   r   4

m = 1

482

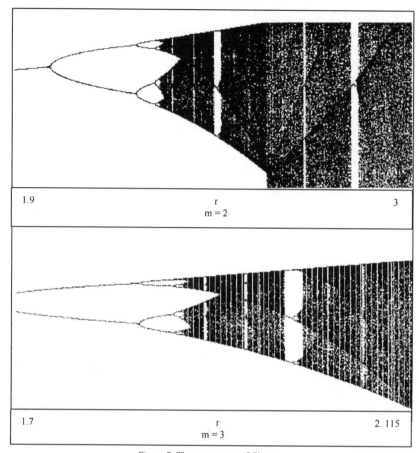

Figure 7. The appearance of Chaos

This is the first one arising from the family of chaotic maps

$$P_{n+1} = rP_n / (1 + P_n{}^m) ,$$

for different integer values of m. Two other ones , for m = 2 and m = 3, are depicted in the above diagram 7 (obtained by means of [2]). Note the dramatic (catastrophic) change of the qualitative behaviour that occurs for a critical value ≈ 2.6 of the parameter where attractors collide, in the 2nd of the above diagrams (cf. [7]). Let us note that there are other maps that produce chaos too, but of a quite different from the above form. Such is e.g. the case of Ricker's map (cf. [6], p.26)

$$P_{n+1} = rP_n exp(-P_n / K).$$

Chaotic behaviour can also appear in the reproduction function of (4). In fact, since flows are often analyzed using their maps, if e.g. K = 1 and $3.8284... \leq r \leq 3.8415...$, a period-3 window occurs for the logistic map

$$P_{n+1} = rP_n(1 - P_n).$$

At the critical value $r = 1 + \sqrt{8} = 3.8284...$ we have a tangent bifurcation (saddle-node bifurcation) and bursts appear that became more frequent as the control parameter is moved farther away from the periodic window, until the system is fully chaotic, a progression known as the *intermittency route to chaos* (cf. e.g. [11], p. 364).

Finally we mention that models concerning tumor growth involving chaos and the logistic map, are recently proposed by Obsemea under the title *"Chaotic Dynamics of tumor Growth and Regeneration"* and by Cordreanu S. and T. under the title *"Suppression of Chaos in some Nonlinear Biological Models"* (cf. [8]).

References

1. Derive 5™, Version 5.05, Texas Instruments.
2. Dynamic Solver, Copyright © J.A. Aguirregabiria, 1992 – 1998.
3. Encarnação, J.L. – Peitgen, H.O. – Sakas, G. – Englert, G. (eds) Fractal Geometry and Computer Graphics. Springer 1992.
4. Hanon,B. – Ruth, M. Dynamic Modeling, with STELLA II software by High Performance Systems, Inc. Springer, 1994.
5. Hargrove,J.L. Dynamic Modeling in the Health Sciences. Springer, 1998.
6. Hoppensteadt, F.C. – Peskin, C.S. Mathematics in Medicine and the Life Sciences. Springer, 1992.
7. http://sprott.physics.wisc.edu/phys505/lect07.htm (p.5/6).
8. International Nonlinear Sciences Conference. Vienna, Feb. 2003.
9. Kaplan,D. – Glass,L. Understanding Nonlinear Dynamics. Springer, 2nd pr. 1998.
10. Nicolis,G. – Prigogine, I. Exploring Complexity. Freeman, 5th pr. 1998.
11. Strogatz, S.H. Nonlinear Dynamics and Chaos. Addison - Wesley, 1994.
12. Thompson, J.R. – Brown, B.W. (eds) Cancer Modeling. Marcel Dekker, 1987.

MEASURING OSTEOPOROSIS USING ULTRASOUND*

JAMES L. BUCHANAN

Department of Mathematics
U.S. Naval Academy, Annapolis, MD
E-mail: jlbuchanan@hotmail.com

ROBERT P. GILBERT

Department of Mathematical Sciences
University of Delaware, Newark, DE
E-mail: gilbert@math.udel.edu

We simulated ultrasonic data to determine whether bones are osteoporotic. Ultrasound allows the clinician to measure various structural coefficients for bone in addition to testing for density. This fact alone means that ultrasound may lead to more accurate assessments of osteoporosis. With regard to the density the ultrasound measurements are in general accurate.

1. Finite element formulation of the problem

In order to determine whether a patient has osteoporosis one usually takes a density measurement for certain critical bones. These measurements are usually made using xray tomography but recently ultrasound measurements have also been used. With this in mind we consider the following experiment where a bone specimen is placed in a water tank and the bone is illuminated by an ultrasound source on one side and the transmitted field observed on the other side. The region occupied by the bone specimen and the water are Ω_b and Ω_w respectively. In Ω_w we have in the two-dimensional case

$$-\nabla^2 P - k_0^2 P = s(x, y, x_0, y_0) \tag{1}$$
$$\nabla P + \rho_w \omega^2 [U_w, V_w] = 0.$$

*This work was supported in part by ONR grant N00014-001-0853 and NSF grant BES-9820813.

for a source located at (x_0, y_0). Multiplying by a test function and applying the divergence theorem gives

$$\int\int_{\Omega_w} (\nabla P \cdot \nabla \psi - k_0^2 P\psi)dA - \int_{\partial\Omega_w} n \cdot \nabla P\psi ds = \int\int_{\Omega_w} s(x, y, x_0, y_0)\psi dA.$$

where n points into the bone. At the water-bone interface the continuity of flux condition gives

$$n \cdot \nabla P = -n \cdot \rho_w \omega^2[U_w, V_w] = -n \cdot \rho_w \omega^2(\beta[U, V] + (1 - \beta)[u, v]).$$

The time-harmonic Biot equations in two dimensions are

$$(\lambda + 2\mu)\partial_{xx}u + \mu\partial_{yy}u + (\lambda + \mu)\partial_{xy}v + Q\partial_{xx}U + Q\partial_{xy}V \qquad (2)$$
$$+p_{11}u + p_{12}U = 0$$
$$(\lambda + 2\mu)\partial_{yy}v + \mu\partial_{xx}v + (\lambda + \mu)\partial_{xy}u + Q\partial_{yy}V + Q\partial_{xy}U$$
$$+p_{11}v + p_{12}V = 0$$
$$Q\partial_x(\partial_x u + \partial_y v) + R\partial_x(\partial_x U + \partial_y V) + p_{12}u + p_{22}U = 0$$
$$Q\partial_y(\partial_x u + \partial_y v) + R\partial_y(\partial_x U + \partial_y V) + p_{12}v + p_{22}V = 0$$

In FEMLAB systems of partial differential equations are written

$$-\partial_{x_j}(c_{\ell kji}\partial_{x_i}u_k + \alpha_{\ell kj}u_k + \gamma_{\ell j}) + \beta_{\ell ki}\partial_{x_i}u_k + a_{\ell k}u_k = f_\ell \qquad (3)$$

with the summation notation convention in effect. For the Biot equations without a source $\alpha, \beta, \gamma, f = 0$ which reduces 3 to

$$-\partial_{x_j}(c_{\ell kji}\partial_{x_i}u_k) + a_{\ell k}u_k = 0. \qquad (4)$$

Multiplying (4) by a test function ϕ and integrating over Ω_b we have the weak form of the equation

$$\int\int_{\Omega_b} (-\partial_{x_j}(c_{\ell kji}\partial_{x_i}u_k) + a_{\ell k}u_k)\phi dA = 0,$$

and from the divergence we have

$$\int\int_{\Omega_b} (c_{\ell kji}\partial_{x_i}u_k\partial_{x_j}\phi + a_{\ell k}u_k\phi) dA - \int_{\partial\Omega_b} n_j(c_{\ell kji}\partial_{x_i}u_k)\phi ds = 0$$

where $n = (n_j)$ is the outward unit normal from Ω_b. For the two dimensional case take $x_1 = x, x_2 = y$ and $u_1 = u, u_2 = v, u_3 = U, u_4 = V$. The stress tensor $T_{j\ell} = c_{\ell kji}\partial_{x_i}u_k$ must be chosen appropriately for the interface conditions. At the water-bone interface continuity of aggregate normal stress, pore fluid pressure, and tangential stress are required, which suggests

$$T = \begin{bmatrix} \sigma_{xx} + \sigma & \sigma_{xy} & \sigma & 0 \\ \sigma_{yx} & \sigma_{yy} + \sigma & 0 & \sigma \end{bmatrix}.$$

where

$$\sigma_{xx} = 2\mu e_{xx} + \lambda e + Q\epsilon, \tag{5}$$
$$\sigma_{yy} = 2\mu e_{yy} + \lambda e + Q\epsilon,$$
$$\sigma_{xy} = \mu e_{xy}, \sigma_{yx} = \mu e_{yx}$$
$$\sigma = Qe + R\epsilon$$

and

$$e_{xx} = \frac{\partial u}{\partial x}, \ e_{xy} = e_{yx} = \frac{\partial u}{\partial y} + \frac{\partial v}{\partial x}, e_{yy} = \frac{\partial v}{\partial y} \tag{6}$$

$$e = \frac{\partial u}{\partial x} + \frac{\partial v}{\partial y}, \epsilon = \frac{\partial U}{\partial x} + \frac{\partial V}{\partial y}. \tag{7}$$

By direct computation

$$\partial_{x_j} T_{j1} = \partial_x(\sigma_{xx} + \sigma) + \partial_y \sigma_{yx} = (\lambda + 2\mu + Q)\partial_{xx}u +$$

$$(\lambda + Q + \mu)\partial_{xy}v + (R + Q)\partial_{xx}U + (R + Q)\partial_{xy}V + \mu\partial_{yy}u$$

$$\partial_{x_j} T_{j3} = \partial_x \sigma = Q(\partial_{xx}u + \partial_{xy}v) + R(\partial_{xx}U + \partial_{xy}V)$$

and similar expressions hold for $\partial_{x_j} T_{j2}$ and $\partial_{x_j} T_{j4}$. Adding $(2)_3$ to $(2)_1$ and $(2)_4$ to $(2)_2$ produces the disired form of the equations

$$(\lambda + 2\mu + Q)\partial_{xx}u + \mu\partial_{yy}u + (\lambda + \mu + Q)\partial_{xy}v + (R + Q)\partial_{xx}U +$$

$$(R + Q)\partial_{xy}V + (p_{11} + p_{12)}u + (p_{12} + p_{22})U = 0$$

$$(\lambda + 2\mu + Q)\partial_{yy}v + \mu\partial_{xx}v + (\lambda + \mu + Q)\partial_{xy}u + (R + Q)\partial_{yy}V$$

$$+(R + Q)\partial_{xy}U + (p_{11} + p_{12})v + (p_{12} + p_{22})V = 0$$

$$Q\partial_x(\partial_x u + \partial_y v) + R\partial_x(\partial_x U + \partial_y V) + p_{12}u + p_{22}U = 0$$

$$Q\partial_y(\partial_x u + \partial_y v) + R\partial_y(\partial_x U + \partial_y V) + p_{12}v + p_{22}V = 0$$

This leads to the following expressions for the stress components

$$T_{11} = c_{1k1i}\partial_{x_i}u_k = c_{1111}\partial_x u + c_{1112}\partial_y u + c_{1211}\partial_x v + c_{1212}\partial_y v$$
$$+ c_{1311}\partial_x U + c_{1312}\partial_y U + c_{1411}\partial_x V + c_{1412}\partial_y V$$
$$= \sigma_{xx} + \sigma = (\lambda + 2\mu + Q)\partial_x u + (\lambda + Q)\partial_y v + (Q + R)\partial_x U + (Q + R)\partial_y V$$
$$\Rightarrow c_{1111} = \lambda + 2\mu + Q, c_{1212} = \lambda + Q, c_{1311} = c_{1412} = Q + R$$

$$T_{21} = c_{1k2i}\partial_{x_i}u_k = c_{1121}\partial_x u + c_{1122}\partial_y u + c_{1221}\partial_x v + c_{1222}\partial_y v$$
$$+ c_{1321}\partial_x U + c_{1322}\partial_y U + c_{1421}\partial_x V + c_{1422}\partial_y V$$
$$= \sigma_{yx} = \mu(\partial_y u + \partial_x v)$$
$$\Rightarrow c_{1122} = c_{1221} = \mu$$

$$T_{12} = c_{2k1i}\partial_{x_i}u_k = c_{2111}\partial_x u + c_{2112}\partial_y u + c_{2211}\partial_x v + c_{2212}\partial_y v$$
$$+ c_{2311}\partial_x U + c_{2312}\partial_y U + c_{2411}\partial_x V + c_{2412}\partial_y V$$
$$= \sigma_{xy} = \mu(\partial_x v + \partial_y u)$$
$$\Rightarrow c_{2112} = c_{2211} = \mu$$

$$T_{22} = c_{2k2i}\partial_{x_i}u_k = c_{2121}\partial_x u + c_{2122}\partial_y u + c_{2221}\partial_x v + c_{2222}\partial_y v$$
$$+ c_{2321}\partial_x U + c_{2322}\partial_y U + c_{2421}\partial_x V + c_{2422}\partial_y V$$
$$= \sigma_{yy} + \sigma = (\lambda + 2\mu + Q)\partial_y v + (\lambda + Q)\partial_x u + (Q + R)\partial_x U + (Q + R)\partial_y V$$
$$\Rightarrow c_{2121} = \lambda + Q, c_{2222} = \lambda + 2\mu + Q, c_{2321} = c_{2422} = Q + R$$

$$T_{13} = c_{3k1i}\partial_{x_i}u_k = c_{3111}\partial_x u + c_{3112}\partial_y u + c_{3211}\partial_x v + c_{3212}\partial_y v$$
$$+ c_{3311}\partial_x U + c_{3312}\partial_y U + c_{3411}\partial_x V + c_{3412}\partial_y V$$
$$= \sigma = Q(\partial_x u + \partial_y v) + R(\partial_x U + \partial_y V)$$
$$\Rightarrow c_{3111} = c_{3212} = Q, c_{3311} = c_{3412} = R$$

$$T_{24} = c_{4k2i}\partial_{x_i}u_k = c_{4121}\partial_x u + c_{4122}\partial_y u + c_{4221}\partial_x v + c_{4222}\partial_y v$$
$$+ c_{4321}\partial_x U + c_{4322}\partial_y U + c_{4421}\partial_x V + c_{4422}\partial_y V$$
$$= \sigma = Q(\partial_x u + \partial_y v) + R(\partial_x U + \partial_y V)$$
$$\Rightarrow c_{4121} = c_{4222} = Q, c_{4321} = c_{4422} = R$$

Consequently

$$c_{11..} = \begin{pmatrix} \lambda + 2\mu + Q & 0 \\ 0 & \mu \end{pmatrix}, c_{12..} = \begin{pmatrix} 0 & \lambda + Q \\ \mu & 0 \end{pmatrix}$$

$$c_{13..} = \begin{pmatrix} Q + R & 0 \\ 0 & 0 \end{pmatrix}, c_{14..} = \begin{pmatrix} 0 & Q + R \\ 0 & 0 \end{pmatrix}$$

$$c_{21..} = \begin{pmatrix} 0 & \mu \\ \lambda + Q & 0 \end{pmatrix}, c_{22..} = \begin{pmatrix} \mu & 0 \\ 0 & \lambda + 2\mu + Q \end{pmatrix}$$

$$c_{23..} = \begin{pmatrix} 0 & 0 \\ Q + R & 0 \end{pmatrix}, c_{24..} = \begin{pmatrix} 0 & 0 \\ 0 & Q + R \end{pmatrix}$$

$$c_{31..} = \begin{pmatrix} Q & 0 \\ 0 & 0 \end{pmatrix}, c_{32..} = \begin{pmatrix} 0 & Q \\ 0 & 0 \end{pmatrix}$$

$$c_{33..} = \begin{pmatrix} R & 0 \\ 0 & 0 \end{pmatrix}, c_{34..} = \begin{pmatrix} 0 & R \\ 0 & 0 \end{pmatrix}$$

$$c_{41..} = \begin{pmatrix} 0 & 0 \\ Q & 0 \end{pmatrix}, c_{42..} = \begin{pmatrix} 0 & 0 \\ 0 & Q \end{pmatrix}$$

$$c_{43..} = \begin{pmatrix} 0 & 0 \\ R & 0 \end{pmatrix}, c_{44..} = \begin{pmatrix} 0 & 0 \\ 0 & R \end{pmatrix}.$$

Also

$$a_{11} = a_{22} = -(p_{11} + p_{12}), a_{13} = a_{24} = -(p_{12} + p_{22})$$

$$a_{31} = a_{42} = -p_{12}, a_{33} = a_{44} = -p_{22}.$$

In FEMLAB interface conditions are of the form $n \cdot (c\nabla u) + q \cdot u = 0$ where for this problem $u = [P, u, v, U, V]^T$. The interface conditions are

- Continuity of flux:

$$n_w \cdot \nabla P + n_w \cdot \rho_w \omega^2 (\beta[U, V] + (1 - \beta)[u, v]) = 0$$

Here n_w points into the bone.
- Continuity of aggregate normal stress:

$$n_b[T_{j1}, T_{j2}] + n_b P = 0$$

since an expansion of the bone induces a compression ($P < 0$) in the water. Here n_b points into the water.
- Continuity of pore pressure:

$$n_b[T_{j3}, T_{j4}] - n_b \beta P = 0.$$

- Vanishing of tangential stress:

Table 1. Parameters for cancellous bovine bone with a porosity of 0.72.

Parameter	Symbol	Value
Pore fluid density	ρ_f	930
Pore fluid bulk modulus	K_f	2.00×10^9
Pore fluid viscosity	η	1.5
Frame material density	ρ_r	1960
Frame material bulk modulus	K_r	2.00×10^{10}
Frame bulk modulus	K_b^*	$3.18 \times 10^9 + 1.01 \times 10^8 i$
Frame shear modulus	μ^*	$1.30 \times 10^9 + 4.14 \times 10^7 i$
Porosity	β	0.72
Permeability	k	5.00×10^{-9}
Structure factor	α	1.10
Pore size parameter	a	4.71×10^{-4}

Table 2. Parameters for cancellous bovine bone with a porosity of 0.81.

Parameter	Symbol	Value
Pore fluid density	ρ_f	930
Pore fluid bulk modulus	K_f	2.00×10^9
Pore fluid viscosity	η	1.5
Frame material density	ρ_r	1960
Frame material bulk modulus	K_r	2.00×10^{10}
Frame bulk modulus	K_b^*	$1.80 \times 10^9 + 5.74 \times 10^7 i$
Frame shear modulus	μ^*	$7.38 \times 10^8 + 2.35 \times 10^7 i$
Porosity	β	0.81
Permeability	k	2.00×10^{-8}
Structure factor	α	1.06
Pore size parameter	a	1.20×10^{-3}

This implies that

$$
q = \begin{pmatrix}
0 & \rho_w \omega^2 (1-\beta) n_{wx} & \rho_w \omega^2 (1-\beta) n_{wy} & \rho_w \omega^2 \beta n_{wx} & \rho_w \omega^2 \beta n_{wy} \\
n_{bx} P & 0 & 0 & 0 & 0 \\
n_{by} P & 0 & 0 & 0 & 0 \\
-n_{bx} \beta P & 0 & 0 & 0 & 0 \\
-n_{by} \beta P & 0 & 0 & 0 & 0
\end{pmatrix}.
$$

2. Biot parameters for bone

We consider specimens of cancellous bone with porosities of $\beta = 0.72$ (normal), $\beta = 0.81$ (osteoporotic) and $\beta = 0.95$ (severely osteoporotic). Tables 1-3 show the Biot parameters for the three specimens.

The parameters were determined as follows:

- the values of ρ_r, ρ_f, K_f and η are from Hosokawa and Otani.

Table 3. Parameters for cancellous bovine bone with a porosity of 0.95.

Parameter	Symbol	Value
Pore fluid density	ρ_f	930
Pore fluid bulk modulus	K_f	2.00×10^9
Pore fluid viscosity	η	1.5
Frame material density	ρ_r	1960
Frame material bulk modulus	K_r	2.00×10^{10}
Frame bulk modulus	K_b^*	$2.57 \times 10^8 + 8.17 \times 10^6 i$
Frame shear modulus	μ^*	$1.05 \times 10^8 + 3.34 \times 10^6 i$
Porosity	β	0.95
Permeability	k	5.00×10^{-7}
Structure factor	α	1.01
Pore size parameter	a	2.20×10^{-3}

- the real parts of K_b^* and μ were calculated using the formulas

$$K_b = \frac{E}{3(1 - 2\nu)}(1 - \beta)^n \qquad (8)$$

$$\mu = \frac{E}{2(1 + \nu)}(1 - \beta)^n$$

used by Hosokawa and Otani with their values $E = 2.2 \times 10^{10}, \nu = 0.32, n = 1.46$ for the Young's modulus and Poisson ratio of solid bone and the empirically determined exponent n.
- there is no guidance on choosing the imaginary parts of K_b^* and μ^*. I used $\operatorname{Im} K_b^* = \ell \operatorname{Re} K_b^*/\pi, \operatorname{Im} \mu = \ell \operatorname{Re} \mu^*/\pi$ with a value for the log decrement $\ell = 0.1$ which is typical of those used in underwater acoustics. There appears to be little sensitivity to this parameter however.
- K_r was calculated from (8) with $\beta = 0$.
- the structure factor was calculated as $\alpha = 1 - r(1 - 1/\beta)$ with $r = 0.25$, again following Hosokawa and Otani.
- Table 4 summarizes the estimates for permeability k and pore size a at various porosities.

The values are from Hosokawa and Otani for $\beta = 0.75, 0.81, 0.83$ and McKelvie and Palmer for $\beta = 0.72, 0.95$.

Based on the estimates in Table 4, permeability is approximately log-linear with respect to porosity whereas pore size is linear. Thus interpolation to obtain estimates at other values of porosity is feasible.

Table 4. Estimated values of permeability and pore size at different porosities.

β	k	a
0.72	5×10^{-9}	4.71×10^{-4}
0.75	7×10^{-9}	8.00×10^{-4}
0.81	2×10^{-8}	1.20×10^{-3}
0.83	3×10^{-8}	1.35×10^{-3}
0.95	5×10^{-7}	2.20×10^{-3}

3. Results

As can be seen from the Tables some of the Biot parameters are expected to be about the same for any specimen. The parameters we will manipulate are $\beta, k, a, \operatorname{Re} K_b^*$ and $\operatorname{Re} \mu^*$. $\operatorname{Im} K_b^*$ and $\operatorname{Im} \mu^*$ seem to have little influence and the remaining parameters are nearly the same for all three specimens. Data for the experiment will be simulated by calculating the acoustic field P_{ij}^{meas} at a series of positions along horizontal or vertical lines a distance s from the bone specimen using the finite elements method with three mesh refinements. The Nelder-Mead simplex method will be used to find a set of parameters $\beta, k, a, \operatorname{Re} K_b^*, \operatorname{Re} \mu^*$ which minimizes the objective function

$$\sum_{i=1}^{N} \left\| P_{i.}^{meas} - P_{i.}^{calc}(\beta, k, a, \operatorname{Re} K_b^*, \operatorname{Re} \mu^*) \right\|_2 \qquad (9)$$

where $1 \leq N \leq 4$ ranges over the number of different horizontal and vertical lines along which measurements are taken. The field values P_{ij}^{calc} will be calculated using the finite elements method with two mesh refinements. The fact that the parameters are of very different magnitudes creates difficulties with the Nelder-Mead method and thus the parameters that the method manipulates will be scaled to be order of magnitude one. $(\widehat{\beta}, \widehat{k}, \widehat{a}, \widehat{\operatorname{Re} K_b^*}, \widehat{\operatorname{Re} \mu^*}) = (\beta, k/10^{-8}, a/0.001, \operatorname{Re} K_b^*/10^9, \operatorname{Re} \mu^*/10^9)$.

We consider bone specimens with porosities $\beta = 0.72, 0.78, 0.84, 0.90$. The values of the five parameters are given in Table 4.

The values for permeability k and pore size a were determined by interpolation. The number and position of the measurements requires more exploration. For the moment we will assume 19 measurements along each of the vertical lines $x = 0.25$ (between source and bone) and $x = 0.95$ (behind bone).

For the simulated experiment we guess that the bone is severely osteoporotic ($\beta = 0.90$) and assume that the actual bone has one of the porosities $\beta = 0.90, 0.84, 0.78, 0.72$. We work at a frequency of 12.5 kHz. Tables 6-10 show the results of the inversions. The inversions was very successful for $\beta = 0.84$, reasonably successful for $\beta = 0.78$ and unsuccessful for $\beta = 0.72$. This is not surprising since the objective function space in such problems typically has numerous local minima and thus the greater the discrepancy between the guess and target the greater the chance of being ensnared by an alternative minimum.

Table 10 shows that the inversion for $\beta = 0.72$ is more successful if the parameters for $\beta = 0.84$ are used.

Table 5. Parameters for simulated experiment.

β	k	a	$\mathrm{Re}\,K_b^*$	$\mathrm{Re}\,\mu^*$
0.72	5.00×10^{-9}	4.71×10^{-4}	3.18×10^{9}	1.30×10^{9}
0.78	1.50×10^{-8}	1.05×10^{-3}	2.23×10^{9}	9.14×10^{8}
0.84	4.00×10^{-8}	1.45×10^{-3}	1.40×10^{9}	5.75×10^{8}
0.90	2.00×10^{-7}	2.00×10^{-3}	7.06×10^{8}	2.89×10^{8}

Table 6. Simulated inversion.

Parameter	β	k	a	$\mathrm{Re}\,K_b^*$	$\mathrm{Re}\,\mu^*$
Guess	0.90	2.00×10^{-7}	2.00×10^{-3}	7.06×10^{8}	2.89×10^{8}
Target	0.90	2.00×10^{-7}	2.00×10^{-3}	7.06×10^{8}	2.89×10^{8}
Result	0.0891	9.33×10^{-8}	1.07×10^{-3}	1.01×10^{9}	4.19×10^{8}

Table 7. Simulated inversion.

Parameter	β	k	a	$\mathrm{Re}\,K_b^*$	$\mathrm{Re}\,\mu^*$
Guess	0.90	2.00×10^{-7}	2.00×10^{-3}	7.06×10^{8}	2.89×10^{8}
Target	0.84	4.00×10^{-8}	1.45×10^{-3}	1.40×10^{9}	5.75×10^{8}
Result	0.859	3.23×10^{-8}	1.21×10^{-3}	1.31×10^{9}	6.39×10^{8}

Table 8. Simulated inversion.

Parameter	β	k	a	$\mathrm{Re}\,K_b^*$	$\mathrm{Re}\,\mu^*$
Guess	0.90	2.00×10^{-7}	2.00×10^{-3}	7.06×10^{8}	2.89×10^{8}
Target	0.78	1.50×10^{-8}	1.05×10^{-3}	2.23×10^{9}	5.75×10^{8}
Result	0.802	2.65×10^{-8}	2.11×10^{-3}	1.71×10^{9}	9.36×10^{8}

Table 9. Simulated inversion.

Parameter	β	k	a	$\mathrm{Re}\,K_b^*$	$\mathrm{Re}\,\mu^*$
Guess	0.90	2.00×10^{-7}	2.00×10^{-3}	7.06×10^8	2.89×10^8
Target	0.72	5.00×10^{-9}	4.71×10^{-4}	3.18×10^9	1.30×10^9
Result	0.776	7.17×10^{-8}	3.33×10^{-2}	1.23×10^{10}	5.13×10^9

Table 10. Simulated inversion.

Parameter	β	k	a	$\mathrm{Re}\,K_b^*$	$\mathrm{Re}\,\mu^*$
Guess	0.84	4.00×10^{-8}	1.45×10^{-3}	1.40×10^9	5.75×10^8
Target	0.72	5.00×10^{-9}	4.71×10^{-4}	3.18×10^9	1.30×10^9
Result	0.745	4.23×10^{-9}	6.08×10^{-4}	4.26×10^9	2.37×10^9

The file results_s03sc.txt contains more results of inversion attempts. Note that the inversions "guess 0.84" → "target 0.72" at 10 kHz and "guess 0.90" → "target 0.84" at 15 kHz failed miserably. This may be because 10 kHz is too low to differentiate between the two specimens and 15 kHz results in fields which are too different for two and three mesh refinements, or it may simply be the case that there are lucky and unlucky frequencies.

References

1. M. A. Biot, *Theory of propagation of elastic waves in a fluid-saturated porous solid. Part I: Lower frequency range*, and *Part II: Higher frequency range*, J. Acoust. Soc. Amer. **28** , 168-178, and 179-191 (1956).
2. M. A. Biot, *Mechanics of deformation and acoustic propagation in porous media*, J. Applied Physics **33** , 1482-1498 (1962).
3. Biot, M. A., "General theory of acoustic propagation in porous dissipative media," J. Acoust Soc. Am. 34, 1254-1264 (1962).
4. Buchanan, J.L., R. P. Gilbert and K. Khashanah, "Determination of the parameters of cancellous bone using low frequency acoustic measurements," (submitted) Computational Acoustics, (2002).
5. Buchanan, J.L., R. P. Gilbert and Yongzhi Xu, "Transient reflection and transmission of ultrasonic waves in cancellus bone", Applicable Analysis, (2003).
6. Hosokawa, A. and T. Otani, "Ultrasonic wave propagation in bovine cancellous bone," Acoust Soc. Am. 101, 558-562 (1997).

7. McKelvie, T.J. and S.B. Palmer, "The interaction of ultrasound with cancellous bone," Phys. Med. Biol. 36, 1331-1340 (1991).

8. Williams, J.L. "Ultrasonic wave propagation in cancellous and cortical bone: prediction of some experimental results by Biot's theory," Acoust Soc. Am. 91, 1106-1112 (1992).

THE EFFECT OF AN ELLIPSOIDAL SHELL ON THE DIRECT EEG PROBLEM

G. DASSIOS AND F. KARIOTOU

University of Patras and Hellenic Open University

The human brain is shaped in the form of an ellipsoid with average semiaxes equal to 6, to 6.5 and to 9cm. This is a genuine 3-D shape that reflects the anisotropic characteristics of the brain as a conductive body. Moreover, the brain is settled within a multi-shell environment of membranes, bone and scalp which are all characterized by different conductivities. The effect of this inhomogeniety on the electric potentials which are registered on the surface of the human head, when an electrical activity is taken place inside the brain, is studied in the present work. Actually, the simplified model of one shell surrounding the brain is used to obtain the electric potential inside the conductive core and shell as well as in the non conductive exterior space. The results are obtained by solving an interior, a transmission and an exterior boundary value problem in ellipsoidal geometry and the results are all expressed in terms of elliptic integrals and ellipsoidal harmonics. Reductions of these results to the single homogeneous brain model as well as to the spheroidal shell model are also included. Comparing the results for the exterior electric potentials, which provide the EEG data, both in the presence and in the absence of the shell, we observe that the effect of the shell is to multiply each multipole term by a factor depending crucially on the conductivity of the shell and on the difference of the conductivities as well as of the geometrical parameters specifying the shell and the core. This factor disappears whenever either the conductivities are equal or the volume of the shell diminishes to zero.

1. Introduction

Brain activity consists of the processing and propagation of chemically generated electric signals through a vast neuronic network. This network is built of at least ten billion (10^{10}) neuronic cells in the outermost layer of the brain, interconnected with each other through one hundred trillion (10^{14}) synapses [3]. Being such a complicated and important organ, the brain is protected by shells consisted of fluid, bone and skin that surround it engulfing the brain in their interior. The shells are all characterized by different electrical conductivities with the cerebrospinal fluid being the most conductive part.

The electrochemical signals propagate through the neuronic network in the form of tiny electrical currents that produce electric and magnetic fields detectable in the exterior of the head. The measured electric potentials offer the data of the Electroencephalogram (EEG), the analysis and interpretation of which provides the basis of the most widely used non invasive method for studying the human brain in vivo, known as Electroencephalography.

The interpretation of the EEG data for the purpose of localizing the electrochemical source inside the brain, that produce the externally measured electric potential field, defines the inverse EEG problem. In order to deal with the inverse EEG problem, one has to know exactly the electric potential field that a given source produces, which determines the forward EEG problem.

In modeling the forward EEG problem, certain assumptions have to be made, concerning the physical model that will approximate the electrochemical source and the geometrical model that will be used for the brain-head approximation.

As far as the source is concerned, the most popular model used is that of a dipole current which, as it is considered to be adequate for the purpose of the forward EEG problem [2], it will be the one used also in the work at hand.

Concerning the geometrical model of the conductor, most of the analytical work that has already been done uses a homogeneous sphere [5, 8, 10], a homogeneous spheroid [8, 13], or small perturbations from the homogeneous sphere [9]. The inhomogeneity that the shells of different conductivities provide in the real brain-head system, is taken into account analytically for the layered spherical and layered spheroidal case [8]. The case of a homogeneous ellipsoid has been studied in [6], incorporating the complete anisotropy of the 3-D space that best fits the anatomical model of the human brain [11].

In the work at hand, the effect of inhomogeneity in ellipsoidal geometry is studied. In fact, we consider an ellipsoidal conductor surrounded by a confocal ellipsoidal shell characterized by different conductivity. It is proved that the conductivity factor, multiplying each multipole term in the multipole expansion of the electric potential, is changed dramatically, incorporating the conductivity values as well as the boundaries of each conductivity support. It is clear then that the analytical study of the forward EEG problem is strongly influenced by the presence of a layered conductivity distribution.

Section 2 states the boundary value problems that the electric potential has to solve in the homogeneous ellipsoidal core, in the ellipsoidal shell and in the non-conductive space outside the ellipsoidal shell. The solutions of these boundary value problems in ellipsoidal geometry, expressed in terms of elliptic integrals and ellipsoidal harmonics, form the content of Section 3. Relative expressions in Cartesian and in tensorial form are also included. In

Section 4 the reduction process to the corresponding homogeneous ellipsoidal result is provided. Finally, the corresponding manipulations needed for the reduction to the spherical-shell model are included in Section 5.

2. Statement of the problem

Let S_α and S_b denote the triaxial ellipsoidal surfaces, which in rectangular coordinates are specified by

$$\frac{x_1^2}{\alpha_1^2} + \frac{x_2^2}{\alpha_2^2} + \frac{x_3^2}{\alpha_3^2} = 1, \qquad 0 < \alpha_3 < \alpha_2 < \alpha_1 < +\infty \qquad (1)$$

and

$$\frac{x_1^2}{b_1^2} + \frac{x_2^2}{b_2^2} + \frac{x_3^2}{b_3^2} = 1, \qquad 0 < b_3 < b_2 < b_1 < +\infty \qquad (2)$$

respectively, where $b_i < \alpha_i$, i = 1,2,3, are their semiaxes. The ellipsoids (1) and (2) are confocal and introduce an ellipsoidal system [4] with coordinates ρ, μ, ν and semifocal distances h_1, h_2, h_3, where

$$h_1^2 = \alpha_2^2 - \alpha_3^2 = b_2^2 - b_3^2 \qquad (3)$$

$$h_2^2 = \alpha_1^2 - \alpha_3^2 = b_1^2 - b_3^2 \qquad (4)$$

$$h_3^2 = \alpha_1^2 - \alpha_2^2 = b_1^2 - b_2^2 \qquad (5)$$

while the ellipsoidal coordinates ρ, μ, ν vary in the intervals $[h_2, +\infty)$, $[h_3, h_2]$ and $[-h_3, h_3]$, respectively. In ellipsoidal coordinates, the surface S_α, given in (1), corresponds to $\rho = \alpha_1$ and it represents the boundary of the head. The surface S_b, given in (2), corresponds to $\rho = b_1$ and stands for the boundary of the brain. The interior to S_b space V_b is defined by the interval $\rho \in [h_2, b_1)$ and it is characterized by the conductivity σ_b. The ellipsoidal shell between S_b and S_α, denoted by V_α, corresponds to the interval $\rho \in (b_1, \alpha_1)$ and is characterized by the conductivity σ_α. The exterior to S_α non conductive space V is defined by $\rho \in (\alpha_1, +\infty)$.

At the point $\mathbf{r}_0 \in V_b$ there exists a primary current dipole source with moment $\mathbf{Q}$ and density function defined by

$$\mathbf{J}^P(\mathbf{r}) = \mathbf{Q}\delta(\mathbf{r} - \mathbf{r}_0) \qquad (6)$$

where δ stands for the Dirac measure.

The primary current $\mathbf{J}^P$ generates an electric field $\mathbf{E}$ in the interior conductive space, which in turn induces a volume current with density $\mathbf{J}^V$

$$\mathbf{J}^V(\mathbf{r}) = \sigma_b \mathbf{E}(\mathbf{r}). \qquad (7)$$

Then, the total current density

$$\mathbf{J}(\mathbf{r}) = \mathbf{J}^P(\mathbf{r}) + \sigma_b \mathbf{E}(\mathbf{r}) \qquad (8)$$

is obtained.

The current **J** generates an electromagnetic wave which propagates in the interior as well as in the exterior to the conducting space.

It can be shown that the values of the characteristic parameters of the human brain allow for both the electric and the magnetic field to be considered as quasistatic [3, 10, 12].

Therefore the electric field **E** and the magnetic field **B** satisfy the quasistatic approximation of Maxwell equations [1, 6, 12], and therefore **E** can be represented by an electric potential u, such that

$$E(\mathbf{r}) = -\nabla u(\mathbf{r}). \tag{9}$$

The potential u is the field recorded in any electroencephalogram. In particular, we denote the electric potential in the interior space V_b by u_b, in the ellipsoidal shell V_α by u_α and in the exterior space V by u. Inserting equations (8) and (9) in Ampere's Law [1, 6, 12] we obtain the Poisson equation

$$\Delta u_b(\mathbf{r}) = \frac{1}{\sigma_b} \nabla \cdot \mathbf{J}^P(\mathbf{r}), \qquad \mathbf{r} \in V_b \tag{10}$$

which the interior potential u_b must satisfy in V_b.

In the source-free spaces V_α and V the potentials u_α and u solve the Laplace equation

$$\Delta u_\alpha(\mathbf{r}) = 0, \qquad \mathbf{r} \in V_\alpha \tag{11}$$

$$\Delta u(\mathbf{r}) = 0, \qquad \mathbf{r} \in V. \tag{12}$$

On the surface S_b the following transmission conditions hold

$$u_b(\mathbf{r}) = u_\alpha(\mathbf{r}), \qquad \mathbf{r} \in S_b \tag{13}$$

$$\sigma_\alpha \partial_n u_\alpha(\mathbf{r}) = \sigma_b \partial_n u_b(\mathbf{r}), \qquad \mathbf{r} \in S_b \tag{14}$$

where the outward normal differentiation on S_b is considered.

On the surface S_α the continuity condition demands that

$$u_\alpha(\mathbf{r}) = u(\mathbf{r}), \qquad \mathbf{r} \in S_\alpha \tag{15}$$

$$\partial_n u_\alpha(\mathbf{r}) = 0, \qquad \mathbf{r} \in S_\alpha. \tag{16}$$

In addition, the asymptotic behavior at infinity

$$u(\mathbf{r}) = O\left(\frac{1}{r}\right), \qquad r \to \infty \tag{17}$$

has to be assumed in order for the exterior problem to be well-posed.

The basic notation for the spectral decomposition of the Laplace operator in ellipsoidal coordinates can be found in [1,4,6], where the interior and the exterior ellipsoidal harmonics, $\mathbb{E}_n^m(\rho,\mu,\nu)$, and $\mathbb{F}_n^m(\rho,\mu,\nu)$ respectively, that are used in what follows, as well as useful relations connecting them, can be found.

3. The interior and exterior electric potential

The solution of equation (12), is an exterior harmonic function that assumes the ellipsoidal expansion

$$u(\rho,\mu,v) = \sum_{n=0}^{\infty} \sum_{m=1}^{2n+1} c_n^m \, IF_n^m (\rho,\mu,v), \rho > \alpha_1 \qquad (18)$$

and satisfies automatically the asymptotic condition (17).

Inside the ellipsoidal shell V_α the electric potential u_α solves equation (11) and therefore it assumes the ellipsoidal expansion

$$u_\alpha(\mathbf{r}) = \sum_{n=0}^{\infty} \sum_{m=1}^{2n+1} [d_n^m \, IE_n^m (\rho,\mu,v) + e_n^m \, IF_n^m (\rho,\mu,v)]. \qquad (19)$$

Finally, in the interior space V_b, which includes the primary source $\mathbf{J}^P$, the interior electric potential u_b solves equation (10), and it is given as a superposition of an interior harmonic function $\Phi(\mathbf{r})$ and the function

$$V(\mathbf{r}) = -\frac{1}{4\pi\sigma_b} \mathbf{Q} \cdot \nabla_r \frac{1}{|\mathbf{r}-\mathbf{r}_0|} = \frac{1}{4\pi\sigma_b} \mathbf{Q} \cdot \nabla_{r_0} \frac{1}{|\mathbf{r}-\mathbf{r}_0|} = \frac{1}{4\pi\sigma_b} \mathbf{Q} \cdot \frac{\mathbf{r}-\mathbf{r}_0}{|\mathbf{r}-\mathbf{r}_0|^3} \qquad (20)$$

which is a particular solution of (10).

Using the ellipsoidal expansion of the interior harmonic function $\Phi(\mathbf{r})$

$$\Phi(\mathbf{r}) = \sum_{n=0}^{\infty} \sum_{m=1}^{2n+1} f_n^m \, IE_n^m (\rho,\mu,v) \qquad (21)$$

the interior electric potential u_b assumes the form

$$u_b(\mathbf{r}) = \frac{1}{4\pi\sigma_b} \mathbf{Q} \cdot \nabla_{r_0} \frac{1}{|\mathbf{r}-\mathbf{r}_0|} + \sum_{n=0}^{\infty} \sum_{m=1}^{2n+1} f_n^m \, IE_n^m (\rho,\mu,v). \qquad (22)$$

The ellipsoidal expansion of the fundamental solution of the Laplace operator for $\rho > \rho_0$ is given in [7] as

$$\frac{1}{|\mathbf{r}-\mathbf{r}_0|} = \sum_{n=0}^{\infty} \sum_{m=1}^{2n+1} \frac{4\pi}{2n+1} \frac{1}{\gamma_n^m} IE_n^m (\rho_0,\mu_0,v_0) IF_n^m (\rho,\mu,v) \qquad (23)$$

where γ_n^m are the normalization constants of the surface ellipsoidal harmonics. Applying properly the gradient operator on (23), we obtain the following form for u_b

$$u_b(\mathbf{r}) = f_0^1 + \sum_{n=1}^{\infty} \sum_{m=1}^{2n+1} \left[f_n^m + \frac{1}{\sigma_b \gamma_n^m} (\mathbf{Q} \cdot \nabla_{r_0} IE_n^m (\rho_0,\mu_0,v_0)) I_n^m (\rho) \right] IE_n^m (\rho,\mu,v) \qquad (24)$$

In (24) we have further expressed the exterior ellipsoidal harmonics in terms of the corresponding interior ones, by means of the elliptic integral I_n^m [1,4,6]. Expansion (24) holds for $\rho > \rho_0$, therefore it holds true on both boundaries S_b and S_α. In (18), (19) and (24) we have expressed all the potentials in terms of ellipsoidal harmonics and therefore the application of the surface conditions (13)-(16) is straightforward. Furthermore, the homogeneity of (14) and (16) in ∂_n allows for the replacement of the normal derivative ∂_n with the ρ-derivative ∂_ρ.

Introducing (18), (19) and (24) in the boundary conditions (13)-(16) and using the orthogonality property of the surface ellipsoidal harmonics, the constants c_n^m, d_n^m, e_n^m, f_n^m are determined as the solutions of a sequence of linear algebraic systems. Long but straightforward calculations lead to the following expressions for the interior potential fields

$$u_\alpha(\mathbf{r}) = d_0^1 + \sum_{n=1}^{\infty} \sum_{m=1}^{2n+1} \frac{\mathbf{Q} \cdot \nabla IE_n^m(\mathbf{r}_0)}{\sigma_b \gamma_n^m b_2 b_3 G_n^m}[H_n^m(\alpha_1) - I_n^m(\rho)]IE_n^m(\rho,\mu,\nu)$$

(25)

and

$$u_b(\mathbf{r}) = \frac{1}{4\pi\sigma_b}\mathbf{Q} \cdot \frac{\mathbf{r} - \mathbf{r}_0}{|\mathbf{r} - \mathbf{r}_0|^3} + d_0^1$$
$$+ \sum_{n=1}^{\infty} \sum_{m=1}^{2n+1} \frac{\mathbf{Q} \cdot \nabla IE_n^m(\mathbf{r}_0)}{\sigma_b \gamma_n^m b_2 b_3 G_n^m}[H_n^m(\alpha_1) - I_n^m(\rho)(1 + b_2 b_3 G_n^m)]IE_n^m(\rho,\mu,\nu)$$

(26)

where the following notation is introduced

$$H_n^m(\alpha_1) = I_n^m(\alpha_1) - \frac{1}{E_n^m(\alpha_1)E_n^{m'}(\alpha_1)\alpha_2\alpha_3}$$

(27)

$$G_n^m = -\frac{\sigma_\alpha}{\sigma_b}\frac{1}{b_2 b_3} + \left(1 - \frac{\sigma_\alpha}{\sigma_b}\right)E_n^m(b_1)E_n^{m'}(b_1)[H_n^m(\alpha_1) - I_n^m(b_1)].$$

(28)

The form (25) holds for $\rho \in (b_1, \alpha_1)$ while the form (26) holds for $\rho \in (h_2, b_1)$. Moreover, applying (20) and (23) in (26) we obtain the ellipsoidal expansion of the potential $u_b(\mathbf{r})$ for $\rho \in (\rho_0, b_1)$, where ρ_0 is the ρ-coordinate of $\mathbf{r}_0$

$$u_b(\mathbf{r}) = d_0^1 + \sum_{n=1}^{\infty} \sum_{m=1}^{2n+1} \frac{\mathbf{Q} \cdot \nabla IE_n^m(\mathbf{r}_0)}{\sigma_b \gamma_n^m b_2 b_3 G_n^m}[H_n^m(\alpha_1) - I_n^m(b_1)$$
$$+ b_2 b_3 G_n^m(I_n^m(\rho) - I_n^m(b_1))]IE_n^m(\rho,\mu,\nu).$$

(29)

The value of u_α on S_α provides the Dirichlet data

$$u_\alpha(\mathbf{r}) = d_0^1 - \sum_{n=1}^{\infty} \sum_{m=1}^{2n+1} \frac{\mathbf{Q} \cdot \nabla IE_n^m(\mathbf{r}_0)}{\sigma_b \gamma_n^m b_2 b_3 G_n^m} \frac{E_n^m(\mu)E_n^m(\nu)}{E_n^{m'}(\alpha_1)\alpha_2\alpha_3}, \qquad \mathbf{r} \in S_\alpha$$

(30)

for the determination of the exterior potential $u(\mathbf{r})$, $\rho > \alpha_1$. Straightforward calculations lead to the evaluation of the constants c_n^m and hence, the exterior electric potential for $\rho > \alpha_1$ reads

$$u(\mathbf{r}) = d_0^1 \frac{I_0^1(\rho)}{I_0^1(\alpha_1)}$$

$$+ \sum_{n=1}^{\infty} \sum_{m=1}^{2n+1} \frac{\mathbf{Q} \cdot \nabla I\!E_n^m(\mathbf{r}_0)}{\sigma_b \gamma_n^m b_2 b_3 G_n^m} [H_n^m(\alpha_1) - I_n^m(\alpha_1)] \frac{I_n^m(\rho)}{I_n^m(\alpha_1)} I\!E_n^m(\rho, \mu, \nu).$$

$$(31)$$

In the sequel, we are going to work further on the last result (31), since the exterior potential is what it is registered in an electroencephalogram. In order to make it more recognizable and practical in use, we analyze further the coefficient $[\sigma_b \gamma_n^m b_2 b_3 G_n^m]^{-1}[H_n^m(\alpha_1) - I_n^m(\alpha_1)]$. Actually, in view of (27) and (28) we obtain

$$H_n^m(\alpha_1) - I_n^m(\alpha_1) = -\frac{1}{E_n^m(\alpha_1) E_n^{m'}(\alpha_1) \alpha_2 \alpha_3} \tag{32}$$

and

$$\sigma_b b_2 b_3 G_n^m =$$
$$-\sigma_\alpha + (\sigma_b - \sigma_\alpha) \left[b_2 b_3 E_n^m(b_1) E_n^{m'}(b_1) \left(I_n^m(\alpha_1) - I_n^m(b_1) \right) - \frac{b_2 b_3 E_n^m(b_1) E_n^{m'}(b_1)}{\alpha_2 \alpha_3 E_n^m(\alpha_1) E_n^{m'}(\alpha_1)} \right] \tag{33}$$

Replacing (32) and (33) into (31) we arrive at the following form for the exterior potential

$$u(\mathbf{r}) = d_0^1 \frac{I_0^1(\rho)}{I_0^1(\alpha_1)}$$

$$+ \sum_{n=1}^{\infty} \sum_{m=1}^{2n+1} \frac{1}{C_n^m} \frac{(\mathbf{Q} \cdot \nabla_{r_0} I\!E_n^m(\mathbf{r}_0))}{\gamma_n^m \alpha_2 \alpha_3 E_n^m(\alpha_1) E_n^{m'}(\alpha_1)} \frac{I_n^m(\rho)}{I_n^m(\alpha_1)} I\!E_n^m(\rho, \mu, \nu) \tag{34}$$

where the coefficient C_n^m is defined by

$$C_n^m = \sigma_\alpha + (\sigma_\alpha - \sigma_b) \left[b_2 b_3 E_n^m(b_1) E_n^{m'}(b_1) \left(I_n^m(\alpha_1) - I_n^m(b_1) \right) - \frac{b_2 b_3 E_n^m(b_1) E_n^{m'}(b_1)}{\alpha_2 \alpha_3 E_n^m(\alpha_1) E_n^{m'}(\alpha_1)} \right] \tag{35}$$

It is worth noticing that this coefficient C_n^m describes the effect of the conductivity term in (34), incorporating the geometry as well as the physics of the different conductivity supports.

Elaborating further on expression (34) by using the interior Lamé functions and the interior ellipsoidal harmonics in terms of Cartesian coordinates and by calculating the action of the gradient operator on $I\!E_n^m$

502

and on E_n^m, we obtain the following analytic form of u expressed in Cartesian coordinates in terms of elliptic integrals

$$u(\rho,\mu,\nu) = d_0^1 \frac{I_0^1(\rho)}{I_0^1(\alpha_1)} + \frac{3}{4\pi\alpha_1\alpha_2\alpha_3} \sum_{m=1}^{3} \frac{Q_m x_m}{C_1^m} \frac{I_1^m(\rho)}{I_1^m(\alpha_1)}$$

$$-\frac{5}{8\pi\alpha_1\alpha_2\alpha_3(\Lambda-\Lambda')} \sum_{m=1}^{3} Q_m x_{0m} \left[\frac{1}{C_2^1} \frac{I_2^1(\rho)}{I_2^1(\alpha_1)} \frac{IE_2^1(\mathbf{r})}{\Lambda(\Lambda-\alpha_m^2)} - \frac{1}{C_2^2} \frac{I_2^2(\rho)}{I_2^2(\alpha_1)} \frac{IE_2^2(\mathbf{r})}{\Lambda'(\Lambda'-\alpha_m^2)} \right]$$

$$+\frac{15}{4\pi\alpha_1\alpha_2\alpha_3} \sum_{\substack{i,j=1 \\ i\neq j}}^{3} \frac{Q_i x_{0j} x_i x_j}{C_2^{6-i-j}(\alpha_i^2+\alpha_j^2)} \frac{I_2^{i+j}(\rho)}{I_2^{i+j}(\alpha_1)} + O(el_3). \tag{36}$$

The notation $O(el_3)$ in (36) denotes ellipsoidal terms of order greater or equal to three.

References

1. G. Dassios and F. Kariotou, "Magnetoencephalography in Ellipsoidal Geometry", *Journal of Mathematical Physics*, Vol. **44**, No 1, pp. 220-241, 2003

2. D.B. Geselowitz, "On Bioelectric Potentials in an Inhomogeneous Volume Conductor", *Biophysical Journal*, **7**, pp. 1-11, 1967.

3. M. Hämäläinen, R. Hari, R.J. Ilmoniemi, J. Knuutila and O. Lounasmaa, "Magnetoencephalography – Theory, Instrumentation, and Applications to Noninvasive Studies of the Working Human Brain", *Rev. Mod. Phys.*, **65**, pp. 413-497, 1993.

4. E.W. Hobson, *"The Theory of Spherical and Ellipsoidal Harmonics"*, Chelsea, N.Y. 1955.

5. R. J. Ilmoniemi, M. S. Hämäläinen and J. Knuutila, "The Forward and Inverse Problems in the Spherical model", pp. 278-282, in *Biomagnetism: Applications and Theory*, edited by Harold Weinberg, Gerhard Stroink, and Toivo Katila, Pergamon Press, New York, 1985.

6. F. Kariotou, "Electroencephalography in Ellipsoidal Geometry", JMAA (in press).

7. T. Miloh, "Forces and moments on a triaxial ellipsoid in potential flow", *Israel J. Techn.*, **11**, pp. 63-74, 1973.

8. J.C. de Munck, "The Potential Distribution in a Layered Anisotropic Spheroidal Volume Conductor", *J. Appl. Phys.*, **64**, pp. 464-470, 1988.

9. G. Nolte, T. Fieseler and G. Curio, "Perturbative Analytical Solutions of the Magnetic Forward Problem for Realistic Volume Conductors", *J. Appl. Phys.*, **89**, pp. 2360-2369, 2001.

10.   J. Sarvas, "Basic Mathematical and Electromagnetic Concepts of the Biomagnetic Inverse Problem", *Phys. Med. Biol.*, **32**, pp. 11-22, 1987.

11.   W.S. Snyder, M.R. Ford, G.G. Warner and H.L. Fisher, Jr., "Estimates of Absorbed Fractions for Monoenergetic Photon Sources Uniformly Distributed in Various Organs of a Heterogeneous Phantom", *Journal of Nuclear Medicine*, Supplement Number, 3, August 1969, Volume 10, Pamphlet No. 5, Revised 1978.

12.   A. Sommerfeld, *"Electrodynamics"*, Academic Press, 1952.

13.   W.X. Wang, "The Potential for a Homogeneous Spheroid in a Spheroidal Coordinate System: I. At an exterior point", *J. Phys. A. Math. Gen.*, **21**, 4245-4250, 1988.

ULTRASONIC BACKSCATTER AND ATTENUATION OF CANCELLOUS BONE

D. DELIGIANNI, K. APOSTOLOPOULOS

Biomedical Engineering Laboratory, Department of Mechanical

Engineering & Aeronautics, University of Patras, Rion 26500, Greece

E-mail: deligian@mech.upatras.gr

In this work the ultrasonic attenuation and the backscatter coefficient of bovine cancellous bone were determined with through transmission and pulse echo measurements correspondingly. A scattering model, based on sound speed fluctuations in a binary mixture, was applied to predict the determined ultrasonic properties as functions of porosity and scatterer size. Good agreement in the magnitudes of these properties between theory and experiment were observed. Linear relationships between the exponent of the frequency dependence of the backscatter coefficient and the porosity were found for scatterer sizes larger than 150 μm..

1. Introduction

The bone strength is the ultimate indicator of bone quality, but it is not directly measurable in vitro. Although bone mass shows a strong correlation with compressive strength, 25-30% of the observed variance in strength is due to other factors such as bone architecture and microstructure. Ultrasound is currently being assessed as an alternative method of evaluating bone quality, following reports that it provides information about structure in addition to density. Despite the diagnostic utility, the fundamental mechanisms underlying the interaction between ultrasound and cancellous bone are not well understood presently.

The two most thoroughly investigated properties have been ultrasonic velocity and attenuation [5,6,10]. These parameters have demonstrated correlation with bone mineral density (BMD), which is an important predictor of fracture risk in degenerative diseases such as osteoporosis.

Ultrasonic scattering occurs when an ultrasonic wave encounters inhomogeneities that possess a different acoustic impedance than the surrounding medium. In general, and particularly in soft tissues, ultrasonic backscatter is known to provide information regarding size, shape, number density and elastic properties of scatterers [2,3,11,13]. In cancellous bone applications, trabeculae can be regarded as possible scattering sites due to high contrast in acoustic properties between mineralized trabeculae and marrow [1,12,16].

In this study the results of in vitro measurements of ultrasonic attenuation and backscatter from bovine cancellous bone are presented. The anisotropy of these ultrasonic properties and the mechanisms underlying ultrasonic scattering in cancellous bone are investigated.

In terms of modeling, attention to date has concentrated on Biot's theory [17], which describes wave propagation in a porous fluid-saturated solid in terms of frame and fluid, and the coupling between the two. Biot's theory, in spite of some success in predicting sound speed in bovine bone, substantially underestimates the attenuation of the fast wave in bone [8]. This suggests that scattering of sound, which is not considered in Biot's theory, may be an important additional attenuation source. Strelitzki et al.[15], as well as Fry and Barger [4], proposed a scattering model, developed for soft tissue and employing velocity fluctuations and a binary mixture law, for the investigation of the frequency dependence of attenuation and the dependence of attenuation due to scattering and backscatter coefficient on porosity and scatterer size. This model, extended to take into account the density, along with the velocity, fluctuations, was used to fit our experimental data.

2. Methods

2.1. Experimental technique
Forty bovine cancellous bone cubes (16 mm x 16 mm x 16 mm) were interrogated in all three perpendicular directions (craniocaudal (CC), anteroposterior (AP) and mediolateral (ML)) by immersion with one (pulse-echo) or two (through transmission) broadband transducers with a center frequency of 1 MHz. RF signals were digitized at 35 MHz. From through transmission measurements ultrasonic speed of sound and attenuation were calculated. Attenuation versus frequency was least-squares fit to a linear function over the range from 200 to 600 kHz. The function was characterized by its value at 500 kHz and the slope of the resulting line (BUA).

Analysis of the pressure wave backscattered from the sample was performed to calculate the backscatter coefficient versus frequency. The transducer output was gated appropriately to permit analysis of a 5 mm long region. The frequency–dependent backscatter coefficient is derived following the method of Roberjot et al. [13]. The measurement method uses a substitution technique, in which the signal scattered from the region being tested is compared to the signal from a standard reflecting target (a steel plate). The deconvolution of the electromechanical response of the measuring equipment from the backscattered data was computed by performing the log-spectral subtraction of the tissue spectra from the calibration spectrum of the standard reflecting target. This apparent backscatter coefficient was compensated for three sources of error: the signal loss caused by bone attenuation, the frequency dependence of the volume insonified by the tansducer and the effect of the gating function.

A linear regression analysis was applied to the logarithm of the measured backscatter coefficient versus the logarithm of the frequency (over the bandwidth 0.2 to 1.2 MHz) to calculate the power term m. This term is known to estimate the characteristic dimension of the scattering structure from backscatter coefficient measurements [3,9,13]. The apparent density (the ratio of the dehydrated, defatted tissue mass to the total specimen volume) of the samples was determined. Images of slices (1 mm thick) from all three orientations of insonification of each sample were taken. The mean trabecular thickness was measured by a suitable software (3-D Calculator).

2.2. Estimation of attenuation due to scattering

For the loss in the transmitted signal due to scatter, it was assumed that scattering be proportional to the mean fluctuation in sound speed (μ^2) and density (ρ^2), with its calculation based on the treatment of materials as immiscible mixtures. Applying the equation of Sehgal (1993) [14] for a two component mixture, the net fluctuation γ^2 is defined as:

$$\left\langle \mu^2 \right\rangle = \phi(1-\phi)(1-\phi+\phi(\frac{c_b}{c_m})^2)\frac{(c_b-c_m)^2}{c_b^2}$$

$$\left\langle \delta^2 \right\rangle = \phi(1-\phi)(1-\phi+\phi(\frac{\rho_b}{\rho_m})^2)\frac{(\rho_b-\rho_m)^2}{\rho_b^2}$$

$$\left\langle \gamma^2 \right\rangle = \mu^2 + \delta^2 \sin^4\frac{\alpha}{2}$$

where: c_b, c_m, ρ_b, ρ_m are the sound speeds and densities in solid bone and marrow respectively, and φ is the porosity. Assuming multiple scattering within a continuum model, the attenuation coefficient due to scattering α_{sc} (dB cm-1), in a medium composed of identical scatterers and with an exponential autocorrelation function, is [15]:

$$\alpha_{sc} = 8.686\frac{4\left\langle \gamma^2 \right\rangle k^4 \alpha^3}{1+k^2\alpha^2}$$

where $k=2\pi/\lambda$ is the wavenumber and α the scatterer size. This equation is valid for both Rayleigh ($k\alpha<<1$) and diffracting scatterers ($k\alpha \cong 1$). The scatterer size may correspond to the mean trabecular thickness, which has been measured from 90 to 300 μm in our bovine cancellous bone specimens. The backscatter coefficient, σ (cm^{-1} sr^{-1}), which represents the fraction of the incident intensity scattered at 180^0 to the propagation direction, is [12]:

$$\sigma = \frac{\langle \gamma^2 \rangle k^4 \alpha^3}{2\pi} \left[\frac{1}{1+k^2\alpha^2} + \frac{3}{(1+9k^2\alpha^2)^2} \right]$$

3. Results

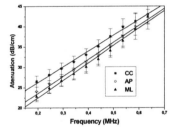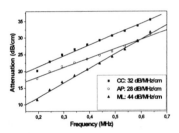

Fig 1 Left: Attenuation as function of frequency in the three orientations of all specimens. Linear fits to the data and standard errors are also shown. Right: Attenuation as function of frequency in all three orientations of a single specimen.

Figure 1 shows the attenuation as function of frequency in the three orientations of insonification of all specimens and of a single specimen.

BUA, which is the slope of the function of attenuation (per length unit) with the frequency variation, in the CC orientation was 38.66 dB/cm/MHz, in the AP orientation, 41.70 dB/cm/MHz and in the ML orientation, 42.84 dB/cm/MHz. Statistically significant differences were found only between CC and ML orientations. The small values of standard error are noticeable. Anisotropy was much more profound in individual specimens.

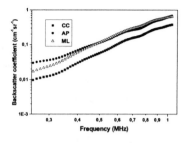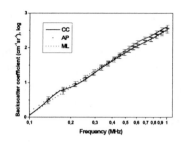

Fig.2. Left: Backscatter coefficient as function of frequency in the three orientations of all specimens. Linear fits to the data and standard errors are also shown. Right: Backscatter coefficient as function of frequency in a single specimen

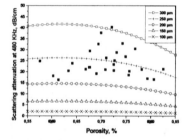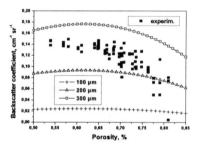

Fig. 3. Predicted and experimental values of attenuation (left) and backscatter coefficient (right) at 500 kHz as functions of porosity and scatterer size. The scatterer sizes used in the theoretical calculations were the measured values of the mean trabecular thickness.

Backscatter coefficient displayed a behavior similar to that of the attenuation. The mean values for the three orientations of all specimens disclosed isotropic behavior. In contrast to attenuation, the backscatter coefficient at 500 kHz was slightly lower for CC orientation. The densities of the 40 specimens ranged from 0.33 to 0.81 g/cm^3, with mean density 0.55 g/cm^3 and standard deviation 0.12 g/cm^3. In this range, the attenuation decreased with increasing density with an exponential decay (data not shown).

The mean trabecular thickness was found to be 90-300 μm. Predicted acoustic parameters, attenuation and backscatter coefficient, increased with increasing porosity, reaching maxima at porosities of 60-75% and decreased thereafter (Fig. 3). Both acoustic parameters tended to increase as scatterer size increased from 100 to 300 μm, regardless of the porosity. Good agreement in both, attenuation and backscatter coefficient, magnitudes between theory and experiment was found. However, the values of both ultrasonic parameters were accumulated between the theoretical curves, corresponding to scatterer sizes 200-300 μm. The porosity dependence of none ultrasonic parameter could be predicted by the theory and, in the range of densities examined, no increase in ultrasonic properties with increasing porosity was observed.

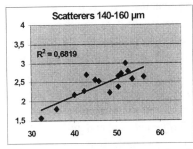 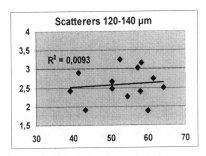

Fig. 4. The slope of the linear function of the logarithm of the measured backscatter coefficient versus the logarithm of the frequency, m, as function of the porosity (%) for a. mean scatterer size 140-160 μm (left), and b. mean scatterer size 120-140 μm (right),

The power terms m of the backscatter coefficient curve were calculated in order to investigate their relation to the scatterer size (which is the characteristic dimension). The terms m were plotted as functions of the porosity for the following grouped scatterer sizes: 80-120 μm, 120-140 μm, 140-160 μm, 160-180 μm, >180 μm. It was found that, for scatterer sizes ≤ 140 μm, no linear relation with porosity was observed. For scatterer sizes ≥ 140 μm and for each of the above groups, very good linear relations were found with porosity (R2>0,50). According to the scattering theory, these terms are independent from porosity.

4. Discussion

A simple theoretical model for acoustic scattering has been applied to bovine cancellous bone to predict BUA and backscatter coefficient. Absorption terms were not incorporated. Limitations were the assumptions of weak scattering, small velocity fluctuations, identical scatterers and an exponential spatial autocorrelation function. Applying the model in our data, density fluctuations were taken into account. The scattering model used yielded predicted acoustic properties, which were comparable with the experimental data both in magnitude and, in qualitative trends, with porosity. More detailed theoretical and experimental studies are required for understanding the interaction of ultrasound with bone.

The attenuation results are consistent with the finding by Glüer et al. [5], that the attenuation slope is larger along the axis of compressive trabeculae compared with the two perpendicular axes in bovine cancellous bone. Over the bandwidth 0.2-0.6 MHz, used for our analysis, the frequency variation of attenuation was perfectly linear.

In human calcaneous bone an aproximate cubic dependence of backscatter coefficient at diagnostic frequencies has been found [1,15]. Since attenuation in calcaneous is widely reported to vary approximately as frequency to the first power. Frequency dependence of backscatter coefficient in bovine cancellous bone, varied from $f^{1.5}$ (almost as attenuation does) to $f^{3.5}$, depending on the porosity and the scatterer size. Our results were obtained in vitro from fresh bovine specimens with relatively high bone mineral density, unlike low densities of human calcaneous bone.

Various groups are using the frequency dependent backscatter coefficient to characterize scatterer sizes in biological tissue [1,2,9], particularly in soft tissue. Sparse scatterer concentrations are generally assumed in relating scattering parameters to this tissue property. For dense scattering media, like bovine cancellous bone, the experimental data showed that the frequency dependent backscatter coefficient changed with the volume fraction and the size of the scatterers. In conclusion, ultrasonic backscatter may represent a useful new approach for the clinical assessment of bone.

References

1. S. Chaffaï, V. Roberjot, F. Peyrin, G. F. Berger, and P. Laugier, *J. Acoust. Soc. Am.* **108(5)**, 2403 (2000).
2. J.-F. Chen, J.A. Zagzebski, and E.L. Madsen, *J. Acoust. Soc. Am.* **96(4)**, 2556 (1994).
3. J. J. Jr. Faran, *J. Acoust. Soc. Am.* **23**, 405 (1951).
4. F. J. Fry, and J.E. Barger, *J. Acoust. Soc. Am.* **63**, 1576 (1978).
5. C. C. Gluer, C. Y. Wu, M. Jergas, S. A. Goldstein, and H. K. Genant, *Calcif. Tissue Int.* **55**, 46 (1996).
6. S. Han, J. Y. Rho, and I. Ziv, *Osteopor. Int.* **6**, 291 (1996).
7. B. K. Hoffmeister, S. A. Whitten, and J. Y. Rho, *Bone* **26**, 635 (2000).
8. A. Hosokawa, and T. Otani, *J. Acoust. Soc. Am.* **101**, 558 (1997).
9. M. F. Insana, R. F. Wagner, D. G. Brown, and T. J. Hall, *J. Acoust. Soc. Am.* **87(1)**, 179 (1990).

10. C. M. Langton, S. B. Palmer, and R. W. Porter, *Eng. Med.* **13**, 89 (1984).
11. R. K. Morse, and K. U. Ingard, *Theoretical Acoustics* (McGraw-Hill, New York, 1968).
12. P.H.F. Nicholson,. R. Strelitzki, R. O. Cleveland, and M. L. Bouxsein, *J. Biomechanics* **33**, 503 (2000).
13. V. Roberjot, L. S. Bridal, P..Laugier, and G. Berger, *IEEE Trans. Ultras. Ferroelec. Freq. Con.* **43**, 970 (1996).
14. C. M. Sehgal, *J. Acoust. Soc. Am.* **94**, 1944 (1993).
15. R. Strelitzki, P.H.F. Nicholson, and V. Paech, *Physiol. Meas.* **19**, 189 (1998).
16. K. A. Wear, *J. Acoust. Soc. Am.* **107(6),** 3474 (2000).
17. J. L. Williams, *J. Acoust. Soc. Am.* **91**, 1106 (1992).

CLASSIFYING PATTERNS RELATING TO THE EARLY DEVELOPMENT OF POSTTRAUMATIC STRESS DISORDER USING PRINCIPAL COMPONENTS ANALYSIS

BETHANY KNORR[1], METIN AKAY[1], THOMAS A. MELLMAN[2]

[1] *Thayer School of Engineering,* [2] *Dartmouth Medical School, Department of Psychiatry*

Introduction

Posttraumatic stress disorder (PTSD) is defined by the *Diagnostic and Statistical Manual of Mental Disorders, Fourth Edition* (*DSM-IV*) as the disorder associated when "The person has experienced an event that is outside the range of usual human experience and that would be markedly distressing to almost anyone." [1] Such events that have been studied in the past as having caused this disorder include combat, assaults, and accidents, among other situations in which a person feels that their life is in danger. Most adults in the United States have been exposed to trauma and approximately 8% of them have met criteria for PTSD. Actually, a high percentage of the population having experience such a traumatic event have been diagnosed with PTSD shortly after the event. However, that percentage drops dramatically within the first year of the event occurrence. [2] Among those patients whose symptoms do not subside, remission is unusual in cases where PTSD duration exceeds this one year period. [3] Therefore, there has been considerable interest in understanding early determinants of the disorder.

Symptoms such as those previously referred to are outlined and described by the *DSM-IV*. The symptoms include but are not limited to the following problems. The patients put forward an effort to avoid any thoughts or feelings associated with the trauma or arousing recollections. They are unable to recall important aspects of the event. The patients have a sense of foreshortened future. Irritatbility or outpursts of anger are experienced. They have difficulty concentrating and going about their usual activities. [1]

Perhaps the most disruptive of all the symptoms are those that occur during sleep, which often disrupt sleep, thereby affecting the daily mental and physical restoration for the patient. In particular, such sleep symptoms include recurrent, distressing recollections and dreams of the traumatic event. These dreams can either be direct replays of the event during which the patient feel as if he or she is reliving the trauma, or they can be any disturbing nightmare that disrupts the normal sleep cycle. [1] It has been previously reported that subjects developing PTSD have trauma replicating dreams and fragmented patterns of rapid eye movement (REM) sleep [4]. These sleep disturbances are prominent in patients suffering from PTSD and lead to increased arousal. However, studies are mixed as to the extent of this sleep disruption. [4,5] Sleep improvement can be an indicator of PTSD since as sleep ability recovers it is usually correlated with an overall reduction of PTSD symptoms. Sleep, therefore, appears to play an important role in the genesis and maintenance of PTSD.

Rapid eye movement (REM) sleep is the stage that is most affected by the mental debilitation of PTSD. This stage of sleep is defined as an acitive sleep period characterized by intense brain activity and bursts of rapid eye movements. Dreaming occurs in the REM stage; therefore it is the stage that is disrupted by violent nightmares. Healthy REM sleep and dreaming may facilitate learning and aid emotional adaptation, unlike the non-REM (NREM) sleep periods which aid in physical resoration. In the past, sleep studies of chronic PTSD have demonstrated increased REM activity. [6]

In order to measure activity and stress during these sleep stages, it is useful to consider the autonomic nervous system (ANS). The components of the system include the parasympathetic nervous system and the sympathetic nervous system. The former system is activated during times of rest. It controls body functions such as pupil constriction, a decrease in heart rate, and an increase in digestion. The parasympathetic nervous system is activated, therefore, when the body is relaxed. The sympathetic nervous system, however, is activated during times of stress when a 'fight or flight' reaction is necessary. It controls pupil dilation, an increase in heart rate and a secretion of norepinephrine and epinephrine from the adrenal medulla. This system is activated during a traumatic event. [7]

The variability of a patient's heart rate is a noninvasive physiological measure that has been useful in the diagnosis of some diseases. Traditionally, heart rate variability (HRV) has been analyzed in the diagnosis of myocardial infarction and other coronary diseases [8,9]. In addition, HRV analysis has been recently proposed as an index of psychiatric disorders, including depression, panic, schizophrenia [10], and most recently, PTSD [11]. The most simple manner of analyzing the HRV signal is to compare means and standard deviations between subjects. This type of examination, however, more sophisticated techniques are need in order to develop an autonomic nervous system indicator.

Spectral analysis may be used as such an analysis. This technique converts the HRV time series to a frequency domain in order to distinguish between the intensity of low frequency and that of high frequency signal elements. In the past, spectral analysis has been used as an effective indicator of autonomic nervous system activation. [8,12,13] It has been shown that the low frequency component of the signal corresponds to elements of both the parasympathetic and sympathetic nervous systems; whereas, the high frequency component corresponds to only the parasympathetic nerous system activity. Therefore, the ratio of low frequency to high frequency activity can accurately quantify the intensity of sympathetic activation. [8,12]

We have shown that spectral analysis is effective to differentiate between subjects with different intensities of PTSD. Specifically, we distinguish between subjects that experience long term PTSD, denoted PTSD + subjects, and those subjects whose PTSD symptoms cease shortly after the traumatic event has occurred, denoted PTSD - subjects. We speculated that increased sympathetic activation during REM sleep could be an indicator of prolonged posttraumatic stress disorder. Using autoregressive techniques to calculate the power spectrum of the HRV signal, we were able to successfully differentiate PTSD + subjects due to their increased sympathetic activation quantified by the low frequency to high frequency ratios from PTSD - subjects. [14,15]

In this study, we propose to use an unbiased pattern recognition methodcalled a principal components analysis (PCA) to classify the HRV patterns recorded from the PTSD + and - subjects. The PCA was chosen since it is an orthonormal decomposition method, and it has been widely used for the classification, the dimensionality reduction, and the noise reduction of biological signals. [16] By conducting this analysis, it is possible to extract only the fundamental features of the signal so that we can capture the essence of the physiological differences between our two subject groups.

METHODS
Patient recruitment
Subjects were recruited from the Dartmouth Hitchcock Medical Center Trauma Service and the University of Miami, Ryder Trauma Center. The included subjects were those people who had recently experience life threatening traumatic events including accidents. Life threatening, in this

case, is characterized by some degree of fear, helplessness, or horror. These patients must also have been able to recall most of the events leading to the injury and must have been alert at the time of initial admission to the hospital. In addition, the subject must be alert at the start of the assessment.

Participation exclusion criteria included the following: clinical signs suggesting traumatic brain injury, pain exceeding a moderate level or interfering with sleep, intoxication at the time of injury, previous psychiatric disorder, and the use of any drug affecting the central nervous system. Subjects without these conditions were assessed longitudinally for PTSD. Those individuals meeting the criteria for the disorder specified in the *Diagnostic and Statistical Manual of Mental Disorders - Fourth Edition* (DSM-IV) [10] and were willing to undergo sleep recordings, received sleep recordings, or polysomnography (PSG), within a month of the traumatic incident. In order to classify each subject as PTSD + or PTSD -, a re-evaluation of the the subject's PTSD status was conducted 6 weeks after the PSD, approximately 2 months after the traumatic event.

In total, 19 subjects (13 male, 6 female) participated in this study. Of these 19 participants, 9 continued to experience PTSD symptoms at a two month follow-up interview (denoted PTSD +) and 10 did not continue to experience these symptoms at the follow-up (denoted PTSD -).

Signal processing

Electrocardiogram (ECG) signals were sampled at a rate of 85 Hz (Miami), 100 Hz (Miami), and 128 Hz (Dartmouth Hitchcock). Epochs for the particular sleep stages were visually scored from the sleep recordings by an expert using the standard criteria. Five minute intervals were extracted from periods of REM, NREM, and wake-like sleep, specifically, the 5 min period directly after falling asleep, the first NREM period, the first REM period, the last REM period, the NREM period preceding the last REM period, and the 5 min period just before waking.

In order to extract the HRV time sequence, the ECG data were resampled at a rate of one half of the original sampling frequency and digitally filtered with order 5 and a cutoff frequency of 20 Hz. Then the RR time intervals were manually estimated and resampled at a rate of 2 Hz using the spline function in order to ensure equally spaced intervals in the HRV time series. In order to perform necessary calculations in our analysis, we cut each signal from approximately 580 components to 512 components.

Analysis

We conducted a principal components analysis (PCA) in order to examine our data for pattern classification. PCA is used to reduce the dimensionality of data. This reduction allows us to extract the principle elements of the signal, taking out any insignificant noise. In the situation, the remaining components can be analyzed and an accurate study of the signal can be conducted. PCA has been used in the past to quantify biological signals including electrocardiogram [16], electroencephalogram [17], and magnetoencephalogram [18].

This mehod of analysis is implemented in the following way. Using our HRV signals, we formed a collection of matrices composed of all signals for the PTSD + subjects (512 x 40), a matrix composed of the PTSD + REM signals, a matrix composed of the PTSD + NREM signals, a matrix composed of the PTSD + wake signals, a matrix composed of all PTSD – subject signals (512 x 44), a matrix composed of PTSD – REM signals, a matrix composed of PTSD – NREM signals, and a matrix composed of PTSD – wake signals. We then calculated covariance matrix C for each of these groups of signals, with the matrix comprised of m signals

$$C_{ij} = \frac{1}{n-1} \sum_{n=1}^{m} \left(x_{in} - \bar{x}_i \right)\left(x_{jn} - \bar{x}_j \right)$$

The eigenvalues λ and eigenvectors ω of the covariance matrix were then caluclated

$$C\omega = \lambda\omega$$

Subsequently, we calculated the principal components pc for each signal

$$pc_{np} = \sum_{i=1}^{512} x_{ni}\omega_{ip}$$

In order to determine to what dimension we should reduce our signal, in other words how many principal components to use, we considered the percent energy contribution for each of the components.

$$\frac{total}{energy} = \sum_{i=1}^{512} \lambda_i$$

$$\frac{\% \quad energy}{of \quad pc_p} = \frac{\lambda_p}{total \quad energy}$$

We must test the reliability of the principal components to account for the main elements of the signal.

$$\hat{x} = pc_{np}\omega_{pi}$$

First, we used all of the principal components to reconstruct the signal. We the reconstructed the signal with only the principal components that accounted for 95 percent of the energy in the signal in order to verify the prediction ability of these values.

These same principal components were used to find patterns with the ability to classify the PTSD + and PTSD - signals. The percent energies of each of these principal components were compared between the two groups. In particular, the means of the percent energies associated with the first principal components for each of the PTSD + subjects were compared to those of the PTSD - subjects. The same was done with the second principle component and so on through those components providing 95 percent of the energy.

The means of the relative energies for each of these principal components were statistically analyzed to compare means between the matrices for the PTSD + and the PTSD - subjects. A one way analysis of variance was used for the analysis, and the significance level was set at $p < 0.05$.

Results

In this study, we are interested in classifying the HRV signals in time domain using the PCA method. Figures 1. and 2. show the HRV signals recorded from PTSD - and PTSD + subjects, respectively. It is obvious from these two time series signals that the HRV signal of the PTSD + subject show much more variability than the series of the PTSD - subject. Therefore, the further quantifications of these data have been done using PCA.

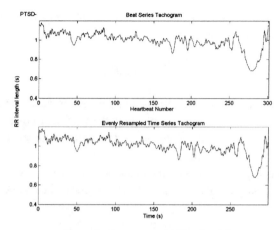

Figure 1. Heart rate variability signal for a
representative PTSD – subject as a function of
heartbeat and as a function of time.

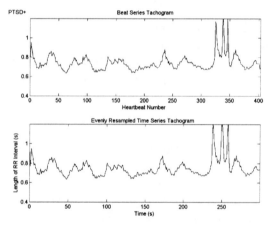

Figure 2. Heart rate variability signal for a
representative PTSD + subject as a function of
heartbeat and as a function of time.

PCA Reliability

First, we tested the reliability of the principal components on the HRV data.
Each signal was reconstructed using the principal components to account
for 95 percent of the signal energy so that we could test the ability of these
principal components to accurately reproduce the backbone of the each of
the signals. For each of the data matrices, at least the first six principal
components were needed to capture this 95 percent of the energy; therefore,
these six values from each matrix were used for comparison reasons.

Figures 3. and 4. show the incredible accuracy of the reconstructions for both a representative PTSD + subject and a representative PTSD - subject.

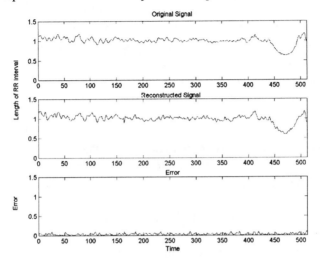

Figure 3. Reconstruction of the HRV signal of a PTSD + subject using the first six principal components. This figures shows the original HRV signal, its reconstruction and the difference between these.

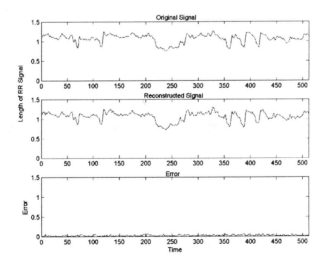

Figure 4. Reconstruction of the HRV signal of a PTSD –subject using the first six principal components. This figures shows the original HRV signal, its reconstruction and the difference between these.

520

Classification

We calculated the eigenvalues, eigenvectors and PCA as detailed in the methods section for each of the data matrices described, providing us with a four data sets depending on sleep stage for each of the PTSD + and PTSD - subject group. These values allowed us to calculate the energy associated with each principal component and the number of components needed to capture the essence of the HRV signal.

The energy associated with the first principal component were compared for each of the aforementioned data matrices. The averages are summarized in Figure 5. As we can see in Figure 5a., the PTSD + patients had a consistently higher associated energy than the PTSD - subjects for PC1. We can also note that this difference is more prominent during the sleep stages and less noticeable during wake. The opposite, however, is true for PC2, PC3, PC4, PC5, and PC6. As shown in Figures 5b., 5c., 5d., 5e., and 5f., the energy associated with the subsequent principal components are higher for the PTSD - subjects than for the PTSD + patients during sleep, again with not much of a difference in wake.

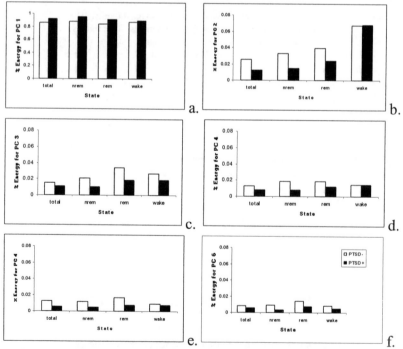

Figure 5. Percent energies associated with each of the first six principal components for matrices comprised of the HRV signals during total sleep, NREM sleep, REM sleep, and wake stages for PTSD + and PTSD – subjects. Figures a., b., c., d., e., and f. represent PC1, PC2, PC3, PC4, PC5, and PC6, respectively.

Statistical Analysis

To test the significance of the results, we used a one way analysis of variance, and our data produced a significant result. Specifically, the principal components of the PTSD - subjects are lower than the same values for PTSD + patients. We achieved this significant result for the first, fourth, fifth, and sixth principal components (PC1, PC4, PC5, PC6) with a p values less than 0.05. The second and third principal components (PC2, PC3) failed to give a significant result with p values between 0.05 and 0.1.

Discussion

In general our new findings implementing PCA to distinguish between PTSD + patients and PTSD - subjects confirm our previous finding which implemented spectral analysis. Previously, we were able to distinguish between these two groups based on their sympathetic activation. In the current study we need to pinpoint the physiological characteristics that account for the differences in energies associated with principal components.

We have noted that the first principal component for the PTSD + subjects carries more signal energy than that of the PTSD - subjects, 92 percent in total compared to 86 percent. Since the percent sum of the associated energies must add to 100, the energies associated with the subsequent principal components must be less for the PTSD + subjects than for the PTSD - subjects. This detail is confirmed by our result. This finding suggests that the HRV signals of the PTSD + subjects are less complex than those signals of the PTSD - subjects since only one principal component is needed to represent it in the time domain. However, for the PTSD - subjects, more principal components are needed to represent the HRV signal since the data is more complex and has more dimensionality and scales in time domain compared to the PTSD +, which is considered to be a sign of the high complexity. [19] We speculate that the first principal component may be representative of the sympathetic activity and dominant for the PTSD - subjects Note that these trends are more apparent in sleep than during wake, which is expected due to the extent of sleep problems observed by patients with PTSD. [6]

In conclusion, the principal components analysis reveals important psysiological characteristics in patients suffering from PTSD. This signal decomposition technique allows us to classify these patients and to distinguish between those subjects denoted PTSD + and those subjects we consider PTSD -. This process, therefore will allow us to predict the extent of time over which a patient will experience PTSD.

Acknowledgement

This research was supported by grant MH54006 from NIMH to Dr. Mellman and a pilot grant. The authors would also like to thank the DHMC Surgical Trauma Service for their clinical support.

References

[1] *Diagnostic and Statistical Manual of Mental Disorders - Fourth Edition* (DSM-IV), American Psychiatric Association, Washington D.C., 1994.

[2] ptsd book with all the symptoms and stuff

[3] R.C. Kessler, A. Sonnega, E. Bromet, M. Huches, C.B. Nelson, "Posttraumatic stress disorder in the National Comorbidity Survey," *Arch. Gen. Psychiatry*, vol. 52, pp. 1048-1060, 1995.

[4] T.A. Mellman, V. Bustamante, A.I. Fins, W.R. Pigeon, and B. Nolan, "Rapid eye movement sleep and the early development of posttraumatic stress disorder," unpublished.

[5] P. Lavie, "Sleep disturbances in the wake of traumatic events," *N. Engl. J. Med.*, vol. 345, pp. 1825-1832, 2001.

[6] T.A. Mellman, D. David, V. Bustamante, J. Torres, and A. Fins, "Dreams in the acute aftermath of trauma and their relationship to PTSD," *J. Traumatic Stress*, vol. 14, pp. 235-241, 2001.

[7] D.S. Goldstein, *The Autonomic Nervous System in Health and Disease*, Marcel Dekker, New York, 2001.

[8] S. Akselrod, D. Gordon, F.A. Ubel, D.C. Shannon, A.C. Barger, and F.J. Cohen, "Power spectrum analysis of heart rate fluctuation: A quantitative probe of beat-to-beat cardiovascular control," *Science*, vol. 213, pp. 220-222, 1981.

[9] S. Policker and I. Gath, "A model for power spectrum estimation of the heart rate signal," *Proc. IEEE Conv. Electr. Electron. Eng.*, Piscataway, NJ, IEEE, 1996.

[10] V.K. Yeragani, R. Pohl, R. Berger, et al., "Decreased heart rate variability in panic disorder patients: A study of power spectral analysis of heart rate," *Psychiatry Res.*, vol. 46, pp. 89-103, 1993.

[11] H. Cohen, M. Kotler, M. Matar, Z. Kaplan, H. Miodownik, and Y. Cassuto, "Power spectral analysis of heart rate variability in posttraumatic stress disorder patients," *Biological Psychiatry*, vol. 41, pp. 627-629, 1997.

[12] A.M. Bianchi, L.T. Mainardi, C. Meloni, S. Chierchia, and S. Cerutti, "Continuous monitoring of the sympatho-vagal balance through spectral analysis: Recursive autoregressive techniques for tracking transient events in heart rate signals," *IEEE Engineering in Medicine and Biology Magazine*, vol. 16, pp. 64-73, 1997.

[13] Task Force of the European Society of Cardiology and the North American Society of Pacing and Electrophysiology, "Heart rate variability: Standards of measurement, physiological interpretation, and clinical use," Circulation, vol. 93, pp. 1043-1065, 1996.

[14] B.R. Knorr, M. Akay, and T.A. Mellman, "Heart Rate Variability during Sleep and the Development of PTSD following Traumatic Injury," Proceedings of the 25th Annual Conference of the IEEE-EMBS, Cancun, Mexico, Sept.17-21, 2002.

[15] T.A. Mellman, B.R. Knorr, W.R. Pigeon, J.C. Leiter, M. Akay, "Heart rate variability during sleep and the early development of PTSD," Biological Psychiatry, in press.

[16] R.B. Panerai, A.L.A.S. Ferreira, and O.F. Brum, "Principal component analysis of multiple noninvasive blood flow derived signals," IEEE Trans. on Biomed. Eng., vol. 35, pp.533-538, 1998.

[17] A.C.K. Soong and Z.J. Koles, "Principal-component localization of the sources of the background EEG," IEEE Trans. on Biomed. Eng., vol. 42, pp. 59-67, 1995.

[18] J.C. Mosher, P.S. Lewis, and R.M. Leahy, "Multiple dipole modeling and localization from spatio-temporal MEG data," IEEE Trans. on Biomed. Eng., vol. 39, pp. 541-557, 1992.

[19] M. Akay, Detection and Estimation of Biomedical Signals, Academic Press, 1996.

KNOWLEDGE-BASED SYSTEMS FOR ARRHYTHMIA DETECTION AND CLASSIFICATION

V.P. OIKONOMOU, M.G. TSIPOURAS, D.I. FOTIADIS

Unit of Medical Technology and Intelligent Information Systems, Dept. of Computer Science, University of Ioannina,
Biomedical Research Institute – FORTH and
Michailideion Cardiology Center, GR 45110, Ioannina, Greece

D.A. SIDERIS

Div. of Cardiology, Medical School, Univ. of Ioannina,
Michailideion Cardiology Center and
Biomedical Research Institute – FORTH, GR 45110, Ioannina, Greece

In this paper two knowledge-based methods for arrhythmia detection and classification using ECG recordings are described, which utilize different information of the ECG signal. The first uses features of the ECG signal (R wave, QRS duration, P wave, RR interval, PR interval, PP interval, QRS similarity and P wave similarity), which are fed into a decision-tree like knowledge-based system. The system can classify all types of arrhythmias. The second is based on the utilization of the RR-duration signal only. Initially, rules based on medical knowledge are used for arrhythmic beat classification and the results are fed into a deterministic automato for arrhythmic episode detection and classification. The system can be used for the classification of limited types of arrhythmia due to the fact that only limited information is carried by the RR-duration signal.

1. Introduction

Arrhythmia is an irregular or abnormal single heartbeat or a group of heartbeats. Arrhythmias can affect the heart rhythm and rate causing irregular rhythms, slow or fast heartbeat [1]. Respiratory sinus arrhythmia (RSA) is a natural periodic variation in heart rate, corresponding to respiratory activity. Arrhythmias can take place in a healthy heart and be of minimal consequence, but they may also indicate a serious problem and lead to stroke or sudden cardiac death [2,3]. Therefore, automatic arrhythmia detection and classification are critical in clinical cardiology, especially if it can be performed in real time. The electrocardiogram (ECG) and its features are used for arrhythmia detection and classification.

Automated methods can help the expert to make a decision and save time. Most of those methods are based on the extraction of specific features of the ECG signal and classification of arrhythmias. The arrhythmia detection methods include sequential hypothesis test [4], time – frequency analysis [5], wavelet analysis [6,7], multifractal analysis [8] and methods based on neural networks [9,10,11]. Some attempts have been made for the classification of arrhythmias [12,13,14].

It is the aim of this work to illustrate the use of knowledge – based systems in arrhythmia detection and classification from ECGs. Two systems are presented. The first uses the ECG signal and its features along with a decision tree like rule-based algorithm. The second is based on the RR duration signal and utilizes a beat-by-beat classification followed by an episode classification schema.

2. Materials and methods

2.1. *ECG signal analysis*

The first method consists of two stages: preprocessing and classification. In the first stage features of the ECG signal such as QRS complex, P wave, RR interval and PP interval are extracted. In addition, the similarity of consecutive QRS complexes and P waves is examined. The extraction of QRS complex is achieved using an algorithm proposed by Tompkins [15] and the P wave using the method described in [16]. The first algorithm has been validated and its performance is quite high. The second is difficult to be validated since no annotated database for P wave exists. The similarity of QRS complexes depends on the amplitude, duration and shape of the waveform. Initially, amplitude and duration are examined and if a waveform satisfies those criteria, its shape is classified to one of the six classes shown in Fig.1. A similarity measure (correlation coefficient) is used for P wave

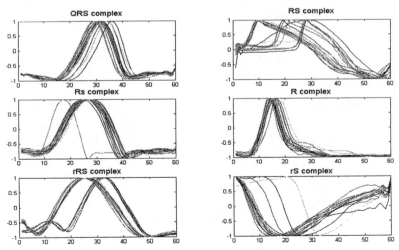

Figure 1. QRS complex shape classes.

similarity.

The above features are used in the second stage, the classification stage. Our classifier is a decision tree like algorithm, which is based in rules, provided by medical experts. The tree consists of four major modules. Each module leads to the classification of some types of arrhythmia. The tree is

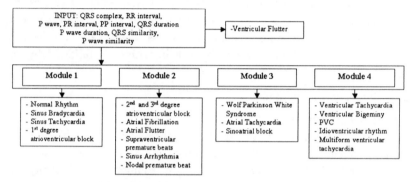

Figure 2: Decision tree like arrhythmia classifier.

shown in Fig. 2.

In Fig. 3 part of the module 3 is presented which leads to classification of arrhythmic episodes of PVC (Premature Ventricular Contraction), couples of PVC and Ventricular Tachycardia.

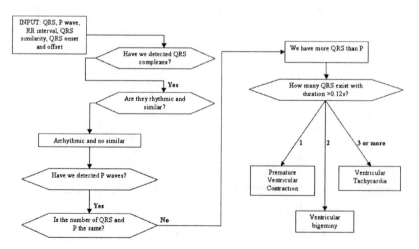

Figure 3: Module 3 details.

2.2. *RR duration signal analysis*

RR duration signal analysis consists of three stages: preprocessing, arrhythmic beat classification and arrhythmic episode detection and classification. Initially, QRS detection is performed on the ECG signal [15] and the RR-interval signal is constructed by measuring the time intervals between successive R waves.

The extracted RR interval signal is used for arrhythmia beat classification. A set of rules, provided by medical experts, are used for the classification. The classification is performed in a 3 RR interval sliding window, classifying the middle RR interval. The RR intervals are classified in four categories: (1) normal sinus beats (N), (2) premature ventricular contractions (PVC), (3) ventricular flutter/fibrillation (VF) and (4) 2° heart block (BII). We assume that a beat not belonging to one of the above arrhythmic categories is classified as normal (category 1). The algorithm starts with window i consisting of the $RR1_i$, $RR2_i$ and $RR3_i$ intervals. The three rules used in our approach are given in Fig. 4.

```
1.    Initialization
    RR2ᵢ from window i is classified as normal (category 1)
2.      Rule 1 - Flutter/fibrillation classification beats
    a. If RR2ᵢ < 0.6 sec and 1.8*RR2ᵢ < RR1ᵢ then
        i.RR2ᵢ is classified in category 3.
        ii.The RR2ₖ of all windows k = i+1, i+2, … i+n with
           (RR1ₖ < 0.7 and RR2ₖ < 0.7 and RR3ₖ < 0.7) or
           (RR1ₖ + RR2ₖ + RR3ₖ < 1.7) are classified in category 3.
    b. If the number of intervals that are continuously classified
       in category 3 is less than 4 then they all are classified in
       category 1 and the algorithm returns to window i.
3.      Rule 2 - Premature ventricular contractions
    If ((1.15*RR2ᵢ < RR1ᵢ) and (1.15*RR2ᵢ < RR3ᵢ)) or
       ((|RR1ᵢ - RR2ᵢ|<0.3) and ((RR1ᵢ<0.8) and (RR2ᵢ<0.8)) and
       (RR3ᵢ>1.2*mean(RR1ᵢ, RR2ᵢ)) or
       ((|RR2ᵢ - RR3ᵢ|<0.3) and ((RR2ᵢ<0.8) and (RR3ᵢ<0.8)) and
       (RR1ᵢ>1.2*mean(RR2ᵢ, RR3ᵢ))
    then RR2ᵢ is classified in category 2.
4.      Rule 3 - Heart block beats
    If (2.2 < RR2ᵢ < 3.0) and
       (|RR1ᵢ - RR2ᵢ| < 0.2 or |RR2ᵢ - RR3ᵢ| < 0.2)
    then RR2ᵢ is classified in category 4
5.    Update window
    a.    i = i + 1
    b.    Go to step 1
```

Figure 4: Beat classification algorithm

A deterministic automato, shown in Fig. 5, is used for arrhythmic episode detection and classification. The results from the beat classification stage are fed into the automato, which detects and classifies six types of arrhythmic episodes: (i) ventricular bigeminy, (ii) ventricular trigeminy, (iii) ventricular couplets, (iv) ventricular tachycardia, (v) ventricular flutter/fibrillation and (vi) 2° heart block.

For each episode a minimum length must be reached. The length for ventricular bigeminy is 5 beats (PVC-N-PVC-N-PVC), for ventricular trigeminy 7 beats (PVC-N-N-PVC-N-N-PVC), for ventricular tachycardia 3 beats, for ventricular flutter/fibrillation 3 beats and for 2° heart block 2 beats. If more than one rhythm type occurs the one that started first prevails.

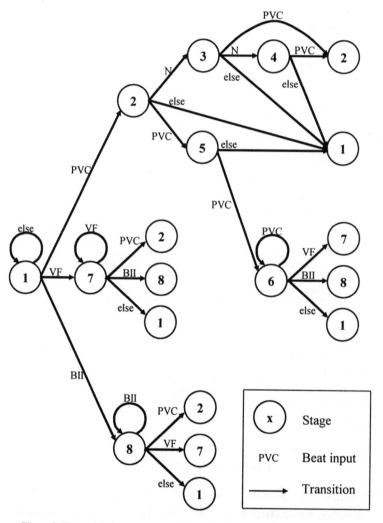

Figure 5: Deterministic automato used for arrhythmic episode detection and classification.

3. Results

The following measures are used for the evaluation of the proposed methods:

$$\text{Sensitivity} = \frac{\text{\# of beats correctly classified in category}}{\text{total \# of beats in category}}, \tag{1}$$

$$\text{P.Predictivity} = \frac{\text{\# of beats correctly classified in category}}{\text{total \# of beats classified in category}}, \tag{2}$$

$$\text{Total Performance} = \frac{\sum\limits_{\text{categories}} \text{correctly classified beats}}{\text{total beats}}. \tag{3}$$

We evaluated the first method in the classification of PVC, PVC couples and Ventricular Tachycardia (VT) from the remainder arrhythmias. The 100 series of the MIT-BIH arrhythmia database [17] was used for evaluation of the method. The obtained sensitivity and the positive predictivity for PVC is 96,89% and 99,53%, respectively, for VT 100% and 1,39%, respectively, and for PVC couples 95,08% and 53,79%, respectively.

The dataset used for the evaluation of the second method was created using all beats from all records of the MIT-BIH arrhythmia database [17] excluding: 2 beats at the start and 2 beats at the end of each record, all beats annotated as A, a, J, S, F, e, j and E and all beats included in atrial flutter or atrial fibrillation episodes because they do not belong to the types of arrhythmia which can be classified by our method. The results for sensitivity and positive predictivity are 98.98% and 99.09%, respectively, for N classified beats, 87.27% and 86.54%, respectively, for PVC classified beats, 98.76% and 97.15%, respectively, for VF classified beats and 99.05% and 89.85%, respectively, for BII classified beats. The total performance is 98.20%.

The arrhythmic episode detection and classification approach was evaluated using the beat classification obtained in the second stage. Sensitivity and positive predictivity are 97.47% and 77.44%, respectively, for ventricular couplets, 90.95% and 81.38%, respectively, for ventricular bigeminy, 73.49% and 85.92%, respectively, for ventricular trigeminy, 81.69% and 98.31%, respectively, for ventricular tachycardia, 100% and 85.71%, respectively, for ventricular flutter/fibrillation and 100% and 83.33%, respectively, for 2° heart block.

4. Discussion

The use of knowledge-based methods in the detection and classification of arrhythmic episodes in ECG recordings has been demonstrated. The methods are based on rules, which are provided by medical experts. The proposed methods indicate high efficiency and compare well with other existing methods.

The first method utilizes all the available information in the ECG signal. However, preprocessing is required to extract several features, which might be a difficult task in the presence of noise. This can be avoided using the second proposed method, which requires only QRS detection. In the second method only the RR interval signal is used therefore only selected types of arrhythmia can be detected and classified.

However, the methods can be combined and work effectively in the detection and classification of all types of arrhythmia. The major advantage of the proposed knowledge-based methods is that they work in real-time and can be easily implemented in hardware to be provided for the clinician as an add-on for existing ECG equipment. We believe that the parameters of the methods need some tuning and the performance can be even higher.

5. Conclusions

We have developed two knowledge-based methods for arrhythmia detection. The first method utilizes the ECG signal and features extracted from it while the second method is based on the RR interval signal. The first method can detect all types of arrhythmia but some of the ECG features that are used are difficult to be extracted (e.g. P wave). The second method is very efficient, but it is limited to specific types of arrhythmias, due to the fact that the RR interval monitors the ventricular activity. A hybrid approach, combining both methods, could be used to achieve better results for classification of all types of arrhythmia.

References

1. E. Sandoe and B. Sigurd. *Arrhythmia - A guide to clinical electrocardiology* (Publishing Partners Verlags GmbH, Bingen, 1991).
2. L. Goldberger and E. Goldberger. *Clinical Electrocardiography.* (The Mosby Company, Saint Louis, 1977).
3. D.A. Sideris. *Primary Cardiology.* (Scientific Editions Grigorios K Parisianos, Athens, 1991) (in Greek).

4. N.V. Thakor, Y.S. Zhu and K.Y. Pan, *Ventricular tachycardia and fibrillation detection by a sequential hypothesis testing algorithm, IEEE Trans Biom Eng 37* 837-843 (1990).

5. V.X. Afonso and W.J. Tompkins, *Detecting ventricular fibrillation, IEEE Eng Med Biol 14* 152-159 (1995).

6. A.S. Al-Fahoum and I. Howitt, *Combined wavelet transformation and radial basis neural networks for classifying life-threatening cardiac arrhythmias, Med Biol Eng Comp 37* 566-573 (1999).

7. L. Khadra, A.S. Al-Fahoum and H. Al-Nashash, *Detection of life-threatening cardiac arrhythmias using wavelet transformation, Med Biol Eng Comp 35* 626-632 (1997).

8. Y. Wang, Y.S. Zhu, N.V. Thakor and Y.H. Xu, *A short-time multifractal approach for arrhythmia detection based on fuzzy neural network, IEEE Trans Biom Eng 48* 989-995 (2001).

9. R.H. Clayton, A. Murray and R.W.F. Campbell, *Recognition of ventricular fibrillation using neural networks, Med Biol Eng Comp 32* 217-220 (1994).

10. T.F. Yang, B. Device and P.W. Macfarlane, *Artificial neural networks for the diagnosis of atrial fibrillation, Med Biol Eng Comp 32* 615-619 (1994).

11. K. Minami, H. Nakajima and T. Toyoshima, *Real-time discrimination of ventricular tachyarrhythmia with Fourier-transform neural network, IEEE Trans Biom Eng 46* 179-185 (1999).

12. Z. Docur and T. Olmez, *ECG beat classification by a hybrid neural network, Comp Meth Prog Biomed 66* 167-181 (2001).

13. S. Osowski and T.H. Linh, *ECG beat recognition using Fuzzy Hybrid Neural Network, IEEE Trans Biom Eng 48* 1265-1271 (2001).

14. M.G. Tsipouras, D.I. Fotiadis and D. Sideris, *Arrhythmia classification using the RR-interval duration signal, in A. Murray, ed., Computers in Cardiology* (IEEE, Piscataway, 2002) 485-488.

15. J. Pan, W.J. Thompkins, *A Real-Time QRS Detection Algorithm, IEEE Trans Biom Eng 32* 230-236 (1985)

16. K. Sternickel, *Automatic pattern recognition in ECG time series, Comp Methods and Programs in Biomed 68* 109-115 (2002).

17. MIT-BIH Arrhythmia Database CD-ROM (Harvard-MIT Division of Health Sciences and Technology, Third Edition, 1997).

RADIAL DISPLACEMENT UNDER JOINT LOAD

GEORGE PAPAIOANNOU

Bone and Joint Center, Henry Ford Health System, Detroit, MI
Department of Biomedical Engineering
Catholic University of America, Washington DC.

C. DEMETROPOULOS, J. GUETTLER, K. JURIST
William Beaumont Hospital MI

D. FYHRIE, S. TASHMAN, K.H. YANG
Bioengineering Center, Wayne State University

Little is known about *in situ/in vivo* meniscal loads and displacements in the knee during strenuous activities. We have developed a method that combines dynamic biplane high-speed radiography (DRSA) and subject-specific finite element models for studying *in situ/in vivo* meniscal behavior. The purpose of this study was to validate this approach for predicting meniscal displacement under joint compression, using a cadaver model with a new knee simulator.

1. Introduction

Anterior cruciate ligament (ACL) deficiency, a common orthopedic condition, results in substantial joint instability. ACL injury alters joint mechanics and is associated with a high incidence of degenerative knee osteoarthritis (OA) and subsequent increased risk for meniscal damage. Animal models have shown a clear connection between mechanical instability and progressive OA. Thus, it is logical to assume that, if ACL reconstruction surgery restores knee stability, then it should reduce the risk of subsequent OA. However, a similar or higher incidence of knee OA has been reported in ACL-deficient individuals who underwent ligament reconstruction, as compared to those who did not. These studies have often been criticized (e.g. retrospective, uncontrolled, possible selection bias). Furthermore, to date there is no clear evidence of a long-term protective effect from these procedures. A review of the literature on the incidence and natural history of ACL injury reveals a high percentage of knee ligament injuries associated with strenuous human tasks, such as sports activities, falls, motor vehicle accidents and other activities[1][13][14][15]. The incidence of ACL injuries has been estimated at approximately 0.5 per

1,000 adults per year in the US and increasing yearly, particularly in women. By conservative estimates, approximately 50-75 thousand ACL reconstructions are performed per year in the U.S., and this number is increasing. With combined surgical/rehabilitation costs of approximately $12,000 per procedure in 1999, annual costs of these procedures could exceed $1 billion in the near future. Clinical studies of surgical effectiveness of reconstruction have been variable and inconclusive, indicating uncertain and unreliable outcomes. Rehabilitation difficulties, patello-femoral problems and loss of motor control all contribute to inconsistencies in restoring normal joint kinematics.

Other factors may contribute to OA development in the ACL reconstructed joint, such as possible damage to other tissues at the time of injury.[15, 16][17, 18][19, 20][21][22]. The incidence of meniscus pathology in the chronic ACL-deficient knee may be as high as 90%[15][23]. Injuries to the meniscus result in over 750,000 surgeries each year in the US. Recent studies suggest that even if no visible meniscus tear is found at surgery, patients have a high risk of developing definite tears as early as 26 months after ACL reconstruction. It is evident from all these studies that ACL reconstruction fails to prevent long term joint degeneration. We suggest failure of the reconstruction to restore normal knee kinematics is the primary contributor to progressive knee joint degeneration. Our studies will also provide a measure of the role of meniscus pathomechanics associated with ACL reconstruction and how this affects OA development.

2. Methods

A single left knee from a cadaveric specimen (male, age 31yrs) was prepared for placement in a load application system (Figure 1a). An arthroscopic procedure was employed so that soft tissues and joint capsule remained intact. Tantalum beads (0.8mm) were placed along the line of anterior cruciate ligament, (n=7), the menisci (n=18), the tibia (n=3) and femur (n=3). Specimens were fixed in cups, aligned with the load application system at 15° flexion and placed it within the biplane radiography system (Figure 1b). The in-situ 3D kinematics of all tantalum beads was measured using dynamic radiography (DRSA) at 250 frames/s (dynamic accuracy ±0.1mm). Meniscal kinematics and knee motion were assessed during a series of uniaxial compressive loading protocols (0-1kN at a rate of 10mms^{-1} to 1ms^{-1}). A K-scan pressure sensor (Tecscan Corporation, S. Boston, MA) was then fixed surgically to the tibia plateaus using an arthoscopic procedure [1]. After appropriate sensor calibration the loading protocol was repeated to evaluate the pressure distribution profile in the articular surfaces. This allowed for the evaluation of the effect of the sensor on the soft tissue kinematics and further calibration of the model by comparing experimental and model results. To construct the model a helical CT system was used to scan the knee in 0.5 mm increments. Two plastic tubes filled with solution of Cupric Sulfate with paramagnetic properties were fixed in the femur and tibia. This along with the tantalum beads allowed for co-registration of bony(CT) and soft tissue (MRI) geometry. MRI measurements were performed using a GE 3 T system. We acquired scans of coronal T2, a sagittal proton density weighted imaging using a fast spin echo sequence and sagittal spoiled gradient echo imaging. Combination of the above imaging protocols allowed for geometry description of meniscal and articular cartilage. The mesh was improved using topology-meshing enhancement software (ANSA Beta Systems Detroit MI)[12] to include the complete internal and external boundaries of the cortical shell and to accelerate the meshing process of cortical and cancellous bone. Element thickness varied from 2-7mm for bone and 1-4mm for cartilage, with solid eight-node elements (Figure 2a,b). The model included both the cortical and trabecular bone of the femur and tibia, articular cartilage of the femoral condyles and tibial plateau, both the medial and lateral menisci with their horn attachments, the transverse ligament, the anterior cruciate ligament, and the medial collateral ligament [1-6]. The material properties were chosen within the range of material properties available in the literature and from our previous studies [1-12]. Cancellous and compact bones were modeled using an elastic-plastic material law with variations of the elastic coefficients depending on the anatomical location. Cartilage was modeled using an elastic material law. The constitutive relation of the menisci treated the tissue as transversely isotropic and linearly elastic. The surface to surface tangential contact algorithm (ABAQUS/standard) was applied for the surface interaction in the knee joint with friction formulation and a friction coefficient of 0.001. Boundary conditions for the simulation were determined from skeletal kinematics (from DRSA) and the loads on the femur (from a 6-axis load cell at the proximal femur interface). The kinematics were applied using the displacement of the tantalum markers. Displacement history of four representative meniscal markers were chosen for comparison with corresponding nodes in the FE model. Relative displacements of marker pairs were expressed as percentages of the pre-test distances, determined from 20 frames of kinematics data acquired prior to the start of the loading protocol. The distance between different markers was expressed as percentage elongation normalized in this way. Displacements between FE model elements at approximately the same location as the implanted makers were determined at 5 time points during the simulation, and compared to the corresponding measured marker displacements.

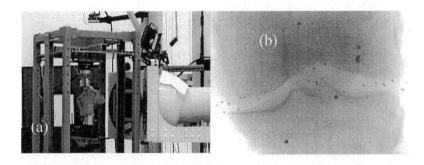

Figure 1 (a). The knee experimental set-up inside the biplane dynamic radiography system. The loading system is capable of controlling planar movement and can apply load up to 3kN at 1.5ms^{-1}. (b) Menisci and ACL tantalum markers in coronal view at the DRSA system-The Kscan Tecscan pressure sensor is not visible in the radiographic image.

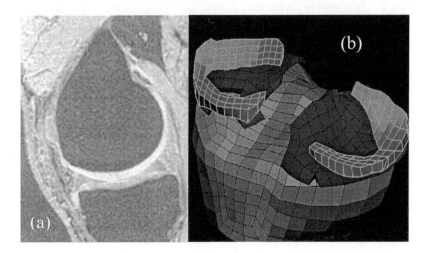

Figure 2 (a). MRI information for cartilage and meniscus geometry. (b) Meniscus mesh (4x4mm density) added in the FE model.

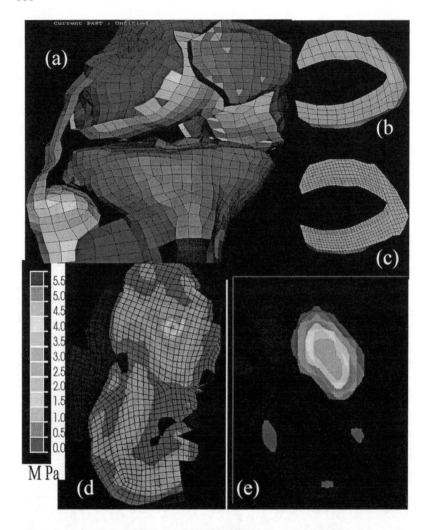

Figure 3 (a). The knee joint FE model. (b)Medial Meniscus mesh (4x4mm density) versus (c)high density mesh (1x1mm) added in the FE model. Cartilage pressure calculated by the model for 1kN axial load (d) predicts the directly measured pressure using the tecscan sensor (e).

3. Results

The experiment permitted acquisition of the menisci kinematics, estimations of radial expansion and direct measurement of articular pressure in the different loading sequences. Motion between the medial and lateral menisci was greater than displacement within the meniscal bodies, reaching a peak of approximately 1.5%. (Figure 4) The gross radial medial-lateral and anterior-posterior meniscal displacement due to pure joint compression of 1 kN was comparable to the predicted model deformations (Figure 3d,e). Insertion of the Tecscan sensor had a minimal effect (<2%) on measured marker displacements. The FE solution was considered converged for an average element size of 1 mm by 1 mm (Figure 3a,b,c). The FE predicted contact force of 550 N in the medial and 320 N in the lateral tibial condyle. Using this mesh size, finite element solutions for the meniscus, indicated that the contact variables changed by more than 7% when we introduced the tecscan in the FE model.

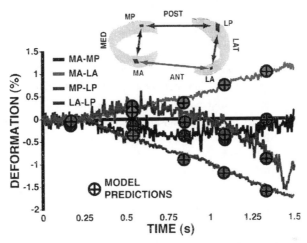

Figure 4 Deformation between meniscal markers and model predictions (corresponding closest element node motion): MA: Medial anterior, MP medial-posterior, LP lateral posterior, LA lateral anterior

4. Discussion

The study presented a method for estimating *in situ* meniscal radial expansion using skeletal kinematics combined with a patient specific FE model. The model was validated using high accuracy 3D kinematics of markers placed directly in the menisci. Good agreement between model predictions and experimental measurement were found, suggesting that the model should be useful for assessing *in vivo* meniscal deformation. We concluded that even in a very controlled uniaxial compression loading condition exclusion of the K-scan sensor from the model can result in relatively large errors in contact and cartilage stress that are not reflected in the change in meniscal kinematics. This local validation enhances the fidelity of the FE knee model and its applicability in the study of articular stress.

Acknowledgments: Dan Alexander MD for the surgical assistance. George Galaitsis and John Skarakis from ANSA Beta Systems MI Detroit for their generous donation and help with ANSA CAE software.

References

1. G. Papaioannou, D. Fyhrie, S. Tashman, K.H Yang, "Knee joint finite element modeling with patient specific properties: Model validation for contact analysis using high accuracy dynamic radiography kinematics data", IN PRESS *Journal of Medical Engineering & Physics* (2003)
2. P. Beillas, G. Papaioannou, K.H. Yang, S. Tashman, "A new method to investigate in-vivo knee behavior using a finite element model of the lower limb" IN PRESS *Journal of Biomechanics,* (2003).
3. I. Jonkers, A. Spaepen, G. Papaioannou, C. Steward "Contribution of Forward Simulation Techniques to the understanding of Muscle Activation Patterns in Mid-stance phase of Gait" *Journal of Biomechanics*, 2002, 35:5:609-619
4. G. Papaioannou, P. Beillas, W. Anderst, K.H. Yang, S. Tashman, "A new method to investigate in-vivo knee behavior using a finite element model of the lower limb",*49th Annual ORS meeting* 2003 New Orleans LA, U.S.A.-*Semifinalist in Orthopaedic Research Society's New Investigator Recognition Awards –NIRA- competition*
5. G. Papaioannou, W. Anderst, D. Fyhrie, S. Tashman, "ACL reconstruction increases in vivo, dynamic joint contact forces: an experimentally driven subject-specific FE analysis", *49th Annual ORS meeting* (2003) New Orleans LA, U.S.A.
6. G. Papaioannou, D. Fyhrie, S. Tashman, "Effects Of Patient-Specific Cartilage Geometry On Contact Pressure: An In-Vivo Finite Element Model Of ACL Reconstruction"*50th Annual ORS meeting* (2004) San Fransisco, U.S.A.
7. G. Papaioannou, S. Tashman F. Nelson, "Morphology Proportional Differences In The Medial And Lateral Compartment Of The Distal Femur", *50th Annual ORS meeting* (2004) San Fransisco, U.S.A.
8. G. Papaioannou, C. Demetropoulos, J. Guettler, K. Jurist, D. Fyhrie, S. Tashman, K.H. Yang, "Osteochondral Defects In The Human Knee With Evaluation Of Defect Size On Cartilage Rim Stress: In-Situ

Study For Finite Element Model Validation"*50th Annual ORS meeting* (2004) San Fransisco, U.S.A.

9. G. Papaioannou, W. Anderst, Y. Yeni, D. Fyhrie, S. Tashman, "Menisci Displacement Under Joint Load: A Subject Specific Finite Element Study With In-Situ Validation"*50th Annual ORS meeting* (2004) San Fransisco, U.S.A.

10. G. Papaioannou, K.H Yang, D. Fyhrie, S. Tashman, "Validation Of A Subject Specific Finite Element Model Of The Human Knee Developed For In-Vivo Tibio-Femoral Contact Analysis" (Nominated For Orthopaedic Research Society's New Investigator Recognition Awards –NIRA- Competition) *50th Annual ORS meeting* (2004) San Fransisco, U.S.A.

11. S. Tashman, D. Collon, K. Anderson, P. Kolowich, W. Anderst, G. Papaioannou, "Increase In Functional Graft Length Over Time After ACL Reconstruction: Effects Of Femoral Tunnel Position" *50th Annual ORS meeting* (2004) San Fransisco, U.S.A.

12. ANSA User's Guide . BETA-CAE Systems. DETROIT MI (2003).

13. Daniel, D.M., et al., "Fate of the ACL-Injured Patient: A Prospective Outcome Study". *Am. J. Sports Med.,* 1994. **22**(5): p. 632-644.

14. Fink, C., et al., "The treatment of fresh anterior cruciate ligament ruptures in relation to age and level of sports activity". Schweiz Z *Med Traumatol,* (1994)(1): p. 26-9.

15. Caborn Dn and Johnson Bm, The natural history of the anterior cruciate ligament-deficient knee. A review. [Review] *[59 refs]. Clinics in Sports Medicine,* (1993). **12**(4): p. 625-36.

16. Bray, R.C., Dandy, D. J., Meniscal lesions and chronic anterior cruciate ligament deficiency. Meniscal tears occurring before and after reconstruction. *J Bone Joint Surg Br,* (1989). **71**(1): p. 128-30.

17. Arnoczky, S.P., Animal Models for Knee Ligament Research, in *Knee Ligaments: Structure, Function, Injury, and Repair,* D.M. Daniel, W.H. Akeson, and J.J. O'Conner, Editors. (1990), Raven Press: New York. p. 401-417.

18. Biswal, S., Hastie, T., Andriacchi, T. P., Bergman, G. A., Dillingham, M. F., Lang, P., Risk factors for progressive cartilage loss in the knee: a longitudinal magnetic resonance imaging study in forty-three patients. *Arthritis Rheum,* (2002). **46**(11): p. 2884-92.

19. Duncan J. B., et al., Meniscal injuries associated with acute anterior cruciate ligament tears in alpine skiers. *Am J Sports Med.,* (1995). **23**(2): p. 170-172.

20. Podskubka, A., et al., [Arthroscopic reconstruction of the anterior cruciate ligament using the transtibial technique and a graft from the patellar ligament--results after 5-6 years]. *Acta Chir Orthop Traumatol Cech,* (2002). **69**(3): p. 169-74.

21. Rodkey, W.G.S., J. R. Li, S. T., A clinical study of collagen meniscus implants to restore the injured meniscus. *Clin Orthop,* (1999)(367 Suppl): p. S281-92.

22. Kobayashi, K., et al., Meniscal tears after anterior cruciate ligament reconstruction. *J Nippon Med Sch,* (2001). **68**(1): p. 24-8.

23. Shelbourne, K.D. and B.P. Rask, The sequelae of salvaged nondegenerative peripheral vertical medial meniscus tears with anterior cruciate ligament reconstruction. *Arthroscopy,* (2001). **17**(3): p. 270-274.

FINITE ELEMENT MODELING OF THE OSTEOCHONDRAL DEFECTS IN THE HUMAN KNEE: INFLUENCE OF DEFECT SIZE ON CARTILAGE RIM STRESS AND LOAD REDISTRIBUTION TO ADJACENT CARTILAGE

GEORGE PAPAIOANNOU

Bone and Joint Center, Henry Ford Health System, Detroit, MI

Department of Biomedical Engineering, Catholic University of America, Washington DC.

C. DEMETROPOULOS, J. GUETTLER, K. JURIST

+William Beaumont Hospital MI

D. FYHRIE, S. TASHMAN, K.H. YANG

Bioengineering Center, Wayne State University

The purpose of this study was to determine the influence of osteochondral defect size on defect rim stress concentration, and load redistribution to adjacent cartilage over the weight-bearing area of the medial and lateral femoral condyles in the human knee. Subsequent finite element (FE) computer modeling of these defects was designed to determine the deformation of the underlying cartilage and to further analyze the development of rim stress.

1. Introduction

Previous studies have shown that focal articular surface defects in the human knee may progress to degenerative arthritis [5]. Although the risk of defect progression to degenerative arthritis is multifactorial [6], defect size is important. To date, the "critical" or "threshold" defect size at which biomechanical forces become potentially damaging to adjacent cartilage has not been clearly identified. As a result, the size at which articular surface restoration needs to be considered has not been well defined. It appears that the treatment of articular surface lesions, when defining treatment according to defect size, is not based on any firm biomechanical data. The purpose of our study was to determine the effect of defect size on defect rim stress concentration and load redistribution to adjacent cartilage over the weightbearing area of the medial and lateral femoral condyles in the human knee. Based on our experimental/modeling technique of cartilage loading around osteochondral defects of different sizes, we sought to determine if a "threshold" or "critical" size exists at which biomechanical forces may become potentially damaging to adjacent cartilage.

2. Methods

Eight fresh frozen cadaveric knees were used to study the rim stresses around osteochondral defects of the femoral condyles. Knees were mounted in 30 degrees of flexion on an Instron materials tester (Instron Corporation, Canton, MA). A K-Scan pressure sensor (Tekscan Corporation, South Boston, MA) was inserted between the femoral condyles and the meniscus to dynamically measure pressure distribution during loading. Each intact knee was loaded to a maximum value of 700 N under load control. Maximum load was maintained for 5 seconds prior to unloading of the specimen. Loading and pressure mapping history were recorded for analysis. Following the testing of the intact knee, full thickness circular defects were created (Fig 1) using a series of custom made coring devices to produce a range of defects. Defect sizes tested included 5 mm, 8 mm, 10 mm, 12 mm, 14 mm, 16 mm, 18 mm and 20 mm. Each knee was progressively tested with each defect size from smallest to largest. To construct the model a helical CT system was used to scan the knee specimen in 0.5 mm increments at 0 degrees of flexion. Two plastic tubes filled with solution of Cupric Sulfate with paramagnetic properties were fixed in the femur and tibia. This along with the tantalum beads allowed for co-registration of bony geometry with soft tissue geometry. The slices were manually digitized for both bone geometry and cortical shell thickness. MRI measurements were performed using a GE 3 T system. We acquired standard anatomic scans which included: coronal T2, a sagittal proton density weighted imaging using a fast spin echo sequence and sagittal spoiled gradient echo imaging. We meshed cancellous bone (thickness 2-7mm) and cartilage (thickness 1-4mm) with solid eightnode elements. Artificial defects were meshed on the model to match a 12 mm defect (Fig. 1). The model included both the cortical and trabecular bone of the femur and tibia, articular cartilage of the femoral condyles and tibial plateau, both the medial and lateral menisci with their horn attachments, the transverse ligament, the anterior cruciate ligament, and the medial collateral ligament and is described in detail in Papaioannou et al. 2003[1][2][3]. The material properties were chosen within the range of material properties available in the literature and from our material properties atlas described in the following session.

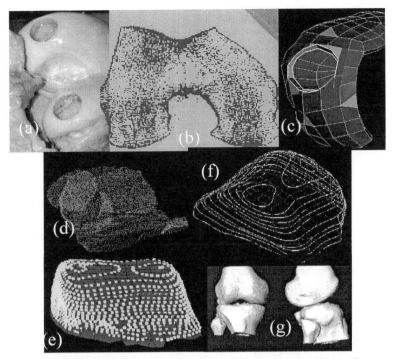

Figure 1. (a) Circular Defect shown at cadaveric femur. (b)Thresholding of CT data slice using ImageJ software and (c)Femoral condyle FE mesh with defect. (d) All the contours from the co-registered CT and MR slices, (e) Imposing topological information from reference mesh that also contains density/material properties information, (f) solid modeling using ANSA[4] beta system software, (g) Surface model of knee joint based on specimen CT for initial model positioning.

2.1 Mechanical properties of human bone from imaging data

Numerous mechanical bone properties studies have been performed over the last three decades. The main body of work concludes that human bone is a highly heterogeneous and anisotropic material. It can be compared to composite materials. It presents to main different tissue formations: a) one with high porosity, trabecular bone and b) one consisting of more compact material the cortical bone. Their material properties can be mapped based on anatomical location. Bone was initially assumed to have an orthotropic behavior whose stiffness matrix containing elastic constants were defined by Eq 1. By reversing the stiffness matrix, the compliance matrix allowed the elastic properties in the three axes of symmetry of the crystal to be determined (Eq. 2.)

$$[C_{ij}] = \begin{bmatrix} C_{11} & C_{12} & C_{13} & 0 & 0 & 0 \\ C_{21} & C_{22} & C_{23} & 0 & 0 & 0 \\ C_{31} & C_{32} & C_{33} & 0 & 0 & 0 \\ 0 & 0 & 0 & C_{44} & 0 & 0 \\ 0 & 0 & 0 & 0 & C_{55} & 0 \\ 0 & 0 & 0 & 0 & 0 & C_{66} \end{bmatrix} \quad (1)$$

$$[S_{ij}] = \begin{bmatrix} \dfrac{1}{E_1} & -\dfrac{\nu_{21}}{E_2} & -\dfrac{\nu_{31}}{E_3} & 0 & 0 & 0 \\ -\dfrac{\nu_{12}}{E_1} & \dfrac{1}{E_2} & -\dfrac{\nu_{32}}{E_3} & 0 & 0 & 0 \\ -\dfrac{\nu_{13}}{E_1} & \dfrac{\nu_{23}}{E_2} & \dfrac{1}{E_3} & 0 & 0 & 0 \\ 0 & 0 & 0 & \dfrac{1}{G_{23}} & 0 & 0 \\ 0 & 0 & 0 & 0 & \dfrac{1}{G_{31}} & 0 \\ 0 & 0 & 0 & 0 & 0 & \dfrac{1}{G_{12}} \end{bmatrix} \quad (2)$$

Were: E_i : Young's moduli in the direction i, G_{ij} : Shear moduli in plane i-j, ν_{ij} : Poisson's ratio stress and strain respectively in directions i,j.

Experimentally, several studies used an ultrasonic transmission technique for assessing bone material properties. Different wave propagation techniques have been used for velocities measurements of bone tissue. Bulk velocities have been measured for cortical bone, and then elastic constants have been obtained. Cancellous bone is porous and highly heterogeneous compared to cortical bone. Homogeneous volume was assumed and bar waves have been used, allowing to assess directly the elastic properties [7]. We used here information from an atlas of mechanical properties of different types of bone (femur, tibia, mandible, humerus, patella, lumbar spine) from eight unembalmed subjects (7 males, 1 female, mean age 60 years (45-68))[7]. Typical range of values of mechanical properties for cortical bone and cancellous bone [8]are summarized in Table 1.

2.2 Models with specimen specific geometric and material properties from CT data

We have investigated the relationships between CT number derived from CT imaging techniques and mechanical properties of bone from previously reported studies. The CT number characterizes a linear coefficient of attenuation of X-ray within the tissue. For the CT scan, the pixel values are represented by an empirical number called CT number expressed in Houndsfield Units (HU) and is defined by the following empirical equation (3):

$$CT(HU) = 1000 \frac{CT - CT_{water}}{CT_{water} - CT_{air}} \quad (3)$$

$$CT(HU) = 1000 \frac{\mu - \mu_{water}}{\mu_{water} - \mu_{air}}$$

μ is the linear coefficient of attenuation of X-ray within the tissue (cm^{-1}). CT number value is depending on acquisition parameters, typical values are 0 and -1000 for water and air respectively. Table 2 summarized some predictive relationships between elastic properties and density and CT numbers of the proximal tibia (upper extremity of the tibia). Tissue characterization from the CT with this method is reflecting significantly the material properties. A custom made program mesh matching properties has been developed allowing the matching of the material properties distribution measured on the stack of CT image data and their assignment to the elements properties. The program consists in matching the measured mean CT number of a region of interest with the characteristics geometric properties of the elements of the mesh. Superimposing of this material properties "topology" information on a generic mesh is therefore possible (Figure 1). It should be noted that for the three (3D) dimensional mesh, it is necessary to match the technical acquisition parameters of the CT data with that of the protocol used to get the predictive relationships and the three-dimensional mesh. Quantitative measurements performed on the CT images required a good 'density' resolution. In order to achieve that, the technical protocol of acquisition of the images should be optimized. Influence of these parameters on direct measurements of CT number have been quantified and showed 10% of variation for the range of 0 to 1200 CT HU. This should be considered when predictive relationships are used to predict elastic properties from literature. Tissue characterization is not reflecting significantly the material properties when MR imaging is used. In fact, the gray level of the region of interest is related to the intensity of proton density, which varies significantly with the acquisition parameters. When predictive relationships are not appropriate, the reviewed database from literature is used. The FE model was then based on layer reconstruction procedure using unidirectional (interior-superior) CT-scan images. While locating the position of the elements back to the CT-scan image we could correct for contour and material properties mismatch between volumetric and meshed information. The pixel gray value of these elements can be approximated to the nearest element located within the contours of the segments (femur and tibia). The average CT gray value of all the pixels contained in the cubic volume and within the contours of the segments was calculated and related to the material properties. For example, our final goal was to differentiate locally the material properties for the whole model based on the review material properties atlas and the assumptions mentioned earlier. For more details on the FE model refer to Papaioannou et al. 2003 [1]. We imported in the model the pressures measured from the pressure maps on the cartilage of the tibia plateau from K-scan readings and

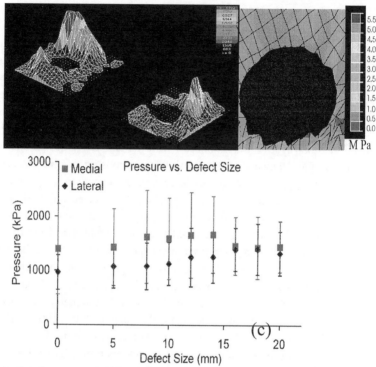

calculated stress and deformation at the locations of interest (Fig 2).

Figure 2. (a) Tekscan pressure sensor output showing stress concentration around a 12 mm defect and (b) FE model calculation of rim stress/deformation on medial condyle. (c) Graph demonstrating the relationship between Peak pressure and Defect size. Note that peak pressure values did not increase significantly as the defects were enlarged, but remained relatively constant.

3. Results

Experimental results demonstrated that pressure distribution around defect sizes 8 mm and smaller were dominated by the effects of the meniscus. Peak pressure values did not increase significantly as the defects were enlarged, but remained relatively constant (Figure 2). However, when defects of 10 mm and greater were created, a peak rim stress was observed at an average distance of 2.64 mm +/- 1.63 mm on the medial condyle and 2.90 mm +/-1.51 mm on the lateral condyle. The model predicted the rim stress topology (fig 2) for the defect size of 12 mm.

4. Discussion

Altered loading secondary to focal articular surface defects in weightbearing areas of the knee may have important implications relating to the long-term integrity of cartilage adjacent to these defects and risk for progression to arthritis. Our biomechanical model suggests that a threshold effect does occur at which rim stress concentration becomes a factor around osteochondral defects. Although the risk of lesion progression to arthritis is certainly multifactorial, rim stress concentration and altered loading may cause degeneration of adjacent cartilage. Our biomechanical data may be used as a size reference to guide clinical decision-making when choosing whether or not to treat osteochondral lesions in weightbearing areas of the knee. Combining high-quality static 3D imaging, and a knee FE model with experimental measurements of contact pressure allowed prediction of the effect that different defect sizes have on cartilage rim stress and pressure distribution in a human cadaveric knee. The model converges for a mesh density of 4x4mm by overestimating slightly the maximum pressure while reproducing the stress topology. Co-registration of multiple imaging modalities increased topological and geometric model fidelity. We also concluded that, even in a very controlled uniaxial compression loading condition, pressure distribution around defect sizes 8 mm and smaller were dominated by the effects of the meniscus. We are now acquiring all of the elements required for a patient specific modeling approach (e.g. CT, MRI, *in vivo* dynamic RSA) Thus, this validated method will enable prediction of patient specific *in vivo* articular stress in the effort to estimate defects on cartilage continuity.

Acknowledgments

George Galaitsis and John Skarakis from ANSA Beta Systems MI Detroit for their generous donation and help with ANSA CAE software.

Bone	cortical	cancellous
E_{axial} (Gpa)	14-27 20	0.011-3.12 0.961
$E_{tangential}$ (Gpa)	7-17 12	0.023-1.5 0.341
E_{radial} (Gpa)	7-16 11	0.024-1.5 0.301
ρ (kg/m3)	1545-2118 1840	55-744 257
CT (HU)	1270-1835 1560	-72-512 143
σ_{ult} (MPa)	<*>	0.11-11 3

Table 1: Range of values of mechanical properties obtained from eight subjects. Minimum, maximum and median values are given. <*>Ultimate strength of cortical specimen was not measured, but literature review gave typically values around 150MPa

RELATIONSHIPS	
$E_{axial} = 0.51\rho^{1.37}$	$R^2 =0.96$
$E_{tangential} =0.06\rho^{1.55}$	$R^2 =0.90$
$E_{radial}=0.06\rho^{151}$	$R^2 =0.89$
$E_{axial} = 296 + 5.2$ CT	$R^2 =0.79$
$\rho =114 + 0.916$CT	$R^2 =0.80$

Table 2. Predictive relationships for the proximal tibia. Young's modulus is expressed in MPa and density in kg/m^3 and CT number in HU.

References

1. G. Papaioannou, D. Fyhrie, S. Tashman, K.H Yang, "Knee joint finite element modeling with patient specific properties: Model validation for contact analysis using high accuracy dynamic radiography kinematics data", IN PRESS *Journal of Medical Engineering & Physics* (2003)
2. G. Papaioannou, P. Beillas, W. Anderst, K.H. Yang, S. Tashman, "A new method to investigate in-vivo knee behavior using a finite element model of the lower limb",*49th Annual ORS meeting* 2003 New Orleans LA, U.S.A.-*Semifinalist in Orthopaedic Research Society's New Investigator Recognition Awards –NIRA- competition*
3. G. Papaioannou, K.H Yang, D. Fyhrie, S. Tashman, "Validation Of A Subject Specific Finite Element Model Of The Human Knee

Developed For In-Vivo Tibio-Femoral Contact Analysis" (Nominated For Orthopaedic Research Society's New Investigator Recognition Awards –NIRA- Competition) *50th Annual ORS meeting* (2004) San Fransisco,

4. ANSA User's Guide . BETA-CAE Systems. DETROIT MI (2003).

5. K. Messner, W. Maletius, "The long-term prognosis of severe damage to weightbearing cartilage in the knee". *Acta Orthop Scand 67: 165-168,* (1996)

6. T. Minas "The role of cartilage repair techniques, including Chondrocyte transplantation, in focal chondral knee damage". *AAOS Instructional Course Lectures* 48: 629-643, (1999)

7. M.C. HoBaTho, R.B. Ashman, "Atlas of mechanical properties of human cortical and cancellous bone", in *In vivo assessment of bone quality by vibration and wave propagation techniques*, Van der Perre G., L. G., Borgwardt A Part II , ACCO, Leuven, 1992, pp. 7-38.

8. T.L. Donahue, M.L. Hull, M.M. Rashid, "A finite element model of the human knee joint for the study of tibio-femoral contact", *J Biomech Eng* 124 (3), 273-80., 2002.

FINGERPRINT VERIFICATION BASED ON IMAGE PROCESSING SEGMENTATION USING AN ONION ALGORITHM OF COMPUTATIONAL GEOMETRY *

M. POULOS

Dept. of Informatics University of Piraeus,
P.O. BOX 96,
49100 Corfu, Greece
E-mail: marios.p@usa.net

A. EVANGELOU

Dept. of Exp. Physiology University of Ioannina,
P.O. BOX 1186,
45110 Ioannina, Greece
E-mail: evagel@uoi.gr

E. MAGKOS

Dept. of Archives and Library Sciences,
Palea Anaktora
49100 Corfu, Greece
E-mail: emagos@unipi.gr

S. PAPAVLASOPOULOS

Dept. of Archives and Library Sciences,
Palea Anaktora
49100 Corfu, Greece
E-mail: sozon@ionio.gr

In this study, we applied a digital image processing system using the onion algorithm of Computational geometry to develop fingerprint verification. This method may be characterized as an alternative method to the used minutiae extraction algorithm proposed by Ratha et al. The proposed algorithm is also compared to a well-known commercial verification algorithm that is based on Ratha's algorithm. In the experimental part the results of the above comparison showed that the proposed method yields correct positive and correct negative verification scores greater than 99%.

---

*This work is supported by University of Piraeus.

550

1. Introduction

In this paper the problem of fingerprint verification via the Internet is investigated. Specifically, the method that is used for the above purpose is based on a traditional finger scanning technique, involving the analysis of small unique marks of the finger image known as minutiae. Minutiae points are the ridge endings or bifurcations branches of the finger image. The relative position of these minutiae is used for comparison, and according to empirical studies, two individuals will not have eight or more common minutiae. [1,2]. A typical live-scan fingerprint will contain 30-40 minutiae. Other systems analyze tiny sweat pores on the finger that, in the same way as minutiae, are uniquely positioned. Furthermore, such methods may be subject to attacks by hackers when biometric features are transferred via Internet [3].

In our case we developed a method that addresses the problem of the rotation and alignment of the finger position. The proposed method is based on computational geometry algorithms (CGA). The advantages of this method are based on a novel processing method using specific extracted features, which may be characterized as unique to each person. These features depend exclusively on the pixels brightness degree for the fingerprint image, in contrast to traditional methods where features are extracted using techniques such as edge and ridge - minutiae points detection. Specifically, these feature express a specific geometric area (convex layer) in which the dominant brightness value of the fingerprint ranges.

For the testing of the accuracy of the proposed method we selected a well-known commercial verification algorithm that is based on Ratha's algorithm. The fingerprint data used in the testing procedure was received from the database of a commercial company. [a]

Thereinafter, we tested the CGA method against Ratha's algorithm with regard to correct positive and negative verification procedures. Finally, the statistical results of both methods were evaluated.

2. Method

In brief, the proposed method is described in the following steps:

(1) *Pre-processing stage* — The input image is made suitable for further processing by image enhancement techniques using Matlab [4].

---

[a]is available free on the Internet :
http://www.neurotechnologija.com/download.html

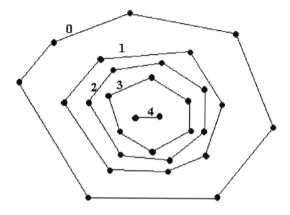

Figure 1.　Onion Layers of a set of points (coordinate vector).

(2) *Processing stage* — The data, which comes from step 1, is submitted
to specific segmentation (data sets) using computational geometry
algorithms implemented via Matlab (see Figure 1).

(3) *Meta-processing stage (during registration only)* — The smallest
layer (convex polygon) of the constructed onion layers is isolated
from the fingerprint in vector form (see Figure 2).

(4) *Verification stage* — This stage consists of the following steps:

    (a) An unknown fingerprint is submitted to the proposed pro-
cessing method (Steps 1 and 2), and a new set of onion layers
is constructed.

    (b) The referenced polygon that has been extracted during the
registration stage is intersected with the onion layers and the
system decides whether the tested vector identifies the onion
layers correctly or not.

(5) *Evaluation of the algorithm in comparison to Ratha's algorithm* —
The above procedure is repeated using a well-known commercial
verification algorithm that is based on Ratha's algorithm.

2.1. Pre-processing stage of CGA method

In this stage a fingerprint image, which is available from any of the known image formats (tif, bmp, jpg, etc), is transformed into a matrix (a two-dimensional array) of pixels [5]. Consider, for example, the matrix of pixel values of the aforementioned array. Then the brightness of each point is proportional to the value of its pixel. This gives the synthesized image of a bright square on a dark background. This value is often derived from the output of an A/D converter. The matrix of pixels, i.e. the fingerprint image, is usually square and an image will be described as N x N m-bit pixels [6,7], where N is the number of points along the axes and m controls the number of brightness values. Using m bits gives a range of 2 m values, ranging from 0 to 2 m -1. Thus, the digital image may be denoted as the following compact matrix form:

$$f(x,y) = \begin{bmatrix} f(0,0) & f(0,1) & \ldots & f(0,N-1) \\ f(1,0) & f(1,1) & \ldots & f(1,N-1) \\ \vdots & \vdots & \vdots & \vdots \\ f(N-1,0) & f(N-1,1) & \ldots & f(N-1,N-1) \end{bmatrix} \quad (1)$$

The coordinate vector of the above matrix is:

$$\mathbf{S} = [f(x,y)] \quad (2)$$

Thus, a vector $1 \times N^2$ of dimension is constructed, which is then used in the next stage [8].

2.2. Processing stage of CGA method

Proposition: We considered that the set of brightness values for each fingerprint image contains a convex subset, which has a specific position in relation to the original set. This position may be determined by using a combination of computational geometry algorithms, which is known as Onion Peeling Algorithms [9] with overall complexity *O(d*n log n)* times.

Implementation: We consider the set of brightness values of a fingerprint image to be the vector **S** (eq.2). The algorithm starts with a finite set of points $\mathbf{S} = \mathbf{S_0}$ in the plane, and the following iterative process is considered. Let $\mathbf{S_1}$ be the set

$$S_0 - \partial H(S_0) : S,$$

minus all the points on the boundary of the hull of **S** . Similarly, define

$$S_{i+1} = S_i - \partial H(S_i).$$

The process continues until the set is (see Figure 1). The hulls are called the layers of the set, and the process of peeling away the layers is called onion peeling for obvious reasons (see Figure 1). Any point on is said to have onion depth, or just depth . Thus, the points on the hull of the original set have depth 0 (see Figure 1).

2.3. *Meta-processing stage of CGA method*

In our case we consider that the smallest convex layer that has depth 3 (see Figure 1) carries specific information, because this position gives a geometrical interpretation of the average of the fingerprint brightness [5]. This feature may be characterized as unique to each fingerprint because the two (2) following conditions are ensured:

(i) The selected area layer is non-intersected with another layer.
(ii) The particular depth of the smallest layer is variable in each case.

Thus, from the proposed fingerprint processing method two (2) variables are extracted: the area of the smallest onion layer and the depth of this layer, which is a subset of the original fingerprint set S values.

2.4. *Verification stage of CGA method*

In this stage we tested the subset S_{xy} against a new subset set N_{xy}, which came from the processing of another set **N**. This testing takes place at the following 3 levels (see Figure 2). Subset S_{xy} is cross-correlated with subset N_{xy}.

(i) The depths of the iterative procedure, from which the subsets were extracted, are compared.
(ii) The intersection between subset N_{xy} convex layer and one of set **S** onion layers is controlled.

Furthermore, it is considered that subset N_{xy} identifies set **S** as the parent onion layers when:

(i) The cross-correlation number of subset S_{xy} N_{xy} is approximately 1
(ii) The intersection [11] between the convex layer of subset N_{xy} and one of the onion layers of set **S** is 0.

Otherwise, subset $\mathbf{N_{xy}}$ does not identify set $\mathbf{S}$ as the parent onion layers.

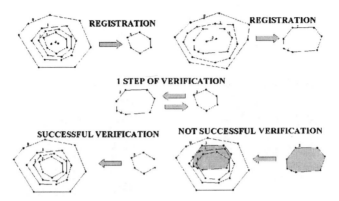

Figure 2. Theoretical presentation of the registration and verification stages of two (2) onions' layers.

2.5. *Verification stage based on Ratha's algorithm*

Fingerprint verification based on Ratha's algorithm is a technique [11,12] to assign a fingerprint into one of the several pre-specified types previously described. Fingerprint verification can be viewed as a coarse level matching of the fingerprints. An input fingerprint is first matched at a coarse level to one of the pre-specified types and then, at a finer level, it is compared to the subset of the database containing that type of fingerprint only. We have developed an algorithm to classify fingerprints into five classes, namely, whorl, right loop, left loop, arch, and tented arch. The algorithm separates the number of ridges present in four directions (0 degree, 45 degree, 90 degree, and 135 degree) by filtering the central part of a fingerprint with a bank of Gabor filters. This information is quantized to generate a FingerCode, which is used for classification [13,14].

3. Experimental Part

In this experiment forty-eight (48) index-finger prints belonging to six (6) individuals (6x8=48) were tested. More specifically, each index-finger print of an individual was tested against the other seven (7) in its group and the forty (40) prints of the other five individuals. In total 2256 or $2 * C_2^8 = \frac{8!}{2!*(8-2)!} = 2256$ verification tests for each of the two methods took place.

3.1. *Pre-processing stage*

In our experiment, each of the recorded fingerprints in TIFF format is represented by a complete 255 × 255 image matrix (equation 1), which came from a converting quantization sampling process implemented via the *imread.m* Matlab function.

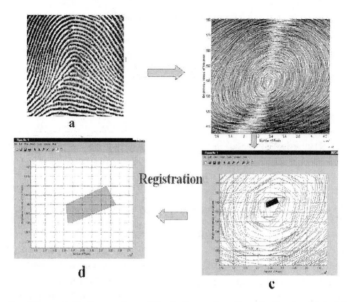

Figure 3.   The analytical procedure of the feature extraction of a fingerprint in 4 frames.

(i) Each pixel of the used fingerprint consists of 8 bits, therefore m=8 and the gray levels of brightness range between 0 and 255.

(ii) The dimension of the created compact matrix $f(x, y)$ of equation 1 is **S** and the coordinate vector is respectively.

3.2. *Processing stage*

The coordinate vector, which was extracted in the pre-processing stage, is submitted to further processing. In particular, the onion layers of vector S are created according to the computational geometry algorithm (figure 3a), which was described in Section 2.2. Thus, a variable number of layers (convex polygons) were extracted for each fingerprint case. In this case, the created onion consisted of 944 layers (convex polygons), and the number of vertices of the smallest internal layer was five (5). Furthermore, the average of vector value **S** in this example was 140,67.

3.3. *Meta-processing stage*

As can be seen in figure 3d the area that encloses the smallest internal layer contains the aforementioned average value. In other words, the area of this layer may be characterized as a specific area in which the dominant brightness value of the fingerprint ranges.

3.4. *Verification stage*

In this stage, it is assumed that the referenced polygon A, must lead to a rejection decision. Then we applied the aforementioned VERIFICATION conditions in order for the system to decide whether polygon B is correctly identified or not. The final decision of this system is that the tested fingerprint is not identified correctly for the following reasons:

Table 1. Positive and Negative Fingerprint Verification Scores using Ratha's and Onion Algorithms.

Individuals	Ratha's						Onion					
A	45	1	0	1	0	0	48	0	0	0	0	0
B	1	44	1	0	0	0	0	47	1	0	0	0
C	1	1	45	1	1	1	1	0	48	0	0	0
D	0	1	1	44	1	0	0	1	0	46	0	0
E	1	0	0	0	44	0	0	0	1	0	47	0
F	0	0	0	1	0	45	1	0	0	0	0	46

(i) The depth of the smallest referenced layer (polygon) was 944 in contrast to that of the tested vector that was 677 respectively.

(ii) The layer of the tested polygon intersected the other layers.

Especially in the negative correct verification case the final decision of the system depended on the position and the sizes of the final characteristic polygons.

3.5. *Ratha's algorithm*

In this case we used a well-known commercial verification algorithm that is based on Ratha's algorithm.

4. Result

In this experiment forty-eight (48) index-finger prints belonging to six (6) individuals (6x8=48) and called A, B, C, D, E and F, were tested. More specifically, each index-finger print of an individual was tested against the other seven (7) in its group and the forty (40) prints of the other five individuals. In total 2256 verification tests for each of the two methods took place.

4.1. *Statistical evaluation*

As can be seen from the diagonal scores on in the above table (1) the correct positive verification test score for the Ratha method, 156/168=0.93 or 93% and for the CGA method is 165/168=0.98 or 98%. Furthermore, the correct negative verification score for the Ratha algorithm is 933/940=0,99 or 99% and the correct negative verification for the CGA method is 938/940 or approximately 100%. In contrast, the false positive verification scores for the Ratha method is 7% and for the CGA method 2%. At this point it is to be noted that the false results of the Ratha method were yielded when the tested image had variations for rotation reasons. On the other hand, the CGA false results were yielded for those tested fingerprint specimens that were not complete.

5. Conclusion

From the results of the experiment it is ascertained that the proposed method, bearing in mind security considerations, can be used for accurate and secure fingerprint verification purposes because the proposed feature extraction is based on a specific area in which the dominant brightness value of the fingerprint ranges. Moreover, the proposed method promisingly allows very small false acceptance and false rejection rates, as it is based on specific segmentation. It has to be noted that biometric applications will gain universal acceptance in digital technologies only when the number of false rejections / acceptances approach zero. The results of this comparison showed that the proposed method yields correct positive and correct negative verification scores greater than 99%. In particular, the proposed CGA method produced extremely reliable results even in cases where the tested fingerprints were complete specimens yet the position or pressure applied was not consistent [15]. The computational complexity of the proposed algorithm may also be characterized as extremely competitive.

References

1. A. K. Jain, A. Ross and S. Pankanti, Fingerprint matching using minutiae and texture features, *Proc. International Conference on Image Processing ICIP, Thessalonica, GR*, 281-285 (2001).
2. D. Maio, D. Maltoni, Direct gray-scale minutiae detection in fingerprints, *IEEE Transactions on PAMI* **19(1)**, 27-40 (1997).
3. L. O'Gorman, Fingerprint verification, in Biometrics, *S.: Kluwer Acadenic Publishers* (1999).
4. T. Poon , P. Banerjee, Contemporary Optical Image Processing With Matlab, *Hardcover: Elsevier Science Ltd* (2001).
5. R. Bracewell, Two-Dimensional Imaging, *NJ: Prentice - Hall, Upper Sandle River* (1995).
6. M. Nixon, A. Aguado, Feature Extraction and Image Processing, *GB: Newnes-Oxford* (2002).
7. R. Gonzales, R. Woods, Digital Image Processing, *NJ: Prentice - Hall, Upper Sandle River* (2002).
8. M. Spiegel, Theory and Problems of Vector Analysis, *London: McGraw-Hill* (1974).
9. J. O'Rourke, Computational Geometry in C, *NY: Cambridge University Press* (1993).
10. J. O'Rourke, J. Chien, C. Olson, and T. Naddor,A new linear algorithm for intersecting convex polygons, *Comput. Graph. Image Proces.* **19 (4)**, 384-391 (1982).
11. C. Calabrese, The Trouble with Biometrics, *Login* **24(4)**, 56-61 (1999).
12. G. Hachez, F. Koeune and J.J. Quisquater, Biometrics, Access Control, Smart Cards: a Not So Simple Combination, *Proc. of the 4th Working Conference on Smart Card Research and Advanced Applications (CARDIS 2000), Bristol, GB* 273-278 (2000).
13. B. Schneier, Applied Cryptography, Protocols, Algorithms and Source Code in C, *GB: Elsevier, 2nd Edition* (1996).
14. A.K. Jain, L. Hong, R. Bolle and S. Pankanti, System and method for deriving a string-based representation of a fingerprint image *US Patent* **12(6)**, 487-496 (2002).
15. A.K. Jain, L.Hong, R.Bolle and S. Pankanti, Determining An Alignment Estimation Between Two (Fingerprint) Images *US Patent* **11(6)**, 314-319 (2001).

MODELS FOR GENE REGULATORY NETWORKS:
A REVERSE ENGINEERING APPROACH

GEORGE C. SAKELLARIS, DIMITRIOS I. FOTIADIS

Unit of Medical Technology & Intelligent Information Systems, Dept of Computer Science, University of Ioannina & Biomedical Research Institute – FORTH, GR 45110 Ioannina GREECE

In this paper we demonstrate the use of Boolean approach in modeling conditional independencies between deterministic and stochastic variables in genetic networks and in the study of the dynamic nature of interacting genes.

1. Introduction

Experimental literature on biomolecular processes provides us with the challenge of multifunctionality, implying regulatory networks as opposed to isolated, linear pathways of causality. Questions which have traditionally been posed in the singular are now being addressed in the plural [1]:

- What are the functions of this gene? Which genes regulate this gene?
- Which genes are responsible for this disease?
- Which drugs will treat better this disease?

System biology is the field of biology, which aims at system-level understanding of biological systems. Most biological phenomena are caused by an ensemble of biochemical entities including mRNA, proteins, small molecules (such as hormones) and ions. Genetic network analysis exploits massively parallel measurements in order to determine the regulatory interactions between genes and their results in various states. Recently, there has been much interest in reverse engineering genetic networks using time series data [1].

There are two proper abstractions that motivate the production of modeling frameworks and data analysis tools. These abstractions are based on two principles: The first is the genetic information flow, which defines the mapping from sequence space to functional space. The genome contains the information for the construction of complex molecular features of the organism, as it is reflected in the process of development.

The second principle is the study of complex dynamic systems in order to move from states to trajectories and attractors. When addressing biological functions, we usually refer to functions in time, i.e. causality and dynamics. On a molecular level, function is manifested by the behavior of complex networks. The dynamics of these networks resemble trajectories of state transitions. The concept of attractors is what really provides meaning to these trajectories, i.e. the attractors are the high dimensional dynamic molecular representations of stable phenotypic structures such as differentiated cells and tissues [1].

The measurement of the expression level of many genes in parallel, as they change over time and react to external stimuli, has become feasible recently due to the development of new technologies such as DNA microarrays. The amount of data obtained from these experiments is enormous. Hence, the challenge is to find automated ways of extracting meaningful information from the expression data, to infer regulatory mechanisms and to reveal the function of genes. To achieve this, various methods are used such as clustering of gene expression data, discrete genetic network inference models, probabilistic genetic network inference methods, and artificial neural networks.

In this paper we consider learning via Boolean Networks and apply them to the inference of genetic networks. We use time series data in order to recover component interactions and sub-networks that correspond to regulatory networks in the cell. The underlying assumption is that some genes depend on others, while others exhibit independence or conditional independence.

2. Materials and Methods

In this section, we describe the methodology for inferring the genetic network using experimental data. Our modelling approach is based on boolean Boolean network. The proposed methodology consists of four stages (fig. 1)

2.1. Dataset

We have utilized the dataset produced by Cho et. al. [3]. The data describes the progression through the eukaryotic cell cycle that is known to be both regulated and accompanied by periodic fluctuations in the expression levels of numerous genes. The experimental work [3] reports the genome-wide characterization of mRNA transcript levels during the cell cycle of the budding yeast S. cerevisiae, which it was found to have a cell cycle–dependent periodicity for 416 of the 6220 monitored transcripts. More than 25% of the 416 genes were found directly adjacent to other genes in the genome that displayed induction in the same cell cycle phase. A short description of the experimental conditions is given in Table 1.

Short Description	17 time points of expression data for synchronized yeast cells
Conditions	YPD + adenine; grown to 8e6 cells/ml at 37 degrees C for 165 minutes, then at 25 degrees
Long Description	Cell-cycle arrested yeast cells are synchronized by being held at a restrictive temperature and then released at t=0. RNA expression via Affymetrix chips is measured every 10 minutes until t=160. Data up to t=40 reflects temperature-induced as well as cell-cycle effects.

Table 1. Experimental conditions [3]

The result of the experiments is an expression data matrix of dimension $N \times M$ consisting of a dataset for N genes whose expression ratios are measured under M different perturbation conditions (Fig. 1). Normally, this matrix contains gene expression levels measured relative to a common baseline condition in logarithmic scale (base 10) with a well-defined dynamic range (e.g. [-2, 2]). The associated data error matrix contains the estimated errors of the data. The logarithmic expression ratio and the error for gene i measured in experiment j are denoted by x_{ij} and σ_{ij}, respectively. In most cases, a reduced data matrix and its associated data error matrix both of dimension $N \times M$ are obtained by applying some sort of significance filtering on the original dataset. This type of filtering is usually mentioned as "significance cuts".

2.2. Preprocessing

According to the "significance cuts" the criterion $|x_{ij}| \geq x_{th}$ applied to the logarithmic ratios, for $i = 1,..., N$ and $j = 1,..., M$ where the thresholds x_{th} must be specified. If a pair (i, j) meets the criterion above, gene i is said to be significant in experiment j. Gene i is selected if there are more experiments in which this gene is significant than a threshold number. Likewise, experiment j is selected if there is a number of genes larger than a threshold which is significant in this experiment. As a result of this choice, the original data matrix of dimension $N \times M$ is reduced to a data matrix of dimension $n \times m$ with $n \leq N$ and $m \leq M$.

In addition to the "significance cuts" another technique has been chosen to repair the missing values that appear in an expression data matrix. A typical microarray consists of thousands of spots and not all of these yield usable data. Sometimes chip imperfections cause defective spots. The KNN-impute algorithm is used to repair the data matrix, filling the missing data values. It does so by looking through the entire dataset for spots with similar expression profiles and then calculating the missing data as an appropriately weighted sum. KNN finds the closest K neighbors in the matrix to row r using the Euclidean distance. The method also returns the actual distances. Any row with missing values is not used [4]. The method also returns the actual distances. Any row with missing values is not used in the computation.

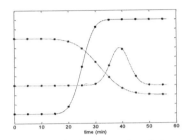

Figure 1 The expression levels of three genes, during a seven step experiment

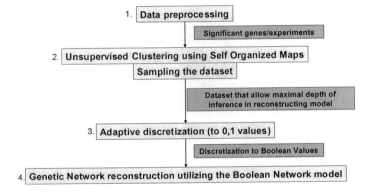

Figure 2 Stages of the proposed method.

2.3. *Unsupervised Clustering using SOM*

We have used Self Organized Maps (SOM)[5,6], an unsupervised clustering algorithm, to produce clusters with co-expressed genes that have related functions. We then create datasets that will allow maximal depth of inference in reconstructing the model.

SOM is quite a unique kind of neural network in the sense that it constructs a topology preserving mapping from the high-dimensional space into map units in such a way that relative distances between data points are preserved. The map units, or neurons, usually form a two-dimensional regular lattice where the location of a map unit carries semantic information. The SOM can thus serve as a clustering tool of high-dimensional data. It is also easy to visualize it due to its typical two-dimensional shape. A real vector describes the set of input samples $x_i \in R^n$ (the observation vectors), where i is the index of the sample, (the discrete-time coordinate). Each node i in the map contains a model vector $m_i(t)$, which has the same number of elements as the input vector. The model vector learning is achieved according to the following formula,

$$m_i(t+1) = m_i(t) + h_{c(x),i} * (x_i(t) - m_i(t)) \tag{1}$$

where $h_{c(x),i}$ is the neighborhood function chosen to be Gaussian.

$$h_{c(x),i} = a(t) * \exp(-\frac{\| r_i - r_c \|}{2\sigma^2(t)}), 0 < a(t) < 1 \tag{2}$$

and r_c is the location of unit c.

The map size has dimensions 45x23. U-Matrix, (fig. 3) is a metric that visualizes the relative distances between neighboring model vectors.

U-matrix

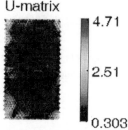

4.71

2.51

0.303

Figure 3 U.matrix describes the relative distances between model vectors

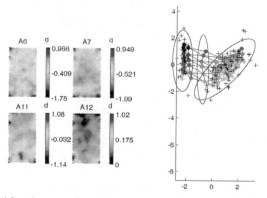

Figure 4 Sample maps and a principle component projection for a sample subset of the dataset

Dataset sampling follows clustering The criterion that is used to determine the number of randomly selected genes depends on the size of each cluster the number of total genes and the number of clusters that have been produced (small sets of co-expressed genes are ignore). The dataset sampling is achieved according to the following formula:

$$\# genes = \frac{size(Cluster)}{\# total\_genes}$$

Finally, clusters of co-expressed genes having related functions are selected in order to construct an accuracy model describing the relations between these genes.

2.4. *Modeling using Boolean Networks*

A Boolean network [9] is a 3-tuple $<G, V, F>$, where $G = <V, E>$, is a directed graph, V is a set of Boolean variables in a one-to-one correspondence with V, and $\{F : f_v : V \rightarrow \{0,1\} \mid v \in V\}$ is a set of Boolean functions. The arguments of f_v are the parents of v in the directed graph G (Fig. 5). When a node v has in-degree zero, the value for the Boolean variable v to 0 or 1 is usually fixed.

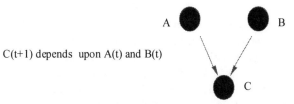

C(t+1) depends upon A(t) and B(t)

Figure 5 A simple Boolean network where component C depends on the state of B and A.

According to the Boolean logic we have to assume that for each gene there are two possible states: on and off. Our aim in that step is to descritize (assign the value 0 to 1) the expression levels of each gene for 17 time steps. A threshold t_i (fig 6.) different for each gene is used, which depends on

- gene's behavior during the experiment
- the expression level of other genes that describe the entropy of the system at the current time

We have propose an adaptive descritezation algorithm adjusting a different threshold for each time series data,

$$th_{it} = \left\{ T : x_{it} > th_{it-1} + a * \arg\left(\sum \frac{d\exp_j}{dt} \right) * \frac{dx}{dt} \right\} \text{ and } th_{i0} = \frac{1}{N} \sum_{t=0}^{n} x_{it}$$

where th_{it} is the threshold for gene i at time t.

Genes correspond to elements in a Boolean network, the wiring of the elements to one another correspond to functional links between genes, and the rules determine the result of a signaling interaction given a set of input values. Genes are idealized as being either on or off, resulting in binary elements interacting according to Boolean rules.

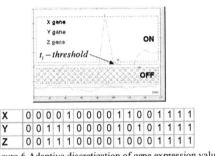

X	0	0	0	0	1	0	0	0	0	1	1	0	0	1	1	1	1
Y	0	0	1	1	1	0	0	0	0	1	0	1	0	1	1	1	1
Z	0	0	1	1	1	0	0	0	1	0	0	0	1	1	1	1

Figure 6 Adaptive discretization of gene expression values

We have used a simple but very effective algorithm (REVEAL) for inferring the relations between genes using experimental data. The concept behind this algorithm is that analyzing the mutual information between input states and output, we are able to infer the sets of input elements controlling each element or gene in the network. The REVEAL algorithm uses the mutual information measures to extract the wiring relationships from state transition tables. These directly lead to the look-up tables of the rules.

The REVEAL algorithm is based on the following criterion: if X is an input data stream and Y an output data stream then If $M(X,Y) = H(Y)$ then X exactly determines $Y M(X,Y) = H(X) + H(Y) - H(X,Y)$. is the mutual information is a natural measure of the dependence between random variables and $H(X) = -\sum_i p_i \log p_i$ is the Shannon entropy declaring the uncertainty, where p_i is the probability of observing a particular event within a given sequence.

The REVEAL algorithm performs well for networks of low k (number of inputs per gene). For higher value of k, the algorithm does not work well. The Boolean network models are based on the notion that biological networks can be represented by binary, synchronously updating switching networks. In real biological system, however, variables change continuously in time. Asynchronous Boolean networks can better approximate this behavior which captures the structure of logical switching networks. The REVEAL algorithm is limited to synchronous Boolean networks.

3. Results

The results of the method were compared with experimental data taken by Simon et al. 2001 [10]. In the experiments a dataset of 259 genes has been used and the produced clusters include 132, 54 and 73 genes, (Table 2).

Dataset	Cl1	Cl2	Cl3	Total
#genes	132	54	73	259
Annotated components	Cln1,Cln2,Ndd1	Swi4/Swi6	Mbp1,Ndd1	

Table 2 Data subsets produced by SOM

The produced network consists of 121 interactions, 87 of them were verified during experiments [10]
The results of our experiments indicate that the proposed methodology is very promising. However, it is plagued by false positives and noisy data. We have determined also several transcription factors. Some problems that have occurred caused by the limited number of connectivity between genes (k=4).

4. Conclusions

We have described a model for a genetic network modeling with Boolean approach. The specificity of the output network is improved using clustering algorithms which generate classes of co-expressed genes with similar functions. In this way we improved the modeling of gene regulatory networks and allowed maximal depth of inference in reconstructing model using datasets produced by unsupervised clustering techniques. The assumption that a gene can expressed only in two states (0, 1) may constraint the reliability of our model. It is highly possible that Boolean networks will be effective only for a very specific class of biological networks. An approach based on probabilistic reasoning could produce more reliable models and for that reason Bayesian Networks and Dynamic Bayesian Networks are currently tested for genetic network reconstruction.

References

1. P. D'haeseleer, S. Liang, and R. Somogyi, Genetic network inference: from co-expression clustering to reverse engineering, *Bioinformatics*, vol. 16, no. 8, pp. 707--726, 2000.
2. Z. Szallasi Genetic Network Analysis in Light of Massively Parallel Biological Data Acquisition. *Pacific Symposium on Biocomputing* 1999.
3. R. J. Cho, M. J. Campbell, E. A. Winzeler, L. Steinmetz, A. Conway, L. Wodicka, T. G. Wolfsberg, A. E. Gabrielian, D. Landsman, D. J. Lockhart, and R. W. Davis. A genomewide transcriptional analysis of the mitotic cell cycle. *Molecular Cell*, vol.2, July 1998.
4. O. Troyanskaya, M. Cantor, G. Sherlock, P. Brown, T. Hastie, R. Tibshirani, D. Botstein and R. B. Altman. Missing value estimation methods for DNA micorarrays. *Bioinformatics* 520-525 2001.
5. T. Kohonen, E. Oja, O. Simula, A. Visa, and J. Kangas, Engineering applications of the self-organizing map. *Proceedings of the IEEE*, 1358-1384, 1996
6. T. Kohonen, Self-Organizing Maps. *Springer*, Berlin 1995
7. S. Liang, REVEAL, A General Reverse Engineering Algorithm For Inference Of Genetic Network Architectures, *Pac. Symp. Biocomput.* 1998
8. Z. Szallasi, Genetic Network Analysis in Light of Massively Parallel Biological Data Aqcuisition, Pac. Symp. Biocomput. 1999
9. S. Kauffman, The Origins of Order, Oxford University Press. 1993
10. I. Simo, J Barnett, N Hannett, CT Harbison, NJ Rinaldi, TL Volkert, JJ Wyric, J Zeitlinger, DK Gifford, TS Jaakkola, RA Young "Serial regulation of transcriptional regulators in the yeast cell cycle", 2001 http://web.wi.mit.edu/young/cellcycle/

HISTORICAL TOPICS ON FUZZY SET THEORY IN BIOMEDICAL ENGINEERING

R. SEISING

Department of Medical Computer Sciences,

Section on Medical Expert and Knowledge-Based Systems, University of Vienna Medical School, Spitalgasse 23, A-1090 Vienna, Austria, e-mail:
rudolf.seising@akh-wien.ac.at

CH. SCHUH

Department of Medical Computer Sciences,

Section on Medical Expert and Knowledge-Based Systems, University of Vienna Medical School, Spitalgasse 23, A-1090 Vienna, Austria

K.-P. ADLASSNIG

Department of Medical Computer Sciences,

Section on Medical Expert and Knowledge-Based Systems, University of Vienna Medical School, Spitalgasse 23, A-1090 Vienna, Austria and Ludwig Boltzmann Institute for Expert Systems and Quality Management in Medicine, Spitalgasse 23, A-1090 Vienna, Austria

The beginning of biomedical engineering was a result of the rapid accumulation of data from medical research. This knowledge explosion in medicine stimulated the speculation that computers could be used to help in the field of medical diagnosis. In the 1950s biomedical investigators used automatic data processing techniques to study correlations of symptoms and signs with diseases. By intensive collaboration between physicians and mathematicians respectively electrical engineers or computer scientists both biology and medicine became, to a certain extend, quantitative sciences. In the second half of the 20th century medical knowledge was also stored in computer systems. To assist physicians in medical decision-making and in their patient care medical expert systems have been constructed that use the theory of fuzzy sets. The present article gives a brief history of applied fuzzy set theory in biomedical engineering, especially "fuzzy relations" and "fuzzy control". We present two Viennese systems representing these concepts: the "fuzzy version" of the Computer-Assisted DIAGnostic System (CADIAG) which was developed at the end of the 1970s, and a fuzzy knowledge-based control system, FuzzyKBWean, which was established as a real-time application based on the use of a Patient Data Management System (PDMS) in the intensive care unit (ICU) in 1996. Systems which combine medical knowledge-bases with a PDMS will include immense expert information for the assistance of physicians in their daily diagnostic and therapeutic routine work. Such systems are able to manage huge medical data because of the increasing quality of storage and database techniques since the last decade of the 20th century.

1. Introduction

Since the 1960s biomedical engineering plays an increasing role in medical practice. The beginning of biomedical engineering was a result of the rapid accumulation of data from medical research. This knowledge explosion in medicine stimulated the speculation that computers could be used to help in the field of medical diagnosis. In the 1950s biomedical investigators used automatic data processing techniques to study correlations of symptoms and signs with diseases. Computerized medical diagnosis was developed by intensive collaboration between physicians and mathematicians respectively electrical engineers or computer scientists. In the 1960s and 1970s various approaches to computerized diagnosis arose using Bayes rule [1, 2], factor analysis [3], and decision analysis [4]. On the other side artificial intelligence approaches came into use, e.g., DIALOG (*Diagnostic Logic*) [5] and PIP (*Present Illness Program*) [6], which were programs to simulate the physicians reasoning in information gathering and diagnosis using databases in form of networks of symptoms and diagnoses.

We use the term "symptom" for any information about the patient's state of health, such as patient history, signs, laboratory test results, ultrasonic results, and X-ray findings. Based on this information a physician has to find a list of diagnostic possibilities for the patient. To master this process he had to study many relationships of obligatory or facultative proving or excluding symptoms for diagnosis in books and journals and in his practical experience. This information about relationships that exist between symptoms and symptoms, symptoms and diagnoses, diagnoses and diagnoses and more complex relationships of combinations of symptoms and diagnoses to a symptom or diagnosis are formalizations of what is called medical knowledge.

2. Computers in Biomedical Engineering

After the Second World War biomedical investigators in the USA began to use automatic data processing techniques to study correlations of signs and symptoms with diseases. Martin Lipkin and James D. Hardy tried to classify and correlate all the data of a medical case (diagnosis of hematological diseases) with the use of a mechanical apparatus [7].

In order to record data from which the diagnoses of hematological diseases had been made, standard textbooks were consulted to choose 26 diseases and to list all characteristics of each disease. Storage and sorting of these medical data were first performed with the use of marginal punch cards. The periphery of these cards had been divided into numbered spaces, and a single hole was punched in each of the spaces. Each of the 138 available spaces had the significance of an item of information, and each card represents a given body of information. One margin of the card consisted of data derived from the case history and a second side consisted of data derived from physical examination. Information relating to peripheral blood examination was assigned to a third side, and bone marrow examination and other laboratory work to the fourth side.

The information of a given disease represented by one single card was transferred to the card in the following way. Where a given positive finding had been previously listed for a disease, a triangular wedge was punched in the appropriate space on the disease card.

"Thus, if one wished to find all the diseases characterized by a single item found under physical examination, for example 'large spleen', one would place this set of cards front to back and place a metal or plastic rod into the hole to which that item had been assigned. When the rod was raised, cards representing diseases characterized by a large spleen would fall, because the triangular wedge would have been punched into that space. Cards without the wedge would be raised" ([7], pp.116 f).

In April 1955, when Lee Browning Lusted, one of the later well known founders of medical informatics and medical decision making, was an instructor at the University of California School of Medicine and assistant at the University of California Hospital, he wrote in the *New England Journal of Medicine*: "Members of the medical profession and electronic engineers have shown increasing interest in the field of medical electronics, with the result that electronic instrumentation has contributed to recent advances in some fields of medicine. Greater application of electronic instrumentation to medical problems will result from teamwork between the engineer and physician" ([8], p. 584).

In 1959, Lusted and Robert S. Ledley, an electrical engineer at George Washington University, Washington D. C. and the later founding president of the National Biomedical Research Foundation (NBRF), published their Science paper *Reasoning Foundations of Medical Diagnosis* [9] where they introduced a logical analysis of the medical diagnosis process. They saw logical concepts inherent 1) in medical knowledge, that is the physicians knowledge about relationships between the symptoms and the diseases, 2) in the signs and symptoms presented by a particular patient, which give further information associated with this patient, and 3) in the final medical diagnosis itself. Ledley and Lusted applied the symbolism associated with the propositional calculus of symbolic logic:

Let x, y, ... represent 'attributes' a patient may have (e.g. x represents the sign 'fever', y represents the disease 'pneumonia'). Let corresponding capital letters X, Y, ... represent statements about these attributes. For example: Y represents the statement "The patient has attribute y.", then the *negation* $\neg Y$ represents the statement "The patient does *not* have attribute y." The combination $X \cdot Y$ represents the combined statement "The patient has the attribute x *and* the attribute y.", and the combination $X+Y$ represents the combined statement "The patient has attribute x *or* attribute y *or* both." The statement "If the patient has attribute x *then* he has attribute y" is symbolized by $X \Rightarrow Y$. Ledley and Lusted defined with only two attributes, symptoms (S) and diseases (D):

$S(i)$ means "The patient has symptom i."	$i = 1, ..., n,$
$D(j)$ means "The patient has disease j."	$j = 1, ..., m.$

From a diagnostic textbook they took abstract example statements:

IF a patient has disease 1 *and not* disease 2,
THEN he cannot have symptom 2. $\qquad D(1) \cdot \neg D(2) \Rightarrow \neg S(2)$

IF a patient has *either or both* of the symptoms,
THEN he must have one *or* both of the diseases. $S(1) + S(2) \Rightarrow D(1) + D(2)$

To consider, in general, more than two attributes, and more complicated expressions Ledley and Lusted used 'Boolean functions' $f(X, Y, ...)$.

After the section on logical concepts Ledley and Lusted continued with a section on probabilistic concepts. They argued that in many cases our medical knowledge is not exact but in the form "If a patient has disease 2, then there is only a certain chance that he will have symptom 2 — that is, say, approximately 75 out of 100 patients will have symptom 2. [...] Since »chance« or »probabilities« enter into »medical knowledge«, then chance, or probabilities, enter into the diagnosis itself" ([9], p. 13). Six years later, it seemed that Lusted had given up the program to use methods of exact mathematics in medicine; in his contribution to a volume on *Computers in Biomedical Research* [10] he was agreeing with a very new claim: "Research on medical diagnosis has served to emphasize the need for better methods of collecting and coding medical information and to demonstrate the inadequacy of conventional mathematical methods for dealing with biological problems. In a recent statement Professor L. A. Zadeh (1962) summed up the situation as follows: »In fact, there is a fairly wide gap between what might be regarded as 'animate' system theorists and 'inanimate' system theorists at the present time, and it is not at all certain that this gap will be narrowed, much less closed, in the near future. There are some who feel that this gap reflects the fundamental inadequacy of the conventional mathematics — the mathematics of precisely-defined points, functions, sets, probability measures, etc. — for coping with the analysis of biological systems, and that to deal effectively with such systems, which are generally orders of magnitude more complex than man-made systems, we need a radically different kind of mathematics, the mathematics of fuzzy or cloudy quantities which are not describable in terms of probability distributions. Indeed, the need for such mathematics is becoming increasingly apparent even in the realm of inanimate systems, for in most practical cases the *a priori* data as well as the criteria by which the performance of a man-made system is judged are far from being precisely specified or having accurately-known probability distributions.«" ([11], p. 321).

3. Fuzzy Sets in Medicine

About two years after his demand quoted by Lusted, Zadeh founded the theory of fuzzy sets. In contrast to usual set theory an object can not only be an element of a set (membership value 1) or not an element of this set (membership value 0) but it can also have a membership value between 0 and 1. Therefore he defined fuzzy sets by their *membership function* μ which is allowed to assume any value in the interval [0,1] instead of their *characteristic function* which is defined to assume the values 0 *or* 1 only (fig. 1) [12].

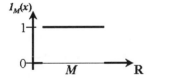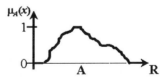

Figure 1: characteristic function of a crisp set M ; membership function of a fuzzy set A.

Relating to fuzzy sets A, B in any universe of discourse X, Zadeh defined *equality*, *containment*, *complementation*, *intersection*, and *union* (for all $x \in X$):

- $A = B$ if and only if $\mu_A(x) = \mu_B(x)$,
- $A \subseteq B$ if and only if $\mu_A(x) \leq \mu_B(x)$,

- ¬ A is the complement of A if and only if $\mu_{\neg A}(x) = 1 - \mu_A(x)$,
- $A \cup B$ if and only if $\mu_{A \cup B}(x) = \max(\mu_A(x), \mu_B(x))$,
- $A \cap B$ if and only if $\mu_{A \cap B}(x) = \min(\mu_A(x), \mu_B(x))$.

The space of all fuzzy sets in X becomes a Boolean algebra; thus, a propositional logic with fuzzy concepts constitutes "fuzzy logic".

In 1969, Zadeh suggested applications of fuzzy sets in medical science: "A human disease, e.g., Diabetes, may be regarded as a fuzzy set in the following sense. Let $X = \{x\}$ denote the collection of human beings. Then diabetes is a fuzzy set, say D, in X, characterized by a membership function $\mu_D(x)$ which associates with each human being x his grade of membership in the fuzzy set of diabetes" ([13], p. 205).

Four years later, Zadeh defined *fuzzy* relations: If $L(A \times B)$ is the set of all fuzzy sets in the Cartesian product $A \times B$ of crisp sets A and B, then a fuzzy relation is a subset of $L(A \times B)$ [14]. Having three sets A, B, and C, to compose fuzzy relations $Q \subseteq L(A \times B)$ and $R \subseteq L(B \times C)$ to get another fuzzy relation $T \subseteq L(A \times C)$, Zadeh introduced the combination rule of a *max-min-composition*: $T = Q \cdot R$ is defined by the following membership function

$$\mu_T(x,y) = \max_{y \in B} \min\{\mu_Q(x,y); \mu_R(y,z)\}, \ x \in A \ y \in B \ z \in C.$$

Medical Knowledge Network. A Database for Computer Aided Diagnosis was the title of Alonzo Perez-Ojedas master thesis in 1976. The basic conception of his work was the representation of "medical knowledge" using an associative model of the human memory. He designed a prototype system to be used in the search for an adequate strategy to simulate an approximate reasoning model in medical decision-making. Some examples are ([15], p. 3.2):

- A runny nose is *almost always* present in a common cold.
- Acute pyelonephritis *usually* presents bladder irritation and infection.
- Acute pyelonephritis presents *occasionally* fever, or chills, and malaise.

The diseases *common cold* and *acute pyelonephritis* are presented by the abbreviations D_1 and D_2 and *runny nose, fever, bladder irritation, infection, chills,* and *malaise* by S_1 to S_6. Therefore the network of medical knowledge could be graphically constructed by elementary knots and arcs. However, Perez-Ojeda modeled the relations (*usually, occasionally,* and *almost always*) by mathematical probability modifiers:

$$D_1 \xrightarrow{\text{almost always}} S_1$$

$$D_2 \xrightarrow{\text{usually}} S_3 \text{ AND } S_4$$

$$D_2 \xrightarrow{\text{occasionally}} (S_2 \text{ OR } S_5) \text{ AND } S_6$$

Figure 2: Examples of elements of the network of medical knowledge.

A more far-reaching concept of modeling relationships between symptoms and diseases was introduced in the 1970s by Elie Sanchez ([16]), "to investigate medical aspects of fuzzy relations at some future time" ([17], p. 47). In 1979 he introduced the relationship between symptoms and diagnoses by the concept of 'medical knowledge': "In a given pathology, we denote by S a set of symptoms, D a set of diagnoses and P a set of patients. What we call "medical knowledge" is a fuzzy relation, generally denoted by R, from S to D expressing associations between symptoms, or syndromes, and diagnoses, or groups of diagnoses" ([18], p. 438). Sanchez adopted Zadeh's max-min-compositional rule as an inference mechanism. It accepts fuzzy descriptions of the patient's symptoms and infers fuzzy descriptions of the patient's diseases by means of the fuzzy relationships described earlier. If a patient's symptom is S_i then the patient's state in terms of diagnoses is a fuzzy set D_j with the following membership function:

$$\mu_{D_j}(d) = \max_{s \in S} \min\{\mu_{S_i}(s); \mu_R(s,d)\}, \quad s \in S, d \in D.$$

$\mu_R(s,d)$ is the membership function of the fuzzy relation "medical knowledge".

With P, a set of patients, and a fuzzy relation Q from P to S, and by 'max-min composition' we get the fuzzy relation $T = Q \cdot R$ with membership function

$$\mu_T(p,d) = \max_{s \in S} \min\{\mu_Q(p,s); \mu_R(s,d)\}, p \in P, s \in S, d \in D.$$

In the nineteen-sixties and -seventies, the Department of Medical Computer Sciences of the University of Vienna Medical School at the Vienna General Hospital envisaged the development of a computer-assisted diagnostic system that did not use stochastic methods [19]. By intensive collaboration between physicians and mathematicians, and engineers a first computer-assisted diagnostic system based on two-valued logic arose in 1968 [20]. In 1976, the second generation of the system was developed on the basis of three-valued logic. Here, in addition to symptoms and diagnoses being considered to be 'present' or 'absent', 'not examined' or 'not investigated' symptoms and 'possible' diagnoses are also included. For this system known as CADIAG-I (*Computer-A*ssisted *DIAG*nosis, version I), the following relationships between symptom (S_i) and disease (D_j) have been defined:

OP: S_i is *obligatory occurring and proving* for D_j.
E: S_i forces *obligatory exclusion* of D_j.
FP: S_i is *facultative occurring and proving* for D_j.
ON: S_i is *obligatory occurring and not proving* for D_j.
FN: S_i is *facultative occurring and not proving* for D_j.
NK: A relationship between the symptom and the disease is *not known*.

With three-valued logic these relationships could be expressed in the form of three-valued logic operators: the symptom's values could be *present* (1), *absent* (0), or *not investigated* (½), whereas the possible diagnoses' values could be *present* (1), *absent* (0), or *possible* (½).

A third version of the computer-assisted diagnostic system, the "fuzzy version" CADIAG-II, appeared in 1980. In Klaus-Peter Adlassnig's *Fuzzy Logical Model of Computer-Assisted Medical Diagnosis* [21], all symptoms $S_i \in S$ are considered to be fuzzy sets of different universes of discourse X with membership functions $\mu_{S_i}(x)$, for all $x \in X$, indicating the strength of x's affiliation in S_i, while all diagnoses $D_j \in D$ are considered to be fuzzy sets in the set P of all patients under consideration, with $\mu_{D_j}(p)$ assigning the patient p's membership to be subject to D_j.

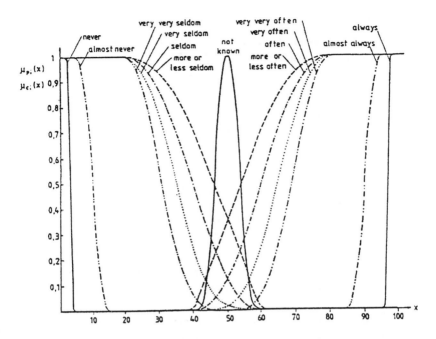

Figure 4: Membership functions of the fuzzy sets *occurrence o (former presence p)*and
confirmability (former conclusiveness c) ([21], p. 145)

Adlassnig considered two relationships between symptoms and diseases, namely *occurrence* — how often does S_i occur with D_j ? — and *confirmability* — how strongly does S_i confirm D_j? — ([22], p. 225). These functions could be determined by linguistic documentation by medical experts and database evaluation by statistical means or a combination of both. In both ways to determine these fuzzy relationships, *occurrence* and *confirmability*, they have been defined as fuzzy sets. When physicians had to specify these relationships by only giving answers like *always, almost always, very often, often, unspecific, seldom, very seldom, almost never*, and *never*, they chose membership functions of fuzzy sets. In the case of medical databases, the membership functions' values of *occurrence* and *confirmability* could be defined as relative frequencies.

Thus, in CADIAG-II, the fuzzy relationships between symptoms (or symptom combinations) and diseases are given in the form of rules with associated fuzzy relationship tupels (*frequency of occurrence o, strength of confirmation c*); their general formulation is ([23], p. 262):

IF antecedent THEN consequent WITH (o, c)

CADIAG-II was very successful in partial tests, e.g., in a study of 400 patients with rheumatic diseases, it elicited the correct diagnosis in 94.5 % ([23], p. 264). More results can be found in [22, 23].

4.   Fuzzy Control and Medical Decisions

The earliest applications of the theory of fuzzy sets are fuzzy controllers that work rather different than conventional controllers. Using the "rule of max-min composition" as an inference rule, Sedrak Assilian and Ebrahim Mamdani developed the concept of *fuzzy control* in the early 1970s [24, 25]. To describe the inputs and outputs of the system, fuzzy sets are used and to describe the fuzzy control strategy, fuzzy-IF-THEN rules are used. If the controller further adjusts the control strategy without human intervention, it is *adaptive*. The adaptive *fuzzy* controller, invented by Assilian and Mamdani, is known as the *self-organizing fuzzy controller*. An adaptive controller is a controller with adjustable parameters and a mechanism for adjusting the parameters [26]. Despite the lack of a formal definition, an adaptive controller has a distinct architecture consisting of two loops: a control loop and a parameter adjustment loop (see fig. 5).

In other words, fuzzy control is based on an I/O function that maps each very low-resolution quantization interval of the input domain into a very low-low resolution quantization interval of the output domain. As there are a few fuzzy quantization intervals covering the input domains, the mapping relationship can be very easily expressed using the IF-THEN formalism. (In some applications this leads to a simpler solution in less designing time.) The overlapping of these fuzzy domains and their usually linear membership functions will eventually allow a rather high-resolution I/O function between crisp input and output variables to be achieved. Mamdani's development of fuzzy controllers in 1974 [25] gave rise to the utilization of these fuzzy controllers in ever-expanding capacities.

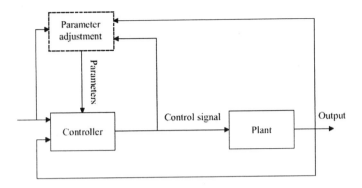

Figure 5: Adaptive control system.

The branch of fuzzy control is being implemented in medical application systems since the 1990's, as real-time applications are being adequately executed by computers since this time. Fuzzy control techniques have recently been applied in various medical processes, such as pain control [27] and blood pressure control [28]. Scientists and physicians at the University of Vienna Medical School and the Vienna General Hospital established the fuzzy knowledge-based control system FuzzyKBWean as a real-time application, based on the use of a *Patient Data Management System* (PDMS) in the intensive care unit (ICU) in 1996.

In other words, fuzzy control can be best applied to production tasks that heavily rely on human experience and intuition, and which therefore rule out the application conventional control methods. The use of PDMS in an ICU since 1992 has made it possible to apply fuzzy control applications in real-time in this medical field.

Mechanical ventilation is such an example. One purpose of mechanical ventilation is to achieve optimal values of arterial O_2-partial pressure (pO_2) and arterial CO_2-partial pressure (pCO_2) while ensuring careful handling of the lung:

- $FiO_2 < 60$ (else oxygen toxicity)
- low inspiratory pressures $P_I < 35$ (else barotrauma)
- small shear forces equivalent to small tidal volumes (else volume trauma)
- prevent atelectasis formation (else shear forces at reopening)

In addition, the patient has to be carefully handled in order to avoid cardiac failure and respiratory muscle fatigue. Both of these conditions have to be observed if the heart rate or the respiratory rate increases. The main physiological input parameters of the weaning system are pO_2 and SpO_2.

580

For instance, the *Biphasic Positive Airway Pressure* (BIPAP) controlled mode is an integrated mode of ventilation of Evita ventilators (Evita, Dräger, Lübeck, Germany). This mode allows spontaneous inspiration during the whole respiratory cycle and thus permits a very smooth and gradual transition from controlled to spontaneous breathing. Ventilatory adjustments are based on two pressure levels: inspiratory pressure P_I (PIP) and expiratory pressure P_E (PEEP); on two durations, inspiration time t_I and expiration time t_E, as well as on the fraction of inspired O_2 (F_IO_2). Within this mode, five parameters can be adjusted: P_I, P_E, t_I, t_E, and F_IO_2.

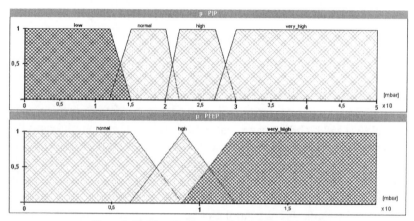

Figure 6: Membership functions for P_E (PIP) and P_I (PEEP).

FuzzyKBWean is a system for weaning patients from artificial ventilation [29]. Although many such expert systems have been described, only a few have been tested in clinical patient care. For example, studies of computer-controlled optimization of positive end-expiratory pressure and computerized protocols for the management of adult respiratory distress syndrome were explored by East and Bohm [30]. A computerized ventilator weaning system for postoperative patients was tested by Strickland and Hasson [31] and Schuh et al. [29].

The procedure for weaning a patient with respiratory insufficiency from mechanical ventilation is a complex control task and requires expertise based on long-standing clinical practice. Fuzzy knowledge-based weaning (Fuzzy-KBWean) is a fuzzy knowledge-based control system that proposes stepwise changes in ventilator settings during the entire period of artificial ventilation at the bedside in real time.

Information is obtained from a PDMS operating at the ICU with a time resolution of one minute. The system is used for postoperative cardiac patients at the Vienna General Hospital. A large part of the explicitly given and implicitly available medical knowledge of an experienced intensive care specialist could be transferred to the fuzzy control system. Periods of deviation from the target are shorter using FuzzyKBWean.

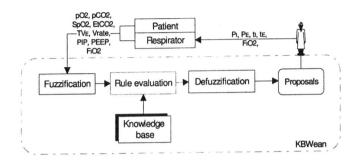

Figure 7: The FuzzyKBWean control process.

5. Outlook

In medicine, two fields of fuzzy applications were developed in the nineteen-seventies: computer assisted diagnostic systems and intelligent patient monitoring systems. Both developments of Zadeh's "rule of max-min composition", namely fuzzy relations and fuzzy control, have been applied in these areas.

For obvious reasons, the available body of medical data (on patients, laboratory test results, symptoms, and diagnoses) will expand in the future. As mentioned earlier, computer-assisted systems using fuzzy methods will be better able to manage the complex control tasks of physicians than common tools. Most control applications in the hospital setting have to be performed within critical deadlines. Decisions have to be made locally and promptly. This is a setting that requires a local hospital intranet rather than the possibilities of the world-wide internet. Using current web technology, integrated systems of both types of fuzzy systems described above can be easily implemented as internet and intranet applications.

582

References

1. Wardle, A., Wardle, L.: Computer Aided Diagnosis — A Review of Research. *Methods of Information in Medicine*, 3, 1976, pp. 174-179.
2. Woodbury, M. A.: The Inapplicabilities of Bayes Theorem to Diagnosis. *Proc. Fifth Int. Conf. on Medical Electronics*. Liege, Belgium. Springfield, Ill., Charles C. Thomas, 1963, pp. 860-868.
3. Überla, K.: Zur Verwendung der Faktorenanalyse in der medizinischen Diagnostik. *Methods of Information in Medicine*, 2, 1965, pp. 89-92.
4. Ledley, R. S., Lusted, L. B.: Medical Diagnosis and Modern Decision Making. In: Bellman, R. (ed.): *Mathematical Problems in the Biological Sciences*, Proceedings of Symposia in Applied Mathematics, 14, pp. 117-158, American Mathematical Society, Providence, R.I., 1962.
5. Pople, H. E., Myers, J. D., Miller, R. A.: DIALOG: A Model of Diagnostic Logic for Internal Medicine. 4. International Joint Conference on Artificial Intelligence, Tblis: USSR, Sept. 3-8, 1975, pp. 848-855.
6. Parker, S. G., Gorry, G. A., Kassiner, J. P., Schwartz, W. B.: Towards the Simulation of Clinical Cognition: Taking a Present Illness by Computer. *American Journal of Medicine*, June 1976, Vol. 60, pp. 981-996.
7. Lipkin, M., Hardy, J. D.: Mechanical Correlation of Data in Differential Diagnosis of Hematological Diseases. *Journal of the American Medical Association*, 166 (1958), pp. 113-125.
8. Lusted, L. B.: Medical Electronics. *The New England Journal of Medicine*, April 7, 1955, pp. 580-585.
9. Ledley, R. S., Lusted, L. B.: Reasoning Foundations of Medical Diagnosis. *Science*, 3. July 1959, Vol. 130, Nr. 3366, pp. 9-21.
10. Lusted, L. B.: Computer Techniques in Medical Diagnosis. In: Stacy, R. W., Waxmann, B. D. (Eds): *Computers in Biomedical Research* Volume I, New York, London: Academic Press 1965, pp. 319-338.
11. Zadeh, L. A.: From Circuit Theory to System Theory. *Proceedings of the IRE*, Vol. 50, 1962, p. 856 ff
12. Zadeh, L. A.: Fuzzy Sets. *Information and Control*, 8, 1965, pp. 338-353.
13. Zadeh, L. A.: Biological Applications of the Theory of Fuzzy Sets and Systems. *The Proceedings of an International Symposium on Biocybernetics of the Central Nervous System*. Little, Brown and Company: Boston 1969, pp. 199-206.

14. Zadeh, L. A.: Outline of a New Approach to the Analysis of Complex Systems and Decision Processes. *IEEE Transactions on Systems, Man, and Cybernetics, SMC-3*, No. 1, January 1973, pp. 28-44.

15. Perez-Ojeda, A.: *Medical Knowledge Network. A Database for Computer Aided Diagnosis*. Master Thesis, Department of Industrial Engineering, University of Toronto. 1976.

16. Sanchez, E.: *Equations de Relations Floues*. Thèse Biologie Humaine, Faculté de Médecine de Marseille. 1974.

17. Sanchez, E.: Resolution of Composite Fuzzy Relation Equations. *Information and Control, 30*, 1976, pp. 38-48.

18. Sanchez, E.: Medical Diagnosis and Composite Fuzzy Relations. Gupta, M. M.; Ragade, R. K.; Yager R. R. (Eds.): *Advances in Fuzzy Set Theory and Applications*. Amsterdam: North-Holland 1979, pp. 437-444.

19. Adlassnig, K.-P., Grabner, G.: The Viennese Computer Assisted Diagnostic System. Its Principles and Values. *Automedica*, 1980, Vol. 3, pp. 141-150.

20. Spindelberger, W. und Grabner, G.: *Ein Computerverfahren zur diagnostischen Hilfestellung*. In: K. Fellinger (Hrsg.): *Computer in der Medizin – Probleme, Erfahrungen, Projekte*. Wien: Verlag Brüder Hollinek 1968, pp. 189-221.

21. Adlassnig, K.-P.: A Fuzzy Logical Model of Computer-Assisted Medical Diagnosis. *Methods of Information in Medicine, 19*, 1980, pp. 141-148.

22. Adlassnig, K.-P.: CADIAG-2: Computer-Assisted Medical Diagnosis using Fuzzy Subsets. Gupta, M. M., Sanchez, E. (Eds.): *Approximate Reasoning in Decision Analysis*. New York: North-Holland Publ. Company, 1982, pp. 219-242.

23. Adlassnig, K.-P.: Fuzzy Set Theory in Medical Diagnosis. *IEEE Transactions on Systems, Man, and Cybernetics, SMC-16*, No. 2, March/April 1986, pp. 260-265.

24. Assilian, S., Mamdani, E.: Learning Control Algorithms in Real Dynamic Systems. *Proc.4th International IFAC/IFIP Conference on Digital Computer Applications to Process Control*, Zürich, March 1974.

25. Mamdani, E.: Application of Fuzzy Algorithms for Control of Simple Dynamics Plant. *Proceedings of the IEEE, 121*(12), 1974, pp. 1585-1888.

26. Åström, K. J., Wittenmark, B.: *Adaptive Control*. Englewood Cliffs, NJ, Addison-Wesley, 2nd ed., 1995.

584

27. Carollo, A., Tobar, A., Hernandez, C.: A Rule-Based Postoperative Pain Controller: Simulation Results. *International Journal Bio-Medical Computing, 33*, 1993, pp. 267-276.
28. Ying, H., McEachern, M., Eddleman, D. W., Sheppard, L. C.: Fuzzy Control of Mean Arterial Pressure in Postsurgical Patients with Sodium Nitroprusside Infusion. *IEEE Transactions on Biomedical Engineering, 39*, 1992, pp. 1060-1069.
29. Schuh Ch., Hiesmayr M., Katz E., Adlassnig K.-P., Zelenka, Ch., Klement E. P.: Integration of Crisp and Fuzzy - Controlled Weaning in an ICU PDMS (FuzzyKBWean). *Proceedings of the World Automation Congress - WAC' 98*, Albuquerque, TSI Press, 1998, pp. 583-588.
30. East, T., Bohm, S.: A Successful Computerized Protocol for Clinical Management of Pressure Control Inverse Ratio Ventilation in ARDS Patients. *Chest, 101*, 1992, pp. 697-710.
31. Strickland, J., Hasson, J.: A Computer-Controlled Ventilator Weaning System. *Chest, 100*, 1991, pp. 1096-1102.

A NEW NEURAL NETWORK-BASED METHOD FOR FACE DETECTION IN IMAGES AND APPLICATIONS IN BIOINFORMATICS[&]

IOANNA-OURANIA STATHOPOULOU AND GEORGE A. TSIHRINTZIS

Department of Informatics, University of Piraeus,Piraeus 185 34,Greece

The rapid and successful detection of a face in an image is a prerequisite to a fully automated face recognition system. In this paper, we present a new neural network–based face detection system, which arises from the outcome of a comparative study of two neural network models of different architecture and complexity. The fundamental difference in the construction of the two models lies in the need to address the face detection problem either by using a general solution based on the full-face image or by composing the solution through the resolution of specific characteristics of the face. The proposed algorithm is based on the assumption that there exists the contrast in brightness between specific regions of the human face. The proposed neural network system is reliable and of reduced error rate. Specifically, we show that the second approach, even though more complicated, exhibits better performance in terms of detection and false - positive rates. Moreover, it can detect successfully faces that are slightly rotated out of the image plane. Although, we apply the proposed method to face detection, the same approach can be followed in similar problems arising in bioinformatics and computer vision technologies.

1. Introduction

The goal of face detection is to determine whether or not there are any faces in the image and, if so, return the location and extent of each face. This problem is quite challenging because faces are non-rigid and have a high degree of variability in size, shape, color and texture. Furthermore, variations in pose, facial expression, image orientation and conditions add to the level of difficulty of the problem.

To address this problem, a number of works have appeared in the literature [e.g., 1-9] in which three main approaches can be identified. These approaches use (1) correlation templates, (2) deformable templates and (3) image invariants, respectively. In the first approach, we compute a difference measurement between one or more fixed target patterns and candidate image locations and the output is thresholded for matches. The deformable templates of the second approach are similar in principle to

---

[&] Partial support for this work were provided by the University of Piraeus Research Center.

correlation templates, but not rigid. To detect faces, in this approach we try to find mathematical and geometrical patterns that depict particular regions of the face and fit the template to different parts of the images and threshold the output for matches. Finally, in approaches based on image invariants, the aim is to find structural features that exist even when pose, viewpoint and lighting conditions vary, and then use them to detect faces.

Most of the aforementioned methods limit themselves to dealing with human faces in frontal view and suffer from several drawbacks: (1) they cannot detect a face that occupies an area smaller than 50 * 50 pixels or more than 3 faces in complex backgrounds or faces in images with defocus and noise problems or faces in side view and (2) they cannot address the problem of partial occlusion of mouth or wearing sunglasses. Although there are some research results that can address two or three of the aforementioned problems, there is still no system that can solve all of them.

In this paper, a new and efficient human face detection system is proposed that combines artificial neural networks and image invariants approaches. More specifically, the paper is organized as follows: in Section 3, we present the proposed face detection algorithm, which is based on the artificial neural network structures of Section 4. In Section 5, we make an evaluation of the performance of the two networks, while in Section 6 we present some conclusions and point to future work.

2.   Proposed Face Detection Algorithm

Our proposed algorithm falls within the third approach mentioned above. Specifically, we defined certain image invariants and used them to detect faces by feeding them into an artificial neural network. These image invariants were found based on the 14*16 pixel ratio template, proposed by P. Sinha, [10,11], which we present below.

2.1. *P. Sinha's Template*

The method proposed by P. Sinha [10,11] combines template matching and image invariant approaches. P. Sinha aimed at finding a model that would satisfactorily represent some basic relationships between the regions of a human face. To be more specific, he found out that, while variations in illumination change the individual brightness of different parts of faces (such as eyes, cheeks, nose and forehead), the relative brightness of these parts remains unchanged. This relative brightness between facial parts is captured by an appropriate set of pairwise brighter - darker relationships between sub-regions of the face.

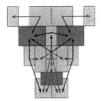

We can see the proposed template in the adjacent Figure 1, where we observe 23 pairwise relationships represented by arrows. The darker and brighter parts of the face are represented by darker and brighter shades of grey, respectively.

Figure 1: The P. Sinha's Template

Our proposed face detection algorithm is built on this model. We preprocess a candidate image in order to enhance the relationships mentioned above and then, use the image as input to an artificial neural network. Then, the neural network will determine whether or not there is a face in the image.

2.2. The Algorithm

The main goal is to find the regions of the candidate image that contain human faces. The proposed system uses Artificial Neural Networks and operates in two modes: training the neural network and using it to detect faces in a candidate image.

To train the neural network, we used a set of 285 images of faces and non-faces. We tried to find images of non-faces that are similar to human faces, so some of the non-face images contained dog, monkey and other animal "faces". These images where gathered from sources of the World Wide Web [12] and preprocessed before entered into the neural network.

To detect faces in a candidate image we apply a window, which scans the entire image, and preprocess each image region, the same way we preprocessed the images of the training set. Specifically, our algorithm works as follows:

1. We load the candidate image. It can be any 3- dimensional (color) image
2. We scan through the entire image with a 35*35 pixel window. The image region defined by the window constitutes the "window pattern" for our system, which will be tested to determine whether it contains a face. We increase the size of the window gradually, so as to cover all the possible sizes of a face in a candidate image.
3. We preprocess the "window pattern":
    3.1 We apply Histogram Equalization techniques to enhance the contrast within the "window pattern".

3.2 We compute the eigenvectors of the image using the Principal Component Analysis and the Nystrom Algorithm [13,14,15,16] to compute the normalized cuts.

3.3 We compute three clusters of the image using the k-means algorithm and color each cluster with the average color.

3.4 We convert the image from colored to grayscale (2D).

4. We resize the processed image into a dimension of 20*20 pixels and use it as input to the artificial neural network, which we present in the next section.

We summarize the basic stages of image processing in Table 1.

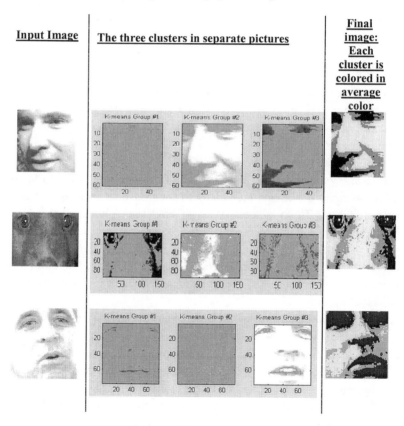

Table 1: The stages of the preprocess of the widow pattern

3. The Artificial Neural Network Structures

To classify window patterns as "faces" or "non-faces", we developed two different artificial neural networks, which are presented next.

3.1. *The First Artificial Neural Network*

This network takes as input the entire window pattern and produces a two-dimensional vector output. The network consists of three hidden layers of thirty, ten and two neurons respectively, as in Figure 2, and the input (window pattern) is of dimension 20-by-20 pixels. The neural network classifies the window pattern as 'face' or 'non-face'. The output vector is of dimension 2-by-1 and equals to [1;0] if the window pattern represents a face or [0;1], else wise.

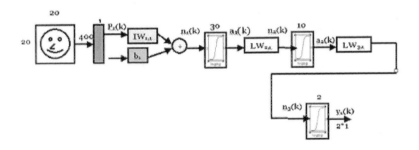

Figure 2: The First ANN's Structure

3.2. *The Second Artificial Neural Network*

The second neural network has four hidden layers with one, four, four and two neurons, respectively, and is fed with the input data: (1) the entire "window pattern" (20-by-20 pixels), (2) four parts of the "window pattern", each 10-by-10 pixels and (3) another four parts of the "window pattern", 5-by-20 pixels. Each of the three types of inputs is fed into different hidden layers of the network. The first, second, and third sets of inputs are fed into the first, second, and third hidden layer, respectively, while the output vector is the same as for the first network. Clearly, the first network consists of fewer hidden layers with more neurons and requires less input data compared to the second.

590

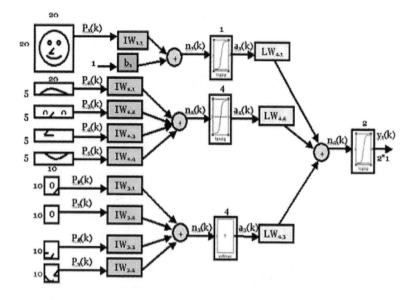

Figure 3: The Second ANN's Structure

4. Performance Evaluation

To train these two networks, we used a common training set of 285 images of faces and non-faces. During the training process, the first and second networks reached an error rate of 10^{-1} and 10^{-10}, respectively.

We required that, for both networks, the output vector value be close to [1;0] when the window pattern represented a face and [0;1] otherwise. This means that the output vector corresponds to the degree of membership of the image in one of the two clusters: "face-image" and "non-face-image". Some results of the two neural networks can be seen in Table 2. The first network, even though it consisted of more neurons than the second one, did not detect faces in the images to a satisfactory degree, as did the second network. On the other hand, the execution speeds of two networks are comparable. Therefore, the second network was found superior in detecting faces in images. Some results of the face detection system are depicted in Figure 4.

FACE IMAGES			
Input image	**Pre-processed window pattern**	**First ANN's output**	**Second ANN's output**
		[0.5 ; 0.5]	[0.947; 0.063]
		[0.6 ;0.4]	[1 ; 0]
		[0.5 ; 0.5]	[0.9717;0.0283]
NON-FACE IMAGES			
		[0.5 ;0.5]	[0 ; 1]
		[0.5 ;0.5]	[0 ; 1]
		[0.5 ; 0.5]	[0 ; 1]

Table 2: Results of the two neural networks for various images

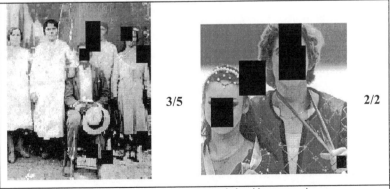

Figure 4: Application of the proposed algorithm on two images.
The first image is an old one of low quality. The system managed to the detect 3 out of the 5 faces. The second image is of better quality and our system detected 2 out of 2 faces.

5. Summary, Conclusions and Future Work

The algorithm that we presented in the present work detected the faces in images satisfactorily. Occasionally, errors (e.g., failure to detect all faces in an image) occurred, especially with faces whose characteristics were not so clear. We believe, however, that the system performance will improve by increasing the training set of the network.

We plan to extend our work in the following three directions: (1) We will improve our system by using a training set that covers a wider range of poses and cases of low quality of images. (2) We will extend our system so as to implement the ability of facial expression analysis. (3) The approach presented in here will be followed in similar problems that arise in bioinformatics and computer vision systems, where the goal is to detect objects with similar characteristics.

References

[1]   A. J. Colmenarez and T.S. Huang, "Face detection with information-based maximum discrimination", Computer Vision and Pattern Recognition, 782–787 (1997).

[2]   G. Yang and T.S. Huang, "Human face detection in a complex background", Pattern Recognition, 27(1): 53–63 (1994)

[3]   S.Y. Lee, Y.K. Ham, R.H. Park, "Recognition of human front faces using knowledge-based feature extraction and neuro-fuzzy algorithm", Pattern Recognition 29,1863-1876 (11) (1996)

[4]   T.K. Leung, M.C. Burl, and P. Perona, "Finding faces in cluttered scenes using random labeled graph matching", *Fifth International Conference on Computer Vision, pages 637–644, Cambridge, Massachusetts, IEEE Computer Society Press* (1995).

[5]   H.A. Rowley, S.Baluja and T. Kanade, "Rotation Invariant Neural Network-Based Face Detection" CMU-CS-97-201 (1997)

[6]   H.A. Rowley, S.Baluja and T. Kanade, "Neural Network-based face detection", IEEE Transactions on Pattern Analysis and Machine Intelligence, 20(1) (1998).

[7]   P. Juell and R. Marsh, "A hierarchical neural network for human face detection", Pattern Recognition 29 (5), 781-787. (1996)

[8]   C. Lin and K. Fan, "Triangle-based approach to the detection of human face", Pattern Recognition 34, 1271-1284 (2001).

[9] K.K. Sung, T. Poggio, "Example-based learning for view-based human face detection", *Proceedings on Image Understanding Workshop, Monterey, CA, 843-850* (1994).

[10] P. Sinha, "Object Recognition via image-invariants"

[11] M.–H. Yang, N. Ahuja, "Face Detection and Gesture Recognition for Human- computer Interaction", Kluver Academic Publishers

[12] Gender Classification (Databases)
http://ise0.stanford.edu/class/ee368a_proj00/project15/intro.html
http://ise0.stanford.edu/class/ee368a_proj00/project15/append_a.html

[13] C. Fowlkes, S. Belongie and J. Malik, " Efficient Spatiotemporal Grouping Using the Nystrom Method"

[14] S. Belongie, "Notes on Clustering Pointsets with Normalized Cut" (2000).

[15] J. Shi and J. Malik, "Normalized Cuts and Image Segmentation", IEEE Transactions On Pattern Analysis and Machine Intelligence, Vol. 22(8) (2000).

[16] Image Segmentation using the Nystrom Method
http://rick.ucsd.edu/~bleong/

AUTOMATED DIAGNOSIS AND QUANTIFICATION OF RHEUMATOID ARTHRITIS USING MRI

EVANTHIA E. TRIPOLITI and DIMITRIOS I. FOTIADIS

Unit of Medical Technology and Intelligent Information Systems, Department of Computer Science, University of Ioannina and Institute of Biomedical Research - FORTH, GR 451 10 Ioannina, Greece

MARIA ARGYROPOULOU

Department of Radiology, Medical School, University of Ioannina, GR 451 10 Ioannina, Greece

Rheumatoid arthritis (RA) is a common chronic inflammatory joint disease which is characterized by persistent synovitis and frequent development of cartilage and bone deteriorations. A major problem in the management of RA is the lack of reliable methods for monitoring joint inflammation, disease progression, and therapeutic response. In this work we propose an automated method for the diagnosis and quantification of rheumatoid arthritis using magnetic resonance images of the hand. The technique relies on the use of fuzzy C-mean algorithm for image segmentation. The number of clusters is determined by minimizing the appropriate validity index. Each vector, used as an input, has four features related to the intensity and position of pixels. A defuzzification process follows the segmentation. Small objects and objects corresponding to vessels are eliminated. The method is evaluated using data from twenty three patients. The obtained sensitivity and positive predictive rate of the proposed method are quite high.

1. Introduction

Today medical imaging technology provides the clinician with a number of complementary diagnostic tools such as x-ray, computer tomography (CT), magnetic resonance imaging (MRI) and positron emission tomography (PET). Routinely these images are interpreted visually and qualitatively by expert radiologists. The conventional methods such as clinical examination, laboratory tests, and radiography (which only shows preceding synovitits), are neither sensitive not specific for diagnosis of rheumatoid arthritis and do not allow detailed evaluation of synovial inflammation [1].

Over the past few years, evidence accumulated shows that MRI is an advanced medical imaging technique providing rich information about the human soft tissues anatomy. MRI facilitates dimensional measurements by eliminating magnification and morphological distortion caused by projectional radiography and increases sensitivity for detecting bone erosions by eliminating superimposition of overlapping structures, which can obscure abnormalities on radiographs. Also, it allows direct visualization of marrow edema/inflammation, synovial tissue and effusion, articular cartilage, tendons and tendon sheaths and intra-articular and peri-articular ligaments, which cannot be detected by x-rays. This makes whole organ assessment of the joint possible. These advantages can be leveraged in clinical trials of rheumatoid arthritis to extend the scope of structural assessments and increase discriminative power [1].

However, the amount of data is far too much for manual analysis/interpretation, and this has been one of the major obstacles in the efficient use of MRI. For this reason, automated or semi-automated computer aided image analysis techniques are necessary. Those are based on the segmentation of MR images into different tissue classes. One of the main problems in image segmentation is uncertainty. Some of the sources of this uncertainty include noise, imprecision in computations and vagueness in class definitions. Traditionally, probability theory was the primary mathematical tool used to deal with uncertainty problems; however, the possibility concept introduced by the fuzzy set theory was gained popularity in modeling and propagating uncertainty in imaging applications [2].

We have developed an automated method for the diagnosis and quantification of rheumatoid arthritis. The method is based on the segmentation of MR images using a fuzzy C-mean schema, which is based on four features related to the image intensity and position of pixels. This is an innovative feature of our system since the fuzzy C-mean algorithm is a common tool in the analysis of MR images of brain using only the image intensity. Segmentation is applied in each MR slice and is followed by a defuzzification process and removal of small objects which correspond to vessels. The findings of the analysis are used as an input in a 3D reconstruction procedure to quantify rheumatoid arthritis. The method is evaluated using a dataset of 279 images.

2. Materials and Methods

2.1. *Dataset*

The MR images were acquired using a Philips Gyroscan ACS-NT imager with fat suppression T1 weighted gadolinium enhanced, spin echo sequence. The mechanism that was used for fat suppression was Spectral Presaturation with Inversion Recovery (SPIR). The hand and wrist was imaged into 12 coronal slices 3mm thick and 0.3mm spaced. The field of view was 21.8x21.8cm and the resulting images were coded into a 256x256 matrix.

The parameters that were used are TR/TE=590ms/15ms, and flip angle 90°. We collected 279 images from 23 patients, which have been annotated for the presence of rheumatoid arthritis by two experts.

2.2. *Image analysis*

The proposed method consists of four stages: preprocessing, determination of image clusters, segmentation and postprocessing.

2.2.1. *Preprocessing*

In this stage each pixel of the image is described by four features, which are: intensity, the average intensity of a 3x3 window whose center is the pixel, the Euclidean distance of the pixel from the center of the wrist and a feature indicating the region where the pixel belongs (wrist, tendon sheaths, metacarpophalangeal joints, proximal and distal interphalangeal joints). Initially, the region of the hand is determined and the coordinates of the region of the three anatomical points of the hand are computed.

The informative marks are removed setting the intensity of the pixels, located in certain areas, equal to zero. All pixels with intensity less than 25 are set to zero. This results in a binary image. Neighboring white pixels with connectivity eight are grouped together to form objects corresponding to the hand region or to motion artifacts. The largest obtained object corresponds to the hand region, whose empty areas are filled using a flood fill algorithm [7]. The perimeter of the hand area can be found easily and it is used for the computation of three anatomical points (subchondral plate of the distal end of the radius - A, the base of the third metacarpal bone - B, distal margin of the third metacarpal bone - C).

The determination of point A is based on the fact that at this point the length of the line segment, which is defined as the distance between the corresponding edges of the perimeter of the hand and is parallel to the x axis of the image, decreases and remains constant around the wrist. The coordinates of point A (A_x, A_y), which is located in the center of the line segment are determined. The determination of points B and C is based on anatomical observations reported in [6]. According to those the carpal height (i.e. the distance from the base of the third metacarpal to the distal articular surface of the radius along the axis of the third metacarpal) ranges from 26.4 - 40.5 mm for mature humans (ages from 25-60). We locate two points that their Euclidean distances from point A are equal to 26.4 mm (point B_1) and 40.5 mm (point B_2), respectively. The length of the line segment connecting B_1 and B_2 is computed. Point B is determined when this length decreases abruptly and then increases again. The coordinates of $B(B_x, B_y)$ are the coordinates of the center of this line segment. The length of the third metacarpal bone (i.e. the distance between point B and point C) ranges from 55.3 - 73.2 mm [6]. We compute the two points with Euclidean distances from point B 55.3 mm (point C_1) and 73.2 mm (point C_2), respectively. The line segment connecting C_1 and C_2 is drawn. Point C corresponds to a position with less than four intersections of this line segment with the perimeter of the hand, when this is moved parallel to its initial position.

The position of a pixel in the hand region is described by its fourth feature. For a pixel belonging to the wrist this feature is equal to one, to the metacarpophalangeal joints this feature is equal to 2, to the proximal and distal interphalangeal joints this feature is equal to 3, to the tendon sheaths this feature is equal to four, otherwise is set equal to zero.

The wrist area is defined as the ellipse with center located at

$$center_x = B_y + \frac{|B_y - A_y|}{2} - 5,$$

$$center_y = B_x + \frac{|B_x - A_x|}{2} \text{ if } B_x < A_x, \qquad (1)$$

$$center_y = A_x + \frac{|B_x - A_x|}{2} \text{ if } B_x > A_x.$$

The minor axis of the ellipse is given as:

$$minor\_axis = A_y - B_y + 30, \qquad (2)$$

and the major axis is the width of the wrist.

The metacarpophalangeal joint regions are defined as circles with centers satisfying the following conditions:

$$center_{x1} = C_x - 2 * \frac{maxdist}{5} \text{ and } C_y < center_{y1} < C_y + 35,$$

$$center_{x2} = C_x - \frac{maxdist}{5} \text{ and } C_y < center_{y2} < C_y + 35,$$

$$center_{x3} = C_x \text{ and } center_{y3} = C_y, \qquad (3)$$

$$center_{x4} = C_x + \frac{maxdist}{5} \text{ and } C_y < center_{y4} < C_y + 35,$$

$$center_{x5} = C_x + 2 * \frac{maxdist}{5} \text{ and } C_y < center_{y5} < C_y + 35,$$

where maxdist is the maximum distance between the corresponding edges of the perimeter of the hand, in the x-direction of the hand. The centers of the circles must lie on the upper half of an ellipse whose center is given as:

$$\text{center}_x = \text{edge}_l + \frac{\text{maxdist}}{2},$$
$$\text{center}_y = C_y + 35, \tag{4}$$

where edge_l is the x coordinate of the left edge of the perimeter of the hand, where the maximum distance was computed. The major axis of the ellipse is given as:

$$\text{major\_axis} = \text{maxdist}, \tag{5}$$

and the minor axis is given as:

$$\text{minor\_axis} = C_y + 70. \tag{6}$$

The area of the distal and proximal interphalangeal joints is defined as the pixels whose coordinates satisfy the following conditions:

$$1 < x < 255,$$
$$\text{firstpix}_y < y < C_y - 75, \tag{7}$$

where firstpix_y is the pixel of the perimeter of the hand region with the smallest y coordinate (Fig. 1). The tendon sheaths area includes the pixels not belonging to one of the above areas and are located between metacarpophalangeal and wrist joints.

2.2.2. Optimal number of clusters

The fuzzy C-mean (FCM) algorithm is well known for its efficiency in clustering large datasets [3]. Although it requires several parameters, the most significant one affecting its performance is the number of clusters c. Different choices of c may lead to different clustering results. Thus, the estimation of the optimal cluster number during the clustering process is important. We determined the number of clusters using an improved version of the method proposed in [4], where instead of the mean intra cluster distance we use the fuzzy mean intra cluster distance. In this way we can find the optimal cluster number with the smallest value of $\upsilon_{SV}(\cdot)$

for $c = 2$ to c_{max}. The validity index $\upsilon_{SV}(\cdot)$ is given as the sum of the normalized under and over partition measure functions $\upsilon_{uN}(c, V, X)$ and $\upsilon_{oN}(c, V)$, respectively, where V is the prototype matrix and X the dataset. The under partition measure function is the mean of fuzzy mean intra cluster distance over the cluster number c. The over partition measure function is given as the ratio of the cluster number to minimum inter cluster distance [4].

2.2.3. *Fuzzy C-mean algorithm*

The standard FCM objective function for partitioning a set of N unlabeled column vectors in R^p into c clusters is given as:

$$J_m = \sum_{i=1}^{c} \sum_{k=1}^{N} u_{ik}^m \| x_k - v_i \|^2, \tag{8}$$

where p is the number of features in each vector x_k, $1 \le k \le N$ and $\{v_i\}_{i=1}^{c}$ are the prototypes of the clusters. In our case p is equal to four and N is equal to 65536. u_{ik} represents the membership value of feature vector x_k in cluster i. The following conditions must hold [5].

$$0 \le u_{ik} \le 1 \qquad \forall i, k,$$

$$\sum_{i=1}^{c} u_{ik} = 1 \qquad \forall k, \tag{9}$$

$$0 < \sum_{k=1}^{N} u_{ik} < N \quad \forall i.$$

The set of values satisfying the above conditions can be arranged in a matrix format $U[c \times N]$. The parameter m is the weighting exponent on each fuzzy membership, it determines the amount of fuzziness of the resulting classification and is equal to two. The FCM objective function is minimized when high membership values are assigned to pixel data that is close to the centroid of its particular class. Low membership values are assigned to pixel data located far from the centroid. The output of the algorithm is the matrix U.

The defuzzification process follows to determine the crisp clusters. We form as much images as the number of clusters. Each column vector x_k is assigned to a cluster (image) with maximum membership value. From the resulting images we select the one with maximum intensity.

2.2.4. *Postprocessing*

In this module we eliminate all the objects not corresponding to inflammation such as vessels, taken into account their size, shape and position. The intensity of pixels of vessels that are close to the perimeter of the hand region, is set equal to zero using morphological operators. The dilation operation (with a squared structure element whose width is 2 pixels) is applied three times on the image of the perimeter and then the resulted image is projected on the image that depicts the inflammation and the vessels. One important characteristic of some vessels is that they have line shape. We compute the eccentricity for all objects of image and we remove those with eccentricity equal to one. This results to the elimination of line shaped vessels. We eliminate also objects with size less than 10 pixels, which do not belong to a region of interest (wrist, metacarpophalangeal, proximal and distal interphalangeal and tendon sheaths). The small vessels which exist in the tendon sheath region are eliminated. Since the inflammation that progresses in this region is larger in size compared to the vessels, we remove all small objects that their distance from all larger objects is more than 15 pixels.

The image that results from the previous step depicts the inflammation and perhaps some remaining vessels. Then, the inflammation can be quantified summing the volumes from all slices.

3. Results

The method was evaluated using 279 images corresponding to 23 patients (18 women, age 23-75 and 5 men, age 34-61). The regions identified are shown in Table 1.

Table 1. Results

# of patients	# of images	# of regions defined by the physician	True Positives	False Negatives	False Positives
23	279	466	443	23	98

The obtained sensitivity is 96.78% and the positive predictive rate is 82.15%. Segmented images of the hand are shown in Fig. 1.

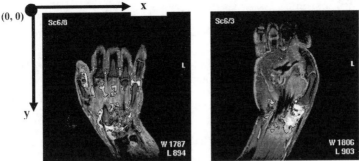

Figure 1. Segmented images of hand

In some cases, the method does not detect all regions of interest since they correspond to old inflammation and are clustered wrongly (Fig. 2).

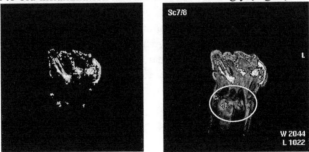

Figure 2. (a) Image after FCM, (b) Missed area.

4. Conclusions

We have developed a method for segmentation and quantification of rheumatoid arthritis using coronal spin echo MR images of hand. The method relies on the use of fuzzy C-mean algorithm for the segmentation. The method was evaluated using 279 images and the computed sensitivity and positive predictive rate were quite high. The method can be used for monitoring joint inflammation, disease progression and therapeutic response.

References

1. Klarlund, M., Ostergaard, M., Jensen, K. E., et al., *Ann. Rheum. Dis.*, 59, 512-528 (2000).
2. N. A. Mohamed, M. N. Ahmed and A. A. Farag, "Modified fuzzy C-mean in medical image segmentation", *Proc. of IEEE-EMBS*, vol 20, part 3, pp 1377- 1380 (1998).
3. J.C. Bezdek, *Pattern recognition with fuzzy objective function algorithms*, New York (1981).
4. Do-Jong KIM, Yong-Woon PARK, Dong-Jo PARK, "Novel Validity Index for Determination of the Optimal Number of Clusters", *IEIC TRANS. INF. & SYST.*, vol.E84-D, no.2 (2001).
5. M.T. Firgueiredo, J.M. Leitao, "Bayesian Estimation of Ventricular Contours in angiographic Images", *IEEE Trans. Med. Imag.*, vol. 11, pp. 416-429 (1992).
6. FA Schuind, RL Linscheid, KN An, EY Chao, "A normal data base of posteroanterior roentgenographic measurements of the wrist", *The Journal of Bone and Joint Surgery*, vol. 76, no. 9, 1418-1429 (1992).
7. Foley, J.D., van Dam, A., Feiner, S.K., Hughes, J.F., 1990., Computer Graphics: Principles and Practice, 2nd edn., Addison-Wesley, Reading, MA, 1174pp.

COMPUTATIONAL TIME-REVERSAL IMAGING VIA SUBSPACE METHODS *

G.A. TSIHRINTZIS

Department of Informatics, University of Piraeus, Piraeus 18534, GREECE
E-mail: geoatsi@unipi.gr

The problem of estimating the location of one or more coherent point scatterers from backscattered wideband (narrowband) multi-static data is formulated as a subspace problem of three (two)-way arrays. Its solution is readily provided by application of well-known high-resolution subspace algorithms.

1. Introduction

Time-reversal of wave pulses has been utilized as a self-correcting method for ameliorating the degradation in focusing that wave pulses suffer as they propagate through spatially inhomogeneous lossless media [1,2,3,4,5,6,7,8,9,10]. In this technique, the invariance of the wave equation under a time-reversal operation[a] is exploited in a three-step procedure which allows focusing on a target embedded in a lossless inhomogeneous medium. First, a pulse is transmitted through the inhomogeneous medium to the target, which scatters off the pulse. Next, the backscattered wave pulse is recorded by appropriate devices at the receiver. In the third step, the received wave is reversed in time and transmitted to the target through the medium. The time-reversed pulse focuses on the target[b] It has been shown both theoretically [1] and experimentally [4,5,6,7,8,9,10] that in the case of a weakly scattering medium that satisfies the first Born approximation, the time-reversal procedure compensates exactly for the distortion caused by the medium inhomogeneity. It has also been shown that under relatively mild conditions, the time-reversal procedure eliminates distortions caused by the

---

*Partial support for this work was provided by the University of Piraeus Research Center.
[a]In simple terms, this invariance implies that if the function $\psi(x, t)$ satisfies (is a solution to) the wave equation, so does the time-reversed function $\psi(x, -t)$.
[b]In the case of more than one targets in the medium, the above procedure has to be repeated several times for better focusing.

medium inhomogeneities even for stronger scatterers which do not satisfy the conditions for validity of the first Born approximation but rather the entire corresponding Born series converges [2].

Recently, there has been significant interest in developing computational signal processing methodologies that somehow create the effect of imaging targets via a time-reversal process without actually performing the three-step procedure of pulse transmission, recording and re-transmission. Such an attempt [11] combined the time-reversal concepts with well-known vector subspace methodologies for processing discrete-time data, namely the MU-SIC algorithm, for the purpose of target detection and identification from discrete backscattered narrowband multi-static measurements.

In the present paper, the formulation of the work in [11] is reviewed and its results are extended in two directions: (i) the case of wideband multistatic measurements is considered and the problem is posed as that of estimating the independent components of a three-way array and (ii) the assumption of validity of the first Born approximation is relaxed to the assumption of a convergent Born series. More specifically, the paper is organized as follows: In Section 2, the basic mathematical formalism of the time-reversal process is reviewed. In Section 3, the key findings of computational time-reversal with vector subspace methods are discussed and the present contribution is outlined. Finally, in Section 4, we summarize the paper and point to future work in this area.

2. Mathematical Foundation of Time-Reversal

A. Definitions
The time-reversal process is founded on the invariance of the wave equation under a time-reversal operation. Indeed, if $\psi(x,t)$ satisfies the wave equation

$$(\nabla^2 - \frac{1}{c^2}\frac{\partial^2}{\partial t^2})\psi(x,t) = 0, \tag{1}$$

the appearance of only second-order temporal derivatives in the equation implies that the time-reversed function $\psi(x,-t)$ satisfies it as well.

It is interesting to indicate the corresponding expression in the domain of the temporal Fourier transform of any solution $\psi(x,t)$ of the wave equation Eq.(1). Let

$$\psi(x,t) = \int_{-\infty}^{\infty} d\omega \, \Psi(x,\omega)e^{-i\omega t}. \tag{2}$$

From the properties of the Fourier transform

$$\psi(x, -t) = \int_{-\infty}^{\infty} d\omega \, \Psi^*(x, \omega) e^{-i\omega t}. \tag{3}$$

Eqs.(2-3) indicate the mathematical equivalence of the time-reversal process to the process of *phase conjugation* which is well-known in optics [1,2,3]. However, the mathematical equivalence does not automatically imply an equivalence in the physical realization of the two processes, as time-reversal is realized with relatively simple devices called *time-reversal mirrors*, while phase conjugation requires more complex non-linear devices [6].

B. Focusing with Time-Reversal

Consider an antenna array embedded in a lossless inhomogeneous medium. The antenna is excited by the transmit signal vector, which has a temporal Fourier transform $\underline{e}^0(\omega)$, and emits a pulse into the medium which is reflected back towards the antenna. The corresponding receive signal vector has temporal Fourier transform which is indicated as $\underline{r}^0(\omega)$ and relates to the transmit signal vector Fourier transform via a transfer matrix $\underline{\underline{K}}(\omega)$[c]

$$\underline{r}^0(\omega) = \underline{\underline{K}}(\omega)\underline{e}^0(\omega). \tag{4}$$

The time-reversal process is now applied, i.e. the transmit signal vector is time-reversed (its Fourier transform conjugated) and re-emitted. That is, the transmit signal Fourier transform is set to

$$\underline{e}^1(\omega) = \underline{r}^{0^*}(\omega) = \underline{\underline{K}}^*(\omega)\underline{e}^{0^*}(\omega) \tag{5}$$

and, therefore, the corresponding receive signal Fourier transform becomes

$$\underline{r}^1(\omega) = \underline{\underline{K}}(\omega)\underline{e}^1(\omega) = \underline{\underline{K}}(\omega)\underline{\underline{K}}^*(\omega)\underline{e}^{0^*}(\omega) \equiv \underline{\underline{T}}(\omega)\underline{e}^{0^*}(\omega). \tag{6}$$

The matrix $\underline{\underline{T}}(\omega) \equiv \underline{\underline{K}}(\omega)\underline{\underline{K}}^*(\omega)$ is called the *time-reversal matrix*, which is justified by the role it plays in the mathematical modeling of the time-reversal process.

If the time-reversal process is repeated n times, the relation in Eq.(6) is generalized as

$$\underline{r}^n(\omega) = \underline{\underline{T}}(\omega)^{\frac{n}{2}}\underline{e}^0(\omega), \tag{7}$$

when n is even, and

$$\underline{r}^n(\omega) = \underline{\underline{T}}(\omega)^{\frac{n-1}{2}}\underline{\underline{K}}^*(\omega)\underline{e}^{0^*}(\omega), \tag{8}$$

---

[c]The corresponding time-domain relation is a convolution w.r.t. the temporal variable of the transmit signal vector and the antenna impulse response.

when n is odd.

The time-reversal matrix is hermitian, therefore non-negative definite. Thus, the spectral theorem implies that for even n

$$\underline{\underline{T}}(\omega)^n = \lambda_1^n \underline{\mu}_1 \underline{\mu}_1^T + \cdots + \lambda_p^n + \underline{\mu}_p \underline{\mu}_p^T$$
$$\approx \lambda_1^n \underline{\mu}_1 \underline{\mu}_1^T, \tag{9}$$

where the superscript T denotes the transpose (without conjugation). In Eq.(9), $\lambda_1 > \cdots > \lambda_p$ are the p non-zero (positive) real eigenvalues of the time-reversal matrix and $\mu_1, \ldots, \mu_p$ are the corresponding eigenvectors and the approximate equality holds for a "large" number n of time-reversal iterations[d].

Eq.(9) is indicative of the focusing capabilities of the time-reversal matrix. Indeed, is seen that *in the midst of a number of scatterers, the time-reversal iterations allow selective focusing on the strongest of them.*

3. Computational Time-Reversal

A. Vector Notation

Let the antenna array consist of N dipoles at arbitrary locations Y_j, $j = 1, 2, \ldots, N$, in the three-dimensional space. We assume that the antenna is embedded in a lossless inhomogeneous medium consisting of $M < N$ point scatterers whose strengths at frequency ω are $\tau_m(\omega)$, $m = 1, 2, \ldots, M$. The antenna is excited by a transmit signal vector with Fourier transform $\underline{e}(\omega) = [e_j(\omega)]$. Its jth element, being a dipole, emits the signal with Fourier transform $\psi_j(x, \omega) = G(x, Y_j) e_j(\omega)$, which causes the formation of a scattered wave

$$\psi_j^s(x, \omega) = \sum_{m=1}^{M} G(x, X_m) \tau_m(\omega) G(X_m, Y_j) e_j(\omega), \tag{10}$$

where X_m, $m = 1, \ldots, M$, are the locations of the M scatterers.

In writing Eq.(10), it has been assumed that the scatterers are weak enough for the first Born approximation to hold. Assuming further that no multiple scattering takes place between scatterers, the scattered wave from simultaneous excitation of all the antenna dipoles is the superposition of

---

[d]A corresponding expression can be derived for an odd number n of time-reversal iterations.

the N terms ψ_j^s, i.e.,

$$\psi^s(x,\omega) = \sum_{j=1}^{N} \psi_j^s(x,\omega) = \sum_{j=1}^{N}\sum_{m=1}^{M} G(x,X_m)\tau_m(\omega)G(X_m,Y_j)e_j(\omega), \quad (11)$$

where G is the *inhomogeneous medium Green function*. Therefore, the receive signal Fourier transform at the antenna array will be $\underline{r} = [r_k(\omega)]$ with

$$r_k(\omega) = \psi^s(Y_k,\omega) = \sum_{j=1}^{N} \psi_j^s(Y_k,\omega)$$

$$= \sum_{j=1}^{N}\sum_{m=1}^{M} G(Y_k,X_m)\tau_m(\omega)G(X_m,Y_j)e_j(\omega). \quad (12)$$

Eq.(12) can be written compactly as an equivalent equation in the form of Eq.(4) with the introduction of the transfer matrix $\underline{\underline{K}}(\omega)$

$$\underline{\underline{K}}(\omega) = \sum_{m=1}^{M} \tau_m(\omega)\underline{g}_m(\omega)\underline{g}_m^T(\omega), \quad (13)$$

where $\underline{g}_m(\omega)$ is the *Green function vector*

$$\underline{g}_m(\omega) = [G(Y_k,X_m)]. \quad (14)$$

The time-reversal matrix that corresponds to the transfer matrix in Eq.(13) is readily computed as

$$\underline{\underline{T}}(\omega) = \sum_{m=1}^{M}\sum_{m'=1}^{M} [\tau_m^*(\omega)\tau_{m'}(\omega)\underline{g}_m^{*T}(\omega)\underline{g}_{m'}]\underline{g}_m^*(\omega)\underline{g}_{m'}^T(\omega), \quad (15)$$

B. Vector Subspace-Based Time-Reversal

It was shown in [11] that *the time-reversal matrix in Eq.(15) for an N element array embedded in a three-dimensional lossless homogeneous background medium containing $1 < M < N$ targets will have rank M under very mild conditions. Its rank will be less than M only under some particularly restrictions on the geometrical distributions of the targets, which cannot be realized with non-planar antenna array consisting of $N > 3$ elements.*

Additionally, a vector subspace methodology was developed in [11] for estimating the number $M < N$ of targets and their corresponding locations X_m, $m = 1,\ldots,M$ from knowledge of the time-reversal matrix. The algorithm was equivalent in form to the well-known MUSIC algorithm [12] of vector subspace signal processing and consisted of forming the *pseudo-spectrum*

$$S(x_p) = \frac{1}{\sum_{m=M+1}^{N} |\underline{\mu}_m^{*T}(\omega)\underline{g}_p(\omega)|^2},$$ (16)

where $\underline{\mu}_m(\omega)$ is an eigenvector of the time-reversal matrix corresponding to the zero eigenvalue and $\underline{g}_p(\omega)$ is the Green function vector in Eq.(14) evaluated at a location x_p, i.e., $\underline{g}_p(\omega) = [G(Y_1, x_p), G(Y_2, x_p), \ldots, G(Y_M, x_p)]^T$. As with the usual MUSIC algorithm, when the test location x_p is set equal to each of the target locations X_m, $m = 1, \ldots, M$, the denominator in Eq.(16) equals zero and, therefore, the pseudo-spectrum value is infinite. In practice, the pseudo-spectrum value will always be finite because of noise in the time-reversal matrix measurement and, therefore, the algorithm seeks the M peaks of the pseudo-spectrum and estimates the target locations as the corresponding peak locations.

C. Extensions to Vector Subspace-Based Time-Reversal

The vector subspace-based time-reversal methodologies can extended in two directions: (1) Their domain of validity exceeds that of validity of the first Born approximation and includes stronger scatterers for which the Born series converges. (2) Vector subspace-based time-reversal methodologies are developed for processing wideband signals. These methodologies are based on the formulation and eigenspace decomposition of three-way arrays.

4. Summary, Conclusions, and Future Research

In the present paper, the vector subspace-based formulation of time-reversal was reviewed and relevant results were extended in two directions: (i) the case of wideband multistatic measurements for which the problem of computational time-reversal limaging is posed as that of estimating the independent components of a three-way array and (ii) relaxation of the assumption of validity of the first Born approximation is relaxed to the assumption of a convergent Born series.

In the future, a detailed evaluation will be conducted of the performance of the proposed algorithm. Future relevant research may also follow the avenues of extending the proposed methodologies to extended (non-point) scatterers with possible multiple scattering between them. This and other research is currently under progress and its findings will be announced in the future.

References

1. Agarwal, G.S. and Wolf,E. (1982). Theory of phase conjugation with weak scatterers, J. Opt. Soc. Am., Vol. 72 p.321
2. Agarwal, G.S. and Wolf,E. (1982). Elimination of distortions by phase conjugation without losses or gains, Opt. Comm., Vol. 43, p.446
3. Agarwal, G.S., Friberg A.T., and Wolf,E. (1983). Scattering theory of distortion correction by phase conjugation, J. Opt. Soc. Am., Vol. 73, p. 529
4. Jackson, D.R. and Dowling, D.R. (1991). Phase conjugation in underwater acoustics, J. Acoust. Soc. Am., Vol. 89, p. 171
5. Prada, C., Wu, F. and Fink, M. (1991). The iterative time-reversal mirror: A solution to self-focusing in the pulse-echo mode, J. Acoust. Soc. Am., Vol. 89, p. 1119
6. Fink, m.(1993). Time-reversal mirrors, J. Phys. D: Appl. Phys., Vol. 26, p. 1333
7. Thomas, J.L., Roux, P. and Fink, M. (1994). Inverse scattering analysis with an acoustic time-reversal mirror, Phys. Rev. Letts., Vol. 72, p. 637, 1994
8. Prada, C., Thomas, J.L., and Fink, M. (1995). The iterative time-reversal process: Analysis of the convergence, J. Acoust. Soc. Am., Vol. 97, p. 62
9. Prada, C., Manneville, S., Spoliansky, D., and Fink, M. (1996). Decomposition of the time-reversal operator: Detection and selective focusing on two scatterers, J. Acoust. Soc. Am., Vol. 99, p. 2067
10. Fink, M. (March 1997). Time-reversed acoustics, Physics Today, p. 34
11. Devaney, A.J. (2003). Super-resolution using time-reversal and MUSIC, J. Acoust. Soc. Am., to appear
12. Stoica, P. and Moses, R. (1997). Introduction to Spectral Analysis, Prentice-Hall, New Jersey, USA

STATISTICAL PATTERN RECOGNITION-BASED TECHNIQUES FOR CONTENT-BASED MEDICAL IMAGE RETRIEVAL[*]

GEORGE A. TSIHRINTZIS

Department of Informatics, University of Piraeus, 80 Karaoli & Dimitriou Street, Piraeus 185.34, Greece

AGGELIKI D. THEODOSSI

Department of Informatics, University of Piraeus, 80 Karaoli & Dimitriou Street, Piraeus 185.34, Greece

Content-based image retrieval (CBIR) was proposed in the early 90's to overcome the difficulties associated with traditional manually annotated (*text-based*) image retrieval systems of the 70's and 80's. Specifically, in CBIR, images are indexed by their own visual content, that is by the distributions in them of features such as color, shape and texture (*general features*) or even human faces and fingerprints (*domain-specific features*). Features are represented by vectors of real numbers in a variety of ways, each emphasizing different aspects of them. Occasionally, the need arises for reduction of the dimensionality of the feature vector length in order to speed up the CBIR system performance. To define useful similarity measures for a feature, one needs to perform a study of the statistical distributions of the feature over a sufficiently broad collection of images and model them with useful statistical models. In this paper, we show that the class of alpha-stable distributions provide a valid model for the distribution of visual features in medical images and reflect the distribution of real data better than Gaussian models.

1. Introduction

During the last decade, we became witnesses to a revolution in the field of Information Systems, caused by the rapid expansion of the information volume and the amazing development of the Internet. This translated into a transition from storing text and numeric data in local databases to construction of large digital libraries,[%] [15;14] containing data of a variety of types and requiring the development of new management and retrieval

---

[*] Partial support for this work was provided by the University of Piraeus Research Center

[%] Such libraries include object-oriented system annals, territorial, image and multimedia databases, special purpose scientific databases, meta-data representations, full text systems, active systems, procedure and enterprise models, workflows, and knowledge based systems.

techniques in order to give their end user the ability to review and analyze information at multiple levels.

In this paper, we survey medical content-based multimedia information retrieval systems. More specifically, the paper is organized as follows: In the third section, we investigate traditional techniques of information retrieval from medical databases and libraries. We present workflow diagrams for content-based information retrieval from homogeneous and heterogeneous data in the fourth section of the paper. We study the statistical distribution of visual features in medical images in the fifth section of the paper and, finally, draw conclusions and point to future work in the sixth section.

2. Traditional Techniques of Information Retrieval

Traditional techniques of information retrieval rely on knowledge and data modeling and have been extensively used in object-oriented system annals, territorial, image, and multimedia databases, and special purpose scientific databases. Generally, there are five traditional search engine types with increasingly blurry borders: *full text*, *catalogue-based*, *meta-search*, *specialized* (*topic maps*) search engines and *dynamic clustering*. An important part of all these techniques is data classification and indexing.

In engines that perform full text search, one of the major difficulties is their obscureness to the semantics, that is the different meanings that a word often has in most languages or the use of the same word in different languages. Consequently, full-text searching often provides too many 'hits' for users, making it difficult for them to retrieve the information they need from a lump of irrelevant information [7].

The traditional way of finding information stored in large information sets is via structured catalogues similar to those used in libraries. Clearly, categorization is subjective since different people comprehend in different ways or have different opinions about the category that a document must be assigned to. Additionally, categorization becomes more complicated when it refers not to simple documents, but to multimedia texts, images and other items. Even though, data cataloguing in libraries has become very sophisticated, it is difficult to correlate items in catalogues using different schemes. As a result, efforts to standardize machine-readable cataloguing (MARC) in an internationally acceptable manner are not available.

Meta-search engines allow their users to make a keyword query to more than one search engine either sequentially or simultaneously. They combine the results by comparing them to both catalogue-based and full-text searches and give the user the ability to select the type of searching that will be executed. Additionally, it is possible to restrict searching to terms that occur in the header of an HTML file, newsgroup listing or e-mail message.

Topic maps, e.g., the ISO Topic Maps standard [ISO/IEC 13250:2000], improve the accessibility of information by facilitating and, to some extent, automating the task of providing navigational resources. They are designed to simplify groupware-supported production of data for which navigational aids such as indices, glossaries, tables of contents, lists and catalogues need to be generated.

Clustering has long been used in information retrieval for data analysis and classification and document visualization. It allows the construction of a taxonomy of a set of documents by forming groups of closely-related documents using a distance metric over the documents and viewing them as points in a metric space. Cluster hypothesis, on which this methodology is based, is resumed on the following sentence: *Documents relevant to a query tend to be more similar to each other than to irrelevant documents and hence are likely to be clustered together* [2].

3. Pattern Recognition-Based Information Retrieval

Research into techniques for content-based information retrieval is still in its infancy. In this section of the paper, we investigate how pattern recognition and image analysis techniques can respond to the needs of retrieval of information from multimedia.

A number of problems arise in the retrieval of information from multimedia data that traditional retrieval techniques are not able to cover, namely **multilinguality, indexing, heterogeneity of the data,** etc. These problems arise as consequence of the fact that different types of data, distributed over different Information Systems, require the storage of knowledge in different representation forms. Although the integration of different representation forms has been investigated for the storage and display of multimedia objects to be retrieved, the only solution so far was in the form of links between the objects or parts of them. [3;9;6]

Specifically, there are two different strategies (according to [6]), that lead to *knowledge representation* in databases: (1) *deductive databases*, which combine relational databases with logical knowledge representation techniques and (2) *semantic data models*, which aim at the integration of terminological models and integrity constrains of the databases. Apart from the above two strategies there are also *object oriented data models* that try to combine the previous two.

Contemporary image retrieval Information Systems use techniques borrowed from the field of Pattern Recognition, according to the workflow in the following figure. Here, we present the case of a Neural Network-based System which utilizes Radial Base Functions [8]. The user submits a query image to the system and then a feature extractor (a set of possible features, e.g. color, is included in Table 1) extracts the corresponding heterogeneous feature vector and forwards it to a database searcher. The database searcher compares each query feature vector with the corresponding feature vectors of database images and returns a metric value vector. Metric value vectors are forwarded to a combiner which is based on a Radial Basis Function Neural Network and yields the ranked list of the retrieved images. Next, the system receives user feedback for the results and trains the Radial Basis Function Network according to this feedback. The system returns to the metric value combiner until a positive user judgment is fed back. In Table 1, we have categorized and grouped various techniques for content-based retrieval that utilize techniques from the field of Pattern Recognition [3;4;8;15].

Table 1. Grouping various techniques for content-based retrieval
with Pattern Recognition Techniques.

FEATURE EXTRACTION		HIGH DIMENSIONAL INDEXING			
DOMAIN SPECIFIC	GENERAL	DIMENSION REDUCTION			
DETECTION	COLOR	KL TRANSFORM		CLUSTERING	
RECOGNITION	TEXTURE	EIGEN-IMAGE	PCA$^{\perp}$	RO W- WI SE	COLUMN-WISE
EXPRESSION	SHAPE				

$^{\perp}$ PCA: Principal Component Analysis

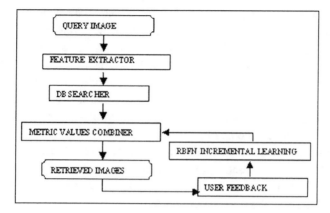

Figure 1. Radial Basis Function Neural Network where metric value vectors are forwarded.

4.   Statistical Feature Distribution

We have performed an extended study of the statistical distributions of visual features in medical images. We found that the class of alpha-stable distributions [12;13;10;11] provides a valid model for the distribution of these features and analyzed the consequences in the design of similarity measures for use in retrieval algorithms and high-dimensional indexing. In practical terms, the predicted probability that the features attain values that deviate from the average
by a large margin is significantly higher than the probability predicted by Gaussian models. This property of alpha-stable models that has been found to reflect the empirical distribution of real data better than Gaussian models.

Specifically, we examined the distribution of Laws's texture energy measure over a set of 54 X-rays of human organs (lungs, head, limbs) that we found in the Internet. Using known techniques, we fitted the feature distribution to an alpha-stable model with parameters: characteristic exponent alpha=0.85, dispersion gamma=7.26, and location parameter delta=0.5. The estimated value of the characteristic exponent indicates a feature distribution that deviates significantly from Gaussianity (corresponding to a characteristic exponent alpha=2) and rather resembles a Cauchy distribution (corresponding to a characteristic exponent alpha=1).

5. Conclusions and Future Work

In this paper, we surveyed medical content-based multimedia information retrieval systems and studied the statistical distributions of visual features in medical images. We found that the class of alpha-stable distributions provides a valid statistical model that reflects the empirical distribution of real data better than Gaussian models. The consequences of these findings on the design of similarity measures and retrieval algorithms are significant and are currently under evaluation. The results of this ongoing work will be presented on a future occasion.

References

[1] L. Amsaleg and P. Gros, "A Robust Technique to Recognize Objects in Images and the Database Problems it Raises," in: *INRIA Technical Report No 4081*, 2000.

[2] Charikar Moses, Chekuri Chandra, Feder Tomas, Motwani Rajeev, (1997). "Incremental Clustening and Dynamic Information Retrieval", 29th Symposium on Theory of Computing, 626-635.

[3] Day Michael, (1999). "Image retrieval: combining content-based and metadata-based approaches", 2ed UK Conference on Image Retrieval, Newcastle, U.K.

[4] Eakins P. John, (2002). "Towards Intelligent Image Retrieval", Pattern Recognition, Elsevier Vol. 35, 3-14.

[5] C. Faloutsos and K-I Lin, "Fastmap: A Fast Algorithm for Indexing, Data-Mining and Visualization of Traditional and Multimedia Datasets," in: Proceedings of SIGMOD, pp. 163-174, 1995.

[6] Fuhr Norbert, (1991). "An Information Retrieval View of Environmental Information Systems", International Conference on Data Base and Expert Systems Applications, Springer –Verlang, ISBN 3-211-82301-8.

[7] Haddouti Hachim, (1998). "Multilinguality Issues in Digital Libraries". EuroMed Net'98 Conference, Nicosia.

[8] Hyoung K. Lee, Yoo I. Suk, (2000). "Nonlinear Combining of heterogeneous features in content based image retrieval", SPIE^—s International Symposium on Intelligent Systems and Advanced Manufacturing, Vol.4197-32, 288-296, Boston, Massachusetts, USA.

[9] Johnson F.Neil, (1999). "In Search of the Right Image: Recognition and Tracking of Image Databases, Collections and The Internet", Technical Report for Center for Secure Information Systems, George Mason University.

[10] R. Kapoor, A. Banerjee, G.A. Tsihrintzis, and N. Nandhakumar, Detection of Targets in Heavy-Tailed Foliage Clutter Using an Ultra-WideBand (UWB)

Radar and Alpha-Stable Clutter Models, *IEEE Transactions on Aerospace and Electronic Systems*, vol. AES-35, pp. 819-834, 1999.

[11] N. Nandhakumar, J. Michel, D.G. Arnold, G.A. Tsihrintzis, and V. Velten, A Robust Thermophysics-Based Interpretation of Radiometrically Uncalibrated IR Images for ATR and Site Change Detection, *IEEE Transactions on Image Processing: Special Issue on Automatic Target Recognition*, vol. IP-6, pp. 65-78, 1997.

[12] G.A. Tsihrintzis, M. Shao, and C.L. Nikias, Recent Results in Applications and Processing of Alpha-Stable-Distributed Time Series, *Journal of the Franklin Institute*, vol. 333B, pp. 467-497, 1996.

[13] G.A. Tsihrintzis and C.L. Nikias, Alpha-Stable Impulsive Interference: Canonical Statistical Models and Design and Analysis of Maximum Likelihood and Moment-Based Signal Detection Algorithms, in: *C.T. Leondes (ed.), Control and Dynamic Systems: Advances in Theory and Applications*, Volume 78, San Diego, CA: Academic Press, Inc., 1996, pp.341-388.

[14] I. Papadakis and V. Chrissikopoulos, (2001). "Environmental Digital Libraries", *Ecological Protection of the Planet Earth I*, pp. 899-907. (eds. V. Tsihrintzis and P. Tsalidis), Xanthi, Greece, (2001).

[15] Rui Yong, Huang S. Thomas, and Chang Shih-Fu, "Image retrieval: current techniques, promising directions, and open issues", *Journal of Visual Communication and Image Representation*, vol.10, pp. 39-62, 1999.

A MOBILE TUTORING SYSTEM FOR MEDICINE

MARIA VIRVOU

Department of Informatics, University of Piraeus,
Piraeus 18534, Greece

EYTHIMIOS ALEPIS

Department of Informatics, University of Piraeus,
Piraeus 18534, Greece

Medical students have many learning and training obligations, ranging from clinical work at hospitals to reading theory and doing coursework. A similar situation is faced by medical instructors who are usually doctors that have to treat patients on top of their tutoring duties. In view of these obligations, the technology of mobile computing can assist considerably the medical educational process since both students and instructors may have access to educational software applications form anywhere at anytime through handheld devices of mobile phones. In this paper, we describe a mobile tutoring system for medicine, called Mobile Tutor, that has been designed in such a way that it can assist medical education while making use of the latest technological advances in mobile computing and educational software.

1. Introduction

In the last decade, Education has benefited a lot from the advances of Web-based technology. Indeed, there have been many research efforts to transfer the technology of ITSs and authoring tools over the Internet. A recent review (Brusilovsky, 1999) has shown that all well-known technologies from the areas of ITS have already been re-implemented for the Web. Some important assets include platform-independence and the practical facility that is offered to students of learning something at any time and any place.

However, in many cases it would be extremely useful to have such facilities in handheld devices, such as mobile phones rather than desktop or portable computers so that additional assets may be gained. Such assets include device independence as well as more independence with respect to time and place in comparison with web-based education using standard PCs. This is certainly the case for medical education due to the heavily loaded schedule of doctors-instructors and medical students. At the current state, there are not many mature mobile tutoring systems since the technology of mobile computing is quite recent and has not yet been used to the extent that it could. However, there have been quite a lot of primary attempts to incorporate mobile features to this kind of

618

educational technology and the results so far confirm the great potential of this incorporation. For example, Ketamo (2003) reports on an adaptive geometry game for handheld devices that is based on student modelling. Ketamo admits that the system developed was very limited and the observed behaviour was defined as very simple. However, an evaluation study that was conducted concerning this system showed that the learning effect was very promising with the handheld platform. A quite different approach is described in the system called KleOS (Vavoula, & Sharples, 2002), which allows users to organise and manage their learning experiences and resources as a visual timeline. The architecture of KleOS allows it to be used on a number of different platforms including handheld devices. However, unlike Mobile Medical Tutor, none of the above systems deals with the problem of facilitating the human instructor in the educational software management Moreover, Mobile Medical Tutor makes use of the desktop application, which is designed in a way that can support the emotional state of learners (Virvou & Alepis, 2003a; Virvou & Alepis, 2003b).

2. General architecture of Mobile Medical Tutor

In this section the general architecture of Mobile Medical Tutor (MMT) is described. First of all we should emphasise the fact that we have an educational application that is meant to assist the medical educational process.

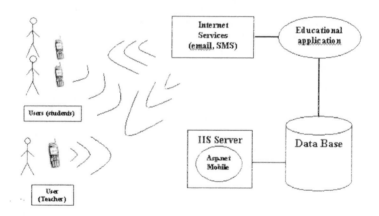

Figure 1: Communication between Instructors, Students and the educational application

For this purpose we have improved the communication of instructors and students by incorporating mobile technology, while the basic concepts of programming for educational purposes are retained.

As we can see in Figure 1, the main architecture of MMT consists of the main educational application, a database, mobile pages and a few Internet services.

In particular, the main application is installed either on a public computer where both students and instructors have access, or alternatively each student may have a copy on his/her own personal computer. Examples of using the main application are shown at Section 4, where the use of MMT by students is described. The underlying reasoning of the system is based on the student modelling process of the educational application. The system monitors the students' actions while they use the educational application and tries to diagnose possible problems, recognise goals, record permanent habits and errors that are made repeatedly. The inferences made by the system concerning the students' characteristics are recorded in their student model that is used by the system to offer advice adapted to the needs of individual students. The database of MMT is used firstly to hold all the necessary information that is needed for the application to run and secondly to keep analytical records of the performance of all the students that use the application.

The mobile pages provide students and instructors with the appropriate interface for their interaction with the main application through their mobile phones or palm tops with mobile support. Mobile pages can contain as many server-side forms as necessary, whereas normal web pages can contain only one. This is quite important because it renders the mobile pages more "powerful" and also it may reduce costs. Mobile controls automatically paginate content according to the device that keeps the paged data on the server until requested by the user.

Finally, Internet services such as e-mail and SMS sending are used in order to deliver messages from and to instructors and students during their interaction with MMT. E-mail messages usually include progress reports of students. In contrast SMS messages are shorter and are used by both instructors and students for many purposes such as to inform about changes in courses and tests, to send scores of particular tests etc.

3. Use of Mobile Medical Tutor by students

Medical students can use MMT to learn through courses that are relevant to their medical studies and to test the knowledge they have acquired. A part of the

620

students' interaction with the system may be made by using mobile devices. Figure 2 illustrates the main form of the educational application on a desktop computer.

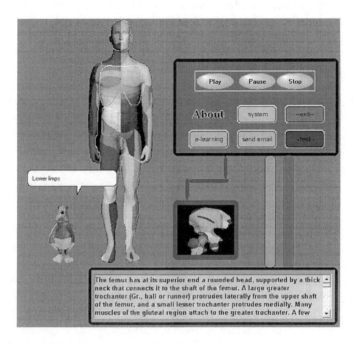

Figure 2. The main form of the application

While using Medical Tutor from a desktop computer, students are able to retrieve information about a particular course. In the example of Figure 2 a medical student is learning anatomy. The information is given in text-form while at the same time an animated agent reads it using a speech engine. The student can choose a specific part of the human body and all the available information will be retrieved from the system's database.

Similarly the student is able to take tests that include questions, answers, multiple-choice, etc, concerning specific parts of the medical theory. The animated agent is present in these modes too to make the interaction more human-like. As we have mentioned, students can use their mobile phone or palm top PC in order to interact with the educational application. They can connect to the mobile pages of the application and after the authentication process they may take tests for specific, relevant to medicine, courses. User authentication is necessary for the creation and improvement of the user model of each student. As we can see in Figure 3 a student named John Burgh is answering a multiple-choice question that corresponds to the upper part of the body and particularly the "head and neck" chapter. When he presses the submit button MMT processes the information and updates the students' short-term user model in the database.

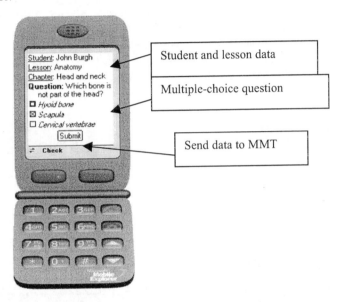

Figure 3. Interaction between a student and MMT through a mobile device

Finally, students may use their mobile device to send short messages through Mobile Medical Tutor to their instructors concerning lessons and tests. Short messages may contain questions about conclusions in the theory or queries about tests that need to be settled. Alternatively email sending through the mobile pages of the system is also supported.

4. Course management by instructors

Teachers are able to connect to the system databases with their wireless device, either mobile phone or mobile Pocket PC simply by entering the corresponding

URL into their device. The URL consists of the IP of the server computer and the name of the mobile ASP.NET page (example: http://195.252.225.118/mobilepage.aspx).

Teachers can use the application to cooperate in the educational process. The teacher and the students are not only able to have easy access to the databases of the application but they can also "communicate" with each other. The communication between teachers and students can be realised in many ways. By using a mobile phone (and thereby connecting to the application's mobile pages) instructors can send short messages via the Short Message Service (SMS), either directly to their students (if they also have mobile phones) or by e-mail. Alternatively instructors can "write" the message to the application's database. In this case, instructors have to declare the name of the receiver and the application will use its audio-visual interface to inform them as soon as they open the application.

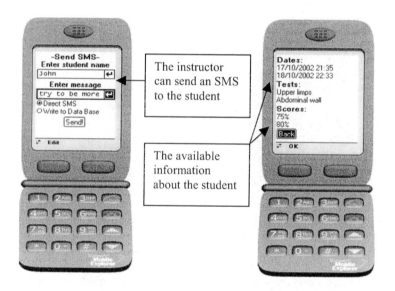

Figure 4 Example: An instructor monitors the progress of a student and sends SMS

An example of an instructor monitoring the progress of a student through a mobile phone and sending an SMS to him is illustrated in Figure 4. Both teachers and students are able to send short messages (SMSs) containing remarks and additional information. The body of the SMS is entered in the "enter message" field and the name of the receiver is written in the "enter student name" field.

5. Conclusions

This paper has described a medical educational system that incorporates mobile technology in order to help students and instructors with many learning and training obligations. MMT has been designed to provide the relatively new mobile facilities while retaining high quality of the educational application with respect to high interactivity personalization and user-friendliness.

This work has shown how mobile devices may be used constructively in medical education by combining existing technologies of educational software with mobile features.

References

1. Brusilovsky, P. (1999) "Adaptive and Intelligent Technologies for Web-based Education." In C. Rollinger and C. Peylo (eds.), Künstliche Intelligenz (4), Special Issue on Intelligent Systems and Teleteaching, 19-25, http://www2.sis.pitt.edu/~peterb/papers/KI-review.html.
2. Ketamo H. (2003) "An Adaptive Geometry Game for Handheld Devices" Educational Technology & Society, 6(1).
3. Vavoula, G. & Sharples, M. (2002) "KleOS: A Personal, Mobile, Knowledge and Learning Organisation System". IEEE International Workshop on Wireless and Mobile Technologies in Education (WMTE'02), pp. 152-155.
4. Virvou M. & Alepis E. (2003a). *Creating tutoring characters through a Web-based authoring tool for educational software*. In Proceedings of the 2003 IEEE International Conference on Systems, Man and Cybernetics, to appear.
5. Virvou M. & Alepis E. (2003b). *Human-like characteristics by speaking animated agents in a web-based tutoring system*. In C. Stephanidis (ed.) Adjunct Proceedings of the 10th International Conference on Human Computer Interaction (HCII'2003), pp. 109-110.

SOME ANALYTICAL ESTIMATES FOR MAGNETIC DRUG TARGETING

P.A. VOLTAIRAS[†] & D.I. FOTIADIS

Department of Computer Science, University of Ioannina
& Biomedical Research Institute-FORTH, Ioannina, GR451 10, Greece

L.K. MICHALIS

Department of Cardiology, University of Ioannina
& Biomedical Research Institute-FORTH, Ioannina, GR451 10, Greece

C.V. MASSALAS

Department of Materials Science, University of Ioannina
& Biomedical Research Institute-FORTH, Ioannina, GR451 10, Greece

The phenomenological ferrohydrodynamic model of the problem of magnetic drug targeting (MDT), that was developed previously (P. A. Voltairas, D. I. Fotiadis and L. K. Michalis, *J. Biomech.*, **35**, 813, (2002)), is extended for a more realistic deformation of the ferrofluid drug drop, due to the blood flow. A suitable "stream" function is proposed along with a perturbation technique.

1. Introduction

Essential for the pharmacological treatment of various diseases is to target, deliver and retain the drug at the suffering area. Many complementary techniques have been studied for that purpose [1]. Among the proposed techniques, MDT assembles many merits, due to its high targeting efficiency and its non-invasive character. The typical apparatus consists of a magnetic fluid drug carrier, that is a fluid or a polymer solution comprised of magnetic nano- or micro-particles (also known as *ferrofluid*), and an external magnetic field. The technique was proposed almost forty years ago and investigated extensively [2-3]. Recently, MDT was associated with *in vivo* and *in vitro* experiments for tumor treatment [4], gene delivery [5] and intraocular retinal repair [6].

In a previous work we proposed a theory for controlling the MDT process and made some preliminary calculations [7]. We assumed that the magnetic drug drop (MDD) had a hemispherical shape in the presence of both the blood stream and the applied magnetic field. This geometric simplification obliged us to impose an additional constraint, in order to link important physical parameters.

[†] Corresponding author. Email address: pvolter@cs.uoi.gr.

As a result, the obtained solution was an upper bound to the exact one. It is the aim of this work to extent our previous study [7] to more realistic MDD deformations, by keeping the three dimensional (3D) character of the boundary value problem (BVP).

2. The Model

2.1. *Problem formulation*

We are concerned with the derivations of conditions for adherence of a MDD on the blood vessel wall. Such conditions, when provided, will help to control the drug targeting process, and thus the accurate release of drug to the suffering area. They correspond to the determination of the geometry and the strength of the applied magnetic field that surpasses the drag that the MDD experiences due to blood flow and results in adhesion, for given biophysical and geometrical parameters (mean blood velocity, drop shape and size, blood vessel geometry, magnetic nano- or micro-particle concentration, etc.). From the fluid dynamics point of view the problem consists of a three dimensional rotational Newtonian viscous two phase flow (MDD and blood) inside a cylindrical tube. A cross section, on the xz-plane, of the 3D geometry of the problem is depicted in Fig. 1.

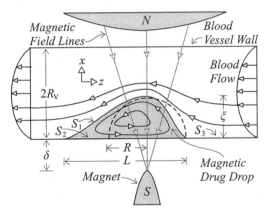

Figure 1. An adherent MDD on the blood vessel wall.

The adherent MDD on the blood vessel wall spreads over a distance L, with height ξ (Fig. 1). The dashed hemisphere of radius R corresponds to the shape that had the MDD in our previous calculations [7]. The coordinate origin is located on the center of the primitive circle on the equatorial yz-plane of a hemispherical drop. Diffusion phenomena are neglected. The blood vessel is considered to be an infinite, rectilinear, rigid, non-porous cylindrical tube, with smooth internal surface of radius $R_V \gg \xi$. The grey conical lines in Fig. 1 correspond to the applied non-uniform magnetic field, with one of the poles located at a distance δ from the blood vessel wall. The magnetophoretic force that experienced the MDD is analogous to the gradient of the magnetic field [7]. Moreover, we consider a diluted, incompressible MDD. This simplifies the magnetostatic field problem considerably, since the magnetophoretic body forces are not present in the hydrodynamic equations and appear only the equivalent surface tractions on the MDD - blood flow interface [7]. Henceforth, bold characters will denote vector fields and the subscripts ( )$_1$ and ( )$_2$ will denote quantities of the magnetic drug and blood regions, respectively. When inertial effects are neglected (low Reynolds number) the blood and MDD flows are described by the Stokes equations

$$\eta_i \nabla^2 \mathbf{u}_i = \nabla p_i , \qquad (1)$$

and the equations of continuity

$$\nabla \cdot \mathbf{u}_i = 0, \qquad (2)$$

for $i = 1, 2$. η is the fluid viscosity; $\mathbf{u}$ is the flow velocity and p is the pressure. Eqs. (1-2) must be supplemented by the magnetostatic potential problem

$$\nabla \cdot \mathbf{B} = 0, \qquad (3)$$

$$\nabla \times \mathbf{H} = 0, \qquad (4)$$

where $\mathbf{B}$ is the magnetic induction and $\mathbf{H}$ the magnetic field. There are three boundaries for the prescribed problem by Eqs. (1-4): the MDD-blood flow interface, S_1; the MDD-blood vessel wall boundary, S_2; and the blood vessel wall boundary, S_3. All three are depicted in Fig. 1. The boundary conditions (BCs) on the interface S_1 are:

$$\mathbf{n} \cdot \mathbf{u}_1 = \mathbf{n} \cdot \mathbf{u}_2 = 0, \qquad (5)$$

$$\mathbf{n} \times [\![\, \mathbf{u} \,]\!] = \mathbf{0}, \qquad (6)$$

$$[\![\, p_{\mathrm{vn}} - p \,]\!] + p_{\mathrm{m}} + p_{\mathrm{n}} = p_{\mathrm{c}}, \qquad (7)$$

$$[\![\, p_{\mathrm{vt}} \,]\!] = 0, \qquad (8)$$

$$\mathbf{n} \cdot [\![\, \mathbf{B} \,]\!] = 0, \qquad (9)$$

$$\mathbf{n} \times [\![\, \mathbf{H} \,]\!] = \mathbf{0}. \qquad (10)$$

where the abbreviation $[\![A]\!] = A_2 - A_1$ is used. Here,

$$p_{vn} = 2\eta\,\mathbf{n}\cdot\partial\mathbf{u}/\partial n, \tag{11}$$

$$p_{vt} = \eta\left(\mathbf{t}\cdot\partial\mathbf{u}/\partial n + \mathbf{n}\cdot\partial\mathbf{u}/\partial t\right), \tag{12}$$

where $\partial/\partial n \equiv \mathbf{n}\cdot\nabla$, $\partial/\partial t \equiv \mathbf{t}\cdot\nabla$, and p_m, p_n and p_c are:

$$p_m = \int_0^H M(H)dH, \tag{13}$$

$$p_n = \mu_0 M_n^2/2 = \mu_0\chi^2 H_n^2/2, \tag{14}$$

$$p_c = \gamma\,\nabla\cdot\mathbf{n}, \tag{15}$$

with $H_n = \mathbf{n}\cdot\mathbf{H}$. The magnetic constitutive law is linear, $M(H) = \chi H$. Here, M is the magnitude of the magnetization vector of the MDD; χ is the constant magnetic susceptibility; μ_0 is the magnetic permeability of vacuum; and γ is the surface tension. On the boundaries S_2 and S_3, the flow satisfies the conditions:

$$\mathbf{u}_1 = \mathbf{u}_2 = \mathbf{0}. \tag{16}$$

Also the finiteness of the flow at the origin and infinity requires:

$$\mathbf{u}_1(r \to 0) = \mathbf{u}_0, \quad \mathbf{u}_2(r \to \infty) = -\mathbf{u}_0, \quad \mathbf{u}_0 = u_0\mathbf{e}_z, \tag{17}$$

where $\mathbf{e}_i$ $(i = x, y, z)$ are the unit base vectors and $u_0 > 0$ is the mean blood velocity. The BVP (1-2, 5-8, 16-17) belongs to the class of free boundary value problems (FBVP), with a stationary interface [8]. We will concentrate on qualitative analysis, by preserving the main biophysical and geometrical characteristics of the flow on a minimum mathematical base. For a diluted MDD the magnetostatic potential problem, Eqs. (3-4, 9-10), is neglected and a point (Coulomb) magnetic source is introduced in the BC (7). For quantitative analysis, it is essential the preservation of the 3D character of the FBVP. The asymmetry in flow, with respect to the x-axis, will be incorporated in the FBVP by considering the MDD-blood flow interface S_1 as a surface of revolution.

2.2. The hemispherical magnetic drop

The FBVP (1-2, 5-8, 16-17) was "solved" in Ref. [7], assuming that the MDD had the hemispherical shape. What we mean by a "solution" is an *almost, mean value to the upper bound* of the exact solution. *Almost*, because the no-slip conditions (16) were not satisfied. *Mean value*, because the BC (7) was satisfied on the average, and *upper bound* because we introduced an additional constraint in order to derive a dependence of the mean blood flow velocity on the applied magnetic field. The validity of the non-slip conditions (16) was also criticized recently [9]. For the case of a hemispherical MDD, the calculation of the flow field is simplified. In spherical polar coordinates

$$(x, y, z) = (r \sin \theta \cos \varphi, r \sin \theta \sin \varphi, r \cos \theta), \tag{18}$$

a velocity field of the form

$$\mathbf{u} = \nabla \times \left(\frac{\Psi(r, \theta) \mathbf{e}_\varphi}{r \sin \theta} \right), \tag{19}$$

satisfies automatically the equation of continuity (2). Eliminating the pressure with cross-differentiation in Eq. (1) yields

$$E^2 \left(E^2 \Psi \right) = 0, \tag{20}$$

where E^2 is the following differential operator:

$$E^2 \equiv \frac{\partial}{\partial r^2} + \frac{\sin \theta}{r^2} \frac{\partial}{\partial \theta} \left(\frac{1}{\sin \theta} \frac{\partial}{\partial \theta} \right). \tag{21}$$

The general non-singular solution of the Eq. (20) can be expressed as a series expansion with eigenfunctions:

$$\Psi_1^n \equiv r^n G_n(\zeta), \quad \Psi_2^n \equiv G_n(\zeta)/r^{n-1}, \tag{22}$$

where G_n are the Gengenbauer functions:

$$G_n(\zeta) = \left(P_{n-2}(\zeta) - P_n(\zeta) \right)/(2n - 1), \tag{23}$$

and P_n are the Legendre polynomials with $\zeta = \cos \theta$. Let us consider only the $n = 2$ mode in the eigenfunction expansion. Then, a solution which satisfies the BCs (5-6, 8) and the conditions for the finiteness of the flow (17), corresponds to a stream function Ψ, of the following form:

$$\Psi_1 = \lambda_3 u_0 \left[1 - r^2/R^2 \right] r^2 \sin^2 \theta/2, \tag{24}$$

$$\Psi_2 = u_0 \left[2\lambda_1 R r - 2\lambda_2 R^3/r - r^2 \right] \sin^2 \theta/2, \tag{25}$$

where

$$\lambda_1 = (3 + 2\gamma_v)/(4 + 4\gamma_v), \quad \lambda_2 = 1/(4 + 4\gamma_v), \quad \lambda_3 = \gamma_v/(2 + 2\gamma_v), \tag{26}$$

and $\gamma_v = \eta_2/\eta_1$ is the viscosity ratio. Substitution of the solution (19, 24-25) into (7), with the pressures p_1, p_2 calculated by Eq. (1), with $p_c \cong 2\gamma/R$, instead of Eq. (15), and for a Coulomb point source applied magnetic field

$$\mathbf{H} = m(\mathbf{r} + \delta \mathbf{e}_x)/(r^2 + \delta^2 + 2\delta x)^{3/2}, \tag{27}$$

where m is the magnetic dipole moment, yields an expression with not similar azimuthal and polar components, which could not be satisfied. Thus, in Ref. [7] we introduced the weaker constraint:

$$\int (\llbracket p_{vn} - p \rrbracket + p_m + p_n - 2\gamma/R) dS = 0, \text{ on } S_1, \tag{28}$$

which corresponds to an average satisfaction of the normal surface tractions along the hemispherical interface S_1. But even then, the traction balance could not lead to a function that depends both on the mean blood velocity u_0, and the applied magnetic field $H_0 = m/\delta^2$. That was a consequence of the high degree of symmetry that it was introduced in our geometrical model, with the applied Coulomb like magnetic field oriented along the x-axis, which could not take into account the deformation that the MDD experiences due to the blood flow. This problem was resolved by projecting the surface tractions along the blood flow direction, and rotating the Coulomb like magnetic field given by Eq. (27) over an angle ω, with respect to the x-axis. That resulted in an additional traction balance condition along the z-axis, written in more compact form:

$$\int (\llbracket \mathbf{T} \rrbracket - 2\gamma/R \mathbf{n}) \cdot \mathbf{e}_z \, dS = 0, \text{ on } S_1, \tag{29}$$

where, $\mathbf{T}$ are the surface tractions with components $T_i = \left(\tau_{ij} + \tau_{ij}^M\right) n_j$ and τ_{ij} and τ_{ij}^M are the stress tensor and the Maxwell's stress tensors, respectively [7].

3. Magnetic Drop Deformation

We consider of primary importance the preservation of the three dimensional character of the flow, in order to obtain quantitative conditions for adherence of the MDD on the blood vessel wall. Therefore, the following analysis aims in no way to be precise, since it is concerned with an asymmetric three dimensional, rotational, two phase viscous flow inside a cylindrical rigid tube. However, the presented arguments provide a method to approach the asymmetry in the flow.

Deviations from the hemispherical shape, could remove both of the above weaknesses in the solution procedure: the additional constraint (29); and the average satisfaction of the BC (7), in the form of Eq. (28). Therefore, we could calculate such deviations from the spherical shape by perturbing the stream function in the following sense:

$$\overline{\Psi}_i = \Psi_i + \varepsilon \Psi_i^*, \quad i = 1, 2, \tag{30}$$

where $0 < \varepsilon \ll 1$, and Ψ_i is given by Eqs. (24-25). The "stream" function $\overline{\Psi}$ resolves all the difficulties if it satisfies the BC (7), along a new interface S_1.

Such an interface could be considered as a surface of revolution, with rotational axis the blood vessel z-axis, that admits the parametric representation:

$$(x, y, z) = (\rho \sin \varphi, \rho \cos \varphi, f(\rho)). \tag{31}$$

Then, due to rotational symmetry, we limit our further computations to a zx-plane problem ($\varphi = 0$). The interface S_1 corresponds to the streamline where the stream function vanishes, and therefore it can be determined by the condition:

$$\overline{\Psi}_i(r, \theta) = 0 \Rightarrow r_i = r_i(\theta), \quad i = 1, 2. \tag{32}$$

Substitution of this polar representation into $\rho = r(\theta) \sin \theta$, solving for $\cos \theta$ and introducing the result in $z = r(\theta) \cos \theta$, yields $z = f(\rho)$ of Eq. (31). But in order for the BCs (5) to be fulfilled, Ψ_i^* should also satisfy the condition for a stationary interface:

$$\partial \overline{\Psi}_i(r, \theta) / \partial r = 0, \quad \text{at} \quad r = r_i(\theta). \tag{33}$$

Moreover, the deformed interface (32), should be the same, regardless of been computed from the interior ($i = 1$), or exterior ($i = 2$) problem. Therefore,

$$r_1(\theta) = r_2(\theta). \tag{34}$$

Finally, the "stream" function Ψ_i^* should also be a solution of Eq. (20) or

$$E^2\left(E^2 \Psi_i^*\right) = 0. \tag{35}$$

The above presented analysis is rigorous, provided all the conditions (32-35) are satisfied. This is not always feasible. Then, the obtained solutions are *weak* solutions, in contrary to the *strong* ones that fulfill Eqs. (32-35). Nevertheless, weak solutions can represent well asymmetric (with respect to the x-axis) MDDs, as long as the perturbational character of our analysis is conserved, $\varepsilon \ll 1$. Calculations related with the satisfaction of the boundary condition (7), will be performed in a forthcoming study. All the "stream" functions that we investigated, are included in the class $\Psi_i^* = R_i(r) \Theta(\cos \theta)$. As an example consider the "stream" function:

$$\Psi_1^* = \Phi_1 r^2 \sin \theta \cos \theta / 2, \quad \Psi_2^* = \Phi_2 \sin \theta \cos \theta / 2 r, \tag{36}$$

with Φ_1, Φ_2 constants. The Eqs. (33-35), are not satisfied. Nevertheless, the interface S_1, that obeys Eqs. (32, 36), has the polar representation:

$$r_1(\theta) = R \sqrt{1 + A_1 \cot \theta}, \quad A_1 = \varepsilon \Phi_1 R / \lambda_3 u_0, \tag{37}$$

and is depicted in Fig. 2. For $A_1 = 0$, S_1 coincides with the hemisphere studied previously [7]. Its singularity corresponds to $L \to \infty$, but the resultant shape approximates better related experimental observations [3]. The fact that $L \to \infty$ should be examined properly in further numerical calculations with the BC (7). In order to conform with our perturbational analysis $A_1 \ll 1$. In Fig. 2, A_1 was on purpose allowed to take values close to unity, for demonstrative reasons.

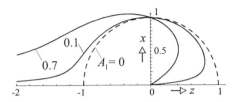

Figure 2. MDD deformation, due to blood flow, calculated from Eq. (37) for varying A_1.

Since the weak solution (36) does not satisfy Eq. (35), the computation of the pressures p_1, p_2 cannot be performed by direct integration of Eq. (1). A good alternative is to consider them as constants, that will be assigned proper values in further numerical calculations with the BC (7).

4. Discussion

We proposed a new method that derives conditions for adherence of a MDD on a blood vessel wall. In order to keep the three dimensional character of the hydrodynamic problem, without confining our analysis to spherical or ellipsoidal MDDs, we considered the shape of the deformed MDD to be a surface of revolution about the blood flow direction. For small deviations from the spherical solution, numerical calculations with the BC (7) should be performed, in order to validate the superiority of the method. The prescribed weak solutions could also be used as input functions for future more sophisticated computer simulations.

References

1. LaVan D.A. et al., *Nature Biotech.*, **21**, 1184 (2003).
2. T. Nakamura et al., *J. Appl. Phys.*, **42**, 1320 (1971)
3. Ruuge E. K. and Rusetski A. N., *J. Mag. Mag. Mat.*, **122**, 335 (1993).
4. C. Alexiou, et al., *Lect. Notes Phys.*, **594**, 233 (2002).
5. C. Plank, et al., *Expert. Opin. Biol. Ther.*, **3**, 745 (2003).
6. D. L. Holligan, et al., *Nanotech.*, **14**, 661 (2003).
7. P. A. Voltairas et al., *J. Biomech.*, **35**, 813 (2002).
8. P. Pelcé, *Dynamics of Curved Fronts*, (Ed.), Academic Press (1988).
9. Q. A. Pankhurst, et al., *J. Phys. D: Appl. Phys.*, **36**, R167 (2003).

Author Index